Methods in Pharmacology and Toxicology

Series Editor
Y. James Kang

For further volumes:
http://www.springer.com/series/7653

TRP Channels in Drug Discovery

Volume I

Edited by

Arpad Szallasi

Transfusion Services, Monmouth Medical Center, Long Branch, NJ, USA

Tamás Bíró

Department of Physiology, Medical and Health Science Center, University of Debrecen, Debrecen, Hungary

Humana Press

Editors
Arpad Szallasi
Transfusion Services
Monmouth Medical Center
Long Branch, NJ, USA

Tamás Bíró
Department of Physiology
Medical and Health Science Center
University of Debrecen
Debrecen, Hungary

ISSN 1557-2153 ISSN 1940-6053 (electronic)
ISBN 978-1-4939-6281-5 ISBN 978-1-62703-077-9 (eBook)
DOI 10.1007/978-1-62703-077-9
Springer New York Heidelberg Dordrecht London

Softcover re-print of the Hardcover 1st edition 2012

Printed on acid-free paper

Humana Press is a brand of Springer
Springer is part of Springer Science+Business Media (www.springer.com)

Preface

With almost 600 reviews, transient receptor potential (TRP) channels arguably represent today's most extensively reviewed pharmacological targets. The literature on TRP channels is vast and still growing: It has exploded from a mere 21 papers in 1995 to over 2,000 in the past 2 years. Yet, even the most studied TRP channels like TRPV1 continue to surprise: as Bernd Nilius points it out in his Introduction, "We are still at the beginning of the beginning."

Over the past decade, both gain- and loss-of-function mutations in TRP channels (so-called TRP channelopathies) have been identified in human disease states ranging from focal segmental glomerulosclerosis (TRPC6) and familial episodic pain syndrome (TRPA1) through brachyolmia and hereditary arthropathy of hand and feet (TRPV4) to mucolipidosis type-4 (TRPML1) and amyotrophic lateral sclerosis and parkinsonism/dementia complex (TRPM7). These findings imply a therapeutic potential for drugs targeting TRP channels in a wide variety of diseases, many with no existing satisfactory treatment options. Indeed, a number of potent, small molecule TRPV1, TRPV3, and TRPA1 antagonists have already entered clinical trials, and many more are in preclinical development.

The TRP superfamily of ion channels in humans is a diverse family of 28 cation channels with varied physiological functions. Their name stems from their similarity on the sequence level to the original *trp* gene from *Drosophila* which, when mutated, resulted in a transient receptor potential in the presence of continued exposure to light. Overall, few generalizations can be made about TRP channels. Most family members share a low level of structural similarity, but some channels are very highly homologous to each other (e.g., TRPC3 and TRPC7; TRPV5 and TRPV6). Many TRP channels form functional channels as homotetramers, though heteromultimerization is not uncommon. The latter phenomenon may have important implications for drug discovery.

Consistent with their diverse structure, TRP channels also serve diverse functions including afferent sensory functions (mechanical, chemical, thermal, noxious, etc.) as well as efferent mechanisms (of growth control, cellular differentiation, vasoregulation, mediator release, etc.). While most family members are cation channels with limited selectivity for calcium, both calcium- (TRPV5 and TRPV6) and sodium-selective (TRPM4 and TRPM5) family members exist. In addition, some TRP channels transport noncanonical cations such as iron (TRPML1), magnesium (TRPV6), or zinc (TRPA1).

Of the 28 TRP channels discovered until today, seven sense hot or warm temperatures (TRPV1 to TRPV4, TRPM2, TRPM4, and TRPM5), whereas two (TRPA1 and TRPM8) are activated by cold. Together, these channels, referred to as "thermoTRPs," cover a wide temperature range with extremes that fall between 10°C (TRPA1) and 53°C (TRPV2). The temperature sensor is believed to be associated with the C terminus. In support of this model, swapping the C-terminal domain of TRPV1 with that of TRPM8 was shown to change the temperature sensitivity of TRPV1 from hot to cold.

Animal data and human genetic studies have shown that TRP channel dysfunction ("TRP channelopathy") can cause various pathological conditions. In fact, the TRML (mucolipin) and TRPP (polycystin) families were named after the human diseases they are associated with (mucolipidosis and polycystic kidney disease, respectively). The founding member of the M (melastatin) family, TRPM1, was identified via comparative analysis of genes that distinguish benign nevi and malignant melanoma. The A (ankyrin) family has only one known member (TRPA1), and its name refers to the unusually high number of ankyrin repeats at the N terminus of the channel protein. Mammalian TRP channels that are most similar to those in *Drosophila* are referred to as canonical (TRPC). Last, the V (vanilloid) family came into existence by expression cloning of the capsaicin receptor TRPV1.

The aim of these volumes is ambitious: They open with a series of "state-of-the-art" minireviews on the most interesting TRP channels (from TRPA1 to TRPV4), followed by a collection of cookbook-like protocol chapters describing various methodologies (ranging from capsaicin inhalation test in humans through rodent models of anxiety to stroke, cancer, diabetes, and experimental colitis models) relevant to TRP channel research. Pain models (TRPs = "Targets for Relief of Pain") were previously detailed in our *Analgesia: Methods and Protocols* volume in the *Methods in Molecular Biology* series. Here, we focus on non-pain models in keeping with the alternative interpretation of TRPs: "Truly Remarkable Proteins."

It is our hope that this book will be useful for graduate students in academic laboratories as well as for scientists developing new drugs at Pharma and clinicians interested in novel drugs in the pipeline.

Long Branch, NJ, USA *Arpad Szallasi*
Debrecen, Hungary *Tamás Bíró*

Contents

Contributors

Karl-Erik Andersson • *Wake Forest Institute for Regenerative Medicine, Wake Forest University, Winston-Salem, NC, USA*

António Avelino • *Department of Experimental Biology, Faculty of Medicine, Oporto University, Porto, Portugal; IBMC – Instituto de Biologia Molecular e Celular, Universidade do Porto, Porto, Portugal*

William Brown • *AstraZeneca R&D Montreal, St-Laurent, QC, Canada*

Lysanne Campeau • *Wake Forest Institute for Regenerative Medicine, Wake Forest University, Winston-Salem, NC, USA*

Ana Charrua • *Department of Urology, Hospital São João, Porto, Portugal; IBMC – Instituto de Biologia Molecular e Celular, Department of Experimental Biology, Faculty of Medicine and Faculty of Nutrition & Food Sciences, Universidade do Porto, Porto, Portugal*

Jun Chen • *Neuroscience Research, Global Pharmaceutical Research and Development, Abbott Laboratories, Abbott Park, IL, USA*

Vladimir Chubanov • *Walther-Straub-Institute of Pharmacology and Toxicology, Ludwig Maximilians-University Munich, Munich, Germany*

Barbara Colsoul • *University of Leuven, Leuven, Belgium*

Célia D. Cruz • *Department of Experimental Biology, Faculty of Medicine, Oporto University, Porto, Portugal; IBMC – Instituto de Biologia Molecular e Celular, University of porto, Porto, Portugal*

Francisco Cruz • *Department of Urology, Hospital São João, Porto, Portugal; IBMC – Instituto de Biologia Molecular e Celular, Universidade do Porto, Porto, Portugal*

Alexander Dietrich • *Walther-Straub-Institute of Pharmacology and Toxicology, Ludwig-Maximilians-University Munich, Munich, Germany*

Matthew A.J. Duncton • *Formerly with Renovis, Inc. (a wholly owned subsidiary of Evotec AG), South San Francisco, CA, USA*

Jonathan T. Eggenschwiler • *Department of Molecular Biology, Princeton University, Princeton, NJ, USA*

Beate Fuchs • *Excellence Cluster Cardio-Pulmonary System, University of Giessen Lung Center, Department of Internal Medicine II, Justus-Liebig-University Giessen, Giessen, Germany*

Claudius Füllhase • *Department of Urology, University Hospital Großhadern, Ludwig-Maximilians-University Munich, Munich, Germany*

Raffaele Gatti • *PharmEste, Ferrara, Italy*

Chandan Goswami • *Institute of Physics, National Institute of Science Education and Research, Bhubaneswar, Orissa, India*

Thomas Gudermann • *Walther-Straub-Institute of Pharmacology and Toxicology,*

Ludwig-Maximilians-University Munich, Munich, Germany
ZSÓFIA HAJNA • *Department of Pharmacology and Pharmacotherapy, Faculty of Medicine, University of Pécs, Pécs, Hungary*
ZSUZSANNA HELYES • *Department of Pharmacology and Pharmacotherapy, Faculty of Medicine, University of Pécs, Pécs, Hungary*
ALEXIS HOFHERR • *Renal Division, Department of Medicine, University Medical Centre Freiburg, Freiburg, Germany; Spemann Graduate School of Biology and Medicine (SGBM) and Faculty of Biology, Albert-Ludwigs-University Freiburg, Freiburg, Germany*
SUN WOOK HWANG • *Korea University Graduate School of Medicine, Seoul, Korea*
HERMANN KALWA • *Brigham and Women's Hospital, Harvard Medical School, Boston, MA, USA*
MIKLOS KECSKES • *University of Leuven, Leuven, Belgium*
PHILIP R. KYM • *Neuroscience Research, Global Pharmaceutical Research and Development, Abbott Laboratories, Abbott Park, IL, USA*
JENNIFER M.A. LAIRD • *AstraZeneca R&D Montreal, St-Laurent, QC, Canada; Department of Pharmacology & Experimental Therapeutics and Alan Edwards Centre for Research on Pain, McGill University, Montreal, QC, Canada*
STEVE MCGARAUGHTY • *Neuroscience Research, Global Pharmaceutical Research and Development, Abbott Laboratories, Abbott Park, IL, USA*
DAVID D. MCKEMY • *Neurobiology Section, Department of Biological Sciences, University of Southern California, Los Angeles, CA, USA*
AURELIE MENIGOZ • *University of Leuven, Leuven, Belgium*
BARBARA A. MILLER • *Departments of Pediatrics and Biochemistry & Molecular Biology, The Pennsylvania State University College of Medicine, Hershey, PA, USA*
EVA MILLQVIST • *Department of Allergology, Institution of Internal Medicine, The Sahlgrenska Academy at University of Gothenburg, Gothenburg, Sweden*
ROMINA NASSINI • *Department of Preclinical and Clinical Pharmacology, University of Florence, Florence, Italy*
BERND NILIUS • *Laboratory of Ion Channel Research, Department of Cellular and Molecular Medicine, KU Leuven, Leuven, Belgium*
CHRISTINA M. PABELICK • *Departments of Anesthesiology and Physiology and Biomedical Engineering, Mayo Clinic, Rochester, MN, USA*
PAMELA PEDRETTI • *Department of Preclinical and Clinical Pharmacology, University of Florence, Florence, Italy*
KOENRAAD PHILIPPAERT • *University of Leuven, Leuven, Belgium*
Y.S. PRAKASH • *Departments of Anesthesiology and Physiology and Biomedical Engineering, Mayo Clinic, Rochester, MN, USA*
OLIVIER RADRESA • *AstraZeneca R&D Montreal, St-Laurent, QC, Canada*
ALEXEY G. RYAZANOV • *Department of Pharmacology, Robert Wood Johnson Medical School – UMDNJ, Piscataway, NJ, USA*
LILLIA V. RYAZANOVA • *Department of Pharmacology, Robert Wood Johnson Medical School – UMDNJ, Piscataway*
BRIJ B. SINGH • *Department of Biochemistry and Molecular Biology, School of Medicine and Health Sciences, University of North Dakota, Grand Forks, ND, USA*

ROBERTO SOLER • *Wake Forest Institute for Regenerative Medicine, Wake Forest University, Winston-Salem, NC, USA; Division of Urology, Federal University of Sao Paulo, São Paulo, Brazil*
MARCELLO TREVISANI • *PharmEste, Ferrara, Italy*
RUDI VENNEKENS • *Laboratory of Ion Channel Research, University of Leuven, Leuven, Belgium*
FABIEN VINCENT • *Pfizer, Inc., Groton, CT, USA*
NORBERT WEISSMANN • *Excellence Cluster Cardio-Pulmonary System, University of Giessen Lung Center, Department of Internal Medicine II, Justus-Liebig-University Giessen, Giessen, Germany*
SUNGJAE YOO • *Korea University Graduate School of Medicine, Seoul, Korea*
DAVID A. ZEEVI • *Department of Medical Genetics, Shaare Zedek Medical Center, Jerusalem, Israel*
STEPHEN ZICHA • *AstraZeneca R&D Montreal, St-Laurent, QC, Canada*

Part I

Introduction

Chapter 1

Introduction to TRPs: A Quest for Novel Drug Targets

Bernd Nilius

Abstract

This introduction lists TRP channels as possible pharmaceutical targets and will also refer to some methodological and still scientific caveats. First, it will be highlighted which scientific problems are still necessary to solve, including permeation, gating regulation. This part leads to the methodological aspect of the book. Second, the well-known problem of evaluating expression of several TRP channels in organs which need to be affected will be shortly discussed. Third, natural compounds as evolutionary important small molecules to modify TRP functions will be added. Last, some important diseases which might have priority as drug targets, focused on neurodegenerative diseases, will be mentioned.

Key words: TPR channels, Methodological caveats, Drug targets, Neurodegenerative diseases

Transient receptor potential (TRP) non-selective cation channels are probably expressed in all organs of our body (Fig. 1), in all cell types but still hide to us many of their specific or modulatory functions. They wait to be understood by the thousands of TRP channel-researchers around the world to be beneficially used for the whole society in many aspects, including targeting human diseases.

Billions of small molecules wait in the drug libraries for potential use in treating diseases or improving the quality of life, presumptuously labeled as serving mankind. Not surprisingly, having all these compounds and synthesizing more and more, the quest for "mankind-serving" targets is a moral and economical issue. And, of course, knowing already many targets, the adventure of a novel potentially interesting gene family encoding target proteins is more than welcome and provides kind of a fascination to both industry and academia. In this "hunting" scheme, TRP channels exactly fit in. They are polymodal, multifunctional, probably serve many cells to do their respective, mostly specialized job, and, especially,

Arpad Szallasi and Tamás Bíró (eds.), *TRP Channels in Drug Discovery: Volume I*, Methods in Pharmacology and Toxicology, DOI 10.1007/978-1-62703-077-9_1, © Springer Science+Business Media, LLC 2012

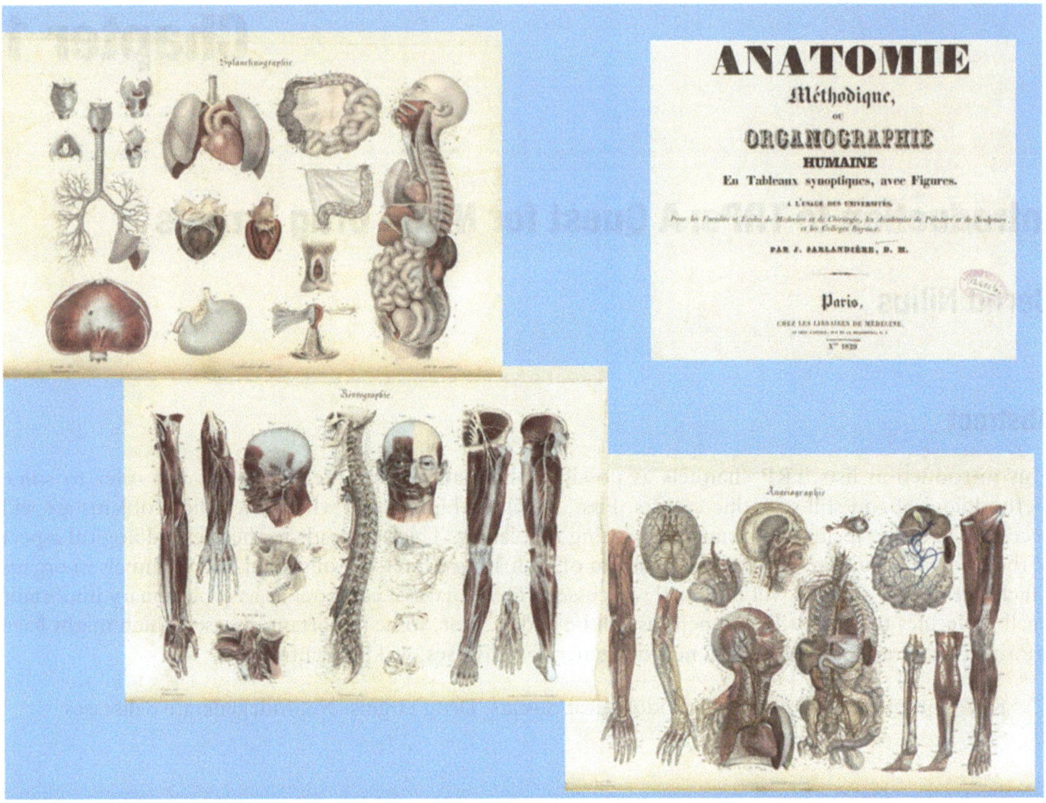

Fig. 1. All organs in our body express TRP channel. An ancient view on the organs of our body Anatomie: Méthodique ou Organographie humaine, from J. Salandiéere, Paris 1829 (from the free collection of Historial Anatomie http://www.nlm.nih.gov/exhibition/historicalanatomies/browse.html#V).

they are unknown and provide still a white area at least in the well-developed research world on ion channels. When the first trp gene was cloned as a protein of unknown function (1), involved in phototransduction in drosophila (2), nobody expected such a revolution in channel physiology. On the contrary, most of the academic world was biased that this might be the key to open the mystery of the still unknown store operated Ca^{2+} entry channels (CRACs, SOCs, SOCEs, CCEs, you name it). Knowing now more about it, STIMs and ORAIs have shown up, we should sensitively consider this too fast enthusiasm as a warning signal! Frankly, we tend to believe that we understand a lot about the 28 mammalian TRPs, but we are still at the beginning of the beginning and are confronted with many mysteries. Just starting, as required for a channel protein, we do not understand yet the regulation of permeation. Is the pore of many TRPs in fact a dynamic structure contributing to gating much more than the known α-pore of classical channels? Do agonists dilate the pore? Might TRP channels possess other pathways for ion conduction as shown for voltage-activated potassium and sodium channels as an ion permeation pathway through the voltage-sensing domain (S1–S4) of the channel, the ω-pore (3–5). Is there only one pore in a TRP homo- or heteromer? Continuing with gating: What creates the gating diversity of TRP channels? Is it the weak voltage dependence? Is it the high sensitivity to membrane lipids such as PIPs? What causes the intriguing activation of some TRP channels by heat or cold? Are all thermoTRPs both cold and heat activated? Is it the high promiscuity of many TRP channels to accessory proteins, and if yes, what are these accessory proteins? We still have very, very little insight into the universe of TRP-interacting proteins, and even have not yet identified a reliable β-subunit of any TRP channel what is so crucial for the much better known siblings of, e.g., the voltage gated cation channels. Needless to mention: TRPs play a crucial role in many diseases (6–11). This is highly convincing for hereditary diseases. It is much also convincing for acquired diseases but more controversial. And even if we agree that TRPs cause hereditary channelopathies, it does not mean that we understand their disease-causing dysfunctions.

Another problem related to tools and methods: many of us just consider TRP channels as kind of less selective Ca^{2+} channels. Importantly, most of the TRP channels have in fact a small Ca^{2+} selectivity and the "fractional Ca^{2+} current" is often less than 5%. We need to measure this "fraction" to check which amount of Ca^{2+} is really contributing to the current (for methods see, e.g., (12, 13). Importantly, the fractional Ca^{2+} current through TRPs might be also regulated! All TRPs trigger depolarization! Therefore, we cannot simply characterize the functional role of a TRP channels by Ca^{2+} imaging. We also need to carefully interpret Flipper results. In addition, a revival of the "single channel measurement" culture as described so brilliantly in the Neher/Sakmann book (14) is a cultural MUST for a TRP (electro)physiologist.

Having said this hopefully not depressing but activating view, a book on methods to approach TRPs as pharmacological targets in acquired and hereditary diseases, supervised and designed by the most effective TRP book editor I have ever met, Arpad Szallasi, is very timely and important, although probably in heavy competition with the many TRP books that appeared already in the last 5 years, especially a book in the same line focused on methods (15).

In this sense, this book is a bolt attempt to label the "unknown" with a sign of a "target". Of course, there are many encouraging examples, but mainly on TRPV1. Ironically, so many functions have been attributed to TRPV1, e.g., in the central nervous system, the development of cerebellar Purkinje cells and hippocampal pyramidal neurons, their contribution to glutamatergic synapse stabilization and function as important player in the neonatal cerebellar cortex and in the refinement of synaptic plasticity. It came as a big surprise that TRPV1 is probably even not expressed in these brain areas (16, 17). This causes another problem? Do we really know where the TRP channels that we wish to target are really located?

We are already in the middle of the problems: We need tools. So far, we have probably relied too much on classical knockout studies. It is time to rely on more intelligent transgenics such as tissue specific knockouts, conditional knockouts, we need more reporter animals for all TRPs, etc. Also, the so widely used knock-down by gene silencing, siRNA, small hairpin RNA methods are often less convincing because of the already mentioned problem of controlling the protein down regulation with sufficiently selective tools. So, we really need better and more specific antibodies. A problem discussed in the book? And as such a wonderful dream for a "TRPist"! If we have tools, small molecular weight inhibitors or activators, acting specifically or at least controllable, we would face as a first requirement revisiting data from knockout experiments. This is a must before we can jump into defining a new drug as targeting a TRP for beneficial reasons.

Even when we claim knowing what TRPs are doing (Ca^{2+} entry and depolarization), we do not know how. The reliable action of channel function and the correct localization of channels rely on innumerous interactions with channel-associated proteins. What do we know about accessory proteins of TRPs and what are the required methods? This book will also refer to this problem. We know signalplexes in which TRPs function or channelosomes (18–20), but even no clear β-subunit is so far defined for a TRP channel (is Pacsin3 such a subunit for TRPV4) (21, 22) At least, targeting TRPs requires our knowledge about associated proteins their dynamic and functional interaction with the channel. Helpful, the recent database on accessory TRP proteins, worthwhile to check regularly http://trpchannel.org/ (see also (23, 24))!

There is another inherent problem: TRPs have a fair and an ugly face. Activation or inhibition can be beneficial for a function in one organ (e.g., block of TRPM4 might counteract heart arrhythmias) but at the same time causes an increased histamine release in mast cells and adrenaline release in chromaffin cells potentiating probably allergic reactions and cause hypertension, respectively. And, block of TRPM4 might be even protective against spinal cord hemorrhage. Similar examples can be given for possible all TRP channels. Therefore, choosing a TRP as drug target has to be critically weighted with effects in other tissues as the targeted cells. This again efforts our knowledge about the functional role of TRPs (25–28). Or, downregulation of TRPC3 may prevent heart hypertrophy, a desired drug target, but also impairs synaptic plasticity by a reduced spine formation, as seen in mental retardation, in particular *Rett Syndrome* (29, 30) and impairs motocontrol (31, 32). TRPV1 is now identified as an antiepileptogenic target. Nerve growth factor (NGF) upregulates TRPV1 expression and triggers epileptogenesis. Also the endocannabinoid anandamide (AEA) activates TRPV1 endogenously and AEA levels are increased in epilepsy. Activation of TRPV1 also triggers apoptotic neuronal death in chronic epilepsy (33). See also that TRPV1, in cooperation with the endocannabinoid system, influences GABAergic and glutamatergic synapses and play a modulatory function on dopamine transmission. Through these mechanisms TRPV1 and endocannabinoids may have an important influence on various neurobiological processes (e.g., control of movement, motivation/reward) and, particularly, on different pathologies affecting these processes like basal ganglia disorders, schizophrenia, and drug addiction (34). But again, is TRPV1 really expressed in the CNS (16, 17)?

Just another example: The "biogenesis" model for MLIV pathogenesis suggests that TRPML1 modulates postendocytic junk delivery to lysosomes by regulating interactions between late endosomes and lysosomes. Effects of TRPML1 loss on hydrolytic activity have a cumulative effect on lysosome function, resulting in a lag between TRPML1 loss and full manifestation of MLIV (35). Disruption of the TRPML1 channel causes also a defective autophagy, which results in oxidative stress and impaired synaptic transmission (36, 37). Thus, TRPML1 is a target for treatment of neuro-degenerative diseases and even Mucolipidosis type IV patients. And what about the iron permeation of TRPML1? A new drug target? Are the respective pathways really cleared up (38, 39)? We have to know the channel function in the correct lipid environment. A very promising example is now the development of patching intracellular organelles, such as lysosomes (40, 41).

Of course, the pharmaceutical industry focuses on small molecules, which are synthesized in vitro and screened with high power methods. Natural compounds teach us a lot about TRP channels. The problem seems to be that they are difficult to synthesize, but is there not a huge intellectual quest for natural compounds? Since more than 5,000 years, mankind fights for spices added to our food. And so many of them are just wonderful TRP channel modulators (42). Interestingly, some natural compounds which have been used in traditional medicine as antidepressants have TRP channels as targets. One of these compounds is *Incensole* acetate which is released by the burning of resin from the *Boswellia* plant has been used for religious and cultural ceremonies for millennia. It activates TRPV3, which is expressed in the brain and causes anxiolytic-like and antidepressive-like behavioral effects (43, 44). *St. John's Wort* has been used medicinally for over 5,000 years. Relatively recently, one of its phloroglucinol derivatives, *hyperforin*, an antidepressive compound, has been identified as effective activator of TRPC6 (45, 46). Hyperforin has been shown to have cognitive enhancing, memory facilitating properties and has probably neuroprotective effects (47). Many more examples can be given (48). However, estimations are done that are less than 1% of all blooming plants have ever been tested for natural compounds which might be useful as medical compounds although all these molecules have past a more than 3.5 billions of years lasting evaluation procedure in evolution!

To close up, again a new book about TRP channels focused on TRPs as drug targets and the required methods (see also (15))! Why is this all so important? I will just close to pin-point one new and exciting field of TRP functions, the brain, which requires so much refined new methods and we need them in our survey on TRPs as drug targets, as recently published (49). TRPs as analgesic targets (the starting point of this explosive worldwide quest for TRP modulators), as targets for bladder diseases, skin diseases, respiratory diseases, heart failure/hypertrophy, but also for hereditary humans diseases (see (10)) has been identified. However, seen the huge social impact, I will focus shortly on the brain. Many discussions have been started about "age as a drug target"! TRPs are again on the stage. But is this hope reliable? Do we have good methods to tackle this exciting perspective? TRPM7 and TRPM2 have been put forward as potential factors in neurodegenerative diseases including Alzheimer´s disease, Parkinson's disease, stroke and diseases associated with oxidant-mediated neuronal damage (50). Especially TRPM2, which is highly expressed in the striatum (caudate nucleus and putamen), is supposed to play a key role in bipolar disorders (BDs) (51, 52). Recent case–control studies implicate TRPM2 conferring risk for bipolar disorder (BD) and genetic variants of TRPM2 have been identified to be coupled with BD supporting a role for this channel in the pathogenesis of this disorder (53) (see for a review (54)). Mutations in TRPM2 have been discovered which result in rapidly inactivated channels which are unable to maintain sustained Mg^{2+} influx (55–58). Via another pathomechanisms, TRPM7 is involved in the *Familial Alzheimers Disease* (*FAD*) which is linked to aberrant $PI(4,5)P_2$ metabolism causing a dysregulation of this channel. The $PI(4,5)P_2$ turnover is affected by the presenilin mutants and the cellular $PI(4,5)P_2$ content correlates with the contents of the 42-residue amyloid ß-peptide (Aß42). Thus FAD mutant presenilin generates toxic Aß42 levels and causes TRPM7 dysfunction (59). Just appeared as a functional channel: TRPM1, the founding member of the TRPM family and until now mainly known as a tumor suppressor and a retinal channel for light perception, could be linked to a complex neurodevelopmental disorder characterized by severe visual impairment, intellectual disability, and refractory epilepsy. This disease is caused by a microdeletion in chromosome 15q13.3 carrying the *trpm1* gene (60). Aging increases also substantially the risk for ischemic brain injury elicits by a sequence of complex biochemical changes including oxidative stress, edema, inflammation, and excitotoxicity. TRPM2 channels are activated by such conditions and downregulation of this channel clearly causes a neuroprotective effect in the cerebral cortex and hippocampus after brain injury, oxidative stress, inflammation and neuronal death (61). The Ca^{2+} activated TRP channel, TRPM4, is involved in the damaging secondary events that accompany brain injury. Changes in capillary permeability result in the extravasation of extracellular fluid, of inflammatory cells and blood, thereby producing cerebral edema, inflammation and progressive secondary hemorrhage. Inhibition of TRPM4 channels provides beneficial effects in animal models of traumatic brain injury and ischemia associated cerebral edema and secondary hemorrhage (62). Therefore, TRPM4 inhibition plays a neuroprotective role, reduces in brain or spinal cord lesion volume and produced a substantial improvement in neurological function (28). But see the already mentioned negative effects! Also TRPM7 is a potential target for neuroprotection after brain injury. Suppressing the expression of TRPM7 in hippocampal CA1 neurons conferred resistance to ischemic cell death, preserved cell function and prevented ischemia-induced deficits in memory (63, 64). Brain injury confers a major burden to Western society and effective treatments are urgently required! But again, what activates TRPM2 and TRPM7 under those conditions? Depletion of intracellular Mg^{2+}, a symptom of ischemic brain injury and a reduction of extracellular Ca^{2+} are both associated with poor neurological outcome and both are conditions which activate TRPM7 and possibly increasing the Ca^{2+} load of neuronal cells. This leads to a secondary injury processes and to cell death (65). Importantly, TRPM7 gene variation might play a role in the risk of ischemic stroke (66). These are all wonderful perspectives! But which mechanisms are involved? How can we really understand that TRP channels are causative in all the sketched diseases? We have to know more and need refined methods! This is the message of this book and a solution of many methodical problems is offered!

This introduction has focused only on a few problems and I have tried to provocatively highlight the need to better methods for an in-depth understanding how TRPs function. An excellent overview on TRP channels as targets for hereditary and acquired diseases is published and is wholeheartedly recommended to all TRPist´s before start reading this book and searching for suitable methods and answers to methodical questions (19)!

References

1. Montell C, Rubin GM (1989) Molecular characterization of the Drosophila trp locus: a putative integral membrane protein required for phototransduction. Neuron 2(4):1313–23
2. Cosens DJ, Manning A (1969) Abnormal electroretinogram from a Drosophila mutant. Nature 224(5216):285–7
3. Gamal El-Din TM et al (2011) Double gaps along Shaker S4 demonstrate omega currents at three different closed states. Channels (Austin) 4:93–100
4. Jurkat-Rott K et al (2010) Sodium channelopathies of skeletal muscle result from gain or loss of function. Pflugers Arch 460(2):239–48

5. Prütting S, Grissmer S (2011) A novel current pathway parallel to the central pore in a mutant voltage-gated potassium channel. J Biol Chem 286:20031–42
6. Abramowitz J, Birnbaumer L (2006) Know thy neighbors: a survey of diseases and complex syndromes that map to chromosomal regions encoding TRP channels. In: Flockerzi V, Nilius B (eds) TRP channels. Springer, New York, pp 377–406
7. Nilius B (2007) TRP channels in disease. Biochim Biophys Acta 1772(8):805–12
8. Nilius B et al (2007) Transient receptor potential channels in disease. Physiol Rev 87:165–217
9. Nilius B, Owsianik G (2010) Channelopathies converge on TRPV4. Nat Genet 42(2):98–100
10. Nilius B, Owsianik G (2010) Transient receptor potential channelopathies. Pflugers Arch 460:437–50
11. Jordt SE, Ehrlich BE (2007) TRP channels in disease. Subcell Biochem 45:253–71
12. Karashima Y et al (2010) Agonist-induced changes in Ca(2+) permeation through the nociceptor cation channel TRPA1. Biophys J 98(5):773–83
13. Samways DS, Egan TM (2011) Calcium-dependent decrease in the single-channel conductance of TRPV1. Pflugers Arch 462: 681–91
14. Sakmann, B. and E. Neher, *Single-Channel Recording*. Second Edition ed. 1995, New York and London: Plenum Press.
15. Zhu MX (2011) TRP channels. In: Eichberg JE (ed) Methods in signal transduction. CRC Press/Taylor & Francis Group, Boca Rota, FL
16. Cavanaugh DJ et al (2011) Trpv1 reporter mice reveal highly restricted brain distribution and functional expression in arteriolar smooth muscle cells. J Neurosci 31:5067–77
17. Cavanaugh DJ et al (2011) Restriction of transient receptor potential vanilloid-1 to the peptidergic subset of primary afferent neurons follows its developmental downregulation in nonpeptidergic neurons. J Neurosci 31: 10119–27
18. Montell C (2005) The TRP superfamily of cation channels. Sci STKE 272:re3
19. Montell C (2003) The venerable inveterate invertebrate TRP channels. Cell Calcium 33(5–6):409–17
20. Ong HL, Ambudkar IS (2011) The dynamic complexity of the TRPC1 channelosome. Channels (Austin) 5:424–31
21. Cuajungco MP et al (2006) PACSINs bind to the TRPV4 cation channel: PACSIN 3 modulates the subcellular localization of TRPV4. J Biol Chem 281:18753–62
22. D'hoedt D et al (2007) Stimulus-specific modulation of the cation channel TRPV4 by PACSIN 3. J Biol Chem 283:6272–80
23. Shin YC et al (2011) TRIP Database: a manually curated database of protein-protein interactions for mammalian TRP channels. Nucleic Acids Res 39:D356–61
24. Kiselyov K et al (2005) Protein-protein interaction and functionTRPC channels. Pflugers Arch 451(1):116–24
25. Mathar I et al (2010) Increased catecholamine secretion contributes to hypertension in TRPM4-deficient mice. J Clin Invest 120: 3267–79
26. Shimizu T et al (2009) TRPM4 regulates migration of mast cells in mice. Cell Calcium 45(3):226–32
27. Vennekens R et al (2007) Increased IgE-dependent mast cell activation and anaphylactic responses in mice lacking the Ca2+-activated non-selective cation channel TRPM4. Nat Immunol 8:312–20
28. Gerzanich V et al (2009) De novo expression of TRPM4 initiates secondary hemorrhage in spinal cord injury. Nat Med 15:185–191
29. Amaral MD, Chapleau CA, Pozzo-Miller L (2007) Transient receptor potential channels as novel effectors of brain-derived neurotrophic factor signaling: potential implications for Rett syndrome. Pharmacol Ther 113(2): 394–409
30. Amaral MD, Pozzo-Miller L (2007) TRPC3 channels are necessary for brain-derived neurotrophic factor to activate a nonselective cationic current and to induce dendritic spine formation. J Neurosci 27(19):5179–89
31. Hartmann J et al (2008) TRPC3 channels are required for synaptic transmission and motor coordination. Neuron 59(3):392–8
32. Hartmann J, Henning HA, Konnerth A (2011) mGluR1/TRPC3-mediated synaptic transmission and calcium signaling in mammalian central neurons. Cold Spring Harb Perspect Biol 3(4).
33. Fu M, Xie Z, Zuo H (2009) TRPV1: a potential target for antiepileptogenesis. Med Hypotheses 73:100–102
34. Fernandez-Ruiz J, Hernandez M, Ramos JA (2010) Cannabinoid-dopamine interaction in the pathophysiology and treatment of cns disorders. CNS Neurosci Ther 16:e72–91

35. Miedel MT et al (2008) Membrane traffic and turnover in TRP-ML1-deficient cells: a revised model for mucolipidosis type IV pathogenesis. J Exp Med 205:1477–90
36. Venkatachalam K et al (2008) Motor deficit in a Drosophila model of mucolipidosis type IV due to defective clearance of apoptotic cells. Cell 135:838–51
37. Venugopal B et al (2009) Chaperone-mediated autophagy is defective in mucolipidosis type IV. J Cell Physiol 219:344–53
38. Dong X-P et al (2008) The type IV mucolipidosis-associated protein TRPML1 is an endolysosomal iron release channel. Nature 455:992–6
39. Kiselyov K et al (2011) TRPML: transporters of metals in lysosomes essential for cell survival? Cell Calcium 50:288–94
40. Dong XP et al (2009) Activating mutations of the TRPML1 channel revealed by proline-scanning mutagenesis. J Biol Chem 284: 32040–52
41. Schieder M et al (2011) Planar patch clamp approach to characterize ionic currents from intact lysosomes. Sci Signal 3:pl3
42. Nilius B, Appendino G (2011) Tasty and healthy TR(i)PS: the human quest for culinary pungency. EMBO Rep 12:1094–1101
43. Moussaieff A, Mechoulam R (2009) Boswellia resin: from religious ceremonies to medical uses; a review of in-vitro, in-vivo and clinical trials. J Pharm Pharmacol 61(10):1281–93
44. Moussaieff A et al (2008) Incensole acetate, an incense component, elicits psychoactivity by activating TRPV3 channels in the brain. FASEB J 22:3024–34
45. Leuner K et al (2007) Hyperforin-a key constituent of St. John's wort specifically activates TRPC6 channels. FASEB J 21(14):4101–11
46. Leuner K et al (2010) Simple 2,4 diacylphloroglucinols as TRPC6 activators – identification of a novel pharmacophore. Mol Pharmacol 77:368–77
47. Griffith TN et al (2010) Neurobiological effects of hyperforin and its potential in alzheimer's disease therapy. Curr Med Chem 17:391–406
48. Vriens J, Nilius B, Vennekens R (2008) Herbal compounds and toxins modulating TRP channels. Curr Neuropharmacol 6(1):79–96
49. Moran MM et al (2011) Transient receptor potential channels as therapeutic targets. Nat Rev Drug Discov 10:601–20
50. McNulty S, Fonfria E (2005) The role of TRPM channels in cell death. Pflugers Arch 451:235–242
51. Uemura T et al (2005) Characterization of human and mouse TRPM2 genes: identification of a novel N-terminal truncated protein specifically expressed in human striatum. Biochem Biophys Res Commun 328:1232–43
52. Xu C et al (2006) Association of the putative susceptibility gene, transient receptor potential protein melastatin type 2, with bipolar disorder. Am J Med Genet B Neuropsychiatr Genet 141:36–43
53. Xu C et al (2009) TRPM2 variants and bipolar disorder risk: confirmation in a family-based association study. Bipolar Disord 11:1–10
54. Chahl LA (2007) TRP's: links to schizophrenia? Biochim Biophys Acta 1772:968–77
55. Hermosura M et al (2008) Altered functional properties of a TRPM2 variant in Guamanian ALS and PD. Proc Natl Acad Sci USA 105:18029–34
56. Hermosura MC, Garruto RM (2007) TRPM7 and TRPM2-Candidate susceptibility genes for Western Pacific ALS and PD? Biochim Biophys Acta 1772:822–35
57. Hermosura MC et al (2002) Dissociation of the store-operated calcium current ICRAC and the Mg-nucleotide-regulated metal ion current MagNuM. J Physiol 539:445–58
58. Hermosura MC et al (2005) A TRPM7 variant shows altered sensitivity to magnesium that may contribute to the pathogenesis of two Guamanian neurodegenerative disorders. Proc Natl Acad Sci USA 102:11510–5
59. Landman N et al (2006) Presenilin mutations linked to familial Alzheimer's disease cause an imbalance in phosphatidylinositol 4,5-bisphosphate metabolism. Proc Natl Acad Sci USA 103:19524–29
60. Lepichon JB et al (2010) A 15q13.3 homozygous microdeletion associated with a severe neurodevelopmental disorder suggests putative functions of the TRPM1, CHRNA7, and other homozygously deleted genes. Am J Med Genet A152A:1300–4
61. Cook NL et al (2010) Transient receptor potential melastatin 2 expression is increased following experimental traumatic brain injury in rats. J Mol Neurosci 42:192–9
62. Simard JM, Kahle KT, Gerzanich V (2010) Molecular mechanisms of microvascular failure in central nervous system injury-synergistic roles of NKCC1 and SUR1/TRPM4. J Neurosurg 113:622–9
63. Rempe DA, Takano T, Nedergaard M (2009) TR(I)Pping towards treatment for ischemia. Nat Neurosci 12:1215–16

64. Sun HS et al (2009) Suppression of hippocampal TRPM7 protein prevents delayed neuronal death in brain ischemia. Nat Neurosci 12:1300–07

65. Cook NL, Van Den Heuvel C, Vink R (2009) Are the transient receptor potential melastatin (TRPM) channels important in magnesium homeostasis following traumatic brain injury? Magnes Res 22:225–34

66. Romero JR, Ridker PM, Zee RY (2009) Gene variation of the transient receptor potential cation channel, subfamily M, member 7 (TRPM7), and risk of incident ischemic stroke. Prospective, nested, case-control study. Stroke 64:791–97

Chapter 2

TRP-Mediated Cytoskeletal Reorganization: Implications for Disease and Drug Development

Chandan Goswami

Abstract

So far the major focus of Transient Receptor Potential (TRP) channels in the context of pathophysiological disorders was centered exclusively on the ionic conductivity mediated by these channels. However, recently the importance of non-ionic functions of TRP channels in different pathophysiological disorders has emerged. Recently several physical and functional interactions of TRP channels with cytoskeletal components have been characterized. These interactions play important roles in executing the non-ionic functions and regulations of TRP channels per se. In the membranous environment, TRP channels form dynamic signaling complexes that include components like microtubule and actin cytoskeleton, other scaffolding and key regulatory components. TRP channels can also regulate the integrity and dynamics of different cytoskeletal systems in complex manner. In many cases, these regulations seem to be independent of Ca^{2+} influx mediated by these channels and thus have immense significance in the context of pathophysiological disorders. In this review, I highlight the importance of TRP channel interactions and multi-directional regulations with cytoskeletal components in detail. This aspect opens up new avenues to target TRP signaling complexes by pharmacological manners. The strategies to target TRP complexes rather than targeting TRP channel solely might be useful for several clinical purposes.

Key words: TRPV, Cytoskeleton, Signalplex, Ion channels, Tubulin, Actin

Abbreviations

4αPDD	4α-phorbol-didecanoate
CIRB domain	Calmodulin- and IP_3R-binding region
CMT2	Charcot–Marie-Tooth disease type 2
DRG neurons	Dorsal root ganglion neurons
eGFP	Enhanced green fluorescence protein
EGTA	Ethylene glycol tetraacetic acid
FRET	Fluorescence Resonance Energy Transfer
5'I-RTX	5'-iodoresiniferatoxin
HUVEC	Human umbilical vein endothelial cell
TRP Channels	Transient receptor potential channels
PC2	Polycystin-2
TRPN	NomPC-like TRP channel

Arpad Szallasi and Tamás Bíró (eds.), *TRP Channels in Drug Discovery: Volume I*, Methods in Pharmacology and Toxicology, DOI 10.1007/978-1-62703-077-9_2, © Springer Science+Business Media, LLC 2012

TRPC	Transient receptor potential canonical
TRPML	Transient receptor potential mucolipin
MAPs	Microtubule-associated proteins
MBP	Maltose-binding protein
NF200	Neurofilament heavy chain 200 KDa
OAG	1-oleoyl-2-acetyl-sn-glycerol
PKC	Protein kinase C
PKD	Polycystic kidney disease
PKCε	Protein kinase Cε sub type
RTX	Resiniferatoxin

1. Introduction

Though the importance of cytoskeleton in different signaling events and cellular functions are well established, the entire arrays of cytoskeletal organization at the sub-membranous region and complexity of the cytoskeleton at the lipid environment have not yet been understood (1, 2). While the presence and function of cortical actin cytoskeleton just beneath the plasma membrane are well established, the events and mechanisms by which membrane proteins and sub-membranous microtubule cytoskeleton regulate each other and execute multiple cellular functions are just emerging (3). In that context, it is important to mention that so far several proteomic studies indicate that components from microtubule cytoskeleton physically interact with several transmembrane proteins such as ion channels, pumps, and receptors (4–7). These reports are also in full agreement with the fact that tubulin, actin, and other cytoskeletal proteins are selectively enriched in several biochemical preparations which represent different membrane fractions (8–11). Even the presence of tubulin and actin is observed in very specialized subset of total membrane fraction, namely in the lipid rafts and/or in the post-synaptic density fraction of the synaptic membranes (9, 12–14). Taken together, both membrane-associated actin and tubulin represent membrane cytoskeleton which is highly relevant for several signaling and provide important platform on which the functions of the membrane proteins are dependent. Surprisingly, the dynamics, biochemical characteristics, regulations, and functions of the sub-membranous microtubule cytoskeleton are still poorly understood (15, 16). However, for simplicity it can be said that a small portion of the microtubule cytoskeletal components are selectively present in the membrane fraction. These components constitute different dynamic complexes that are primarily formed by the transmembrane proteins. Within these complexes, the cytoskeletal components can interact directly or indirectly with the membrane proteins.

Transient Receptor Potential (TRP) channels are a group of newly discovered non-selective cation channels that can be activated by several physical and chemical stimuli and are involved in several physiological functions (17–20). Dysfunction and mis-regulation of TRP channels have been linked with several pathophysiological and genetic disorders (21). So far, research with TRPs has focused mainly on the ionic conductivity mediated by these channels. In spite of several reports demonstrating that TRP channels share co-localization with several cytoskeletal proteins at highly specific sub-cellular locations, the importance of TRP-cytoskeleton cross-talk in the context of structure–function and regulation has been neglected for a long time. Only in recent time, the importance of cytoskeletal proteins in the multi-dimensional regulation of TRP channels has come to light. So far, a handful of reports suggest that cytoskeletal proteins play an important role in the context of structure–function and regulation of TRP channels. In the same context, a number of studies have indicated that TRP channels physically interact with actin and microtubule cytoskeleton as well as with many other proteins at the plasma membrane. Therefore the significance of these interactions is manifested by the functional aspects of TRP channels. All these studies also indicate that cytoskeletal, vesicular and other membrane regulatory proteins interact with TRP channels and form scaffolds at the plasma membrane which can be described as dynamic functional complexes central to many physiological functions. Improper structure–function and regulation of these complexes seem to be intimately associated with the development of diseases and pathophysiological conditions. Therefore these complexes represent potential, specific and novel pharmacological targets.

It is noteworthy to mention that so far crosstalk between TRP channels with cytoskeleton has not been investigated properly though the indication of such crosstalk was reported long back. For example, it has been demonstrated that capsaicin-responsive DRG neurons are devoid of neurofilament 200 kDa protein (Capsaicin-responsive dorsal root ganglion (DRG) neurons cannot be labeled with a monoclonal antibody (RT97) that detects NF200 kDa) though the reason for such specific regulation is still not known (22, 23). Nevertheless, understanding of such fine regulations between TRP channels and cytoskeleton has tremendous importance in the case of several pathophysiological disorders and diseases. In this review I highlight the interaction of different cytoskeletal proteins with TRP channels at several levels and also how these complexes are regulated.

2. Cross-Talk Between TRP Channels and Cytoskeleton: Co-localization and Genetic Interactions

The importance of cytoskeleton in the context of function and regulation of TRP channels came from the common observation that these channels and specific cytoskeletal proteins are co-expressed in some specialized cells. Often these two groups of proteins are located at distinct subcellular structures also, a topic that has been discussed in detail by us recently (24). As these subcellular structures are characterized by the presence of these specialized proteins and/or by intricate cytoskeletal organization, specific localization strongly suggests that TRP channels either interact with some of the cytoskeletal proteins and/or are involved with the development as well as function of these structures. For example, polycystine channel sub type 2 (PC2) co-localizes with polyglutamylated tubulin at the basal bodies/cilia of ciliated epithelial cells present in mouse trachea (25). This co-localization is also in agreement with the involvement of PC2 channel in ciliary function. PC2 forms a complex with pericentrin and this interaction is also required for primary cilia assembly (25). In the same context, both PC1 and PC2 are present in the primary cilium of kidney cells (26). PC2 channel is also localized at the primary cilia of renal epithelial cells (27). *Xenopus* TRPN1 (NOMPC) localizes to the tip of the microtubule-based cilia in epithelial cells (kinocilial bulb) and tip of the inner-ear hair cells (28). There TRPN co-localizes with cytoskeletal components like actin, tubulin, and Cdh23. In *Drosophila melanogaster*, NOMPC (a member of the TRP channel family) localizes to the tubular body and distal cilium of Campaniform and Chordotonal receptor cells and is involved in these ciliary functions (29). In *Drosophila*, TRPN (=NOMPC) localizes at the distal end of mechanosensory cilia also and co-localizes with EYS (an extracellular protein that marks the proximal end of the sensory cilia) (30). TRPC6 localizes in podocytes where it interacts with podocin and nephrin, components that belong to actin cytoskeleton (31). These examples suggest that TRP channels localize at the specific cytoskeleton-enriched structures and share a special relation with the cytoskeletal components.

Like PC, TRPN and TRPC channels, recent results suggest that TRPV channels share physical and functional interactions with cytoskeletal components. TRPV1 and TRPV4 are reported to be localized at the tip of filopodia in both neuronal and non-neuronal cells when expressed ectopically (24, 32, 33). This is also in agreement with the endogenous localization of TRPV1, TRPV4 and other TRP channels at the spines (33–35). Interestingly, expression of TRPV1 induces filopodia that possess a characteristic bulbous head which contains negligible amount of F-actin but accumulates TRPV1 there (24, 32). This phenotype resembles well with the expression of the non-conventional myosin II, III, V, X and XV (24, 36–47). These similarities suggest that overexpression of TRPV1 may alter the function of these myosin motors and execute similar dominant-negative effects. Indeed, changes in the expression pattern as well as distribution of certain cytoskeletal proteins including non-conventional myosin motors after the ectopic expression and/or activation of TRP channels have been reported (32, 48, 49). Ectopic expression of TRPV1 in F11 cells results in altered expression as well as reorganization of non-conventional myosins, namely endogenous myosinIIa and myosinIIIa (32). In agreement with this, observation another study has also confirmed that overexpression of TRPC6 in transgenic mice resulted in an increased expression of beta-myosin heavy chain in cardiac tissues (50).

In many cases, the development and the function of these specialized cells/structures are regulated by both TRP channels and these cytoskeletal proteins. In agreement with this, mutations in either TRP channels or specific cytoskeletal proteins often lead to similar, if not same, phenotype as well as pathophysiological disorders and/or syndromes. Taken together, involvements of these two groups of proteins in common functions and occurrence in same cell (even in the same sub-cellular regions) are highly indicative of physical, functional, and genetic interactions (51). As the examples are too many, it is impossible to cover all these in this review. However, some key examples, like multi-dimensional relation of TRP channels with different motor proteins are described here.

For example, as is the case in many ciliary proteins, mutations either in PC1 and PC2 are also involved in polycystic kidney disease (PKD) and result in defective localization, cilia formation and/or loss of flow-induced Ca^{2+} signaling (26, 52). This agrees with the fact that PC channels are regulated by microtubule-based motor proteins such as KIF3a and KIF3b (53). In a similar context, mutations in either kinesin (Kif1b) or TRPV4 result in similar pathophysiology and development of Charcot–Marie-Tooth disease type 2 (CMT2) disease suggesting a strong genetic link between these two (21, 54–57). Mutations in TRP channels as well as in different non conventional myosin motors are also reported to develop similar pathophysiological disorders and other syndromes like deafness and blindness. For example, both the development and proper function of the stereocilia of hair cells are important for hearing. In normal conditions, the ciliary tips of hair cells contain enriched amounts of endogenous TRP channels as well as several nonconventional myosin motor proteins, indicating that the function of these cells are dependent on these two groups of proteins at the ciliary tips. Indeed, several reports suggest that in the case of deafness, several nonconventional myosin motors (myosin I, IIA, IIIA, VI, VIIA and XV) are important for either development of the stereocilia of hair cells in the inner ear or proper localization of

TRP channels at the tips of these stereocilia (58–60). Reciprocally, mutations and abnormal expression–functions of several TRP channels, namely NompC, TRPML1, TRPML2, TRPML3, TRPV4, TRPV5, and TRPV6 also lead to deafness (30, 61–68).

Like auditory defects, development, polarization of retinal cells and proper trafficking of pigments in the retinal cells are involved in the proper light-sensing mechanisms. In case of blindness, both TRP channels and non-conventional myosins are involved (60). Retrospectively, TRP channel was first discovered by Minke et al. in *Drosophila melanogaster* as the mutant was defective in light-sensing mechanism (69). Indeed, so far several TRP channels have been reported to express in retinal cells. Some of these TRP channels are involved in photo-response and essential for the light sensation as mutation in these TRP channels causes different forms of blindness (70). For example, mutation in TRPM1 is responsible for blindness as it is involved in retinal ON bipolar function (71, 72). In agreement with the involvement in common functions, mutations in myosin motors are also involved in blindness. Mutation in myosin VIIa is involved in the development of "Usher syndrome type 1B" (60). In *Drosophila*, Ca^{2+}-activated myosin V is involved with the closure of the pupil and thus with the light sensation procedure (73). Apart from the genetic interaction, recent reports clearly indicate physical as well as functional interactions between these two groups of proteins. Recently it has been reported that translocation of eGFP-tagged TRP-like channels to the rhabdomeral membrane in *Drosophila* photoreceptors is myosin III dependent (74). A recent proteomic screen has also identified myosin as an interacting protein for TRPC5 and TRPC6 (75). Another study showed that myosin IIa is directly phosphorylated by TRPM7, a cation channel fused to an alpha-kinase (76). In the same notion, a recent proteomic screen has identified the heavy chain of myosin X and cytosolic dynein heavy chain as an interacting protein of TRPC3 (77). All these results suggest that TRP channels and some of the specific cytoskeletal proteins like kinesins and nonconventional myosins are involved in same functions. However, detailed studies are needed to understand these genetic interactions.

3. Physical Interaction of TRP Channels with Microtubule Cytoskeleton

In the last few years, major progress has been made to elucidate and further characterize the physical interaction of TRP channels with components from microtubule cytoskeleton like tubulin, microtubule associated proteins (MAPs) and different motor proteins (3, 6, 78). So far, direct physical interaction of tubulin has been reported for members belonging to TRPV, TRPC, and TRPP

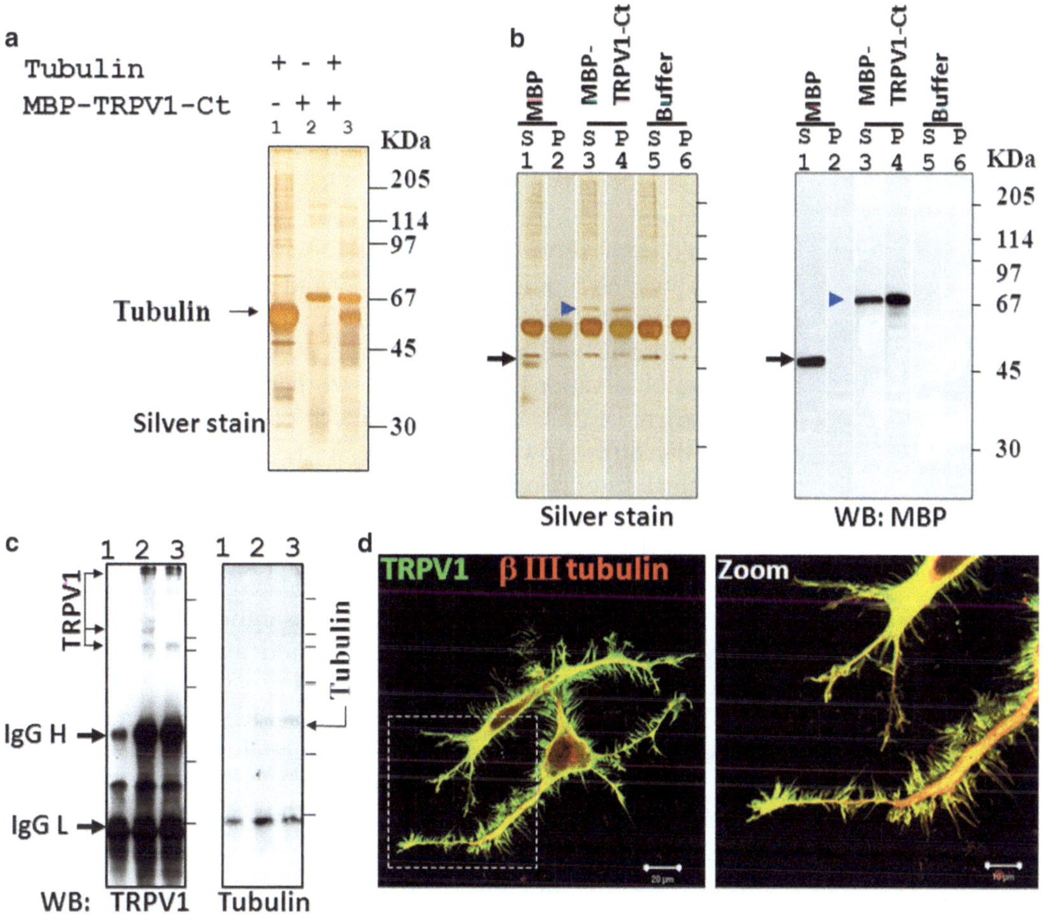

Fig. 1. TRPV1 interacts directly with microtubule cytoskeleton and induces cellular changes upon expression as well as activation. (**a**) MAP-enriched tubulin (lane 1, input) was added to the MBP-TRPV1-Ct coupled with amylase resin (lane 2). A significant fraction of the tubulin interacts with the MBP-TRPV1-Ct. (**b**) The MBP-TRPV1-Ct directly interacts with polymerized microtubules. Purified tubulin dimers were incubated with GTP to form microtubules either in the presence of MBP only (lane 1–2) or in the presence of MBP-TRPV1-Ct (lane 3–4) or in the presence of buffer only (lane 6–7). Polymerized microtubules and bound proteins were subsequently separated from unpolymerized tubulin dimers or unbound proteins by centrifugal separation of pellet (**P**) fraction from supernatant fraction (**S**). Silver-stained gel (*left side*) and anti-MBP western blot analysis reveal specific interaction of MBP-TRPV1-Ct (*blue arrow head*) with polymerized microtubules. MBP alone does not interact with polymerized microtubules and thus do not appear in the pellet fraction. (**c**) Tubulin co-immunoprecipitates with TRPV1. Anti-GFP antibody was used for immunoprecipitation from F11 cells transiently expressing GFP only (lane 1) or GFP TRPV1 (lane 2). Immunocomplexes were probed for TRPV1 (*left side*) and tubulin (*right side*). Presence of tubulin is detected only in lane 2, but not in lane 1. (**d**) Ectopic expression of TRPV1 (*green*) alters cellular morphology and induces multiple filopodial structure. TRPV1 co-localizes with neuron-specific β-tubulin subtype III (*yellow*) in such filopodial structures. Scale bar 20 mm and 10 mm respectively.

channels (3, 6, 7, 75, 79). Here I discuss the details of these interactions.

Probably the best characterization for the interaction of TRP channels with tubulin has been illustrated by TRPV1, alternatively known as capsaicin receptor (Fig. 1). By proteomic analysis, we identified tubulin as a component present in the complex formed with the C-terminal cytoplasmic domain of TRPV1 (80). The interaction was subsequently confirmed by several biochemical approaches including co-immunoprecipitation, microtubule co-sedimentation, direct pull-down assay and cross-linking experiments (49, 80). This interaction is direct as both purified tubulin and the C-terminal cytoplamic domain of TRPV1 tagged with maltose-binding protein (MBP-TRPV1-Ct) can form a stable complex. We identified two short regions located within the C-terminus of TRPV1, namely amino acids 710–730 and 770–797 that can retain tubulin interaction independently (49). In contrast, MBP-TRPV1-Nt failed to interact with tubulin in a direct pull-down assay (80). Based on these observations, it was proposed that the tubulin interaction was restricted within the C-terminal cytoplasmic region of TRPV1 only (49, 80). However, recently it has been shown that the N-terminal cytoplasmic region of TRPV1 can also interact with tubulin (81). This difference might be due to the experimental systems and the procedures used. Taken together, this suggests that there might be more than one region located in TRPV1 that can be involved in tubulin interaction. It might also suggest that TRPV1-tubulin interaction is dynamic and might be involved in the conformational changes.

The C-terminal cytoplasmic region of TRPV1 preferably interacts with the β-tubulin and to a lesser extent also with the α-tubulin (49). The cross-linking experiment revealed that MBP-TRPV1-Ct interacts with β-tubulin quickly and the entire amount of β-tubulin forms a high-molecular weight complex with MBP-TRPV1-Ct within a minute. In contrast, the MBP-TRPV1-Ct interacts with α-tubulin slowly and almost half of the β-tubulin fails to form high-molecular weight complex with MBP-TRPV1-Ct even after an hour. This also suggests a stronger binding of TRPV1 to the plus end rather than the minus end of microtubules as the plus ends of microtubule proto-filaments are decorated with β-tubulin. It is therefore tempting to speculate that TRPV1 may act as a microtubule plus-end-tracking protein (+TIP)(6). However, whether TRPV1 can indeed serve as a plus-end-tracking protein remains to be explored.

A significant understanding about the TRPV1 interaction with tubulin has been derived from sequence analysis of the binding regions. Interestingly, there are two short tubulin-binding stretch sequences that reveal tubulin-binding ability. These two sequence stretches contain highly basic amino acids and contain very high isoelectric points, 11.17 and 12.6 (49). In the context of microtubule interaction, these two regions can act as polycationic stretch sequences that can favor microtubule formation and stabilize them (82–84). Indeed, these two short stretch sequences can also interact with soluble tubulin as well as with polymerized microtubules (49). Interestingly, if assumed to form α-helical conformation, then all the basic amino acids present in these two regions are projected to one side, suggesting potential interactions with negatively charged

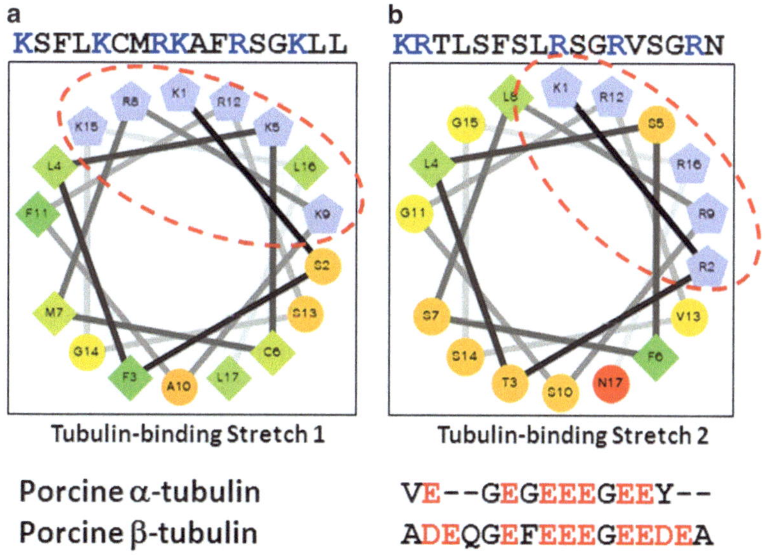

Fig. 2. Tubulin-binding motifs located at the C-terminus of TRPV1 are characterized by the presence of multiple positive charged residues. (**a**) The basic amino acids (indicated in *blue*) are located at one side of the putative helical wheel, where it can interact with the acidic C-terminus of tubulin. (**b**) The extreme C-terminus of both α- and β-tubulin contains highly negatively charged amino acids (indicated in *red*) and is mostly unstructured.

surface. In this context, it is important to note that the C-terminal over-hanging regions of tubulin contain a large number of negatively charged glutamate (E) residues in a stretch characterized as an unstructured region of the tubulin and thus referred to as "E-hook". Due to the presence of several negatively charged residues, the E-hook can be important for the interaction with the TRPV1, especially to these positively charged regions (Fig. 2). In agreement with this, previously we have demonstrated that $\alpha_s\beta_s$-tubulin (protease-digested tubulin dimers that lack approximately 45 amino acids from the C-terminal region including the E-hooks) does not bind to the MBP-TRPV1-Ct (49). More interestingly, in many TRPV channels the distribution of these basic amino acids composing the tubulin-binding regions is conserved even though the overall amino acid conservation is rather limited. This may suggest that tubulin interaction is apparently under high evolutionary pressure and it might be conserved in many TRP channels. Indeed, the C-terminal cytoplasmic region of TRPV4 also reveals interaction with tubulin (33). Similarly, the C-terminal region of the TRPV2 also interacts with purified tubulin (Unpublished observation). However, the exact amino acids of TRPV4 and TRPV2 that are involved in the tubulin interaction have not been determined yet.

It is important to note that different post-translationally modified tubulin, like tyrosinated tubulin (a marker for dynamic microtubules), acetylated tubulin, polyglutamylated tubulin, detyrosinated tubulin, phospho (serine) tubulin and neuron-specific β-III tubulin (all markers for stable microtubules) interact with

MBP-TRPV1-Ct (49). The same phenomenon was also observed with MBP-TRPV4-Ct (33). These results strongly suggest that TRPV channels interact not only with soluble tubulin, but also with assembled microtubules which represent various dynamic states of the microtubules. Indeed, the purified cytoplasmic domain of TRPV1-Ct, TRPV2-Ct, and TRPV4-Ct co-sediment with polymerized microtubules also. In addition to sole binding, MBP-TRPV1-Ct and MBP-TRPV4-Ct exert strong stabilization effect on microtubules. This stabilization effect of the C-terminus of TRPV1 and/or TRPV4 becomes especially apparent under microtubules depolymerizing conditions such as in the presence of nocodazol or increased Ca^{2+} concentrations (33, 80). These observations fit well with the fact that polycations favor microtubule polymerization and stabilization (82).

As TRPV1 represents a non-selective cation channel, the role of increased concentration of Ca^{2+} on the properties of TRPV1–tubulin and/or TRPV1–microtubule complex is of special interest. Tubulin binding to MBP-TRPV1-Ct is sensitive to the presence of Ca^{2+} (80). In contrast, the absence or presence of extra Ca^{2+} has no effect on the interaction of tubulin with MBP-TRPV4-Ct (80). Interestingly, the microtubules formed with MBP-TRPV1-Ct in the presence of Ca^{2+} become 'cold-stable' as these microtubules do not depolymerize further at low temperatures (80). The exact mechanisms by which Ca^{2+} modulates these physio-chemical properties in vitro are not clear. In this regard, it is important to mention that tubulin has been shown to bind two Ca^{2+} ions to its C-terminal sequence (85–88) and thus Ca^{2+}-dependent conformational changes of tubulin may underlie the observed effects of Ca^{2+} (89).

In addition to the interaction with αβ-tubulin dimer with the TRPV channels, there are several reports that suggest that other components of the microtubule cytoskeleton also interact with the TRPV channels. For example, a yeast two-hybrid screen has reported interaction of kinesin 2 and kinesin family member 3B with the TRPV1 (78). Similarly, MAP7 interaction with TRPV4 has been demonstrated (90). This interaction is also mapped down to the C-terminal cytoplasmic domain of TRPV4, especially within the amino acid region 798 to 809. This MAP7 interaction is involved in the surface expression of TRPV4. The biochemical data of direct interaction as well as microtubule stabilization find their correlates in cell biological studies. Ectopic expression of TRPV1 in dorsal root ganglia-derived F11 cells results in co-localization of TRPV1 and microtubules and accumulation of endogenous tyrosinated tubulin (a marker for dynamic microtubules) in close vicinity to the plasma membrane (80). As suggested by its preference to bind to the plus-end-exposed β-tubulin, TRPV1 apparently stabilizes microtubules reaching the plasma membrane and thereby increases the number of pioneering microtubules within the actin cortex. Similarly, TRPV4 co-localizes with microtubules at the plasma membrane (33). Therefore, the stabilization of microtubules in the plasma membrane induces cellular morphological changes. This also explains at least in part why overexpression of TRPV1 induces massive induction of filopodial structures in neuronal as well as in non-neuronal cells. The mechanism for this is currently under investigation and apparently also includes alterations in the actin cytoskeleton. However, co-localization of TRPV1 with tubulin has been observed all along the filopodial stalk and, of note, including the filopodial tips (32, 91). Tubulin and components attributed to stable microtubules (like acetylated tubulin and MAP2a/b) were also observed within these thin filopodial structures (32).

Apart from the TRPV members, few TRPC members have been reported to interact with tubulin. TRPC5 and TRPC6 have sown to form signaling complexes that contain tubulin (75). In a similar manner, TRPC1 also interacts with tubulin and this interaction is involved with the surface expression of the channel (79). A proteomic study has also reported interaction of few microtubule cytoskeletal proteins, namely microtubule-associated protein 2 (MAP2) and cytosolic dynein heavy chain with TRPC3 (77). In addition, physical interaction of TRPC5 with stathmin, a factor that causes disassembly of microtubule cytoskeleton has been demonstrated (92). The interaction with stathmin is important for the neurite extension, growth cone function and also for synaptogenesis. Interestingly, TRPC5 interaction with stathmin plays an important role in the regulation of neurite and filopodial length also. In agreement with other TRP channels, PC1 and PC2 channels also interact with different tubulins (like α-tubulin, β-tubulin, γ-tubulin, acetylated α-tubulin) and the kinesin motor proteins KIF3A and KIF3B (53, 93).

4. Regulation of Microtubule Cytoskeleton by TRP Channels

In the last few years, significant progress has been made to elucidate the regulation of microtubule cytoskeleton by TRP channels. This is not surprising as activation of TRP channels initiates Ca^{2+} signaling as well as many other signaling events. Indeed recent reports suggest that activation of TRP channels is not an all-or-none event and thus these channels can regulate the cytoskeleton by both Ca^{2+}-dependent as well as Ca^{2+}-independent mechanisms (94). Recently we have demonstrated that TRPV channels regulate cytoskeleton in many different manners. For example, activation of TRPV1 by specific agonists like Resiniferatoxin (RTX) or Capsaicin leads to rapid destabilization of microtubules (48, 95). Notably, TRPV1 activation predominantly affects the dynamic microtubules and not the stable microtubules. This conclusion has been drawn mainly due to the observation that majority of the tyrosinated tubulins (marker for dynamic microtubules) but not the acetylated or polyglutamylated tubulins (markers for stable microtubule) appear as soluble tubulin after the activation (48). Similarly, activation of TRPV4 also results in disassembly of microtubules (33). Though the exact molecular factors and pathways involved in this microtubule disassembly are not known, involvement of Cam-Ca^{2+} complex can be speculated (96–98). This is due to the fact that catalytic amounts of Cam-Ca^{2+} complex are known to cause severe microtubule disassembly. However, this TRPV-induced microtubule disassembly can be achieved even in a Ca^{2+}-independent manner, especially under certain conditions. For example, TRPV1 mediated microtubule disassembly can also be achieved in the presence of Ca^{2+} chelators like EGTA and strong inhibitor of TRPV1, like 5'I-RTX (49, 91). Even expression of the N-terminal fragment of TRPV1 (ΔTRPV1-Nt) can cause microtubule disassembly in response to some specific components like estrogen (91). These results strongly suggest that Ca^{2+}-independent pathways are also involved in the microtubule disassembly. However, further studies are needed to dissect these different signaling events in detail.

5. Interaction of TRP Channels with Actin Cytoskeleton

Similar to microtubule cytoskeleton, a large number of studies suggest that TRP channels interact with actin cytoskeleton, both physically and functionally. Often TRP channels physically interact with G-actin and other components associated with actin cytoskeleton. For example, members of TRPV, TRPC and PC channels are reported to form molecular complexes that contain actin and/or related components (33, 53, 93). The interaction of TRP channels with actin cytoskeleton is functionally important and relevant for several reasons. First, actin cytoskeleton is located just beneath the plasma membrane and thus has enough physical proximity to interact with the TRP channels. Next, in many cases, TRP channels are present in specialized subcellular structures like at the spines, filipodial tips, etc., that are characterized by the presence of bundled actin cytoskeleton (32–34). In addition, there are TRP channels (like TRPV4) that are involved in mechanosensation, a complex process and thus are supposed to bridge lipid bi-layer with sub-membranous cytoskeleton (7, 99, 100).

So far TRPV4 represents the best characterized TRP channel in terms of multi-dimensional interaction with actin cytoskeleton. Based on the fluorescence resonance energy transfer (FRET) performed in live cells, it was demonstrated that actin and TRPV4 share a close proximity, possibly a physical interaction between these two (101). This physical interaction is logical as both TRPV4 and actin cytoskeletons are functionally involved in mechanosensation as well as in mechanical pain (102). In agreement with that, recently we have demonstrated that the C-terminus of TRPV4 interacts directly with soluble actin as well as with polymerized actin filaments (33). In addition, presence of α2 integrin, an actin-binding protein in the signaling complex formed by TRPV4 has also been reported (103). These interactions are also in agreement with the fact that TRPV4 is enriched in structures like cilia, filopodia, focal adhesion points, dendritic spines and in lamellipodia, where it can regulate the dynamics of actin cytoskeleton (33, 104–110). In the same notion, involvement of TRPV4 in the intercellular junction formation in keratinocytes has been demonstrated (111). In spite of these studies, the exact location on TRPV4 where actin or other actin cytoskeletal proteins bind with it has not been determined. However, we demonstrated that soluble tubulin competes with soluble actin for binding on MBP-TRPV4-Ct suggesting that both actin and tubulin may bind to the same site located on TRPV4-Ct (33). This also suggests that TRPV4 may have a complex regulatory mechanism that switches it from actin cytoskeleton to microtubule cytoskeleton or vice versa.

In contrast with TRPV4, interaction of other TRPV members with soluble actin cytoskeleton is not well established. As TRPV1 localizes in the actin cytoskeleton-enriched structures like at the filopodial structures and at the dendritic spine, it is expected that TRPV1 interacts with actin cytoskeleton (32, 112). However, so far the direct physical interaction of TRPV1 with actin has not been established. In contrary, it has been shown that the same immune complex of TRPV1 that contains tubulin actually lacks actin (80). Even purified MBP-TRPV1-Ct does not interact with soluble actin in a condition where it interacts with tubulin (49). These results apparently suggest that TRPV1 may not interact directly with actin cytoskeleton. However, it might be possible that TRPV1-actin complex is extremely dynamic in nature, needs full-length TRPV1 (or even tetrameric structures) and difficult to extract in soluble phase. Therefore, bio-chemical methods may not be suitable to confirm the interaction and further live cell imaging studies are needed.

Recently, few TRPC members have also been reported to interact with actin cytoskeleton. In a proteomic screen it has been shown that the signaling complex formed by TRPC5 and TRPC6 contains actin and other actin cytoskeletal associated proteins, namely spectrin and myosins (75). Among all, spectrin seems to be a conserved interacting protein for many TRP channels, especially for TRPC members. In agreement with that, a recent study demonstrated that the C-terminal cytoplasmic part of hTRPC4, specifically amino acid residues 686 to 977 interact with αII- and βV-spectrin in a yeast two-hybrid assay (113). Within this region, the amino acids residues 730–758 of hTRPC4 are critical for the interaction with spectrin (113). This interaction was further confirmed by glutathione S-transferase pulldown and co-immunoprecipitation experiments. This interaction with spectrin is important for the surface expression of TRPC members. Further deletion studies confirmed that amino acids 730–758 of hTRPC4 are critical for the interaction with spectrin. This region contains a coiled-coil domain and is juxtaposed to the Ca^{2+}/calmodulin- and IP_3R-binding region (CIRB-domain) suggesting that the interaction with the cytoskeletal components can have influence on other regulation as well. It is likely that spectrin interacts with TRPC5 also as the same sequence which is important for interaction is present in TRPC5. A recent proteomic study has also reported the interaction of several actin cytoskeletal proteins, namely, spectrin α-chain, spectrin β-chain as well as cofilin-1 as interacting proteins of TRPC3 (77). In the same notion, it has been demonstrated that TRPC4 also interacts with SESTD1, a previously uncharacterized protein that contains a lipid-binding SEC14-like domain and a spectrin-type cytoskeleton interaction domain (114). TRPC4 also interacts with 4.1 protein indicating that TRP channels and cytoskeletal proteins indeed form complex membrane scaffolds. This interaction is also due to a small sequence located at the C-terminus of the TRPC4 which is enriched with positively charged residues. Due to this ionic interaction, association of TRPC4 with membrane cytoskeleton is sensitive to high salt (115). Interaction of PC channels with actin cytoskeleton and associated components like monomeric actin, the actin-related components α-actinin and gelsolin have also been demonstrated (53, 93). However, more studies are needed to identify the entire spectrum of interacting proteins belonging to actin cytoskeleton.

6. Regulation of Actin Cytoskeleton by TRP Channels

Being permeable to Ca^{2+}, activation of TRP channels has the potential to regulate actin cytoskeleton. Indeed several reports suggest that TRP cannels regulate actin cytoskeleton by various manners. Interestingly, the nature of regulation and exact effect of TRPs-mediated regulation of actin cytoskeleton depend on few factors, mainly on the identity of the TRP channels and the cellular system. Best characterization of TRP-mediated regulation of actin cytoskeleton has been demonstrated for TRPV channels, namely for TRPV4. This also fits well with the involvement of TRPV4 in several cellular functions that are also known to require active participation of the actin cytoskeleton. It is noteworthy to mention that TRPV4 is a key molecule involved in mechanical force mediated biological processes. For example, TRPV4 activity is central to cytoskeleton-dependent/mediated regulatory volume decrease of cells, a process where actin-binding proteins contribute to cell volume regulatory ion channel activation (116–120). In the same notion, a recent study demonstrated that disruption of the actin cytoskeleton increases the intracellular mobility of TRPV4-GFP and results in loss of co-localization of TRPV4 with actin (121). Recently it has been reported that TRPV4 regulates the morphology of human umbilical vein endothelial cell (HUVEC) (122). Activation of TRPV4 in this cell line causes rapid retraction and condensation of cells. In a similar manner, a prolonged activation also causes detachment of cells from the plates. Interestingly, these effects can be blocked by the ruthenium red, a TRP channel blocker. This suggests that TRPV4 activation alters the cytoskeletal integrity and dynamics and affects focal adhesion points as well as microtubules (122). Indeed, recent reports also suggest that TRPV4 activation regulates the morphology and migration of neuroendocrine (GN11) cells (123). These effects are in full agreement with what has been seen in F11 cells that express TRPV4 (33). This cell retraction due to TRPV4 activation is partly due to the loss of microtubules which disrupt the balance between anterograde force mediated by microtubule cytoskeleton and the retrograde force mediated by acto-myosin components.

In a similar manner, functional consequence of TRPV1 on the actin cytoskeleton has been shown in some systems. However, the effect of TRPV1 on actin cytoskeleton seems to be different depending on the cellular systems. For example, TRPV1 activation may enhance actin polymerization in some cellular systems whereas other cellular systems may remain unaffected or reveal depolymerization. Indeed it has been shown that capsaicin treatment increases the actin cytoskeleton, and also increases the actin filament content in neutrophils (124). Similarly, the effect of TRPV1 activation on the actin cytoskeleton has been demonstrated in sperm cells also. It has been shown that inhibition of TRPV1 by capsazepine during capacitation leads to the inhibition of actin polymerization in the acrosomal region (125, 126). Another study also demonstrated that activation of TRPV1 in premature spermatozoa promotes actin cytoskeletal depolymerization and a loss of acrosome structure integrity (125, 126). In case of F11 cells, the dorsal root ganglion-derived cells, activation of TRPV1 results in rapid microtubule disassembly but does not cause disassembly of actin or neurofilament cytoskeleton (18).

Other TRP channels also reveal functional interactions with actin cytoskeleton. For example, TRP1 in human platelets (hTRP1) couples with IP3 receptor and this coupling is controlled by actin cytoskeleton as stabilization of the cortical actin cytoskeleton with pharmacological means prevents this coupling (127). This result suggests that the sub-membranous actin filaments act as negative clamp which prevents constitutive coupling between TRP1 and IP3. In the same manner, PC2 channels regulate the morphology of BeWo cells that represent Human Trophoblast Choriocarcinoma (128). In this cell line, PC2 co-localizes at the cytokinetic midbody where the dynamics of actin cytoskeleton is important for the final step of the cell division. In the case of neurons, TRPC1 regulates growth cone dynamics by a fine balance of LIM kinase and slingshot phosphatase activity which in turn regulates ADF/cofilin (129). This process is involved in growth cone attraction and repulsion. Taken together these results suggest that TRP-mediated signaling events are involved in reorganization of actin cytoskeleton. These reorganization effects in turn control many of the cellular functions like acrosomal reaction, fertilization, and functional aspects of neurons and immune cells.

7. Regulation of TRP Channels by Cytoskeletal Components

Recently there are a handful of reports that suggest a feedback regulation of TRP channels by the components of the cytoskeleton. Interestingly these feedback regulations by cytoskeleton can occur at the cellular level as well as at the single molecular level. At the cellular level, there were few studies which demonstrated that the functions of certain TRP channels are dependent on the status of the cellular cytoskeleton. Especially it has been shown that pharmacological modulation of different cytoskeletons results in altered influx of ions via these TRP channels. For example, a microtubule stabilizer drug, namely Taxol, reduces TRPV4-dependent currents while the microtubule-disrupting agents like Colchicine and Vincristine as well as actin cytoskeleton regulating drugs like Phalloidin (a stabilizer) or Cytochalasin B (a destabilizer) do not alter the TRPV4-mediated current (90). In agreement with this, recently we also have demonstrated that pharmacological stabilization of microtubules by applying Taxol results in reduction in the Ca^{2+} influx in response to 4α-phorbol-didecanoate (4αPDD) (an agonist of TRPV4) as measured by Ca^{2+} influx assay in whole cells (33). Interestingly, the degree of reduction in Ca^{2+} influx is much robust in the case of second time application of 4αPDD. Notably, this reduction is independent of the expression or availability of the TRPV4 at the plasma membrane, suggesting that dynamics of the microtubule cytoskeleton can regulate the ion channel function. This hypothesis is also supported indirectly by the whole-cell recordings measuring the TRPV4 activation conducted by heat activation. This is due to the fact that activation of single TRPV4 ion channel in response to heat is possible in whole cell recordings but not in a cell-free inside-out patch clamp experiments, suggesting that in the latter some cellular factor is missing (108). Therefore it can be speculated that components from microtubule cytoskeleton like α- or β-tubulin as well as MAPs might be important for the channel function at the single ion channel level. Similarly, involvement of actin cytoskeleton in the regulation of TRPV4 at the cellular level has been demonstrated. Pharmacological disruption of actin by latrunculin-A results in loss of sensing hypotonicity and the onset of regulatory volume decrease (121). Tubulin interaction seems to control the Ca^{2+} homeostasis via TRPC members present in the ER. It is important to mention that microtubule-based motor proteins indirectly regulate the calcium-selective store-operated currents, a function where TRPC channels are involved (130). This is due to the fact that either stabilization or destabilization of microtubules by pharmacological drugs like Taxol or nocodazole results in altered distribution of cellular organelles as well as availability of the TRPC channels by modulat ing the endoplasmic reticulum-to-plasma membrane coupling events (130, 131). Ionic conductivity mediated by TRPC7 is also regulated by the status of the actin cytoskeleton as disruption of actin cytoskeleton by Cytochalasin-B results in inhibition of OAG-activated and TRPC7-mediated currents (132).

In agreement with all these reports, it has been demonstrated that the activity of the mechanosensitive ion channels in cultured sensory neurons appears to depend largely on the status of the cytoskeleton. Thus, disruption of actin or microtubule cytoskeleton by pharmacological agents greatly reduces the activity of mechanosensitive channels (133). In this regard, it is important to mention that most of these studies involve Ca^{2+} imaging and/or whole cell patch clamp as the read out systems and thus analyzed the effect of cytoskeletal alteration on a population of TRP channels in general. These studies give a partial mechanistic view of cytoskeletal involvement only and do not address the multidirectional regulation of cytoskeleton on the ion channels, especially at the level of single molecules. This is simply due to the fact that availability of the ion channels at the plasma membrane and normal trafficking of these ion channels are limited if the cytoskeleton is disrupted. Therefore it remains to be established whether the modulation of TRP channels can occur through direct interaction with the cytoskeleton.

As mentioned, recent reports suggest that cytoskeletal components can regulate the properties, especially the ionic conductivity and related behavior of TRP channels, at the single channel level also. This is mainly due to the fact that interactions of cytoskeletal components with TRP channels affect other interactions and thus modulate the sensitization–desensitization properties as well as channel opening probabilities. In most cases, sensitization–desensitization of TRP channels can be modulated through phosphorylation–dephosphorylation events. For example, recently we have demonstrated that phosphorylation of MBP-TRPV1-Ct in vitro at S800 position by PKCε is significantly reduced due to the interaction of tubulin to the MBP-TRPV1-Ct (91). This result strongly suggests that microtubule dynamics is an important regulator for the ionic conductivity mediated by TRPV1. In this respect, it is important to mention that S800 is a key position that regulates the sensitization–desensitization of TRPV1 (134). The phosphorylation–dephosphorylation of TRP channels can also be regulated by the Ca^{2+}-dependent and/or independent kinases as well as by the Ca^{2+} influx through the channel itself. Kinases like PKCε and other PKCs are also involved in the sensitization–desensitization of TRP channels. Physical interaction and involvement of Ca^{2+}-binding/sensing proteins like calmodulin and CamKII in the desensitization of TRP channels have been reported (33, 135–141). For example, TRPV1 and TRPV4 interact with Calmodulin and are regulated by CamKII and these interactions are involved in the regulation of ion channels. So it is becoming prominent that TRP channels are modulated by Ca^{2+}-dependent as well as Ca^{2+}-independent mechanisms and the Ca^{2+}-independent regulation of TRP channels is just emerging. An example of Ca^{2+}-independent regulation of TRP channel is the regulation of TRPC by Homer. It has been shown that TRPC mutants lacking the homer-binding site become constitutively active (142). Even, point mutations in the ankyrin repeat region (supposed to be involved in protein–protein interaction) of TRPV4 results in constitutively active or inactive channel (21, 54–56). These examples strongly indicate that other scaffolding proteins and cytoskeletal components can regulate TRP channel though the experimental evidences are still limited.

However, in recent time, very few studies have addressed this problem and attempted to establish a direct modulatory role of the cytoskeleton. The best examples of such studies were performed on TRPP channels (53, 93). Montalbetti and co-workers performed single-channel electrophysiological experiments of polycystin channel 2 (PC2) on reconstituted lipid bilayers. This system arguably eliminates all factors except the channel-associated complex. Interestingly, monomeric actin, the actin-related components α-actinin and gelsolin, tubulin including acetylated α-tubulin, and the kinesin motor proteins (KIF3A and KIF3B) are present in these membranes, possibly due to the direct interaction with PC2 channels (53, 93). Disruption of actin filaments with cytochalasin D or with the actin-severing protein gelsolin activates the channel. This activation can be inhibited by the addition of soluble monomeric G-actin with ATP, which induces actin polymerization. This indicates that actin filaments, but not soluble actin, are an endogenous negative regulator of PC2 channels. Also microtubules regulate PC2 channel function only in opposing manner. Depolymerization of microtubules with Colchicine rapidly inhibits the basal level of PC2 channel activity, whereas polymerization and/or stabilization of microtubules from soluble tubulin with GTP and Taxol stimulates the PC2 channel activity (93). Involvement of the microtubule cytoskeleton in the regulation of PC2 channel has also been described in vivo in primary cilia of renal epithelial cells (27). In that system, addition of microtubule destabilizer (Colchicine) rapidly abolished channel activity, whereas the addition of microtubule stabilizers (Taxol) increased channel activity (27). Similar results were obtained using reconstituted lipid bilayer system, which reveals that both spontaneous activity and the opening probability of TRPP3 ion channels are increased by the addition of α-actinin, demonstrating that this channel can be indeed modulated by cytoskeleton (27). Certainly more studies are needed to explore such regulations at the single level.

8. Importance of Cytoskeletal Reorganization in Pathophysiological Disorders: New Pharmacological Challenges, Future Prospects and Concluding Remarks

The direct importance of TRP channels in Ca^{2+} signaling has been the major focus for a long time and still being investigated by many. However, in the last few years another aspect of TRP channels has emerged: the importance of Ca^{2+}-independent signaling event via TRP channels. In this context, the signaling complexes formed by the TRP channels and the cytoskeletal components at the submembranous compartment are of high importance. Being transmembrane proteins, TRP channels play a central role as scaffolds at the sub-membranous regions on which other components are sequestered, interact among themselves and finally form the functional signalplexes (Fig. 3). Interestingly, cytoskeletal components are not only present in these signalplexes, but also play significant roles to "fix" the entire signalplex in the context of cellular space and time. Therefore, both TRP channels and cytoskeletal components are involved in the "scaffolding" of the signalplexes. As the formation of these signalplexs needs TRP channels as transmembrane proteins only, the formation of the signalplexes is dependent on the availability of the TRP channels at the membrane but is largely independent of their Ca^{2+} channel activity. Interestingly, these components also take part in the regulation of ion channel opening—closure properties to a large extent.

The coordinated regulation of different cytoskeletons and vesicular trafficking by TRP channels has immense implication in the context of pharmacological treatment of pathophysiology and several disorders. For example, TRPV1 plays an inhibitory role in urothelial cancer cell invasion and metastasis by altering the micro tubule cytoskeleton (143). This type of understanding may allow targeting the cytoskeleton of cancer cells via specific TRP channels or vice versa. In that manner, sprouting of neuronal cells, function of immune cells and sperm cells, etc., can be effectively modulated by targeting TRP-cytoskeletal complex per se. This strategy can be effective as expression, localization, function and regulation of TRP channels are specific yet versatile in nature. For example, the complex of TRP channel with β-tubulin III can be specifically targeted in neurons and/or in some specific cancer cells where the expression of β-tubulin III is reported. Thus, different properties of the individual TRP-signalplexes can be used for the pharmacological and clinical purposes.

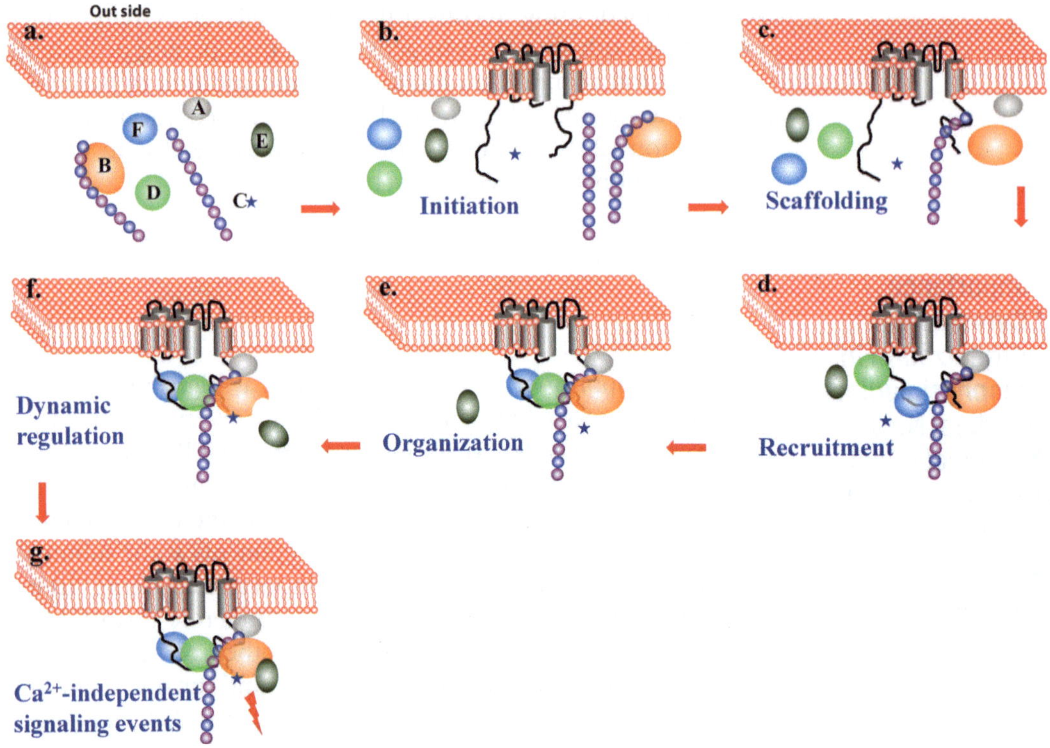

Fig. 3. Ca^{2+}-independent signaling events mediated by TRP channels. As TRP channels are transmembrane in nature, interaction of submembranous cytoskeleton and other scaffolding proteins with these channels initiate formation of scaffolds at the plasma membrane on which several other regulatory factors can associate and get involved in complex signaling events. As this scaffolding act of TRP channels is independent of their ion channel activity, such signaling events are dependent on the presence of TRP channels but independent of the TRP channel-mediated Ca^{2+} influx. These Ca^{2+}-independent signaling events can be described in several distinct steps like initiation (step b), scaffolding (step c), recruitment (step d), organization (step e), dynamic regulation (step f) and signaling events (step g). Though several membrane-associated factors (like A), microtubule-associated factors (like B) and other cytosolic factors (like C, D E and F) are present in the cell, these components cannot form novel signaling complex as these components are either not available at the submembranous region or not properly sequestered there (As indicated in step a). Interaction of TRP channels with cytoskeleton initiates some conformational changes (step b) and results in the formation of novel scaffolds on which some of these key components can sequester (step c). This sequestration of these key factors facilitates recruitment and sequestration of several other cytosolic and membrane-associated factors on the existing TRP complex (Step d). All these associated factors adjust and fine-tune their organization by further conformational changes (step e). The components sequestered in this complex can regulate each other and the entire signaling complex becomes dynamic (f). These complexes can also be further regulated by transient events like by kinase or phosphatase activity (such as by C) and result in some signaling events (step g).

Understanding the molecular mechanism of both Ca^{2+}-dependent and Ca^{2+}-independent signaling events has importance in basic research and also has pharmacological as well as clinical interests. This is especially due to the fact that application of common microtubule-based chemotherapeutics like Taxol and Vinca drugs in cancer patients is known to induce strong neuropathic pain (144–152). Though certain signaling events are involved in these chemotherapeutics-induced pain and hyperalgesia development, the molecular mechanisms behind these pathophysiological symptoms are still largely unknown (102, 153). However, changes in the microtubule orientation, structure and other changes have been reported. In that context, physical and functional interactions of TRP channels with microtubule cytoskeletal proteins is significant. A better understanding of these physical and functional interactions may allow targeting these pathophysiological disorders in a more systemic manner. This may be more useful for the application of microtubule-based chemotherapeutics also. In future more studies must be conducted in these aspects.

Acknowledgement

This review reflects the views and interpretation of the data available in the literature. The author regrets for not being able to include all the scientific works due to space limitations. The author acknowledges the support and intellectual inputs from Mr. Rakesh Majhi and other lab members. CG acknowledges the support previously obtained from Freie University of Berlin and Max Planck Institute of Molecular Genetics, Berlin. Financial support from NISER, Bhubaneswar is appreciated.

References

1. Sheetz MP (2001) Cell control by membrane-cytoskeleton adhesion. Nat Rev Mol Cell Biol 2:392–396
2. Sheetz MP, Sable JE, Döbereiner HG (2006) Continuous membrane-cytoskeleton adhesion requires continuous accommodation to lipid and cytoskeleton dynamics. Annu Rev Biophys Biomol Struct 35:417–434
3. Goswami C, Hucho T (2008) Novel aspects of the submembraneous microtubule cytoskeleton. FEBS J 275:4653
4. Arce CA, Casale CH, Barra HS (2008) Submembraneous microtubule cytoskeleton: regulation of ATPases by interaction with acetylated tubulin. FEBS J 275:4664–4674
5. Roychowdhury S, Rasenick MM (2008) Submembraneous microtubule cytoskeleton: regulation of microtubule assembly by heterotrimeric Gproteins. FEBS J 275:4654–4663
6. Goswami C, Hucho T (2008) Submembraneous microtubule cytoskeleton: biochemical and functional interplay of TRP channels with the cytoskeleton. FEBS J 275: 4684–4699
7. Chen XZ, Li Q, Wu Y, Liang G, Lara CJ, Cantiello HF (2008) Submembraneous microtubule cytoskeleton: interaction of TRPP2 with the cell cytoskeleton. FEBS J 275:4675–4683
8. Feit H, Barondes SH (1970) Colchicine-binding activity in particulate fractions of mouse brain. J Neurochem 17:1355–1364
9. Walters BB, Matus AI (1975) Tubulin in postynaptic junctional lattice. Nature 257: 496–498
10. Zenner HP, Pfeuffer T (1976) Microtubular proteins in pigeon erythrocyte membranes. Eur J Biochem 71:177–184
11. Yoshimura Y, Yamauchi Y, Shinkawa T, Taoka M, Donai H, Takahashi N, Isobe T, Yamauchi T (2004) Molecular constituents of the postsynaptic density fraction revealed by proteomic analysis using multidimensional liquid chromatography-tandem mass spectrometry. J Neurochem 88:759–768
12. Li N, Shaw AR, Zhang N, Mak A, Li L (2004) Lipid raft proteomics: analysis of in- solution digest of sodium dodecyl sulfate-solubilized lipid raft proteins by liquid chromatography-matrix-assisted laser desorption/ionization tandem mass spectrometry. Proteomics 4: 3156–3166

13. Feuk-Lagerstedt E, Movitz C, Pellmé S, Dahlgren C, Karlsson A (2007) Lipid raft proteome of the human neutrophil azurophil granule. Proteomics 7:194–205
14. Foster LJ, De Hoog CL, Mann M (2003) Unbiased quantitative proteomics of lipid rafts reveals high specificity for signaling factors. Proc Natl Acad Sci USA 100:5813–5818
15. Wolff J (2009) Plasma membrane tubulin. Biochim Biophys Acta 1788:1415–1433
16. Stephens RE (1986) Membrane tubulin. Biol Cell 57:95–109
17. Ramsey IS, Delling M, Clapham DE (2006) An introduction to TRP channels. Annu Rev Physiol 68:619–647
18. Clapham DE (2003) TRP channels as cellular sensors. Nature 426:517–524
19. Clapham DE, Runnels LW, Strübing C (2001) The TRP ion channel family. Nat Rev Neurosci 2:387–396
20. Venkatachalam K, Montell C (2007) TRP channels. Annu Rev Biochem 76:387–417
21. Verma P, Kumar A, Goswami C (2010) TRPV4-mediated channelopathies. Channels (Austin) 4(4):319–328
22. Winter J (1987) Characterization of capsaicin-sensitive neurones in adult rat dorsal root ganglion cultures. Neurosci Lett 80: 134–140
23. Wood JN, Winter J, James IF, Rang HP, Yeats J, Bevan S (1988) Capsaicin-induced ion fluxes in dorsal root ganglion cells in culture. J Neurosci 8:3208–3220
24. Majhi R, Sardar P, Goswami L, Goswami C (2011) Right time – right location – right move: TRPs find motors for common functions. Channels (Austin) 5(4):375–381
25. Jurczyk A, Gromley A, Redick S, San Agustin J, Witman G, Pazour GJ, Peters DJ, Doxsey S (2004) Pericentrin forms a complex with intraflagellar transport proteins and polycystin-2 and is required for primary cilia assembly. J Cell Biol 166:637–643
26. Nauli SM, Alenghat FJ, Luo Y, Williams E, Vassilev P, Li X, Elia AE, Lu W, Brown EM, Quinn SJ, Ingber DE, Zhou J (2003) Polycystins 1 and 2 mediate mechanosensation in the primary cilium of kidney cells. Nat Genet 33:129–137
27. Li Q, Montalbetti N, Wu Y, Ramos A, Raychowdhury MK, Chen XZ, Cantiello HF (2006) Polycystin-2 cation channel function is under the control of microtubular structures in primary cilia of renal epithelial cells. J Biol Chem 281:37566–37575
28. Shin JB, Adams D, Paukert M, Siba M, Sidi S, Levin M, Gillespie PG, Gründer S (2005) Xenopus TRPN1 (NOMPC) localizes to microtubule-based cilia in epithelial cells, including inner-ear hair cells. Proc Natl Acad Sci USA 102:12572–12577
29. Liang X, Madrid J, Saleh HS, Howard J (2011) NOMPC, a member of the TRP channel family, localizes to the tubular body and distal cilium of Drosophila campaniform and chordotonal receptor cells. Cytoskeleton (Hoboken) 68:1–7
30. Lee J, Moon S, Cha Y, Chung YD (2010) Drosophila TRPN(=NOMPC) channel localizes to the distal end of mechanosensory cilia. PLoS One 5:e11012
31. Dryer SE, Reiser J (2010) TRPC6 channels and their binding partners in podocytes: role in glomerular filtration and pathophysiology. Am J Physiol Renal Physiol 299:F689–F701
32. Goswami C, Hucho T (2007) TRPV1 expression-dependent initiation and regulation of filopodia. J Neurochem 103:1319–1333
33. Goswami C, Kuhn J, Heppenstall PA, Hucho T (2010) Importance of non-selective cation channel TRPV4 interaction with cytoskeleton and their reciprocal regulations in cultured cells. PLoS One 5(7):e11654
34. Shibasaki K, Murayama N, Ono K, Ishizaki Y, Tominaga M (2010) TRPV2 enhances axon outgrowth through its activation by membrane stretch in developing sensory and motor neurons. J Neurosci 30:4601–4612
35. Yamamoto S, Wajima T, Hara Y, Nishida M, Mori Y (2007) Transient receptor potential channels in Alzheimer's disease. Biochim Biophys Acta 1772:958–967
36. Loudon RP, Silver LD, Yee HF Jr, Gallo G (2006) RhoA-kinase and myosin II are required for the maintenance of growth cone polarity and guidance by nerve growth factor. J Neurobiol 66:847–867
37. Medeiros NA, Burnette DT, Forscher P (2006) Myosin II functions in actin-bundle turnover in neuronal growth cones. Nat Cell Biol 8:215–226
38. Les Erickson F, Corsa AC, Dose AC, Burnside B (2003) Localization of a class III myosin to filopodia tips in transfected HeLa cells requires an actin-binding site in its tail domain. Mol Biol Cell 14:4173–4180
39. Bridgman PC, Dave S, Asnes CF, Tullio AN, Adelstein RS (2001) Myosin IIB is required for growth cone motility. J Neurosci 21: 6159–6169

40. Bridgman PC, Elkin LL (2000) Axonal myosins. J Neurocytol 29:831–841
41. Berg JS, Derfler BH, Pennisi CM, Corey DP, Cheney RE (2000) Myosin-X, a novel myosin with pleckstrin homology domains, associates with regions of dynamic actin. J Cell Sci 113:3439–3451
42. Berg JS, Cheney RE (2002) Myosin-X is an unconventional myosin that undergoes intrafilopodial motility. Nat Cell Biol 4:246–250
43. Sousa AD, Berg JS, Robertson BW, Meeker RB, Cheney RE (2006) Myo10 in brain: developmental regulation, identification of a headless isoform and dynamics in neurons. J Cell Sci 119:184–194
44. Sousa AD, Cheney RE (2005) Myosin-X: a molecular motor at the cell's fingertips. Trends Cell Biol 15:533–539
45. Bohil AB, Robertson BW, Cheney RE (2006) Myosin-X is a molecular motor that functions in filopodia formation. Proc Natl Acad Sci USA 103:12411–12416
46. Bennett RD, Mauer AS, Strehler EE (2007) Calmodulin-like protein increases filopodia-dependent cell motility via up-regulation of myosin-10. J Biol Chem 282:3205–3212
47. Belyantseva IA, Boger ET, Friedman TB (2003) Myosin XVa localizes to the tips of inner ear sensory cell stereocilia and is essential for staircase formation of the hair bundle. Proc Natl Acad Sci USA 100:13958–13963
48. Goswami C, Dreger M, Otto H, Schwappach B, Hucho F (2006) Rapid disassembly of dynamic microtubules upon activation of the capsaicin receptor TRPV1. J Neurochem 96:254–266
49. Goswami C, Hucho TB, Hucho F (2007) Identification and characterisation of novel tubulin-binding motifs located within the C-terminus of TRPV1. J Neurochem 101:250–262
50. Kuwahara K, Wang Y, McAnally J, Richardson JA, Bassel-Duby R, Hill JA, Olson EN (2006) TRPC6 fulfills a calcineurin signaling circuit during pathologic cardiac remodeling. J Clin Invest 116:3114–31126
51. Avraham KB (1997) Motors, channels and the sounds of silence. Nat Med 3:608–609
52. Xu C, Rossetti S, Jiang L, Harris PC, Brown-Glaberman U, Wandinger-Ness A, Bacallao R, Alper SL (2007) Human ADPKD primary cyst epithelial cells with a novel, single codon deletion in the PKD1 gene exhibit defective ciliary polycystin localization and loss of flow-induced Ca2⁺ signaling. Am J Physiol Renal Physiol 292:F930–F945
53. Montalbetti N, Li Q, Wu Y, Chen XZ, Cantiello HF (2007) Polycystin-2 cation channel function in the human syncytiotrophoblast is regulated by microtubular structures. J Physiol 579:717–728
54. Auer-Grumbach M et al (2010) Alterations in the ankyrin domain of TRPV4 cause congenital distal SMA, scapuloperoneal SMA and HMSN2C. Nat Genet 42:160–164
55. Landouré G et al (2010) Mutations in TRPV4 cause Charcot-Marie-Tooth disease type 2C. Nat Genet 42:170–174
56. Deng HX et al (2010) Scapuloperoneal spinal muscular atrophy and CMT2C are allelic disorders caused by alterations in TRPV4. Nat Genet 42:165–169
57. Zhao C et al (2001) Charcot-Marie-Tooth disease type 2A caused by mutation in a microtubule motor KIF1Bbeta. Cell 105:587–597
58. Redowicz MJ (2002) Myosins and pathology: genetics and biology. Acta Biochim Pol 49:789–804
59. Libby RT, Steel KP (2000) The roles of unconventional myosins in hearing and deafness. Essays Biochem 35:159–174
60. Hasson T, Heintzelman MB, Santos-Sacchi J, Corey DP, Mooseker MS (1995) Expression in cochlea and retina of myosin VIIa, the gene product defective in Usher syndrome type 1B. Proc Natl Acad Sci USA 92: 9815–9819
61. Grimm C, Cuajungco MP, van Aken AF, Schnee M, Jörs S, Kros CJ, Ricci AJ, Heller S (2007) A helix-breaking mutation in TRPML3 leads to constitutive activity underlying deafness in the varitintwaddler mouse. Proc Natl Acad Sci USA 104:19583–19588
62. Tabuchi K, Suzuki M, Mizuno A, Hara A (2005) Hearing impairment in TRPV4 knockout mice. Neurosci Lett 382:304–308
63. Sidi S, Friedrich RW, Nicolson T (2003) NompC TRP channel required for vertebrate sensory hair cell mechanotransduction. Science 301:96–99
64. Di Palma F, Belyantseva IA, Kim HJ, Vogt TF, Kachar B, Noben-Trauth K (2002) Mutations in Mcoln3 associated with deafness and pigmentation defects in varitint-waddler (Va) mice. Proc Natl Acad Sci USA 99:14994–14999
65. Nilius B, Voets T, Peters J (2005) TRP channels in disease. Sci STKE 295:re8

66. Puertollano R, Kiselyov K (2009) TRPMLs: in sickness and in health. Am J Physiol Renal Physiol 296:F1245–F1254
67. van Aken AF et al (2008) TRPML3 mutations cause impaired mechano-electrical transduction and depolarization by an inward-rectifier cation current in auditory hair cells of varitint-waddler mice. J Physiol 586:5403–5418
68. Nagata K, Zheng L, Madathany T, Castiglioni AJ, Bartles JR, García-Añoveros J (2008) The varitint-waddler (Va) deafness mutation in TRPML3 generates constitutive, inward rectifying currents and causes cell degeneration. Proc Natl Acad Sci USA 105:353–358
69. Minke B (1977) Drosophila mutant with a transducer defect. Biophys Struct Mech 3:59–64
70. Ribelayga C (2010) Vertebrate vision: TRP channels in the spotlight. Curr Biol 20:R278–R280
71. Koike C, Numata T, Ueda H, Mori Y, Furukawa T (2010) TRPM1: a vertebrate TRP channel responsible for retinal ON bipolar function. Cell Calcium 48:95–101
72. van Genderen MM et al (2009) Mutations in TRPM1 are a common cause of complete congenital stationary night blindness. Am J Hum Genet 85:730–736
73. Satoh AK, Li BX, Xia H, Ready DF (2008) Calcium-activated myosin V closes the Drosophila pupil. Curr Biol 18:951–955
74. Meyer NE, Joel-Almagor T, Frechter S, Minke B, Huber A (2006) Subcellular translocation of the eGFPtagged TRPL channel in Drosophila photoreceptors requires activation of the phototransduction cascade. J Cell Sci 119:2592–2603
75. Goel M, Sinkins W, Keightley A, Kinter M, Schilling WP (2005) Proteomic analysis of TRPC5- and TRPC6-binding partners reveals interaction with the plasmalemmal Na(+)/K(+)-ATPase. Pflugers Arch 451:87–98
76. Clark K et al (2006) TRPM7, a novel regulator of actomyosin contractility and cell adhesion. EMBO J 25:290–301
77. Lockwich T et al (2008) Analysis of TRPC3-interacting proteins by tandem mass spectrometry. J Proteome Res 7:979–989
78. Stein AT, Ufret-Vincenty CA, Hua L, Santana LF, Gordon SE (2006) Phosphoinositide 3-kinase binds to TRPV1 and mediates NGF-stimulated TRPV1 trafficking to the plasma membrane. J Gen Physiol 128:509–522
79. Bollimuntha S, Cornatzer E, Singh BB (2005) Plasma membrane localization and function of TRPC1 is dependent on its interaction with β-tubulin in retinal epithelium cells. Vis Neurosci 22:163–170
80. Goswami C, Dreger M, Jahnel R, Bogen O, Gillen C, Hucho F (2004) Identification and characterization of a Ca^{2+} -sensitive interaction of the vanilloid receptor TRPV1 with tubulin. J Neurochem 91:1092–1103
81. Laínez S et al (2010) GABAA receptor associated protein (GABARAP) modulates TRPV1 expression and channel function and desensitization. FASEB J 24:1958–1970
82. Erickson HP, Voter WA (1976) Polycation-induced assembly of purified tubulin. Proc Natl Acad Sci USA 73:2813–2817
83. Lefèvre J et al (2011) The C terminus of tubulin, a versatile partner for cationic molecules: binding of Tau, polyamines, and calcium. J Biol Chem 286:3065–3078
84. Savarin P et al (2010) A central role for polyamines in microtubule assembly in cells. Biochem J 430:151–159
85. Hayashi M, Matsumura F (1975) Calcium binding to bovine tubulin. FEBS Lett 58:222–225
86. Solomon F (1977) Binding sites for calcium on tubulin. Biochemistry 16:358–363
87. Serrano L, Valencia A, Caballero R, Avila J (1986) Localization of the high affinity calcium-binding site on tubulin molecule. J Biol Chem 261:7076–7081
88. Fong KC, Babitch JA, Anthony FA (1988) Calcium binding to tubulin. Biochim Biophys Acta 952:13–19
89. Soto C, Rodriguez PH, Monasterio O (1996) Calcium and gadolinium ions stimulate the GTPase activity of purified chicken brain tubulin through a conformational change. Biochemistry 35:6337–6344
90. Suzuki M, Hirao A, Mizuno A (2003) Microtubule-associated [corrected] protein 7 increases the membrane expression of transient receptor potential vanilloid 4 (TRPV4). J Biol Chem 278:51448–51453
91. Goswami C, Kuhn J, Dina OA, Fernández-Ballester G, Levine JD, Ferrer-Montiel A, Hucho T (2011) Estrogen destabilizes microtubules through an ion-conductivity-independent TRPV1 pathway. J Neurochem 117:995–1008
92. Greka A, Navarro B, Oancea E, Duggan A, Clapham DE (2003) TRPC5 is a regulator of hippocampal neurite length and growth cone morphology. Nat Neurosci 6:837–845
93. Montalbetti N, Li Q, González-Perrett S, Semprine J, Chen XZ, Cantiello HF (2005) Effect of hydro-osmotic pressure on polycystin-2 channel function in the human syncytiotrophoblast. Pflugers Arch 451:294–303

94. Blumberg PM, Pearce LV, Lee J (2011) TRPV1 activation is not an all-or-none event: TRPV1 partial agonism/antagonism and its regulatory modulation. Curr Top Med Chem 11:2151–2158
95. Han P et al (2007) Capsaicin causes protein synthesis inhibition and microtubule disassembly through TRPV1 activities both on the plasma membrane and intracellular membranes. Biochem Pharmacol 73:1635–1645
96. Job D, Fischer EH, Margolis RL (1981) Rapid disassembly of cold stable microtubules by calmodulin. Proc Natl Acad Sci USA 78:4679–4682
97. Karr TL, Kristofferson D, Purich DL (1980) Calcium ion induces endwise depolymerization of bovine brain microtubules. J Biol Chem 255:11853–11856
98. Lieuvin A, Labbe JC, Doree M, Job D (1994) Intrinsic microtubule stability in interphase cells. J Cell Biol 124:985–996
99. Alenghat FJ, Nauli SM, Kolb R, Zhou J, Ingber DE (2004) Global cytoskeletal control of mechanotransduction in kidney epithelial cells. Exp Cell Res 301:23–30
100. Kaazempur Mofrad MR, Abdul-Rahim NA, Karcher H, Mack PJ, Yap B et al (2005) Exploring the molecular basis for mechanosensation, signal transduction, and cytoskeletal remodeling. Acta Biomater 1:281–293
101. Ramadass R, Becker D, Jendrach M, Bereiter-Hahn J (2007) Spectrally and spatially resolved fluorescence lifetime imaging in living cells: TRPV4-microfilament interactions. Arch Biochem Biophys 463:27–36
102. Dina OA, McCarter GC, de Coupade C, Levine JD (2003) Role of the sensory neuron cytoskeleton in second messenger signaling for inflammatory pain. Neuron 39:613–624
103. Alessandri-Haber N, Dina OA, Joseph EK, Reichling DB, Levine JD (2008) Interaction of transient receptor potential vanilloid 4, integrin, and SRC tyrosine kinase in mechanical hyperalgesia. J Neurosci 28:1046–1057
104. Lorenzo IM, Liedtke W, Sanderson MJ, Valverde MA (2008) TRPV4 channel participates in receptor-operated calcium entry and ciliary beat frequency regulation in mouse airway epithelial cells. Proc Natl Acad Sci USA 105:12611–12616
105. Gradilone SA, Masyuk AI, Splinter PL, Banales JM, Huang BQ et al (2007) Cholangiocyte cilia express TRPV4 and detect changes in luminal tonicity inducing bicarbonate secretion. Proc Natl Acad Sci USA 104:19138–19143
106. Teilmann SC, Byskov AG, Pedersen PA, Wheatley DN, Pazour GJ et al (2005) Localization of transient receptor potential ion channels in primary and motile cilia of the female murine reproductive organs. Mol Reprod Dev 71:444–452
107. Andrade YN, Fernandes J, Vázquez E, Fernández-Fernández JM, Arniges M et al (2005) TRPV4 channel is involved in the coupling of fluid viscosity changes to epithelial ciliary activity. J Cell Biol 168:869–874
108. Watanabe H, Davis JB, Smart D, Jerman JC, Smith GD et al (2002) Activation of TRPV4 channels (hVRL-2/mTRP12) by phorbol derivatives. J Biol Chem 277:13569–13577
109. Waning J, Vriens J, Owsianik G, Stuwe L, Mally S et al (2007) A novel function of capsaicin-sensitive TRPV1 channels: involvement in cell migration. Cell Calcium 42:17–25
110. Shibasaki K, Suzuki M, Mizuno A, Tominaga M (2007) Effects of body temperature on neural activity in the hippocampus: regulation of resting membrane potentials by transient receptor potential vanilloid 4. J Neurosci 27:1566–1575
111. Sokabe T, Fukumi-Tominaga T, Yonemura S, Mizuno A, Tominaga M (2010) The TRPV4 channel contributes to intercellular junction formation in keratinocytes. J Biol Chem 285:18749–18758
112. Goswami C, Rademacher N, Smalla KH, Kalscheuer V, Ropers HH, Gundelfinger ED, Hucho T (2010) TRPV1 acts as a synaptic protein and regulates vesicle recycling. J Cell Sci 123:2045–2057
113. Odell AF, Van Helden DF, Scott JL (2008) The spectrin cytoskeleton influences the surface expression and activation of human transient receptor potential channel 4 channels. J Biol Chem 283:4395–4407
114. Miehe S, Bieberstein A, Arnould I, Ihdene O, Rütten H, Strübing C (2010) The phospholipid-binding protein SESTD1 is a novel regulator of the transient receptor potential channels TRPC4 and TRPC5. J Biol Chem 285:12426–12434
115. Cioffi DL, Wu S, Alexeyev M, Goodman SR, Zhu MX, Stevens T (2005) Activation of the endothelial store-operated ISOC Ca2+ channel requires interaction of protein 4.1 with TRPC4. Circ Res 97:1164–1172
116. Pan Z, Yang H, Mergler S, Liu H, Tachado SD et al (2008) Dependence of regulatory volume decrease on transient receptor potential vanilloid 4 (TRPV4) expression in human corneal epithelial cells. Cell Calcium 44:374–385

117. Becker D, Blase C, Bereiter-Hahn J, Jendrach M (2005) TRPV4 exhibits a functional role in cell-volume regulation. J Cell Sci 118: 2435–2440
118. Liu X, Bandyopadhyay B, Nakamoto T, Singh B, Liedtke W et al (2006) A role for AQP5 in activation of TRPV4 by hypotonicity: concerted involvement of AQP5 and TRPV4 in regulation of cell volume recovery. J Biol Chem 281:15485–15495
119. Cantiello HF, Prat AG, Bonventre JV, Cunningham CC, Hartwig JH et al (1993) Actin-binding protein contributes to cell volume regulatory ion channel activation in melanoma cells. J Biol Chem 268:4596–4599
120. Blase C, Becker D, Kappel S, Bereiter-Hahn J (2009) Microfilament dynamics during HaCaT cell volume regulation. Eur J Cell Biol 88:131–139
121. Becker D, Bereiter-Hahn J, Jendrach M (2009) Functional interaction of the cation channel transient receptor potential vanilloid 4 (TRPV4) and actin in volume regulation. Eur J Cell Biol 88:141–152
122. Willette RN et al (2008) Systemic activation of the transient receptor potential vanilloid subtype 4 channel causes endothelial failure and circulatory collapse: Part 2. J Pharmacol Exp Ther 326:443–452
123. Zaninetti R et al (2011) Activation of TRPV4 channels reduces migration of immortalized neuroendocrine cells. J Neurochem 116: 606–615
124. Wang JP, Tseng CS, Sun SP, Chen YS, Tsai CR, Hsu MF (2005) Capsaicin stimulates the non-store-operated Ca^{2+} entry but inhibits the store-operated Ca^{2+} entry in neutrophils. Toxicol Appl Pharmacol 209:134–144
125. Bernabò N et al (2010) Role of TRPV1 channels during the acquisition of fertilizing ability in boar spermatozoa. Vet Res Commun 34(Suppl 1):S5–S8
126. Bernabò N, Pistilli MG, Mattioli M, Barboni B (2010) Role of TRPV1 channels in boar spermatozoa acquisition of fertilizing ability. Mol Cell Endocrinol 323:224–231
127. Rosado JA, Sage SO (2001) Activation of store-mediated calcium entry by secretion-like coupling between the inositol 1,4,5-trisphosphate receptor type II and human transient receptor potential (hTrp1) channels in human platelets. Biochem J 356:191–198
128. Ramos AJ, Cantero MR, Zhang P, Raychowdhury MK, Green A, MacPhee D, Cantiello HF (2008) Morphological and electrical properties of human trophoblast choriocarcinoma, BeWo cells. Placenta 29:492–502
129. Wen Z, Han L, Bamburg JR, Shim S, Ming GL, Zheng JQ (2007) BMP gradients steer nerve growth cones by a balancing act of LIM kinase and Slingshot phosphatase on ADF/cofilin. J Cell Biol 178:107–119
130. Wu S, Chen H, Alexeyev MF, King JA, Moore TM, Stevens T, Balczon RD (2007) Microtubule motors regulate ISOC activation necessary to increase endothelial cell permeability. J Biol Chem 282:34801–34808
131. Redondo PC, Harper AG, Sage SO, Rosado JA (2007) Dual role of tubulin-cytoskeleton in store-operated calcium entry in human platelets. Cell Signal 19:2147–2154
132. Lemonnier L, Trebak M, Lievremont JP, Bird GS, Putney JW Jr (2006) Protection of TRPC7 cation channels from calcium inhibition by closely associated SERCA pumps. FASEB J 20:503–505
133. Cho H, Shin J, Shin CY, Lee SY, Oh U (2002) Mechanosensitive ion channels in cultured sensory neurons of neonatal rats. J Neurosci 22:1238–1247
134. Mandadi S et al (2006) Increased sensitivity of desensitized TRPV1 by PMA occurs through PKCepsilon-mediated phosphorylation at S800. Pain 123:106–116
135. Numazaki M, Tominaga T, Takeuchi K, Murayama N, Toyooka H, Tominaga M (2003) Structural determinant of TRPV1 desensitization interacts with calmodulin. Proc Natl Acad Sci USA 100:8002–8006
136. Rosenbaum T, Gordon-Shaag A, Munari M, Gordon SE (2004) Ca^{2+}/calmodulin modulates TRPV1 activation by capsaicin. J Gen Physiol 123:53–62
137. Mohapatra DP, Nau C (2005) Regulation of Ca^{2+}-dependent desensitization in the vanilloid receptor TRPV1 by calcineurin and cAMP-dependent protein kinase. J Biol Chem 280:13424–13432
138. Price TJ, Jeske NA, Flores CM, Hargreaves KM (2005) Pharmacological interactions between calcium/calmodulin-dependent kinase II alpha and TRPV1 receptors in rat trigeminal sensory neurons. Neurosci Lett 389:94–98
139. Lishko PV, Procko E, Jin X, Phelps CB, Gaudet R (2007) The ankyrin repeats of TRPV1 bind multiple ligands and modulate channel sensitivity. Neuron 54:905–918
140. Holakovska B, Grycova L, Bily J, Teisinger J (2011) Characterization of calmodulin binding domains in TRPV2 and TRPV5 C-tails. Amino Acids 40:741–748
141. Phelps CB, Wang RR, Choo SS, Gaudet R (2010) Differential regulation of TRPV1, TRPV3, and TRPV4 sensitivity through a

conserved binding site on the ankyrin repeat domain. J Biol Chem 285:731–740
142. Yuan JP et al (2003) Homer binds TRPC family channels and is required for gating of TRPC1 by IP3 receptors. Cell 114:777–789
143. Caprodossi S et al (2011) Capsaicin (CPS) promotes a more aggressive gene expression phenotype and invasiveness in null-TRPV1 urothelial cancer cells. Carcinogenesis 32:686–694
144. Bhalla KN (2003) Microtubule-targeted anticancer agents and apoptosis. Oncogene 22:9075–9086
145. Blagosklonny MV, Fojo T (1999) Molecular effects of paclitaxel: myths and reality (a critical review). Int J Cancer 83:151–156
146. Lee JJ, Swain SM (2006) Peripheral neuropathy induced by microtubule-stabilizing agents. J Clin Oncol 24:1633–1642
147. White JG, Rao GH (2008) Effects of a microtubule stabilizing agent on the response of platelets to vincristine. Blood 60:474–483
148. Röytta M, Raine CS (1985) Taxol-induced neuropathy: further ultrastructural studies of nerve fibre changes in situ. J Neurocytol 14:157–175
149. Gilbar P, Hain A, Peereboom VM (2009) Nail toxicity induced by cancer chemotherapy. J Oncol Pharm Pract 15:143–155
150. Canta A, Chiorazzi A, Cavaletti G (2009) Tubulin: a target for antineoplastic drugs into the cancer cells but also in the peripheral nervous system. Curr Med Chem 16:1315–1324
151. Topp KS, Tanner KD, Levine JD (2000) Damage to the cytoskeleton of large diameter sensory neurons and myelinated axons in vincristine-induced painful peripheral neuropathy in the rat. J Comp Neurol 424:563–576
152. Tanner KD, Levine JD, Topp KS (1998) Microtubule disorientation and axonal swelling in unmyelinated sensory axons during vincristine-induced painful neuropathy in rat. J Comp Neurol 395:481–492
153. Goswami C, Goswami L (2010) Filamentous microtubules in the neuronal spinous process and the role of microtubule regulatory drugs in neuropathic pain. Neurochem Int 57: 497–503

Part II

TRP Channels, from A1 to V4

Chapter 3

TRPA1 in Drug Discovery

Jun Chen, Steve McGaraughty, and Philip R. Kym

Abstract

TRPA1 is one of the few ion channels with human genetic validation for pain. A TRPA1 gain-of-function mutation is linked to familial episodic pain syndrome in humans. This milestone discovery, coupled with a growing preclinical literature implicating TRPA1 in multiple indications, has made TRPA1 an attractive therapeutic target. With extensive investment across the pharmaceutical industry, several novel nonreactive TRPA1 antagonist series have emerged in patents, and two TRPA1 compounds have recently advanced to human clinical trials. A review of the diverse roles for TRPA1 in pain signaling and other indications such as itch and respiratory diseases is presented along with an overview of known small molecule activators and antagonists of the TRPA1 receptor.

Key words: TRPA1 antagonists, Pain, Respiratory disease, Overactive bladder, Itch

1. Introduction

TRPA1, also known as ANKTM1 and p120, belongs to the transient receptor potential (1) superfamily, which consists of a large group of cation channels present in species from yeast to mammals (2, 3). In mammals, there are 28 members from six subfamilies (TRPC, TRPV, TRPM, TRPML, TRPP, and TRPA), playing critical roles in physiological processes ranging from vasorelaxation, fertility, cell growth to sensory function. TRPA1 is the only member of the TRPA subfamily, and distinctively different from other TRP channels with an overall low sequence homology (<40% in transmembrane domains), a large N terminus containing many ankyrin repeat domains (ARD), and sensitivity to a variety of stimuli. Activation of TRPA1 leads to membrane depolarization and increase in intracellular Ca^{2+} which initiate multiple downstream signaling cascades that ultimately control a variety of physiological processes.

Arpad Szallasi and Tamás Bíró (eds.), *TRP Channels in Drug Discovery: Volume I*, Methods in Pharmacology and Toxicology, DOI 10.1007/978-1-62703-077-9_3, © Springer Science+Business Media, LLC 2012

1.1. Expression

Although TRPA1 was originally cloned from cultured human lung fibroblasts, it is most predominantly expressed in primary sensory neurons where it co-localizes with pain markers and peptidergic nociceptors (e.g., TRPV1, CGRP, substance P, and bradykinin receptors) (4–7). Among acutely dissociated DRG neurons, 30% respond to TRPA1 agonists (8). TRPA1 expression level is increased in rodent models of inflammatory and neuropathic pain, as well as in avulsion-injured human DRG (4, 6, 9, 10). Furthermore, TRPA1 expression is increased by nerve growth factors and nociceptive signals (4, 11). Besides its predominant expression in sensory neurons, TRPA1 has been reported to also be expressed in nonneuronal cells and tissues, including skin (12), inner ear (13), urinary bladder (14, 15), stomach (16), and cerebral arteries (17).

1.2. Structure–Function

A functional TRPA1 channel is predicted to be formed by four subunits each consisting of six transmembrane domains, a pore domain between helices 5 and 6, and large intracellular segments of cytoplasmic N- and C termini (Fig. 1). This overall architecture was recently confirmed by a 16 Å resolution electron microscopy structure of mouse TRPA1 (18). TRPA1 is a nonselective cation channel that can open or close in response to conformational changes induced by binding of reactive or nonreactive ligands, changes in levels of intracellular Ca^{2+}, or by other modifications of the protein. Specific domains and amino acids that play an important role in channel gating have been identified by site-directed mutagenesis.

TRPA1 is activated by electrophilic compounds such as allyl isothiocyante (AITC) through covalent modification. Several cysteine and lysine residues in the N terminus have been identified as covalent modification sites, including C415, C422, and Cys622 in mTRPA1 and C621, C641, C665, and K711 in hTRPA1

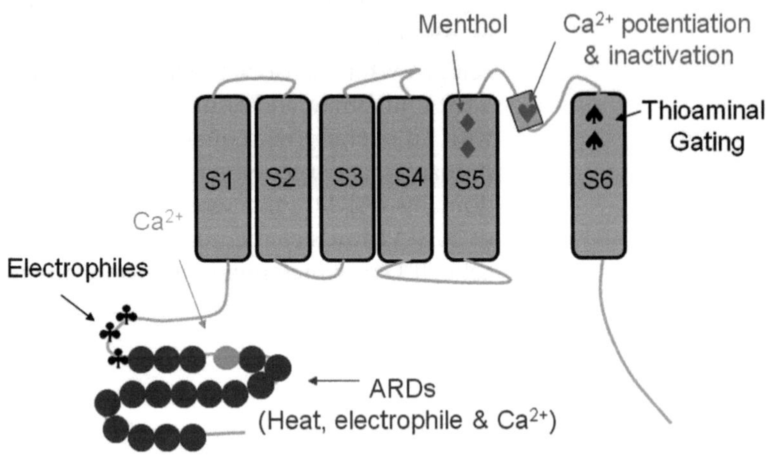

Fig. 1. TRPA1 structural scheme and domains critical for function. See text for abbreviation.

(NM_007332) (19, 20). The quadruple mutation in hTRPA1 (C621S/C641S/C655S/K711Q) was found to reduce, but did not eliminate AITC sensitivity, indicating contributions from other yet unidentified residues (21). Another observation is that only one residue is in equivalent position of human and mouse channels (e.g., C622 in mTRPA1 and C621 in hTRPA1), suggesting that covalent modification may differ across species.

TRPA1 contains a large number (at least 12) of ARD in its N zterminus. ARDs are responsible for the difference in heat sensitivity between Drosophila TRPA1 and hTRPA1, and these domains are also involved in modulating sensitivity to chemical irritants and cytosolic Ca^{2+} (22). The complex regulation of TRPA1 channel gating that is mediated by changes in intracellular Ca^{2+} has been shown to involve several different structural domains. Cytosolic Ca^{2+} activates the channel by binding to residues of an EF-hand domain in the N terminus (e.g., S468 and T470 in human TRPA1) (23, 24). Permeation Ca^{2+} ions potentiate, and then desensitize the channel through binding to an acid residue in the putative selectivity filter (D918 in rat TRPA1) (25). Several residues in the S5 domain (S876/T877 in mouse TRPA1) were identified as critical sites for menthol binding (26). In a previous study, we found that thioaminals covalently modified channel proteins but produced species-specific effects: activation of rat TRPA1, but block of human TRPA1. The opposite gating was attributed to residue difference in several S6 residues: A946 and M949 in rat TRPA1 were shown to be responsible for channel activation, and equivalent residues in human TRPA1 (S943 and I946) resulted in channel block (21). Therefore these residues may constitute a part of the gating machinery of TRPA1.

2. TRPA1 Activators

The most striking property of TRPA1 is its activation by a plethora of stimuli, including electrophilic agonists, nonelectrophilic agonists, and physiological stimuli. This property fulfills its role as a broad chemosensor to changes in external and internal environment.

2.1. Electrophilic Agonists

Many electrophilic compounds present in pungent natural products or chemical irritants activate TRPA1 by reacting with nucleophilic residues in the channel (Fig. 2). Some of the natural products that have shown to activate TRPA1 include allyl isothiocynate (AITC) from mustard oil, cinnamaldehyde from cinnamon, allicin from raw garlic, and isovelleral from the fungus *Lactarius vellereus* (Fig. 2) (5, 27, 28). Their pungency and deterrent power are largely mediated through TRPA1 (29). Additionally, TRPA1 can be activated by a large number of reactive environmental irritants, such as acrolein from cigarette smoke, isothiocyanates from industry waste, CN from tear gas, acetaldehyde from alcohol, hyperchlorite from bleach, and ozone from air pollutants (29–32). In spite of their different structures and origins, these compounds are all electrophilic and activate TRPA1 through a common mechanism, covalent modification of key cysteine, and lysine residues localized in the N terminus of the channel (Fig. 1) (19, 20, 33). An increasing number of potential endogenous ligands have also been identified, including lipid peroxidation products (e.g., 4-hydroxynonenal), cyclopentenone prostaglandin metabolites (e.g., 15-d-PGJ_2), and reactive oxygen species (e.g., H_2O_2 and nitric oxide) and reactive nitrogen species (nitric oxide) (34, 35). These molecules have the potential to nonselectively modify proteins, but their proalgesic and pro-inflammatory effects are largely attributed to TRPA1.

2.2. Nonelectrophilic Agonists

An increasing number of structurally diverse, nonreactive TRPA1 agonists have also been identified, including arachidonic acid, URB597, farnesyl thiosalicylic acid, trinitrophenol, flufenamic acid, d-9-THC, 2-APPB, diclofenac, isoflurane, and PF-4840154 (Fig. 2) (27, 36–40). The apparent lack of chemical reactivity suggests that they activate TRPA1 through noncovalent interactions, but the exact mechanism has not been elucidated.

2.3. Physiological Stimuli

Physiological stimuli that activate TRPA1 include changes in intracellular Ca^{2+}, hypertonicity, and arguably noxious cold. Ca^{2+} is an important regulator of TRPA1 function and has complex effects:

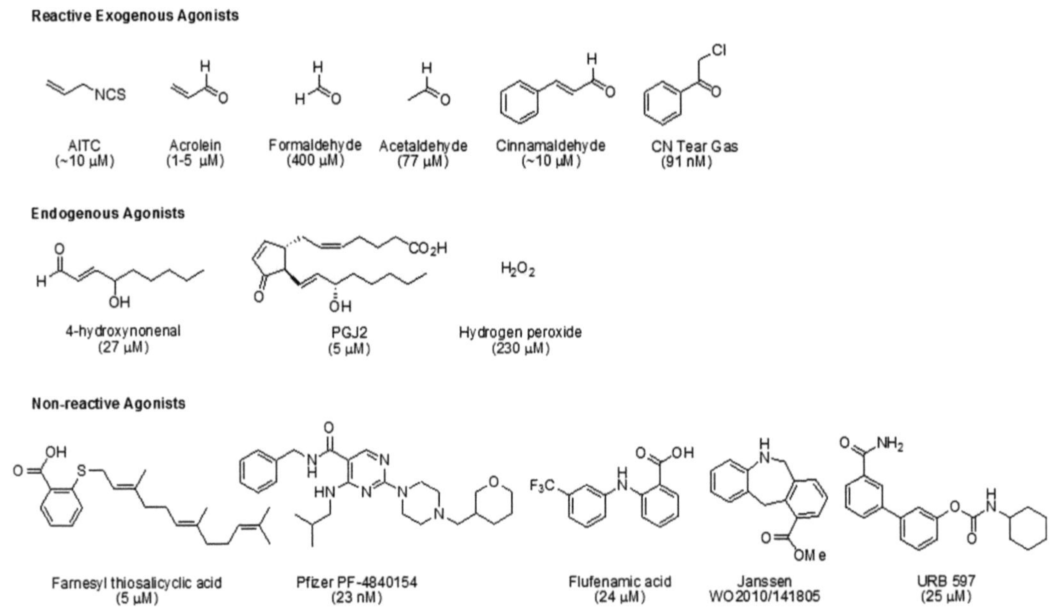

Fig. 2. TRPA1 agonists.

intracellular Ca^{2+} directly activate TRPA1 via binding to an N-terminal EF-hand domain (24); while permeating Ca^{2+} potentiates and desensitizes the channel (25). Hypertonicity activates both recombinant and native TRPA1 in DRG neurons, implicating a possible role of TRPA1 in mechanosensation (41). Bradykinin activates TRPA1 indirectly through the phospholipase C pathway, and its nociceptive effect is reduced in TRPA1-deficient mice (27, 29). TRPA1 can also be activated downstream of Mas-related G protein-coupled receptors, which mediated histamine-independent itch (42). It was reported originally that noxious cold activated TRPA1 (4), but this issue remains controversial (43, 44).

3. Antagonists

To date, several antagonists (HC-030031, AP18, A-967079, and Chembridge-5861528) have been utilized as tools for probing TRPA1 function and therapeutic utility (45–49). The broad interest in TRPA1 antagonists is noted by the large number of patents filed in recent years (>20 patents since 2007).

3.1. Electrophilic Antagonists

Covalent modification is the most important mechanism for TRPA1 activation, but not all electrophilic compounds are TRPA1 agonists. Amgen and Abbott have independently identified a series of reactive, thioaminal containing compounds that function as human TRPA1 antagonists (Fig. 3) (21, 50). The blocking effect is dependent on the electrophilic nature of the compounds, since removing the reactive sulfur atom of the compounds or the key cysteine residues in the channel abolishes inhibition. Interestingly, the same compounds activate the rat channel. Therefore, covalent modification can lead to different functional consequences; whether channels open or close depends not only on their respective proteins (e.g., rat and human TRPA1), but also on the nature of specific chemical adducts formed by electrophilic compounds. These compounds are useful tools in studying TRPA1 structure–function, but the species-specific effects limit their potential for advancement.

In contrast to thioaminals, compounds from an oxime series have more consistent effects across species (46, 49, 51). AP18 was the first reported antagonist within this series, with IC_{50} of 3.1 μM and 4.5 μM on human and rat TRPA1, respectively (46). Abbott independently identified the same chemotype from a high throughput screen (52). Subsequent medicinal chemistry efforts led to the identification of oxime A-967079 (IC_{50}: 67 nM on hTRPA1 and 290 nM on rat TRPA1) that represented the first potent and selective TRPA1 antagonist that demonstrated good oral bioavailability (51). Although the α, β-unsaturated oxime functionality could potentially react with nucleophiles, neither AP18 nor A-967079

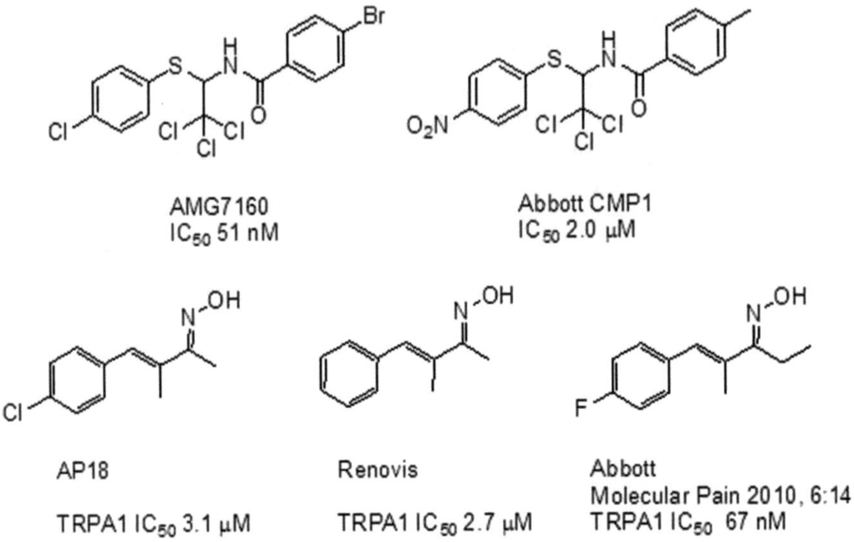

Fig. 3. Electrophilic antagonists.

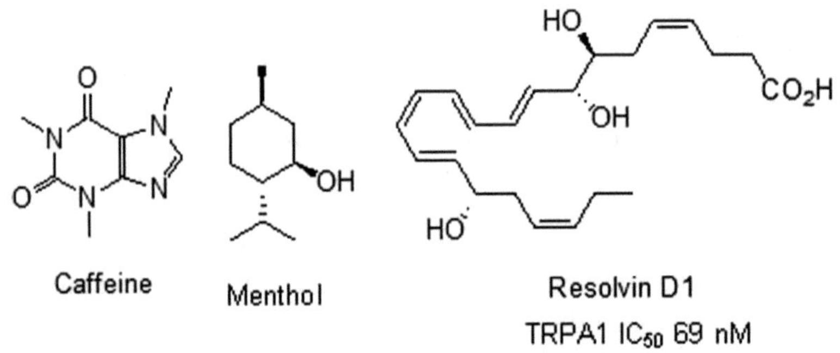

Fig. 4. Natural antagonists.

was found positive in a battery of reactivity assays including glutathione assay, La antigen-based ALARM NMR or ALARM-MS (51). Furthermore, block of TRPA1 by A-967079 was reversible. Taken together, these data suggest that some oximes may not covalently modify the channel, or that they could covalently modify the channel but in a rapidly reversible manner.

3.2. Nonreactive Antagonists

3.2.1. Naturally Occurring Antagonists

Several natural products exhibit inhibitory effects on TRPA1 (Fig. 4). Caffeine suppresses human TRPA1 activity at millimolar concentrations (1). Although it is relatively weak and it activates, rather than blocks mouse TRPA1, caffeine has served as an important starting point for the construction of potent TRPA1 antagonists described in patents from Hydra and Glenmark. Menthol blocks mouse TRPA1 at high concentrations (e.g., 300 μM) and activates at lower concentrations (e.g., 10 μM). It exhibits only an agonist effect on human TRPA1 across a range of concentrations (26). It will be interesting to explore whether menthol can be used as a building block for the discovery of future potent antagonists. Resolvin D1, a naturally occurring anti-inflammatory lipid, blocked mouse TRPA1 with IC_{50} of 69 nM (53). In behavioral testing, Resolvin D1 blocked nociceptive responses evoked by cinnamaldehyde and formalin. Interestingly, Resolvin D1 is even more potent in inhibiting TRPV3 and TRPV4 (IC_{50}: 28.9 nM and 8.1 nM, respectively). It is not clear whether Resolvin D1 interact with these channels directly or through an indirect mechanism. Nonetheless, these results suggest the existence of endogenous TRPA1 antagonists.

3.2.2. Pharmaceutical Nonelectrophilic Antagonists

Hydra Biosciences was the first to patent potent TRPA1 antagonists that incorporate the xanthine alkaloid core that is present in caffeine (Fig. 5). Building off the purine core at the N7 methyl group with an arylated amide functionality led to significant gains in TRPA1 potency (HC-030031 IC_{50} < 5–20 μM, WO2007073505). Further manipulation by changing the aryl group from phenyl to thiazole, and appending a second aryl functionality attached to aliphatic pyrrolidine capping group led to exceptionally potent TRPA1 antagonists (Example 1 IC_{50} 4 nM, WO2010/075353). Transposition of the lipophilic side chain to the C8 position of the purine heterocycle also leads to TRPA1 antagonists of reasonable potency (Example 1 IC_{50} 497 nM, WO2010/132838).

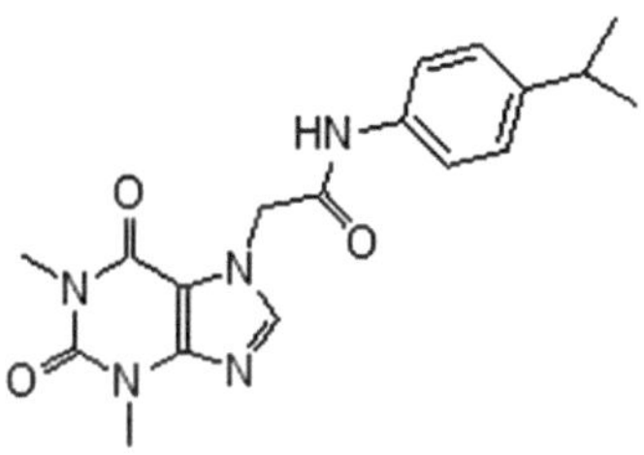

Hydra Biosciences
HC-030031
WO2007073505
TRPA1 IC_{50} < 5,000 - 20,000 nM

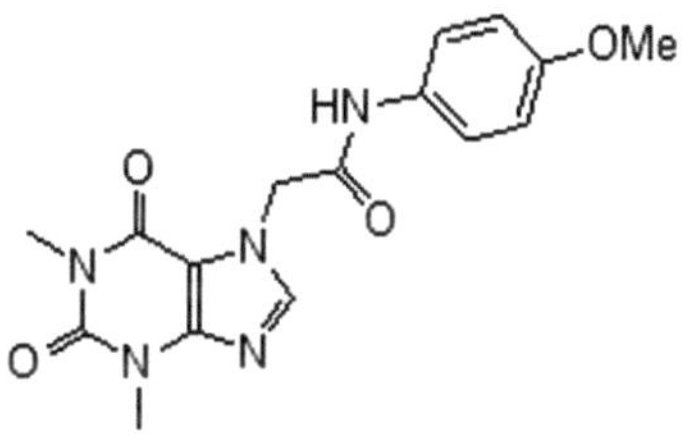

Hydra Biosciences
WO2009002933
TRPA1 IC_{50} < 10,000 nM

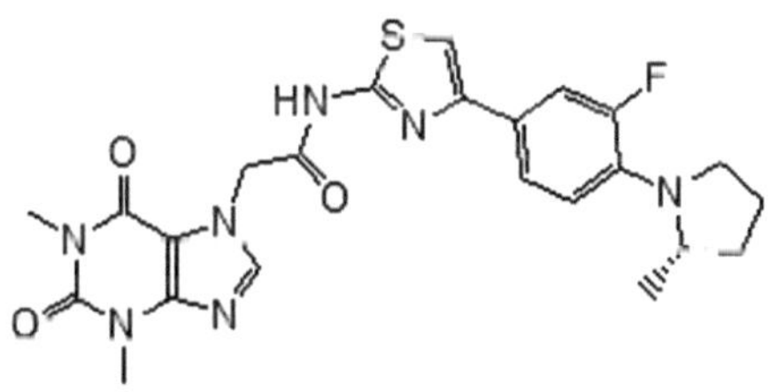

Hydra Biosciences
WO2010/075353
Example 1
TRPA1 IC_{50} 4 nM

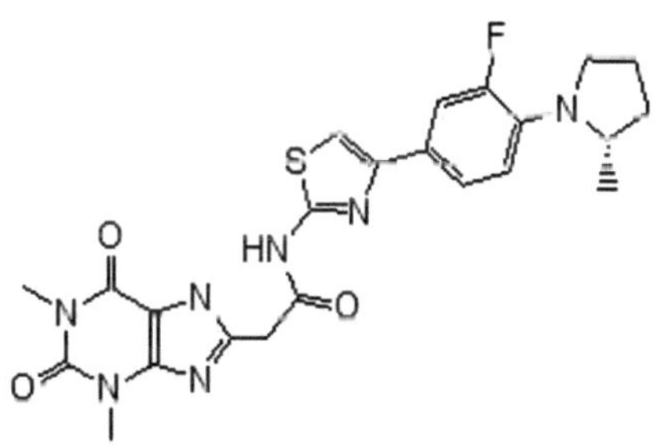

Hydra Biosciences
WO2010/132838
Example 2
TRPA1 IC_{50} 497 nM

Fig. 5. Hydra antagonists.

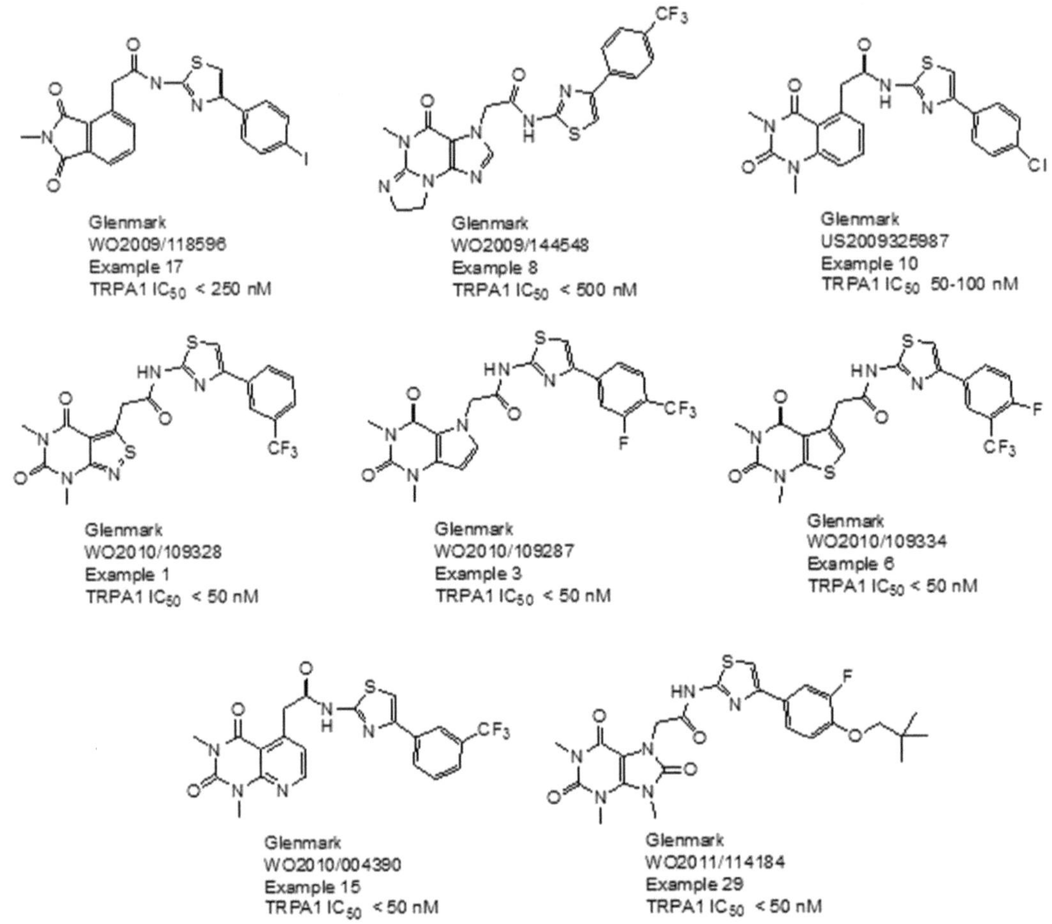

Fig. 6. Glenmark antagonists.

Glenmark Pharmaceuticals also filed patents for TRPA1 antagonists based on the caffeine xanthine alkaloid core (Fig. 6). The Glenmark approach has been focused on modification of the xanthine ring to create several new classes of TRPA1 antagonists. Reversal of the 6-5 purine core to a 5-6 phthalimide system maintained TRPA1 potency (Example 17 IC_{50} < 250 nM, WO2009/118596). In addition, replacement of the bicyclic purine core with the tricyclic imidazopurinone maintains potency at TRPA1 (Example 8 IC_{50} < 500 nM, WO2009/144548). TRPA1 potency is also maintained when the five membered imidazole fragment of the purine core is replaced with a variety of different heterocycles, including phenyl (US2009325987), isothiazole (WO2010/109328), pyrrole (WO2010/109287), thiophene (WO2010/109334), pyridine (WO2010/004390), and imidazolone (WO2011/114184).

TRPA1 antagonists featuring greater structural diversity have been reported by Janssen and Merck (Fig. 7). Janssen has described

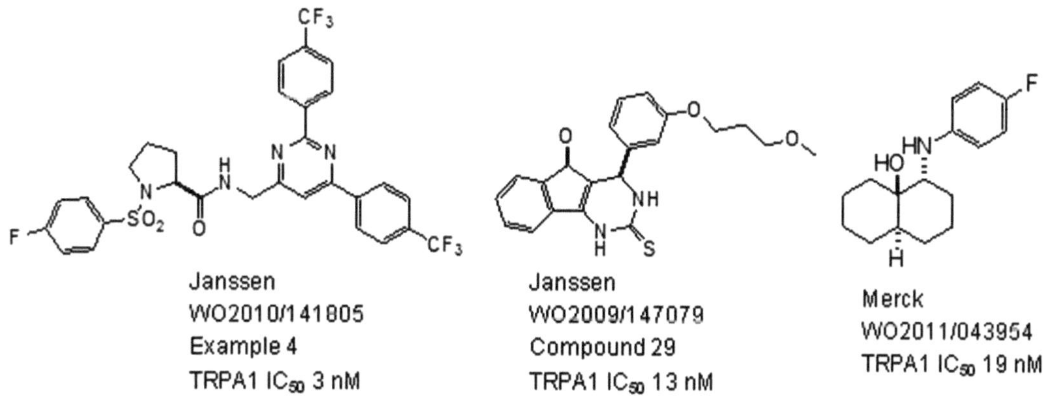

Fig. 7. Janssen and Merck antagonists.

a class of highly potent biaryl pyrimidines that contain a pyrrolidine carboxamide side chain (Example IC_{50} 3 nM, WO2010/141805). They have also described a second lead series based on the tricyclic thioxodihydroindenopyrimidinone core (Compound 29 IC_{50} 13 nM, WO2009/147079). Merck has described a lower molecular weight lead series based on a hydroxy-substituted aminodecalin core (Example, IC_{50} 19 nM, WO2011/043954). This lead series demonstrates more attractive drug-like properties (MW 263, cLog P 4.1) than many previously described antagonists.

4. TRPA1 Involvement in Pain

Evidence for TRPA1 contributions to pain starts with the observation that local application of "natural" TRPA1 agonists trigger spontaneous pain in humans (54) and induce nocifensive behaviors in animals (29). TRPA1 is expressed in small- and medium-sized peptidergic primary afferent somatosensory neurons that express TRPV1 and receptors for calcitonin gene-related peptide, substance P, and bradykinin, which are key mediators/transmitters in nociceptive signaling (5, 7, 29). TRPA1 expression is increased in these sensory neurons in humans with an avulsion injury as well as in animal models of neuropathic and inflammatory pain (7, 10). Recently, a gain-of-function mutation in TRPA1 was linked to familial episodic pain syndrome in humans, providing strong validation of TRPA1 in human pain states (55).

4.1. Chemical Nociception

TRPA1 is a key chemosensor for many pungent natural products and environmental irritants including AITC from horse radish, cinnamaldehyde from cinnamon, allicin from garlic, acrolein from cigarette smoke, and ozone from air pollutants (56, 57). These agents trigger tissue irritation and a burning sensation. TRPA1 is activated indirectly by bradykinin, a mediator of neurogenic inflammation (27). In behavioral experiments, TRPA1$^{-/-}$ mice lost aversion to oral AITC, and had diminished nocifensive responses to hind paw injection of AITC and intraplantar bradykinin (29, 58).

4.2. Mechanical Nociception

Gene ablation (TRPA1$^{-/-}$) studies have examined TRPA1 contributions to mechanical nociception. In uninjured rats, mechanical thresholds were not different between wild type and knockout animals, nor was there an impairment in the development of mechanical hyperalgesia after an inflammatory injury (46, 59). However, mechanical sensitization typically induced by mustard oil injection was not observed in TRPA1$^{-/-}$ mice (59), and a TRPA1 antagonist (AP18) did not reduce mechanical hyperalgesia in inflamed knockouts despite effectiveness in wild types (46). Mechanically activated currents in small diameter primary afferent neurons were also significantly reduced in amplitude in TRPA1$^{-/-}$ animals compared to wild types (60). Another study demonstrated that TRPA1$^{-/-}$ animals had less sensitivity to low-intensity mechanical stimuli compared to wild-type mice, and responses to high-intensity (noxious) mechanical stimulation were appreciably impaired (58). Hypertonic solutions were also found to activate recombinant and endogenously expressed rat TRPA1 (41); Taken together, a possible role for TRPA1 in mechanotransmission was suggested by these studies.

Using the skin-nerve preparation, it was found that TRPA1 in normal skin are necessary for mechanotransmission in several types of primary afferent fibers including slowly adapting C-fibers, Aδ-fiber mechanononociceptors, and slowly adapting Aβ-fibers (61). HC-030031 did not alter the responses of C- and A-fibers to low-intensity mechanical stimulation of normal skin (62); however, it reduced the responsiveness of C-fibers once the stimulus force was increased into the high-intensity range. Similarly, with an in vivo preparation, systemic administration of A-967079 reduced the responses of spinal wide dynamic range (WDR) and nociceptive specific neurons to high-intensity mechanical stimulation in uninjured rats as well as in rats with an osteoarthritic (OA) or inflammatory injury (49). In contrast, the transmission of low-intensity mechanical stimulation was impeded by A-967079 only following an inflammatory injury and was not altered in uninjured animals. The combined data suggest that TRPA1 factors into high-intensity mechanotransmission under both normal and pathological conditions, but shifts to include lower intensity stimulation following an inflammatory injury. The lowering of mechanical thresholds could be linked to TRPA1 up-regulation and/or sensitization by inflammatory mediators such as bradykinin and prostaglandins (27, 33).

The physiological data underscore effects of TRPA1 antagonists on mechanical sensitivity in behavioral assays. Responses to mechanical stimuli in inflamed and neuropathic rats were decreased by systemic administration of HC-030031 and Chembridge-5861528 (47, 48, 63). Intraplantar injection of AP18 attenuated mechanical hyperalgesia in wild type but not in knockout mice (46). Intrathecal delivery of A-967079 or Chembridge-5861528 attenuated secondary mechanical sensitivity following capsaicin or formalin injection (64). However, these data are contrasted by the observations that A-967079 did not attenuate mechanical hypersensitivity in models of neuropathic and inflammatory pain, and TRPA1-specific antisense oligodeoxynucleotides. Thus, although the majority of data points to a role for TRPA1 in mechanosensation in uninjured and sensitized states, some inconsistencies still need to be explained.

4.3. Cold Nociception

The role of TRPA1 in noxious cold sensation is highly controversial (43, 44). Channel activation by noxious cold has been reported in several studies (4, 65), but has been disputed or attributed to an indirect intracellular Ca^{2+}-mediated mechanism (5, 24). TRPA1-deficient mice were shown to be less responsive to acetone-mediated evaporative cooling, cold plate, and tail immersion than wild-type animals (58, 65); while other studies showed normal cold sensitivity in knockout mice (29, 66). Injection of A-967079 or HC-030031 did not produce deficits in the cold plate test in uninjured mice, but attenuated cold allodynia in models of neuropathic pain (51, 67). HC030031 also reversed paclitaxel-mediated cold hyperalgesia in diabetic rats, and this effect may have been related to the increased expression of TRPA1 that was triggered by the overproduction of mitochondrial H_2O_2. TRPA1-specific antisense oligodeoxynucleotides were efficacious in attenuating cold allodyia in inflammatory and neuropathic pain models (7). These data suggest that TRPA1 plays a role in cold allodynia under disease states. It has also been suggested that TRPA1 may play a role in cold sensing in visceral sensory neurons, but not in cutaneous nociceptors (68, 69).

4.4. Heat Nociception

TRPA1 is not considered a heat-activated channel and thus there are limited studies in this area. Genetic ablation of TRPA1 or antagonism of the receptor with HC-030031 does not alter responses to noxious heat stimulation in uninjured animals (47, 58). However, HC-030031 was found to decrease thermal hyperalgesia in the paclitaxel model of chemotherapy-induced neuropathic pain (63). Interestingly, HC-030031 also reduced cold and mechanical hypersensitivity in this model, and this effect was likely mediated through proteinase-activated receptor 2 (PAR2)-related sensitization of neurons containing TRP receptors including TRPA1 (63). Additionally, injection of TRPA1 agonists enhances WDR neuronal responses to heat stimulation in naive rats (70, 71). This effect may not be surprising since topical application of AITC or CA can trigger burning pain and thermal hyperalgesia, which could be attributed to indirect sensitization of TRPV1 (54).

4.5. Spontaneous Nociception

In addition to effects on evoked stimulation, evidence suggests that TRPA1 may be involved in nonevoked discomfort or pain. Nonevoked pain is observed in the majority of patients with chronic injury and is a primary reason for seeking medical care. Spontaneous firing of WDR neurons is elevated in injured animals and likely reflects injury-related sensitization and possibly nonevoked pain (72, 73). Injection of A-967079 reduced “heightened” spontaneous firing of WDR neurons in inflamed rats (49). Interestingly, elevated spontaneous firing of WDR neurons in OA rats was not decreased by injection of A-967079. The differential effect of A-967079 on spontaneous firing in inflamed and OA rats was postulated to be related to the degree of inflammation on the days of neuronal recording.

5. TRPA1 Involvement in Other Diseases

5.1. Respiratory Diseases

Respiratory diseases have become a significant public health burden and unmet medical need. TRPA1 has been linked to airway sensitization, chronic cough, asthma, and Chronic Obstruction of Pulmonary Diseases (COPD) (74–76). The airways are innervated by sensory neurons that express TRPA1 in vagal and trigeminal nerves, which are responsible for transducing noxious stimuli into pain signals, and for mediating protective behaviors (sneezing, cough, and respiratory depression). A plethora of airway irritants activate TRPA1, including acrolein from cigarette smoke, ozone (air pollutants), isocyanate (industrial waste), and tear gas (chemical warfare) (56, 57). Transient TRPA1 activation by exogenous irritants evokes pain and protective reflex (sneezing, cough, and respiratory depression, avoidance), therefore providing a protective alert to eliminate exposure to irritants and limit damage to the body. However, persistent channel activation by endogenous ligands causes or aggravates existing airway irritation. Asthma, COPD, and rhinitis are associated with increased levels of reactive oxygen species (H_2O_2), lipid peroxidation production (e.g., 4HNE), and pro-inflammatory mediators (e.g., bradykinin and nerve growth factors) (77, 78). These agents activate TRPA1 directly (e.g., H_2O_2, 4HNE), indirectly (bradykinin), or increase TRPA1 surface expression (nerve growth factors). In turn, the heightened TRPA1 activity leads to the release of neurotransmitters and cytokines, further promoting neurogenic inflammation (56).

In guinea pig, aqueous extracts from cigarette smoke, acrolein, and crotonaldehyde induce neurotransmitter release, tracheal plasma extravasation, and bronchi contraction. These effects are inhibited by TRPA1 antagonist HC-030031 (56). In mice, airway exposure to oxidants (hypochlorite and H_2O_2) evokes respiratory depression manifested as a reduction in breathing frequency and increased end expiratory pause, which were abolished by genetic deletion (31). In an ovalbumin-induced mouse asthma model, genetic ablation and treatment with HC-030031 significantly reduced the induction of cytokines, chemokines, neurotransmitters, as well as leukocyte infiltration and airway hyperactivity (32). While these results are encouraging, many respiratory diseases are chronic, multifaceted disorders that involve inflammatory pathways. For example, asthma is associated with acute and chronic inflammation of the airway mucosa, as well as airway remodeling. Chronic obstructive pulmonary disease (COPD) is characterized by activated inflammation pathways that lead to destruction of the lung over time. It remains to be seen whether TRPA1 antagonists can alleviate symptoms and reverse disease progression in inflammatory-mediated respiratory diseases, and it is unclear whether a TRPA1 antagonist will offer more benefits than current therapies (e.g., anti-inflammatory drugs and bronchodilators) and other emerging mechanisms. Nonetheless, respiratory diseases represent an exciting frontier to be explored for TRPA1 antagonists.

5.2. Overactive Bladder

The involvement of TRPA1 in bladder activity is implied by its expression pattern and the effect of TRPA1 agonists and antagonists on bladder function. TRPA1 is expressed on C-fiber bladder afferents and urothelial cells (14, 15). The expression of TRPA1 is significantly increased in overactive bladder conditions in both rats and humans (14, 79). AITC and cinnamaldehyde increase basal bladder pressure and reduce voided volume in naive rats (15). Furthermore, injections of either a TRPA1 antisense oligodeoxynucleotide or antagonist (HC-030031) were both shown to attenuate bladder overactivity induced by spinal cord injury (79). Collectively these studies suggest that TRPA1 may be involved in bladder function under normal and pathological conditions.

5.3. Itch

Itch is a major complaint in patients with skin, kidney, diabetic, and neurological disorders (80). Ironically, many medical treatments evoke itch, limiting their clinical usage. For example, patients treated with opioids and local anesthetics frequently experience itch (81). The antimalaria drug chloroquine causes intolerable itch, which is resistant to current treatment of antihistamines. A recent study found that TRPA1 knockout mice exhibited diminished itch response to chloroquine and the endogenous pruritogen BAM8-22 (42). Since chloroquine and BAM8-22 evoke itch through two separate Mar-related G protein-coupled receptors (MrgprA3 and MrgprC11), TRPA1 may be a downstream integrator of multiple histamine-independent pathways.

Besides expression in pruritic sensory neurons, TRPA1 is detected in nonneuronal cells from human skin (12). RT-PCR and western blot demonstrated robust TRPA1 expression in primary culture of epidermal keratinocytes, melanocytes, and fibroblasts.

Immunohistochemisty of human samples localized TRPA1 in the basal layer of the epidermis, dermis, and epithelium of hair follicle. These results suggest that TRPA1 expressed in the skin may directly participate in itch responses.

6. Outlook

The recent finding that a gain-of-function mutation of TRPA1 causes familial episodic pain syndrome in humans is a major milestone for TRPA1 research, establishing TRPA1 as one of two ion channels with human genetic validation for pain (55). This result, coupled with a growing preclinical literature implicating TRPA1 in multiple indications, has made TRPA1 an attractive therapeutic target. With extensive investment across the pharmaceutical industry, several novel nonreactive TRPA1 antagonist series have emerged in patents, and two TRPA1 compounds have recently advanced to human clinical trials. In February 2011, Glenmark announced the entry of the first TRPA1 antagonist into Phase I trials. GRC 17536 is reported to be highly potent (IC_{50} < 10 nM), selective, and orally available. It is claimed to be efficacious in inflammatory and neuropathic pain, and efficacious in airway inflammation from an asthma model. In January 2012, Cubist Pharmaceuticals/Hydra Biosciences announced entry of CB-625 into Phase I trials. CB625 is reported to be a highly potent, selective TRPA1 antagonist with efficacy in surgically induced and inflammatory mediated pain in animal models. Positive outcomes from these and other clinical trials would galvanize additional TRPA1 drug discovery efforts.

References

1. Nagatomo K, Kubo Y (2008) Caffeine activates mouse TRPA1 channels but suppresses human TRPA1 channels. Proc Natl Acad Sci USA 105(45):17373–17378
2. Montell C, Birnbaumer L et al (2002) The TRP channels, a remarkably functional family. Cell 108(5):595–598
3. Clapham DE (2003) TRP channels as cellular sensors. Nature 426(6966):517–524
4. Story GM, Peier AM et al (2003) ANKTM1, a TRP-like channel expressed in nociceptive neurons, is activated by cold temperatures. Cell 112(6):819–829
5. Jordt SE, Bautista DM et al (2004) Mustard oils and cannabinoids excite sensory nerve fibres through the TRP channel ANKTM1. Nature 427(6971):260–265
6. Nagata K, Duggan A et al (2005) Nociceptor and hair cell transducer properties of TRPA1, a channel for pain and hearing. J Neurosci 25(16):4052–4061
7. Obata K, Katsura H et al (2005) TRPA1 induced in sensory neurons contributes to cold hyperalgesia after inflammation and nerve injury. J Clin Invest 115(9):2393–2401
8. Niforatos W, Zhang XF et al (2007) Activation of TRPA1 channels by the fatty acid amide hydrolase inhibitor 3'-carbamoylbiphenyl-3-yl cyclohexylcarbamate (URB597). Mol Pharmacol 71(5):1209–1216
9. Katsura H, Obata K et al (2006) Antisense knock down of TRPA1, but not TRPM8, alleviates cold hyperalgesia after spinal nerve ligation in rats. Exp Neurol 200(1):112–123

10. Anand U, Otto WR et al (2008) TRPA1 receptor localisation in the human peripheral nervous system and functional studies in cultured human and rat sensory neurons. Neurosci Lett 438(2):221–227
11. Schmidt M, Dubin AE et al (2009) Nociceptive signals induce trafficking of TRPA1 to the plasma membrane. Neuron 64(4):498–509
12. Atoyan R, Shander D et al (2009) Non-neuronal expression of transient receptor potential type A1 (TRPA1) in human skin. J Invest Dermatol 129(9):2312–2315
13. Corey DP, Garcia-Anoveros J et al (2004) TRPA1 is a candidate for the mechanosensitive transduction channel of vertebrate hair cells. Nature 432(7018):723–730
14. Du S, Araki I et al (2007) Transient receptor potential channel A1 involved in sensory transduction of rat urinary bladder through C-fiber pathway. Urology 70(4):826–831
15. Streng T, Axelsson HE et al (2008) Distribution and function of the hydrogen sulfide-sensitive TRPA1 ion channel in rat urinary bladder. Eur Urol 53(2):391–399
16. Kondo T, Obata K et al (2009) Transient receptor potential A1 mediates gastric distention-induced visceral pain in rats. Gut 58(10):1342–1352
17. Earley S, Gonzales AL et al (2009) Endothelium-dependent cerebral artery dilation mediated by TRPA1 and Ca^{2+}-Activated K+ channels. Circ Res 104(8):987–994
18. Cvetkov TL, Huynh KW et al (2011) Molecular architecture and subunit organization of TRPA1 ion channel revealed by electron microscopy. J Biol Chem 286(44):38168–38176
19. Hinman A, Chuang HH et al (2006) TRP channel activation by reversible covalent modification. Proc Natl Acad Sci USA 103(51):19564–19568
20. Macpherson LJ, Dubin AE et al (2007) Noxious compounds activate TRPA1 ion channels through covalent modification of cysteines. Nature 445(7127):541–545
21. Chen J, Zhang XF et al (2008) Molecular determinants of species-specific activation or blockade of TRPA1 channels. J Neurosci 28(19):5063–5071
22. Cordero-Morales JF, Gracheva EO et al (2011) Cytoplasmic ankyrin repeats of transient receptor potential A1 (TRPA1) dictate sensitivity to thermal and chemical stimuli. Proc Natl Acad Sci USA 108(46):E1184–1191
23. Doerner JF, Gisselmann G et al (2007) Transient receptor potential channel A1 is directly gated by calcium ions. J Biol Chem 282(18):13180–13189
24. Zurborg S, Yurgionas B et al (2007) Direct activation of the ion channel TRPA1 by Ca^{2+}. Nat Neurosci 10(3):277–279
25. Wang YY, Chang RB et al (2008) The nociceptor ion channel TRPA1 is potentiated and inactivated by permeating calcium ions. J Biol Chem 283(47):32691–32703
26. Xiao B, Dubin AE et al (2008) Identification of transmembrane domain 5 as a critical molecular determinant of menthol sensitivity in mammalian TRPA1 channels. J Neurosci 28(39):9640–9651
27. Bandell M, Story GM et al (2004) Noxious cold ion channel TRPA1 is activated by pungent compounds and bradykinin. Neuron 41(6):849–857
28. Escalera J, von Hehn CA et al (2008) TRPA1 mediates the noxious effects of natural sesquiterpene deterrents. J Biol Chem 283(35):24136–24144
29. Bautista DM, Jordt SE et al (2006) TRPA1 mediates the inflammatory actions of environmental irritants and proalgesic agents. Cell 124(6):1269–1282
30. Bang S, Kim KY et al (2007) Transient receptor potential A1 mediates acetaldehyde-evoked pain sensation. Eur J Neurosci 26(9):2516–2523
31. Bessac BF, Sivula M et al (2008) TRPA1 is a major oxidant sensor in murine airway sensory neurons. J Clin Invest 118(5):1899–1910
32. Caceres AI, Brackmann M et al (2009) A sensory neuronal ion channel essential for airway inflammation and hyperreactivity in asthma. Proc Natl Acad Sci USA 106(22):9099–9104
33. Materazzi S, Nassini R et al (2008) Cox-dependent fatty acid metabolites cause pain through activation of the irritant receptor TRPA1. Proc Natl Acad Sci USA 105(33):12045–12050
34. Trevisani M, Siemens J et al (2007) 4-Hydroxynonenal, an endogenous aldehyde, causes pain and neurogenic inflammation through activation of the irritant receptor TRPA1. Proc Natl Acad Sci USA 104(33):13519–13524
35. Andersson DA, Gentry C et al (2008) Transient receptor potential A1 is a sensory receptor for multiple products of oxidative stress. J Neurosci 28(10):2485–2494
36. Hill K, Schaefer M (2007) TRPA1 is differentially modulated by the amphipathic molecules trinitrophenol and chlorpromazine. J Biol Chem 282(10):7145–7153
37. Maher M, Ao H et al (2008) Activation of TRPA1 by farnesyl thiosalicylic acid. Mol Pharmacol 73(4):1225–1234

38. Matta JA, Cornett PM et al (2008) General anesthetics activate a nociceptive ion channel to enhance pain and inflammation. Proc Natl Acad Sci USA 105(25):8784–8789
39. Hu H, Tian J et al (2010) Activation of TRPA1 channels by fenamate nonsteroidal anti-inflammatory drugs. Pflugers Arch 459(4): 579–592
40. Ryckmans T, Aubdool AA et al (2011) Design and pharmacological evaluation of PF-4840154, a non-electrophilic reference agonist of the TrpA1 channel. Bioorg Med Chem Lett 21(16):4857–4859
41. Zhang XF, Chen J et al (2008) Transient receptor potential A1 mediates an osmotically activated ion channel. Eur J Neurosci 27(3): 605–611
42. Wilson SR, Gerhold KA et al (2011) TRPA1 is required for histamine-independent, Mas-related G protein-coupled receptor-mediated itch. Nat Neurosci 14(5):595–602
43. Caspani O, Heppenstall PA (2009) TRPA1 and cold transduction: an unresolved issue? J Gen Physiol 133(3):245–249
44. Kwan KY, Corey DP (2009) Burning cold: involvement of TRPA1 in noxious cold sensation. J Gen Physiol 133(3):251–256
45. McNamara CR, Mandel-Brehm J et al (2007) TRPA1 mediates formalin-induced pain. Proc Natl Acad Sci USA 104(33):13525–13530
46. Petrus M, Peier AM et al (2007) A role of TRPA1 in mechanical hyperalgesia is revealed by pharmacological inhibition. Mol Pain 3:40
47. Eid SR, Crown ED et al (2008) HC-030031, a TRPA1 selective antagonist, attenuates inflammatory- and neuropathy-induced mechanical hypersensitivity. Mol Pain 4:48
48. Wei H, Hamalainen MM et al (2009) Attenuation of mechanical hypersensitivity by an antagonist of the TRPA1 ion channel in diabetic animals.". Anesthesiology 111(1): 147–154
49. McGaraughty S, Chu KL et al (2010) TRPA1 modulation of spontaneous and mechanically evoked firing of spinal neurons in uninjured, osteoarthritic, and inflamed rats. Mol Pain 6:14
50. Klionsky L, Tamir R et al (2007) Species-specific pharmacology of Trichloro(sulfanyl)ethyl benzamides as transient receptor potential ankyrin 1 (TRPA1) antagonists. Mol Pain 3:39
51. Chen J, Joshi SK et al (2011) Selective blockade of TRPA1 channel attenuates pathological pain without altering noxious cold sensation or body temperature regulation. Pain 152(5): 1165–1172
52. Chen J, Lake MR et al (2007) Utility of large-scale transiently transfected cells for cell-based high-throughput screens to identify transient receptor potential channel A1 (TRPA1) antagonists. J Biomol Screen 12(1):61–69
53. Bang S, Yoo S et al (2010) Resolvin D1 attenuates activation of sensory transient receptor potential channels leading to multiple anti-nociception. Br J Pharmacol 161(3):707–720
54. Namer B, Seifert F et al (2005) TRPA1 and TRPM8 activation in humans: effects of cinnamaldehyde and menthol. Neuroreport 16(9): 955–959
55. Kremeyer B, Lopera F et al (2010) A gain-of-function mutation in TRPA1 causes familial episodic pain syndrome. Neuron 66(5): 671–680
56. Andre E, Campi B et al (2008) Cigarette smoke-induced neurogenic inflammation is mediated by alpha, beta-unsaturated aldehydes and the TRPA1 receptor in rodents. J Clin Invest 118(7):2574–2582
57. Bessac BF, Sivula M et al (2009) Transient receptor potential ankyrin 1 antagonists block the noxious effects of toxic industrial isocyanates and tear gases. FASEB J 23(4):1102–1114
58. Kwan KY, Allchorne AJ et al (2006) TRPA1 contributes to cold, mechanical, and chemical nociception but is not essential for hair-cell transduction. Neuron 50(2):277–289
59. Bautista DM et al (2005) Pungent products from garlic activate the sensory ion channel TRPA1. Proc Natl Acad Sci USA 102(34):12248–12252
60. Kwan KY, Glazer JM et al (2009) TRPA1 modulates mechanotransduction in cutaneous sensory neurons. J Neurosci 29(15):4808–4819
61. Brierley SM, Castro J et al (2011) TRPA1 contributes to specific mechanically activated currents and sensory neuron mechanical hypersensitivity. J Physiol 589(Pt 14): 3575–3593
62. Kerstein PC, del Camino D et al (2009) Pharmacological blockade of TRPA1 inhibits mechanical firing in nociceptors. Mol Pain 5:19
63. Chen Y, Yang C et al (2011) Proteinase-activated receptor 2 sensitizes transient receptor potential vanilloid 1, transient receptor potential vanilloid 4, and transient receptor potential ankyrin 1 in paclitaxel-induced neuropathic pain. Neuroscience 193:440–451
64. Wei H, Koivisto A et al (2011) Spinal transient receptor potential ankyrin 1 channel contributes to central pain hypersensitivity in various pathophysiological conditions in the rat. Pain 152(3):582–591
65. Karashima Y, Talavera K et al (2009) TRPA1 acts as a cold sensor in vitro and in vivo. Proc Natl Acad Sci USA 106(4):1273–1278

66. Knowlton WM, Bifolck-Fisher A et al (2010) TRPM8, but not TRPA1, is required for neural and behavioral responses to acute noxious cold temperatures and cold-mimetics in vivo. Pain 150(2):340–350
67. del Camino D, Murphy S et al (2010) TRPA1 contributes to cold hypersensitivity. J Neurosci 30(45):15165–15174
68. Dunham JP, Kelly S et al (2008) Inflammation reduces mechanical thresholds in a population of transient receptor potential channel A1-expressing nociceptors in the rat. Eur J Neurosci 27(12):3151–3160
69. Fajardo O, Meseguer V et al (2008) TRPA1 channels mediate cold temperature sensing in mammalian vagal sensory neurons: pharmacological and genetic evidence. J Neurosci 28(31):7863–7875
70. Merrill AW, Cuellar JM et al (2008) Effects of TRPA1 agonists mustard oil and cinnamaldehyde on lumbar spinal wide-dynamic range neuronal responses to innocuous and noxious cutaneous stimuli in rats. J Neurophysiol 99(2):415–425
71. Sawyer CM, Carstens MI et al (2009) Mustard oil enhances spinal neuronal responses to noxious heat but not cooling. Neurosci Lett 461(3):271–274
72. Chu KL, Faltynek CR et al (2004) Increased WDR spontaneous activity and receptive field size in rats following a neuropathic or inflammatory injury: implications for mechanical sensitivity. Neurosci Lett 372:123–126
73. McGaraughty S, Chu KL et al (2008) Contributions of central and peripheral TRPV1 receptors to mechanically evoked and spontaneous firing of spinal neurons in inflamed rats. J Neurophysiol 100:3158–3166
74. Bessac BF, Jordt SE (2008) Breathtaking TRP channels: TRPA1 and TRPV1 in airway chemosensation and reflex control. Physiology (Bethesda) 23:360–370
75. Simon SA, Liedtke W (2008) How irritating: the role of TRPA1 in sensing cigarette smoke and aerogenic oxidants in the airways. J Clin Invest 118(7):2383–2386
76. Geppetti P, Patacchini R et al (2010) Cough: the emerging role of the TRPA1 channel. Lung 188(Suppl 1):S63–68
77. Dolovich J, Back N et al (1970) Kinin-like activity in nasal secretions of allergic patients. Int Arch Allergy Appl Immunol 38(4): 337–344
78. Winterbourn CC, Kettle AJ (2000) Biomarkers of myeloperoxidase-derived hypochlorous acid. Free Radic Biol Med 29(5):403–409
79. Andrade EL, Forner S et al (2011) TRPA1 receptor modulation attenuates bladder overactivity induced by spinal cord injury. Am J Physiol Renal Physiol 300(5):F1223–1234
80. Verhoeven EW, Kraaimaa FW et al (2007) Prevalence of physical symptoms of itch, pain and fatigue in patients with skin diseases in general practice. Br J Dermatol 156(6): 1346–1349
81. Atanassoff PG, Brull SJ et al (1999) Enhancement of experimental pruritus and mechanically evoked dysesthesiae with local anesthesia. Somatosens Mot Res 16(4):291–298

Chapter 4

Canonical Transient Receptor Potential Channel Expression, Regulation, and Function in Vascular and Airway Diseases

Brij B. Singh, Christina M. Pabelick, and Y.S. Prakash

Abstract

The physiological function of all cells is uniquely regulated by changes in cytosolic Ca^{2+} levels. Although several mechanisms increase cytosolic Ca^{2+} levels, Ca^{2+} influx across the plasma membrane upon the release of Ca^{2+} from the internal stores is one of the major mechanisms in most nonexcitable cells and in some excitable cells. Such Ca^{2+} channels, which are activated by intracellular Ca^{2+} store depletion are referred to as store-operated Ca^{2+} entry (SOCE) channels and have been shown to be essential for many biological functions including fluid and enzyme secretion, immune regulation, hypertension, pulmonary function, neurosecretion, synaptic plasticity, and vascular diseases. Canonical transient receptor potential (TRPCs) have been proposed as components of the store-operated Ca^{2+} channel (SOCC) which mediates SOCE. TRPC channels are nonselective cation channels, present in a signaling complex where they interact with key proteins critical for their regulation. In this regard, there is increasing recognition that Ca^{2+} entry via SOCE channels plays critical roles in the lung, particularly for vascular and airway function. Indeed, regulation/targeting of TRPC channels appear to be important in normal vascular and airway physiology as well as pathophysiology of lung diseases. In this chapter, we briefly summarize the current state of knowledge regarding expression, regulation, and function of TRPC channels in the vasculature and the respiratory system.

Key words: Calcium signaling, SOCE, TRPC channels, SOCE and lung function

1. Introduction

Stimulation of cells using various agonists or pharmacological agents leads to an increase in intracellular Ca^{2+} ($[Ca^{2+}]_i$) (1, 2). This increase in $[Ca^{2+}]_i$ results from both release of Ca^{2+} from intracellular stores as well as Ca^{2+} influx across the plasma membrane. Increased $[Ca^{2+}]_i$ thereby regulates many fundamental process such as gene expression, cell proliferation and apoptosis, secretion, and muscle contraction (3–5). Although release of intracellular Ca^{2+} stores is critical especially for the initial $[Ca^{2+}]_i$ response especially at the local level, it is the influx of external Ca^{2+} which is always

Arpad Szallasi and Tamás Bíró (eds.), *TRP Channels in Drug Discovery: Volume I*, Methods in Pharmacology and Toxicology, DOI 10.1007/978-1-62703-077-9_4, © Springer Science+Business Media, LLC 2012

essential for the global and sustained response. Thus, Ca^{2+} influx is not only essential for refilling the internal stores, but can also contribute towards regulating/fine tuning a host of critical biological processes.

The first step in this process is always the binding of a hormone or a growth factor to its receptor, which is localized at the plasma membrane. Activation of these receptors regulates breakdown of phosphatidylinositol-4,5-bisphosphate (PIP_2) into inositol 1,4,5-trisphosphate (IP_3) and diacylglycerol (DAG) by the enzyme phospholipase C (PLC) (2, 6). The release of Ca^{2+} from the intracellular stores (endoplasmic reticulum or its specialized counterpart in muscle, the sarcoplasmic reticulum) is generally signaled by the formation of small mediators such as IP_3. In addition, Ca^{2+} influx can be signaled by second messengers such as cyclic nucleotides and DAG; however, in classical messenger pathway, DAG does not appear to be the primary mediator of activated Ca^{2+} entry (Fig. 1). In most instances, the signal for Ca^{2+} entry is somehow derived from IP_3-mediated depletion of Ca^{2+} from intracellular stores, a process called "capacitative Ca^{2+} entry (CCE)" or "store-operated Ca^{2+} entry (SOCE) (7)." Identification of transient receptor potential channel (TRPCs) as nonselective Ca^{2+} channels and Orai as a selective Ca^{2+} channel (discussed in other chapters in this book) opens up new avenues for biological research, since these channels

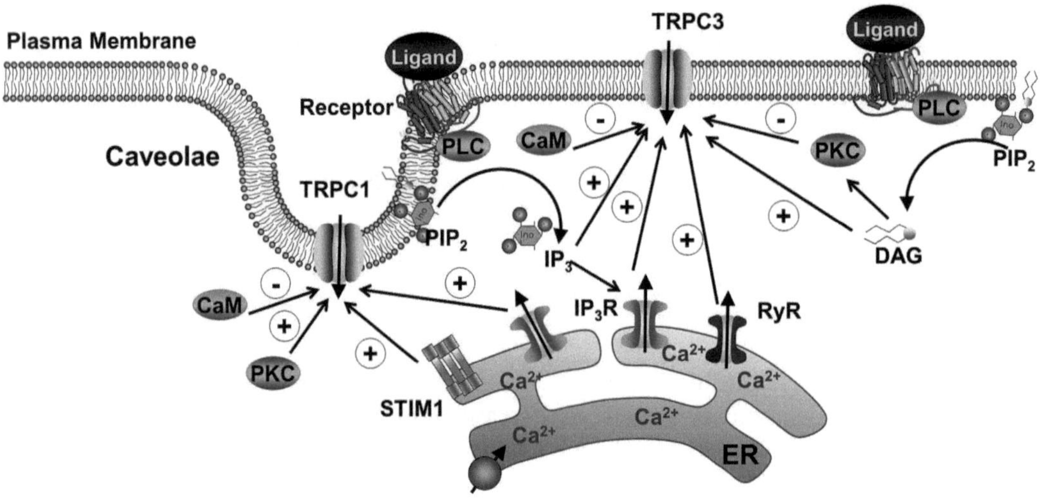

Fig. 1. Schematic representation of TRPC modulation, exemplified by TRPC1 and TRPC3. In response to agonist stimulation, production of IP_3 (via PLC-catalyzed PIP2 degradation) results in release of intracellular Ca^{2+} stores from IP_3 receptor channels. Additional Ca^{2+} release may occur via ryanodine Receptor (RyR) channels. PLC additionally stimulates the DAG-PKC pathway. Enhanced Ca^{2+} release results due to the depletion of intracellular stores, triggers store-operated Ca^{2+}entry that is critically mediated by the Ca^{2+} sensor STIM1 on the ER membrane. Channels such as TRPC1 are stimulated by IP_3R and STIM1 in response to depleted intracellular Ca^{2+} stores. Additional stimulation may involve PKC, while CaM generally is inhibitory. Channels such as TRPC3 are also stimulated by factors such as IP_3, IP_3R, RyR, and additionally by DAG (but not PKC). Thus a number of interactive cascades triggered by agonist stimulation and altered intracellular Ca^{2+} regulation can differentially modulate TRPC activity. Additional factors not shown here, but discussed in the text include RhoA, the cytoskeletal matrix, and other signaling intermediates.

are critical for regulating key biological processes (4, 8–10). In addition, although these channels have been attributed in few diseases to date, it is anticipated that both TRPC and Orai family of ion channels have an extraordinarily important role in basic cellular function as well as in diseases as highlighted in this review. However, in this chapter we will focus our attention on the regulation and functional importance of TRPC channels as it relates to vascular function and disease.

2. TRPC Channel Assembly and Functional Importance of the Signalplex

The TRP family of ion channels comprises of 28 members (divided into TRPC, TRPV, TRPM, TRPML, and TRPA subfamilies). In order to be functional in modulating Ca^{2+} entry, TRPC channel proteins have to be associated with other signaling molecules (11, 12). In *Drosophila*, the TRP channels are associated in a signalplex, which provides a unique platform to modulate its channel activity. The key component of the *Drosophila* signalplex is INAD, which exhibits five PDZ protein domains and interacts with multiple proteins such as calmodulin, PLC, G-protein, and protein kinase C. Similarly, mammalian TRPC channels have also been shown to interact with multiple proteins (STIM1, Orai1, Gαq/11, PMCA, SERCA, IP_3Rs, RyRs Homer, Caveolin1, calmodulin, GPCRs, receptor tyrosine kinases, NFkappaB, PLCg, PLCb, NCX1, Na^+/K^+/ATPase, NHERF, and others (13, 14)). However, a scaffold protein that can be similar to INAD has not yet been identified. Also it is not clear if all these proteins are always attached to TRPC channels, or are recruited in a more dynamic fashion, when the channel is either activated or inactivated.

Recent research has indicated that TRPC channels are assembled into lipid raft domains, which are not only critical for the function of TRPC channels, but are also important in the assembly and retention of TRPC proteins within the plasma membrane. Moreover, interplay between this signalplex could be essential for regulating fundamental biological function exhibited by TRPC channels. Interestingly, mutations in the *Drosophila* INAD not only decrease Ca^{2+} entry, but also alter the localization of all the corresponding interacting proteins. Thus, it could be anticipated that disruption of the mammalian TRPC signalplex could also lead to altered localization of TRPC channels and its associated proteins, thereby disrupting Ca^{2+} signaling as observed in *Drosophila* signalplex. Although multiple reviews have been published on TRPC channels, few of these reviews have explored the implications of TRPC channel interactions with a multitude of proteins. Thus, in this chapter we will focus on a couple of these interactions and discuss the functional implications of such interactions (summarized in Fig. 1).

2.1. Regulation of TRPCs Via Calmodulin

Ca^{2+} entry is critical for cellular responses, but excessive Ca^{2+} influx is toxic to cells. Therefore, Ca^{2+} channels have evolved an intrinsic activation/inactivation mechanism to regulate Ca^{2+} entry. Similar to voltage-gated Ca^{2+} channels, TRPC proteins are also known to interact with a Ca^{2+} sensor protein "calmodulin." We and other investigators have shown that the C-terminus of TRPC is essential for calmodulin binding (15, 16) that is involved in Ca^{2+}-dependent feedback inhibition of the TRPC channels by acting as a Ca^{2+} sensor[15]. Although it has been shown that deletion of this calmodulin-binding domain significantly decreases the inactivation of the TRPC channel, other possibilities cannot be ruled out which are (1) deletion of the calmodulin-binding domain from the C-terminus of TRPC protein could limit/inhibit its interaction with other regulatory proteins necessary for channel inactivation, primarily Homer or other unknown proteins, (2) deletion of the C-terminus could alter the stability of the TRPC1 channel thereby increasing Ca^{2+} entry, and (3) could inhibit TRPC phosphorylation as observed with other TRPV channels (17). Beside identification of the calmodulin-binding domain in the TRPC channels, no studies have been performed to indicate the amino acids critical for this regulation. In voltage-gated channels, calmodulin exerts two opposing effects on individual channels. It can initially promote (facilitation) and subsequently inhibit (inactivation) channel opening, via Ca^{2+}-transduction capabilities of calmodulin (18). Ca^{2+} release-activated Ca^{2+} currents (I_{crac}) were also shown recently to be activated in a similar fashion by calmodulin (19). Calmodulin has four EF hand motifs, which can each bind to a Ca^{2+} ion. Ca^{2+} binding to its amino-terminal lobes initiates channel inactivation while Ca^{2+} binding to the carboxyl-terminal lobe induces facilitation of voltage-gated channels. Since TRPCs functionally interact with calmodulin, it is possible that mechanism(s) similar to voltage-gated or I_{crac} channels could also be present in TRPC channels (mainly TRPC1 and 3). Thus, it would be interesting to study and understand the dual role of calmodulin in TRPCs activation/inactivation. Also, multiple calmodulin-binding sites have been identified on various TRPCs; however, only few of these sites have been shown to have a functional implication. Thus, it could be hypothesized that one or more calmodulin-binding domains present on the TRPC channels could be associated with the retention of calmodulin protein and only upon stimulation the calmodulin could detach from its original binding site and would interact with its regulatory site in order to control Ca^{2+} influx. If such a phenomenon occurs, then inhibition/facilitation of the movement of calmodulin could be used to manipulate Ca^{2+} influx in various pathological conditions and more research is needed to fully understand the role of calmodulin in Ca^{2+} entry.

2.2. TRPC Activation and Interaction with IP_3R and RyR Channels

TRPC channels conduct relatively large Na^+, Ca^{2+}, Ba^{2+}, and Sr^{2+} inward currents and little outward current up to +40 mV (20). Although TRPC channels could be classified into two different groups depending on their mode of activation (store versus receptor operated), in physiological conditions all TRPC channels are activated by G-protein coupled mechanism. It has been also shown that the TRPC3/6/7 subfamily could also be gated by diacylglycerol (DAG) (21). TRPC3, TRPC6, and TRPC7 form a structural and functional subfamily of DAG-sensitive cation channels, coupling receptor/PLC-signaling pathways to cation entry. Although these TRPC proteins (3, 6, and 7) are generally classified as the DAG-responsive subfamily, it is still a highly contentious issue as to whether DAG can be regarded as the physiological activator of native channel complexes, while there is no doubt that all members of the TRPC3/6/7 subfamily can principally be activated by DAG. As deduced from pharmacological inhibition of DAG lipase and DAG kinase, it appears that endogenously generated DAG is sufficient for channel activation (22). Most notably, receptor agonists and DAGs do not display additive effects on TRPC3 and TRPC6 current amplitudes, suggesting that the same TRPC channels are activated by DAG and by PLC-linked receptors and that DAG may be the decisive second messenger generated by PLC. However, so far a direct interaction of DAG with TRPC3/6/7 proteins has not been demonstrated. Further, TRPC3/6/7 activation by C1 domain-containing proteins, such as chimeras, MUNC13s, RasGRPs, and even DAG kinases, cannot be excluded and deserves experimental clarification.

Several studies have shown that TRPC channels interact with IP_3 and ryanodine receptor (IP_3R, RyR) (23–25). TRPC3–IP_3R interaction has been shown to be critical for gating of the TRPC3 channel. Kiselyov et al. (24) demonstrated that depletion of intracellular Ca^{2+} stores activated single channels in the plasma membrane of cells stably transfected with TRPC3. On excision, channel activity was lost but could be restored by IP_3. With more extensive washing, restoration of channel activity required addition of both IP_3 and the IP_3 receptor. These results suggest that expressed TRPC3 is gated by IP_3-liganded IP_3 receptors, consistent with the conformation coupling model (7). Full-length IP_3R coupled to silent TRPC3 channel in intact cells did not activate them until stores were depleted of Ca^{2+}. However, constructs containing the IP_3-binding domain activated TRPC3 channels in unstimulated cells and restored gating of TRPC3 by IP_3 in excised plasma membrane patches (23). Thus, it has been concluded that the N-terminal domain of the IP_3R functions as a gate and is sufficient for the activation of TRPC3. Interestingly, it has been shown that TRPC3 can be activated by ryanodine receptors (RyRs), thus contributing towards gating of the TRPC3 channel (24). Coupling of TRPC3 to IP_3R or RyRs in the same cells was found to be mutually exclusive, suggesting that clustering and segregation of TRPC3–IP_3R and TRPC3–RyR complexes in plasma membrane microdomains was essential for Ca^{2+} entry by different agonists (24, 25). Overall, these previous data demonstrate a complex, multiprotein mediated gating and activation mechanism for TRPC channels.

2.3. Interplay Between Calmodulin, Homer, and Phosphorylation

TRPC1 is known to be phosphorylated by PKC (26), whereas TRPC3 activity is shown to be inhibited by PKC (27). Both pharmacological (PKC antagonist) or genetic (kinase-defective mutant) manipulations show that inhibition in TRPC1 phosphorylation decreased Ca^{2+} entry, indicating that PKC phosphorylation of TRPC1 is an important determinant of Ca^{2+} entry in human endothelial cells. Phosphorylation of TRPC1 and resulting Ca^{2+} entry were found to be functionally essential for the increase in permeability induced by thrombin in confluent endothelial monolayers. In contrast, two functional phosphorylation sites have been identified in TRPC3, which are activated by different protein kinases. Protein kinase G (PKG) inactivates TRPC3 by direct phosphorylation on Thr-11 and Ser-263 sites, whereas protein kinase C (PKC) inactivates TRPC3 by phosphorylation on Ser-712. Point mutations at PKG phosphorylation sites (T11A-S263Q) of TRPC3 reduced phorbol-12-myristate-13 acetate inhibition by inhibiting the PKG activity. Thus, overall it suggests that the inhibitory action of PKC on TRPC3 is partly mediated through PKG, which was observed in multiple cells. Furthermore, OAG-induced $(Ca^{2+})_i$ rise was also inhibited by the addition of PKC and PKG agonists. Importantly, the PMA inhibition on OAG-induced $(Ca^{2+})_i$ rise was significantly reduced by PKG inhibitors, suggesting an important role of PKG in the PMA-induced inhibition of TRPC channels in endothelial cells. Although it is possible that different TRPC channels could be regulated by different kinases, it remains to be seen if other cells that have functional TRPC1–TRPC3 channels are either inhibited or activated via these kinases. The Src kinase was found to be essential for the activation of TRPC3 by muscarinic receptors (28); however, no direct evidence with regard to the phosphorylation of TRPC3 channel was been established. On the other hand, the Src family kinase Fyn does indeed bind to TRPC6 and regulate its activity (29). The interaction involves the SH2 domain of Fyn and the N-terminal region of TRPC6, which result in tyrosine phosphorylation of TRPC6 and showed increased channel activity. TRPC6 activity is also regulated by a CaMK II-dependent phosphorylation (30). However it remains to be seen if the kinase directly interacts with the channel. In addition, the phosphorylation state of the TRPC channels can also be regulated by various phosphates, and although till date no phosphates have been identified that interacts with TRPC channels, it could be speculated that these phosphates could also have a significant role in the regulation of TRPC-mediated channel activity.

The N-terminus of IP_3Rs is able to activate TRPC1, TRPC3, and TRPC4 (16, 31). Interestingly, IP_3R peptides compete with calmodulin for binding to TRPC channels, and addition of calmodulin prolongs the interval between Ca^{2+} release from intracellular stores and activation of Ca^{2+} influx (16, 32). These findings suggest a model in which calmodulin blocks IP_3R-dependent activation of the channel. However, deletion of the calmodulin-binding sequences in TRPC1 did not result in spontaneously active channels as observed in Homer knockout mice, but rather showed an increased Ca^{2+} influx. Since calmodulin binding affinity increases with elevated Ca^{2+}, this mechanism may be important for certain adaptations and be influenced by cytosolic Ca^{2+} or release of Ca^{2+} from intercellular stores. Additionally, since both Homer and calmodulin bind at the C-terminus of TRPC1 channel, there could be competition between the two proteins, and depending on the need and the complexity of the cell, different proteins may be bound to regulate Ca^{2+} influx accordingly. Overall these findings indicate a high degree of complexity of TRPC activation and regulation in biologically relevant systems allowing for multiple modes of regulation. However, further research is needed to confirm these interactions under more dynamic conditions involving activation/inactivation of these TRPC channels.

2.4. Functional Interaction Between STIM1, Orai1, and TRPCs

Using large-scale RNA-interference screens on cellular signaling proteins, two groups have identified STIM proteins (stromal interaction molecule 1 and 2) as the most important regulators of SOCE (33, 34). Identification of STIM, as the missing link involved in transmitting the message of ER store depletion to the PM-SOCC apparatus, provided an answer for the most perplexing mysteries in store-operated Ca^{2+} signaling biology. STIM1 and STIM2 are both single-pass transmembrane proteins with N-terminal Ca^{2+} binding EF-hands (canonical and hidden) and SAM (sterile alpha motif) domain located either extracellular or in the ER lumen. On the other hand, the C-terminus domain of STIM1 is localized to the cytoplasm and is shown to contain the ERM (ezrin–radixin–moesin) and coiled-coil protein interacting domains (35, 36). STIM1 has been shown to be glycosylated and associated with the PM (37–39); however, the precise role of PM STIM1 in SOCE remains unclear (40). In addition, STIM1 is shown to be phosphorylated (37), predominantly at serine residues, and this appears to be important for SOCE regulation during cell division (41, 42). Similarly STIM2, which is about 45% identical with STIM1, has also been identified as a second modulator of Ca^{2+} signaling; however, recent reports indicate that STIM2 is primarily involved in maintaining basal Ca^{2+}, rather than initiating SOCE (40). Functionally, STIM1 in the ER responds to the changes in ER Ca^{2+} levels via reversible binding of Ca^{2+} with its EF-hand domain. Following depletion of ER Ca^{2+}, STIM1 undergoes homotypic interactions and forms oligomeric clusters (puncta) at the ER–PM junctions. The STIM1 puncta at the peripheral ER can thus physically associate with SOCC that are present in the plasma membrane thereby activating SOCE (33, 43–48). Although identification of the ER Ca^{2+} sensor, STIM1, has revolutionized our understanding of SOCE, questions of what the importance of STIM1 puncta is, and how they are stabilized and recruited to sites of ER in opposition to the PM are still unsolved mysteries. Whether the composition of the PM at these junctional sites influences the specificity of STIM1 recruitment is not known. Thus, information on the identity of the ER–PM junction sites, at subplasma membrane regions, where STIM1 clusters translocate and associate with the PM-SOCC to activate SOCE is warranted. Nevertheless, identification of STIM1 as a critical regulator of SOCE channels has provided novel insight into their activation paradigms.

Recent studies have demonstrated an obligatory role of STIM1 in activating TRPC channels and regulating SOCE. In this context, STIM1 activation of TRPC1 channel has been extensively studied. Physiologically, the interaction between STIM1 and TRPC1 is dynamic and reversible and is regulated by the status of ER Ca^{2+} stores such that, upon agonist-mediated store-depletion STIM1–TRPC1 interaction is increased. However, following agonist clearance and subsequent re-filling of the ER stores the STIM1–TRPC1 complex dissociates to a resting state (49). Although the precise control of this interaction cycle is not established, the kinetics underlying the homomultimeric clustering of STIM1 appears to be a prime driving force. STIM1 has been shown to associate with TRPC1 in a complex containing Orai1; however, independently their regulation by STIM1 is distinct, since functional interaction sites for STIM1 are different (50, 51). STIM1 binds to native TRPC channels as efficiently as it binds with ectopically expressed channels. With regard to STIM–TRPC associations, there seems to be a preference. STIM1 interacts with TRPC1, C2, C4, and C5 but does not seem to interact with TRPC3, C6, and C7. It has been suggested that although STIM1 does not directly bind to certain TRPC channels it is still able to influence their function indirectly since, native SOCCs can constitute heteromeric TRPC assemblies. This distinction in the activation mechanism of TRPC channels could provide an explanation as to why TRPC1/C4/C5 are deemed as candidates for store depletion, whereas TRPC3/C6/C7 are mainly believed to be dependent on second messenger system. Similarly, TRPC1 has been shown to functionally interact with Orai1 channels and Orai1 channels have been shown to be regulators of TRPC channels (52, 53). In addition, SOCE via the TRPC channels has been shown to be both Orai1 dependent and Orai1 independent (54, 55), suggesting that positioning of specific channels along with precise interactions with other molecules at cellular compartments could co-ordinate Ca^{2+} entry. Overall, these results suggest that depending on the cell type and the need of

Ca^{2+} signal, these interactions could have additive effect on the regulation of TRPC channels and that Ca^{2+} channels are more complex that requires multiple proteins for the generation of Ca^{2+} currents. In addition, compartmentalization of these key proteins could also play a significant role in these functional interactions and many of the TRPC interacting proteins have been shown to be partitioned in both lipid raft and nonlipid raft fractions (13).

3. Targeting of TRPC Channels to the Plasma Membrane (PM)

TRPC channels have been shown to interact with many regulatory and scaffolding proteins. Thus, it can be anticipated that some of these proteins will play an important role in PM retention as well as targeting of TRPCs to the membrane. Moreover, since TRPC channels have the inherent ability to form multimeric channels (homomeric and heteromeric), selective association between different subtypes can generate a plethora of different channel with very different functional characteristics and thereby can ultimately regulate numerous biological functions. However, in order for TRPC channels to be functional they have to be targeted to the PM and associate with their regulatory proteins.

3.1. TRPC1 Interaction with Caveolin-1 Lipid Rafts

Lipid rafts are planar domains in the PM that are rich in sphingolipids, cholesterol, plasmenylethanolamine, and arachidonic acid giving them a liquid-ordered state that is in stark contrast to the surrounding highly disordered phospholipid bilayer[13]. One biophysical property of lipid rafts is that they are resistant to low temperature detergent solubilization with nonionic detergents. Caveolae are a subclass of lipid rafts containing a family of integral membrane proteins, caveolins that cause them to take the form of flask-shaped invaginations of the PM (9, 13, 56). Lipid rafts and caveolae have been shown to play an important role in the regulation of various cellular functions including organization of cell signaling machinery such as receptor tyrosine kinases (RTKs) and GPCRs, establishment of cell polarity, cholesterol transport, and endocytosis. Since Ca^{2+} signaling cascade, such as PIP_2, $G_{\alpha q/11}$, muscarinic receptors, PMCA, and IP_3R-like protein, and Ca^{2+} signaling events such as receptor-mediated turnover of PIP_2 have also been localized to caveolar microdomains in the plasma membrane, it is likely that caveolae are important in regulating Ca^{2+} influx.

An interesting study showed that agonist-stimulated Ca^{2+} signal originates in specific areas of the PM that are enriched in caveolin-1 and that disruption of lipid raft domains not only disrupts these Ca^{2+} signals, but significantly decreases Ca^{2+} influx. Caveolae have been now identified as the primary sites of Ca^{2+} wave initiation and previous studies have suggested that caveolae function as receptacles for the PM-bound Ca^{2+} signaling machinery (57). Crosstalk between endoplasmic reticulum and caveolae has also been observed in endothelial cells, where IP_3R agonists generate Ca^{2+} waves that initiate at caveolin-1-rich edges of the cell (57). Furthermore, B cell activation is initiated in lipid rafts, also accompanied by release of Ca^{2+} from the IP_3 gated stores and after store depletion STIM1 protein has been shown to associate with caveolar lipid raft domains (49, 58). Thus, lipid rafts could very well serve as platforms for Ca^{2+} signaling, as well as the sites for the localization of various Ca^{2+} channels (13, 56). Interestingly, in addition to G-protein coupled receptors $G_{q/11}$, phospholipase C are known to be present in lipid raft domains, suggesting that proteins necessary for TRPC1 activation are grouped together to effectively coordinate signaling. Thus, it has been proposed that caveolae may regulate spatial organization of Ca^{2+} signaling by contributing to the assembly of Ca^{2+} signaling complex as well as the site of Ca^{2+} entry.

TRPC1 is known to be localized in caveolin-scaffolding lipid raft domains of the PM where it could associate with other signaling proteins mainly IP_3R, Caveolin-1, calmodulin, PMCA, and $G_{\alpha q/11}$(14). Although multiple putative caveolin-binding domains both in the N- and C-terminal regions of the TRPC1 gene have been identified, their functional importance is yet to be determined. However, it has been noted that TRPC1 which lacks the N-terminal Caveolin-1-binding domain does not get trafficked to the PM and shows significant decrease in Ca^{2+} influx. Additionally, this site has been observed to be conserved in all TRPC channels, but it remains to be seen if other TRPC channels also require caveolin-1 for their trafficking or recruitment to the PM. Expression of the mutant Cav1 (lacking its protein scaffolding and membrane anchoring domains) also shows disruption in PM localization of TRPC1 as well as significant decrease in Ca^{2+} influx upon store depletion. This effect was not due to a generalized disruption of PM lipid raft domains, since staining with cholera toxin is retained. Moreover, cholesterol depletion and re-introduction has shown to have a profound effect on TRPC1-dependent endothelin-evoked arterial contraction, implying TRPC1 regulation by cholesterol is critical in vascular pathology (59). Thus, overall, interaction between TRPC1 N-terminus with Caveolin-1 appears to play a critical role in the functional expression of this protein in the PM with TRPC1 function depending on the presence of cholesterol, or at least the integrity of cholesterol-rich caveolae. There is much that is still unknown about the extent that lipid raft domains and lipid raft function affect Ca^{2+} signaling. Some questions that remain unanswered are whether intracellular trafficking of TRPC channels depends to some degree on lipid raft-mediated processes in cells, and whether caveolins are also responsible for the turnover of TRPC1 protein (endocytosis). In this regard, it may not be just

TRPC channels that are modulated by caveolar structures. TRPV channels localized to caveolae/lipid rafts could be activated via endocannabinoids present in lipid rafts. Nonetheless, it is clear that caveolae/lipid raft-mediated pathways are important components of Ca^{2+} signaling and trafficking of TRPC1 channels.

3.2. TRPC3 Interaction with SNARE Complex Proteins

The SNARE superfamily of proteins has become one of the most extensively studied elements of intracellular trafficking. The roles of these proteins in membrane trafficking, docking, and fusion include the establishment of tight membrane contacts and the formation of a scaffold to support the secretory machinery. Among the most important and ubiquitous SNARE proteins are VAMPs, transmembrane proteins located in the vesicular membranes, and SNAP-25, which contains a region of palmitoylated cysteines by which it associates with the PM (60). In addition to other modulators of vesicular trafficking such as PI3-kinase, caveolin-1, PLCγ, and likely PIP_2 itself, TRP channels may also associate with SNARE proteins. Indeed, vesicular trafficking is emerging as a key mechanism for regulating PM expression and function of TRP channels.

A newly discovered mechanism of TRPC channel insertion is exocytotic insertion of these channels into the PM. A number of proteins involved in vesicular trafficking are known to interact with TRP channels, which include stathmin, VAMP2, syntaxin, SNAP, and NSF, all of which could contribute to PM insertion of TRPC channels. An initial yeast two-hybrid screen revealed that the N-terminus of TRPC3 interacts with the SNARE proteins mainly VAMP2 and a-soluble *N*-ethylmaleimide-sensitive factor attachment protein (αSNAP). Moreover, co-immunoprecipitations assays indicate that other SNARE complex protein such as SNAP23, Syntaxin1, and NSF also interact with TRPC3 channels. Interestingly agonist stimulation of cells, but not passive store depletion, resulted in PM translocation of TRPC3 that is dependent on its association with SNARE complex proteins. For example, cleavage of VAMP2 with tetanus toxin significantly decreases PM TRPC3 expression (61). These findings indicate that activation of TRPC3 by agonist stimulation involves exocytosis of subplasmalemmal vesicles containing TRPC3 into the plasma membrane. Moreover, these data indicate that TRPC3 is present in readily available pools of vesicles that can fuse with the PM immediately upon agonist stimulation, enhancing Ca^{2+} influx. Here it is interesting to note that TRPC3 directly interacts with αSNAP, which could be involved in rapid endocytosis of vesicles and thus termination of Ca^{2+} influx.

Another mode of TRPC3 regulation is via PLCg which also interacts with the N-terminus of TRPC3. However, unlike VAMP2, PLCg does not affect PM expression of TRPC3 in response to agonist stimulation. Interestingly, TRPC5 is also shown to be present in vesicles and PLCg is critical for its insertion into the PM

(62). Stimulation of tyrosine kinase growth factor receptors resulted in PM insertion of TRPC5 via PI3K and the Rho GTPase Rac1. Activated Rac1 seems to be critical for the fusion of TRPC5 with the PM. Similarly, TRPC6 translocates to the PM upon stimulation of the GPCR muscarinic M3 receptor and with the lipid 1-oleoyl-2-acetyl-*sn*-glycerol (OAG) (63). Overall, it can be summarized that TRPC channels are present in vesicles and PM insertion may involve a Ca^{2+}-mediated step that requires SNARE proteins or activation of cell surface receptors.

3.3. TRPC Interaction with Cytoskeleton Proteins

A number of studies in different cell types have reported that multiple cytoskeleton-associated mechanisms can regulate activation of TRPC channels. Small molecular GTP-binding proteins that are involved in vesicular trafficking such as Rac1 and Rho-GTPase interact with many TRPC channels. Interaction of TRPs with components of the cytoskeleton (actin) and microtubules (tubulin) could serve as a stable anchor for retaining the channels in specific PM locations. Additions of agents which modify actin and tubulin severely disrupt PM localization of TRPC channels (25, 64). It could be hypothesized that dynamic interactions with cytoskeletal components, mainly actin and microtubules, provide a system for trafficking of channel to and from the PM. Remodeling of these pathways can either promote or limit access of the vesicles to the PM. Cytoskeletal interactions can also relocate required molecular components that are involved in regulating their function or trafficking in close proximity to TRP channels. For example TRPC4 and TRPC5 interaction with spectrin links the channels to the cytoskeleton via ezrin (16).

Myosin light chain kinase is also known to regulate TRPC5 function, although it is not clear whether this involves direct effects on TRPC5 or is mediated via regulation of the cytoskeleton. There is dynamic recruitment of proteins that regulate the cytoskeleton, e.g., RhoA recruitment has been reported during TRPC1 channel activation. Actin polymerizing agents cause internalization of TRPC3. Importantly, cytoskeletal components also interact with PIP_2 and thus changes in PIP_2 levels significantly affect the status of the underlying cytoskeleton. Vesicular trafficking, including internalization as well as exocytosis, is regulated via compartmentalized changes in plasma membrane PIP_2–cytoskeletal interactions. Notably, proteins involved in vesicle fusion as well as ezrin are enriched in lipid raft domains. Since PIP_2 is relatively enriched in lipid raft domains, it is interesting to speculate that these plasma membrane lipid microdomains can coordinate trafficking as well as regulation of TRP channels. Inhibition of actin polymerization by cytochalasin D or latrunculin A completely disorganizes the actin network and fully blocks SOCE. Importantly, store-dependent movement of STIM1 has also been shown to be dependent on microtubule organization (65) .

4. TRPCs and Vascular Function

Expression and potential roles of TRPC channels in regulating vascular structure and function have been relatively well studied. Both vascular endothelial cells and smooth muscle cells express a variety of TRPC channels that appear to serve a number of functions of the vasculature under normal conditions, as well as to contribute to the pathophysiology of several diseases both in the systemic and pulmonary circulations.

4.1. TRPCs in Endothelial Cells

Vascular endothelial cells appear to express all seven members of the TRPC subfamily (66), although understanding of their roles and interactions is far from complete. A large body of work on TRPCs in the endothelium is based on nonprimary endothelial cells where the culturing process may influence the profile of expressed proteins. Furthermore, current knowledge on TRPCs in endothelial cells is based on both systemic and pulmonary vascular endothelium from both human and nonhuman species, where TRPC isoform expression (and even function) may depend on the vascular bed, the species itself, and the environment (normal vs. disease state). Additionally, heterogeneity in TRPC expression even within a vessel, and the oligomerization of TRPC isoforms further confound interpretation. Nonetheless, the roles of TRPC1, 3, 4, and 6 in endothelial cells have been explored to a larger extent than TRPC2 and TRPC7.

TRPC1 expression has been demonstrated in both pulmonary (67–69) and systemic (26, 70–72) endothelial cells, and is generally considered to be involved in Ca^{2+} influx (which may have the downstream effect of regulating vascular permeability by disrupting the adherens junction) (66, 73). In this regard, with the more recent recognition of STIM1 as a key regulator of store-operated Ca^{2+} entry (SOCE) and plasma membrane Orai1 being the putative Ca^{2+} entry channel (74–78), the exact role of TRPCs as influx channels per se has been made more complicated (41). One model suggests a heterodimer of TRPC1 and TRPC4 which then complexes with STIM1 and Orai1 in regulating SOCE (76, 79). However, other studies have independently shown that reduced expression of either TRPC isoform does not influence SOCE (80), while STIM1 suppression (41) does not affect influx via these isoforms. Here, one factor that could potentially explain these discrepant results is the physical location of TRPCs in the plasma membrane, particularly within caveolae (81, 82). For example, Malik and colleagues have demonstrated that the constitutive caveolar protein caveolin-1 helps functionally link TRPC1 with IP_3 receptor channels in the endoplasmic reticulum of pulmonary artery endothelial cells (83), thereby helping regulate Ca^{2+} influx. Whether caveolin-1 interact with STIM1 within this setting to influence influx via TRPC1 has not been examined in vascular endothelium per se, but has been demonstrated in other cell systems (9, 56, 84). The link between TRPC1 and the ER in vascular endothelium appears to be also important in regulating receptor-operated Ca^{2+} influx. For example, in human pulmonary artery endothelial cells, TRPC1-IP_3 receptor interaction regulates thrombin and VEGF-induced Ca^{2+} influx (85, 86). This pathway may play a role in the enhanced vascular permeability in the presence of inflammation, especially in the lung leading to pulmonary edema.

TRPC3 has been found in endothelial cells of several vascular beds (87–93) and is considered an oxidant-sensitive channel (66, 73) that helps regulate redox sensitivity (important in generation of atherosclerotic disease). As with TRPC1, TRPC3 can also form hetero-multimeric channels with TRPC4 in regulating $(Ca^{2+})_i$ (94). Relevant to disease, data from coronary artery endothelial cells suggest that reduction in Ca^{2+} influx due to blunted TRPC3 expression may be important in generation of atheromas (89, 95). Conversely, relevant to malignant hypertension, TRPC3 expression is increased in renal arteries (90), with the assumption that the resultant elevated Ca^{2+} influx somehow impairs renovascular tone and the downstream effects on the endocrine system. In addition to being a component of heteromultimeric influx channels (as earlier), TRPC4 by itself appears to be important as a mediator of SOCE in endothelial cells (but see (80)). For example, endothelium-dependent relaxation of the aorta is impaired in TRPC4 knockout mice, while Ca^{2+} influx in endothelial cells when stimulated with standard agonists such as ACh or ATP (important for downstream nitric oxide generation) is suppressed in the absence of TRPC4 (71). As with TRPC1, TRPC4 is also important for vascular permeability, influencing thrombin-induced Ca^{2+} influx (96, 97). However, what is not known is whether TRPC4 independently regulates SOCE, or only within the setting of interactions with other TRPC isoforms.

Compared to other isoforms, the role of TRPC5 in endothelial cells is somewhat unique. In aortic endothelial cells, TRPC5 translocates to the plasma membrane in response to stimulation by lysophosphatidylcholine, subsequently modulating Ca^{2+} influx (62, 98). Furthermore, one study found that TRPC5 is sensitive to NO (via nitrosylation) (99), but has not been found to be the case in another study (100). Nonetheless, the relevance of TRPC5 may lie in atherosclerotic disease wherein activation of TRPC5 in response to lysophosphatidylcholine prevents endothelial cell migration and thus impairs re-endothelialization of the injured vessel (98). The role of NO in TRPC5 modulation in the setting of atherosclerosis is not known. TRPC6 has been shown to be a mediator of receptor-operated Ca^{2+} influx in endothelial cells of different vascular beds (98, 101, 102), responding to factors such as RhoA (102) and bradykinin (103). Furthermore, TRPC6 appears to be important in angiogenesis, working with TRPC3 in its response to VEGF (101, 104, 105). From the above brief discussion, it is clear that TRPC channels can serve a variety of endothelial cell functions including NO generation, permeability, migration, and angiogenesis. While there is much to learn about TRPCs in vascular disease, it is also clear that modulation of TRPC channels in the endothelium (which may be technically feasible via an intravascular administration of inhibitors or drug delivery mechanisms) should be an important area for drug discovery in cardiovascular disease.

4.2. TRPCs in Vascular Smooth Muscle

The relevance of TRPCs in vascular smooth muscle again lies in their modulation of vascular tone as well as vascular remodeling in the setting of a variety of diseases in both the systemic and pulmonary circulations. As with endothelial cells, multiple TRPC isoforms have been found in the smooth muscles of different vascular beds and appear to again be involved in a variety of functions (66, 73, 106–113).

In contrast to a role in SOCE in endothelial cells, TRPC1 has been generally recognized as modulating receptor-operated Ca^{2+} entry (ROCE) in smooth muscle (114–118). Well-known vaso constrictors such as endothelin-1 (116) and angiotensin II (115) activate Ca^{2+} influx via this latter mechanism using TRPC1. Whether TRPC1 is involved in SOCE per se especially in vascular smooth muscle has not been firmly established. In vitro, TRPC1 overexpression enhances Ca^{2+} influx following store depletion and vasoconstriction of rat pulmonary artery (119) and in cerebral arteries (120). However, data from intact aortic or cerebral arteries argue against a role for TRPC1 in SOCE (121). Regardless, TRPC1 appears to be important in vascular remodeling, e.g., with angiotensin II exposure (122) or with pulmonary artery smooth muscle proliferation (123) relevant to pulmonary hypertension. Interestingly, the association between angiotensin II and smooth muscle growth in coronary artery appears to involve SOCE and interactions with STIM1 (124). These discrepant data are difficult to reconcile unless more information is obtained regarding other proteins (including other TRPC isoforms) that could potentially interact with TRPC1 in modulating Ca^{2+} influx. The role of TRPC3 in vascular smooth muscle as a receptor-operated channel is also different than in endothelial cells, being sensitive to constrictors such as endothelin-1 (88) and mediated by DAG (125, 126). More recent data suggest that TRPC3 may be indirectly involved in vasodilation, by virtue of its inhibition by an NO/PKG pathway, at least in carotid artery (127). As with TRPC1, elevated levels of TRPC3 have also been associated with angiotensin II-associated hypertension, at least in rats (128).

Compared to endothelium, the function of TRPC4 in vascular smooth muscle has been examined only to a limited extent.

As with endothelium, TRPC4 may interact with other TRPC isoforms in modulating Ca^{2+} influx in response to constrictors such as endothelin-1 (e.g., in cerebral artery) (129) or in response to mechanical stimulation (e.g., in aorta) (130). Whether these functions of TRPC4 are important in regulation of vascular tone or in vascular remodeling still remains to be established. However, limited data in the dog suggest that TRPC4 (potentially interacting with other TRPC isoforms) may be important in cerebral vasospasm following subarachnoid hemorrhage (129). The role of TRPC5 has been largely studied in the cerebral circulation (129, 131), where this isoform appears to mediate SOCE, potentially interacting with TRPC1. Furthermore, as in endothelial cells TRPC5 is activated by lysophosphatidylcholine (132, 133). However, rather than modulating vascular tone, TRPC5 appears to be involved in smooth muscle cell proliferation, which may be of relevance in vascular remodeling in the setting of hypertension.

TRPC6 has been relatively well studied in vascular smooth muscle (134–144), initially reported to response to adrenergic stimulation (142), being activated by diacylglycerol. TRPC6 is also activated by a number of vasoconstrictors including angiotensin II (145), vasopressin (141, 143), and serotonin (141). Unlike other isoforms, TRPC6 may be particularly important in reflex vasoconstriction in response to mechanical stretch (146). However, it is unclear whether mechanical stretch directly activates Ca^{2+} influx via TRPC6 or whether this channel is indirectly activated by virtue of other agents (e.g., angiotensin II) which may rise upon mechanical stimulation. Separately, TRPC6 appears to be important in the pulmonary circulation in the setting of enhanced smooth muscle proliferation (147, 148) and has been reported to be upregulated in idiopathic pulmonary hypertension. Even under normal circumstances, TRPC6 is involved in hypoxic pulmonary vasoconstriction (149), an important feature of the pulmonary circuit in maintenance of ventilation-perfusion matching. As with TRPC3, in vascular smooth muscle, TRPC7 appears to be a receptor-operated channel (150), potentially interacting with TRPC3 or TRPC6 in modulating Ca^{2+} influx (66). However, there is currently no information on the role of TRPC7 in normal vascular function or in diseases.

4.3. TRPCs in Vascular Disease

From the brief discussion earlier, it is clear that the vascular expresses a variety of TRPC isoforms all with complex roles that contribute to vascular structure and function. There is now increasing evidence that many of these isoforms, working in combination or independently, contribute to a number of diseases either by virtue of modulating $(Ca^{2+})_i$ or by altering vascular structure. While it is beyond the scope of this chapter to detail the current body of information in this area, some points are highlighted here.

Growth and proliferation of vascular smooth muscle is a key factor in conditions such as atherosclerotic disease, hypertension (both systemic and pulmonary), and vascular restenosis following percutaneous angioplasty or in fistulas (151–156). Here, there is clearly a correlation between elevated $[Ca^{2+}]_i$ levels (especially long term) and enhanced proliferation (155, 157–159). TRPCs, by virtue of being influx channels (be it receptor- or store-operated) can influence $[Ca^{2+}]_i$ and thus contribute to proliferation. For example, as discussed earlier, TRPC1 in coronary, renal, and cerebral arteries and TRPC5 in the venous system have been implicated in the systemic circulation. Pulmonary artery smooth muscle proliferation is associated with elevated TRPC6, while other studies have found elevated TRPC3 to be contributory. From a functional perspective, altered TRPC isoform expression may alter vascular tone or reactivity. Here, TRPC1 and TRPC6, by virtue of modulating vascular smooth muscle membrane potential may contribute to vasoconstriction. Both isoforms are also activated by common vasoconstrictor agonists. Additional contributions (depending on the vascular bed) may come from TRPC1 and TRPC7. Hypoxic pulmonary vasoconstriction involves TRPC6 relevant to pulmonary hypertension. Furthermore, TRPC6 may contribute to myogenic tone, important in autoregulation, especially in the cerebral circulation. Relevant to disease, TRPC1 and TRPC4 may contribute to cerebral vasospasm.

5. TRPCs in the Respiratory System

Compared to relatively detailed examination of TRPCs in the vasculature, there is currently limited information on the role of TRPCs in normal function of the airway or in respiratory diseases not involving the pulmonary vasculature (see (114, 160–163) for review). Akin to vascular endothelium and smooth muscle, airway epithelial cells and airway smooth muscle are key elements regulating structure and function. There is currently very little information on TRPC isoform-regulated regulation of airway epithelial function. Both TRPC4 and TRPC6 are expressed in human bronchial epithelium (164), and TRPC6 expression is increased in epithelia of patients with cystic fibrosis (165), and could potentially interact with the CFTR protein in controlling Ca^{2+} within the epithelium.

A few studies have examined TRPC isoforms in airway smooth muscle and their potential involvement in diseases such as asthma and COPD (see (114, 160, 162, 163, 166–169) for review). Similar to vascular smooth muscle, studies in humans, pigs, and guinea pig all show the expression of TRPC1, TRPC3, TRPC4, TRPC5, TRPC6, and TRPC7 in airway smooth muscle, albeit to different levels (164, 170–174). However, the majority of functional studies have focused on TRPC3. Interestingly, unlike in vascular smooth muscle, TRPC3 and not TRPC1 is the major component of native nonselective cation channels and thus contributes to elevation of $[Ca^{2+}]_i$ in response to agonist stimulation and with store depletion (169, 175). Limited data demonstrate that TRPC3 regulates resting membrane potential in airway smooth muscle, with siRNA-induced TRPC3 suppression resulting in lowering of membrane potential (169, 175). Here, the relative importance of membrane potential in this cell type compared to the vasculature may be different, thus making the functional importance of TRPC3 at baseline also different. In contrast to TRPC3, TRPC1 does not play a role in airway smooth muscle membrane potential. Furthermore, TRPC3 is also involved in $[Ca^{2+}]_i$ responses to bronchoconstrictor agonists such as acetylcholine (169, 171). In addition to allowing Ca^{2+} entry, TRPC3 may also modulate local Ca^{2+} levels, especially in the proximity of the plasma membrane, thus influencing other mechanisms such as Na^+/Ca^{2+} exchange or membrane potential which may have the downstream effect of enhancing influx via L-type Ca^{2+} channels (176). Here, a recent discovery has been the localization of TRPC3 to caveolae in human airway smooth muscle cells (56, 84), which introduces a new level of complexity in terms of TRPC3 regulation that is currently unexplored. Nonetheless, the fact that TRPC3 may work via both store- and receptor-operated mechanisms suggests a role for this protein in maintaining influx over sustained periods following agonist stimulation.

Given the multiple effects of TRPC3 on $[Ca^{2+}]_i$ there has been some exploration of altered TRPC3 expression and function in airway diseases. Here, a few studies have reported increased TRPC3 expression in airways of patients with asthma (169, 175, 177). Separately, the pro-inflammatory cytokine TNFa has been shown to upregulate TRPC3 expression in human airway smooth muscle (171). Furthermore, siRNA suppression of TRPC3 results in blunting of TNFa-induced enhancement of $[Ca^{2+}]_i$ and thus airway contractility (171).

There is evidence that TRPC1 may be important in airway smooth muscle cell proliferation (162), and thus a contributor to airway thickening and hyper reactivity in diseases such as asthma, with TRPC1 suppression attenuating SOCE as well as cell proliferation (123). While TRPC4 has been shown to be involved in pulmonary artery smooth muscle cell proliferation, a similar role in airway has not been demonstrated. Furthermore, it is currently not known whether either of these isoforms works independently or in collaboration in modulating airway cell proliferation, or whether they in fact directly influence mechanisms that control proliferation. Compared to other isoforms, TRPC6 is considered to be important in asthma and COPD (160), although the exact mechanisms are still unclear. Mice lacking TRPC6 show substantially less numbers of inflammatory cells in the airway in response to allergen challenge (178), suggesting a role for this isoform in allergic airway inflammation, relevant to asthma. In this regard, TRPC6 may be involved in the Ca^{2+} response of neutrophils to chemokines such as CXCL1 or CXCL8, as well as neutrophil migration in response to CXCL2 (178). In a similar vein, TRPC6 expression is increased in lung macrophages in COPD patients (179) (while other isoforms may not be substantially altered). TRPC6 is also expressed in airway smooth muscle (171). However, the airways of mice lacking TRPC6 actually show higher airway reactivity to methacholine in models of allergen sensitization and challenge (180), which may be partly explained by a compensatory increase in TRPC3 within the airway. Thus, the role of TRPC6 in the airway remains far from complete.

6. Conclusion

In conclusion TRPC channels are present within signaling microdomains that are necessary for their function. Interaction of TRPC proteins with a multitude of other PM-bound, lipid raft-associated and intracellular proteins could facilitate or inhibit Ca^{2+} entry via TRPCs, thereby contributing to the heterogeneity of spatiotemporal $[Ca^{2+}]_i$ changes necessary for cellular physiology. Further, protein–protein interaction in the trafficking of TRPC channels by different stimuli provides for sophisticated control of influx. Additionally as nonselective channels, TRPCs can contribute towards "cell plasticity" necessary for Ca^{2+}-dependent functions. Interestingly, modulation of TRPC channel upon stimulation could have a significant impact on the regulation of Ca^{2+} signaling, which could be modified in certain pathological conditions. Here, the diverse expression and regulation of TRPCs in cells of the vasculature and airways indicate the importance of these proteins in health and disease (summarized in Table 1). Thus, pharmacological intervention to regulate TRPC channel expression and activity should be a key avenue for treatment of vascular and airway pathologies and pathophysiologies.

Table 1
Summary of TRPC isoform expression and function in vasculature and airway

	TRPC1	TRPC2	TRPC3	TRPC4	TRPC5	TRPC6	TRPC7
Endothelial cells	SOCE Responses to thrombin, VEGF Vascular Permeability and edema		SOCE Redox sensitive Atherosclerosis Malignant hypertension	SOCE Responses to thrombin Vascular permeability and edema		ROCE Sensitivity to RhoA	
Vascular smooth muscle	ROCE Vascular remodeling		ROCE DAG sensitivity Vasodilation? Hypertension	SOCE Vasospasm	SOCE NO sensitivity Cerebral vascular responses Vascular remodeling	Reflex and hypoxic pulmonary vasoconstriction Pulmonary remodeling	
Epithelial cells				Present		Present	
Airway smooth muscle	Present Airway remodeling Asthma		SOCE Modulation of other Ca regulatory mechanisms Asthma	Present	Present	Present Asthma COPD	

Acknowledgments

This work was supported by NIH grants awarded to Brij B Singh (award number RO1 DE 017102-06 and 5P20RR017699), Christina M. Pabelick (R01 HL090595) and Y.S. Prakash (R01 HL088029).

References

1. Bootman MD, Berridge MJ, Roderick HL (2002) Calcium signalling: more messengers, more channels, more complexity. Curr Biol 12:R563–565
2. Berridge MJ, Lipp P, Bootman MD (2000) The versatility and universality of calcium signalling. Nat Rev Mol Cell Biol 1:11–21
3. Berridge MJ, Bootman MD, Roderick HL (2003) Calcium signalling: dynamics, homeostasis and remodelling. Nat Rev Mol Cell Biol 4:517–529
4. Parekh AB, Putney JW Jr (2005) Store-operated calcium channels. Physiol Rev 85:757–810
5. Putney JW Jr (2004) Store-operated calcium channels: how do we measure them, and why do we care? Sci STKE 2004:pe37
6. Putney JW Jr (1993) Excitement about calcium signaling in inexcitable cells. Science 262:676–678
7. Putney JW Jr (1990) Capacitative calcium entry revisited. Cell Calcium 11:611–624
8. Putney JW (2011) Origins of the concept of store-operated calcium entry. Front Biosci (Schol Ed) 3:980–984
9. Ambudkar IS, Brazer SC, Liu X, Lockwich T, Singh B (2004) Plasma membrane localization of TRPC channels: role of caveolar lipid rafts. Novartis Found Symp 258:63–70, discussion 70–64, 98–102, 263–106
10. Birnbaumer L et al (2000) Mechanism of capacitative Ca^{2+} entry (CCE): interaction between IP3 receptor and TRP links the internal calcium storage compartment to plasma membrane CCE channels. Recent Prog Horm Res 55:127–161, discussion 161–122
11. Montell C (2005) The TRP superfamily of cation channels. Sci STKE 2005:re3
12. Montell C (1999) Visual transduction in Drosophila. Annu Rev Cell Dev Biol 15:231–268
13. Pani B, Singh BB (2009) Lipid rafts/caveolae as microdomains of calcium signaling. Cell Calcium 45:625–633
14. Ambudkar IS, Ong HL (2007) Organization and function of TRPC channelosomes. Pflugers Arch 455:187–200
15. Singh BB, Liu X, Tang J, Zhu MX, Ambudkar IS (2002) Calmodulin regulates Ca^{2+}-dependent feedback inhibition of store-operated Ca^{2+} influx by interaction with a site in the C terminus of TrpC1. Mol Cell 9:739–750
16. Zhu MX (2005) Multiple roles of calmodulin and other Ca^{2+}-binding proteins in the functional regulation of TRP channels. Pflugers Arch 451:105–115
17. Niemeyer BA, Bergs C, Wissenbach U, Flockerzi V, Trost C (2001) Competitive regulation of CaT-like-mediated Ca2+ entry by protein kinase C and calmodulin. Proc Natl Acad Sci USA 98:3600–3605
18. DeMaria CD, Soong TW, Alseikhan BA, Alvania RS, Yue DT (2001) Calmodulin bifurcates the local Ca2+ signal that modulates P/Q-type Ca2+ channels. Nature 411:484–489
19. Moreau B, Straube S, Fisher RJ, Putney JW Jr, Parekh AB (2005) Ca^{2+}-calmodulin-dependent facilitation and Ca^{2+} inactivation of Ca^{2+} release-activated Ca^{2+} channels. J Biol Chem 280:8776–8783
20. Zhu X et al (1996) trp, a novel mammalian gene family essential for agonist-activated capacitative Ca^{2+} entry. Cell 85:661–671
21. Minke B, Cook B (2002) TRP channel proteins and signal transduction. Physiol Rev 82:429–472
22. Venkatachalam K, Zheng F, Gill DL (2003) Regulation of canonical transient receptor potential (TRPC) channel function by diacylglycerol and protein kinase C. J Biol Chem 278:29031–29040
23. Kiselyov K, Mignery GA, Zhu MX, Muallem S (1999) The N-terminal domain of the IP3 receptor gates store-operated hTrp3 channels. Mol Cell 4:423–429
24. Kiselyov KI et al (2000) Gating of store-operated channels by conformational coupling to ryanodine receptors. Mol Cell 6:421–431

25. Lockwich T, Singh BB, Liu X, Ambudkar IS (2001) Stabilization of cortical actin induces internalization of transient receptor potential 3 (Trp3)-associated caveolar Ca^{2+} signaling complex and loss of Ca^{2+} influx without disruption of Trp3-inositol trisphosphate receptor association. J Biol Chem 276: 42401–42408
26. Ahmmed GU et al (2004) Protein kinase Calpha phosphorylates the TRPC1 channel and regulates store-operated Ca^{2+} entry in endothelial cells. J Biol Chem 279: 20941–20949
27. Kwan HY, Huang Y, Yao X (2006) Protein kinase C can inhibit TRPC3 channels indirectly via stimulating protein kinase G. J Cell Physiol 207:315–321
28. Vazquez G, Wedel BJ, Kawasaki BT, Bird GS, Putney JW Jr (2004) Obligatory role of Src kinase in the signaling mechanism for TRPC3 cation channels. J Biol Chem 279: 40521–40528
29. Hisatsune C et al (2004) Regulation of TRPC6 channel activity by tyrosine phosphorylation. J Biol Chem 279:18887–18894
30. Shi J et al (2004) Multiple regulation by calcium of murine homologues of transient receptor potential proteins TRPC6 and TRPC7 expressed in HEK293 cells. J Physiol 561:415–432
31. Boulay G et al (1999) Modulation of Ca^{2+} entry by polypeptides of the inositol 1,4, 5-trisphosphate receptor (IP3R) that bind transient receptor potential (TRP): evidence for roles of TRP and IP3R in store depletion-activated Ca^{2+} entry. Proc Natl Acad Sci USA 96:14955–14960
32. Vaca L, Sampieri A (2002) Calmodulin modulates the delay period between release of calcium from internal stores and activation of calcium influx via endogenous TRP1 channels. J Biol Chem 277:42178–42187
33. Liou J et al (2005) STIM is a Ca^{2+} sensor essential for Ca^{2+}-store-depletion-triggered Ca^{2+} influx. Curr Biol 15:1235–1241
34. Roos J et al (2005) STIM1, an essential and conserved component of store-operated Ca^{2+} channel function. J Cell Biol 169: 435–445
35. Zeng W et al (2008) STIM1 gates TRPC channels, but not Orai1, by electrostatic interaction. Mol Cell 32:439–448
36. Worley PF et al (2007) TRPC channels as STIM1-regulated store-operated channels. Cell Calcium 42:205–211
37. Manji SS et al (2000) STIM1: a novel phosphoprotein located at the cell surface. Biochim Biophys Acta 1481:147–155
38. Williams RT et al (2001) Identification and characterization of the STIM (stromal interaction molecule) gene family: coding for a novel class of transmembrane proteins. Biochem J 357:673–685
39. Zhang SL et al (2005) STIM1 is a Ca^{2+} sensor that activates CRAC channels and migrates from the Ca^{2+} store to the plasma membrane. Nature 437:902–905
40. Brandman O, Liou J, Park WS, Meyer T (2007) STIM2 is a feedback regulator that stabilizes basal cytosolic and endoplasmic reticulum Ca^{2+} levels. Cell 131:1327–1339
41. Smyth JT et al (2009) Phosphorylation of STIM1 underlies suppression of store-operated calcium entry during mitosis. Nat Cell Biol 11:1465–1472
42. Yu F, Sun L, Machaca K (2009) Orai1 internalization and STIM1 clustering inhibition modulate SOCE inactivation during meiosis. Proc Natl Acad Sci USA 106:17401–17406
43. Liou J, Fivaz M, Inoue T, Meyer T (2007) Live-cell imaging reveals sequential oligomerization and local plasma membrane targeting of stromal interaction molecule 1 after Ca^{2+} store depletion. Proc Natl Acad Sci USA 104:9301–9306
44. Luik RM, Wang B, Prakriya M, Wu MM, Lewis RS (2008) Oligomerization of STIM1 couples ER calcium depletion to CRAC channel activation. Nature 454:538–542
45. Luik RM, Wu MM, Buchanan J, Lewis RS (2006) The elementary unit of store-operated Ca^{2+} entry: local activation of CRAC channels by STIM1 at ER-plasma membrane junctions. J Cell Biol 174:815–825
46. Muik M et al (2008) Dynamic coupling of the putative coiled-coil domain of ORAI1 with STIM1 mediates ORAI1 channel activation. J Biol Chem 283:8014–8022
47. Wu MM, Buchanan J, Luik RM, Lewis RS (2006) Ca^{2+} store depletion causes STIM1 to accumulate in ER regions closely associated with the plasma membrane. J Cell Biol 174:803–813
48. Xu P et al (2006) Aggregation of STIM1 underneath the plasma membrane induces clustering of Orai1. Biochem Biophys Res Commun 350:969–976
49. Pani B et al (2008) Lipid rafts determine clustering of STIM1 in endoplasmic reticulum-plasma membrane junctions and regulation of store-operated Ca^{2+} entry (SOCE). J Biol Chem 283:17333–17340
50. Kim MS et al (2009) Native Store-operated Ca^{2+} influx requires the channel function of orai1 and TRPC1. J Biol Chem 284: 9733–9741

51. Yuan JP et al (2009) SOAR and the polybasic STIM1 domains gate and regulate Orai channels. Nat Cell Biol 11:337–343
52. Cheng KT, Ong HL, Liu X, Ambudkar IS (2011) Contribution of TRPC1 and Orai1 to Ca^{2+} entry activated by store depletion. Adv Exp Med Biol 704:435–449
53. Salido GM, Jardin I, Rosado JA (2011) The TRPC ion channels: association with orai1 and STIM1 proteins and participation in capacitative and non-capacitative calcium entry. Adv Exp Med Biol 704:413–433
54. Lee KP et al (2010) An endoplasmic reticulum/plasma membrane junction: STIM1/ Orai1/TRPCs. FEBS Lett 584:2022–2027
55. Lee KP, Yuan JP, So I, Worley PF, Muallem S (2010) STIM1-dependent and STIM1-independent function of TRPC channels tunes their store-operated mode. J Biol Chem 285:38666–38673
56. Prakash YS et al (2007) Caveolins and intracellular calcium regulation in human airway smooth muscle. Am J Physiol Lung Cell Mol Physiol 293:L1118–1126
57. Isshiki M et al (2002) Sites of Ca^{2+} wave initiation move with caveolae to the trailing edge of migrating cells. J Cell Sci 115:475–484
58. Guo B, Kato RM, Garcia-Lloret M, Wahl MI, Rawlings DJ (2000) Engagement of the human pre-B cell receptor generates a lipid raft-dependent calcium signaling complex. Immunity 13:243–253
59. Tiruppathi C, Ahmmed GU, Vogel SM, Malik AB (2006) Ca^{2+} signaling, TRP channels, and endothelial permeability. Microcirculation 13:693–708
60. Lin RC, Scheller RH (2000) Mechanisms of synaptic vesicle exocytosis. Annu Rev Cell Dev Biol 16:19–49
61. Singh BB et al (2004) VAMP2-dependent exocytosis regulates plasma membrane insertion of TRPC3 channels and contributes to agonist-stimulated Ca^{2+} influx. Mol Cell 15: 635–646
62. Bezzerides VJ, Ramsey IS, Kotecha S, Greka A, Clapham DE (2004) Rapid vesicular translocation and insertion of TRP channels. Nat Cell Biol 6:709–720
63. Cayouette S, Lussier MP, Mathieu EL, Bousquet SM, Boulay G (2004) Exocytotic insertion of TRPC6 channel into the plasma membrane upon Gq protein-coupled receptor activation. J Biol Chem 279:7241–7246
64. Bollimuntha S, Cornatzer E, Singh BB (2005) Plasma membrane localization and function of TRPC1 is dependent on its interaction with beta-tubulin in retinal epithelium cells. Vis Neurosci 22:163–170
65. Smyth JT, DeHaven WI, Bird GS, Putney JW Jr (2007) Role of the microtubule cytoskeleton in the function of the store-operated Ca^{2+} channel activator STIM1. J Cell Sci 120: 3762–3771
66. Earley S, Brayden JE (2010) Transient receptor potential channels and vascular function. Clin Sci 119:19–36
67. Moore T et al (1998) Regulation of pulmonary endothelial cell shape by Trp-mediated calcium entry. Chest 114:36S–38S
68. Moore TM et al (1998) Store-operated calcium entry promotes shape change in pulmonary endothelial cells expressing Trp1. Am J Physiol 275:L574–582
69. Brough GH et al (2001) Contribution of endogenously expressed Trp1 to a Ca^{2+}-selective, store-operated Ca^{2+} entry pathway. FASEB J 15:1727–1738, official publication of the Federation of American Societies for Experimental Biology
70. Groschner K et al (1998) Trp proteins form store-operated cation channels in human vascular endothelial cells. FEBS Lett 437: 101–106
71. Freichel M et al (2001) Lack of an endothelial store-operated Ca^{2+} current impairs agonist-dependent vasorelaxation in TRP4$^{-/-}$ mice. Nat Cell Biol 3:121–127
72. Paria BC et al (2003) Tumor necrosis factor-alpha induces nuclear factor-kappaB-dependent TRPC1 expression in endothelial cells. J Biol Chem 278:37195–37203
73. Di A, Malik AB (2010) TRP channels and the control of vascular function. Curr Opin Pharmacol 10:127–132
74. Cheng KT, Ong HL, Liu X, Ambudkar IS (2011) Contribution of TRPC1 and Orai1 to Ca^{2+} entry activated by store depletion. Adv Exp Med Biol 704:435–449
75. Hogan PG, Lewis RS, Rao A (2010) Molecular basis of calcium signaling in lymphocytes: STIM and ORAI. Annu Rev Immunol 28:491–533
76. Lee KP et al (2010) An endoplasmic reticulum/plasma membrane junction: STIM1/ Orai1/TRPCs. FEBS Lett 584:2022–2027
77. Putney JW (2010) Pharmacology of store-operated calcium channels. Mol Interv 10:209–218
78. Schindl R et al (2009) Recent progress on STIM1 domains controlling Orai activation. Cell Calcium 46:227–232
79. Yuan JP et al (2009) TRPC channels as STIM1-regulated SOCs. Channels 3:221–225
80. Abdullaev IF et al (2008) Stim1 and Orai1 mediate CRAC currents and store-operated

calcium entry important for endothelial cell proliferation. Circ Res 103:1289–1299

81. Patel HH et al (2007) Increased smooth muscle cell expression of caveolin-1 and caveolae contribute to the pathophysiology of idiopathic pulmonary arterial hypertension. FASEB J 21:2970–2979, Official publication of the Federation of American Societies for Experimental Biology
82. Remillard CV, Yuan JX (2006) Transient receptor potential channels and caveolin-1: good friends in tight spaces. Mol Pharmacol 70:1151–1154
83. Kwiatek AM et al (2006) Caveolin-1 regulates store-operated Ca^{2+} influx by binding of its scaffolding domain to transient receptor potential channel-1 in endothelial cells. Mol Pharmacol 70:1174–1183
84. Sathish V et al (2011) Caveolin-1 in cytokine-induced enhancement of intracellular Ca^{2+} in human airway smooth muscle. Am J Physiol Lung Cell Mol Physiol 301:607–614
85. Mehta D et al (2003) RhoA interaction with inositol 1,4,5-trisphosphate receptor and transient receptor potential channel-1 regulates Ca^{2+} entry. Role in signaling increased endothelial permeability. J Biol Chem 278:33492–33500
86. Jho D et al (2005) Angiopoietin-1 opposes VEGF-induced increase in endothelial permeability by inhibiting TRPC1-dependent Ca2 influx. Circ Res 96:1282–1290
87. Liu CL, Huang Y, Ngai CY, Leung YK, Yao XQ (2006) TRPC3 is involved in flow- and bradykinin-induced vasodilation in rat small mesenteric arteries. Acta Pharmacol Sin 27:981–990
88. Peppiatt-Wildman CM, Albert AP, Saleh SN, Large WA (2007) Endothelin-1 activates a Ca^{2+}-permeable cation channel with TRPC3 and TRPC7 properties in rabbit coronary artery myocytes. J Physiol 580:755–764
89. Smedlund K, Vazquez G (2008) Involvement of native TRPC3 proteins in ATP-dependent expression of VCAM-1 and monocyte adherence in coronary artery endothelial cells. Arterioscler Thromb Vasc Biol 28: 2049–2055
90. Thilo F, Loddenkemper C, Berg E, Zidek W, Tepel M (2009) Increased TRPC3 expression in vascular endothelium of patients with malignant hypertension. Mod Pathol 22:426–430, An official journal of the United States and Canadian Academy of Pathology, Inc
91. Xi Q et al (2008) IP3 constricts cerebral arteries via IP3 receptor-mediated TRPC3 channel activation and independently of sarcoplasmic reticulum Ca^{2+} release. Circ Res 102:1118–1126
92. Balzer M, Lintschinger B, Groschner K (1999) Evidence for a role of Trp proteins in the oxidative stress-induced membrane conductances of porcine aortic endothelial cells. Cardiovasc Res 42:543–549
93. Hill AJ et al (2006) A TRPC-like non-selective cation current activated by alpha 1-adrenoceptors in rat mesenteric artery smooth muscle cells. Cell Calcium 40:29–40
94. Eder P et al (2007) Phospholipase C-dependent control of cardiac calcium homeostasis involves a TRPC3-NCX1 signaling complex. Cardiovasc Res 73:111–119
95. Kwan HY, Huang Y, Yao X (2007) TRP channels in endothelial function and dysfunction. Biochim Biophys Acta 1772:907–914
96. Tiruppathi C et al (2002) Impairment of store-operated Ca^{2+} entry in TRPC4(-/-) mice interferes with increase in lung microvascular permeability. Circ Res 91:70–76
97. Tiruppathi C, Minshall RD, Paria BC, Vogel SM, Malik AB (2002) Role of Ca^{2+} signaling in the regulation of endothelial permeability. Vascul Pharmacol 39:173–185
98. Chaudhuri P et al (2008) Elucidation of a TRPC6-TRPC5 channel cascade that restricts endothelial cell movement. Mol Biol Cell 19:3203–3211
99. Yoshida T et al (2006) Nitric oxide activates TRP channels by cysteine S-nitrosylation. Nat Chem Biol 2:596–607
100. Wong CO, Sukumar P, Beech DJ, Yao X (2010) Nitric oxide lacks direct effect on TRPC5 channels but suppresses endogenous TRPC5-containing channels in endothelial cells. Pflugers Arch 460:121–130
101. Ge R et al (2009) Critical role of TRPC6 channels in VEGF-mediated angiogenesis. Cancer Lett 283:43–51
102. Singh I et al (2007) Galphaq-TRPC6-mediated Ca^{2+} entry induces RhoA activation and resultant endothelial cell shape change in response to thrombin. J Biol Chem 282: 7833–7843
103. Leung PC et al (2006) Mechanism of non-capacitative Ca^{2+} influx in response to bradykinin in vascular endothelial cells. J Vasc Res 43:367–376
104. Cheng HW, James AF, Foster RR, Hancox JC, Bates DO (2006) VEGF activates receptor-operated cation channels in human microvascular endothelial cells. Arterioscler Thromb Vasc Biol 26:1768–1776
105. Yao X, Garland CJ (2005) Recent developments in vascular endothelial cell transient

receptor potential channels. Circ Res 97: 853–863
106. Albert AP, Saleh SN, Large WA (2009) Identification of canonical transient receptor potential (TRPC) channel proteins in native vascular smooth muscle cells. Curr Med Chem 16:1158–1165
107. Beech DJ (2005) Emerging functions of 10 types of TRP cationic channel in vascular smooth muscle. Clin Exp Pharmacol Physiol 32:597–603
108. Brayden JE, Earley S, Nelson MT, Reading S (2008) Transient receptor potential (TRP) channels, vascular tone and autoregulation of cerebral blood flow. Clin Exp Pharmacol Physiol 35:1116–1120
109. Gonzalez-Cobos JC, Trebak M (2010) TRPC channels in smooth muscle cells. Front Biosci 15:1023–1039, a journal and virtual library
110. House SJ, Potier M, Bisaillon J, Singer HA, Trebak M (2008) The non-excitable smooth muscle: calcium signaling and phenotypic switching during vascular disease. Pflugers Arch 456:769–785
111. Landsberg JW, Yuan JX (2004) Calcium and TRP channels in pulmonary vascular smooth muscle cell proliferation. News Physiol Sci 19:44–50, An international journal of physiology produced jointly by the International Union of Physiological Sciences and the American Physiological Society
112. Remillard CV, Yuan JX (2006) TRP channels, CCE, and the pulmonary vascular smooth muscle. Microcirculation 13:671–692
113. Wamhoff BR, Bowles DK, Owens GK (2006) Excitation-transcription coupling in arterial smooth muscle. Circ Res 98:868–878
114. Guibert C, Ducret T, Savineau JP (2011) Expression and physiological roles of TRP channels in smooth muscle cells. Adv Exp Med Biol 704:687–706
115. Saleh SN, Albert AP, Peppiatt CM, Large WA (2006) Angiotensin II activates two cation conductances with distinct TRPC1 and TRPC6 channel properties in rabbit mesenteric artery myocytes. J Physiol 577:479–495
116. Tai K et al (2008) Agonist-evoked calcium entry in vascular smooth muscle cells requires IP3 receptor-mediated activation of TRPC1. Eur J Pharmacol 583:135–147
117. Dietrich A, Kalwa H, Gudermann T (2010) TRPC channels in vascular cell function. Thromb Haemost 103:262–270
118. Tang C, To WK, Meng F, Wang Y, Gu Y (2010) A role for receptor-operated Ca^{2+} entry in human pulmonary artery smooth muscle cells in response to hypoxia. Physiol Res 59:909–918
119. Kunichika N et al (2004) Overexpression of TRPC1 enhances pulmonary vasoconstriction induced by capacitative Ca^{2+} entry. Am J Physiol Lung Cell Mol Physiol 287: L962–969
120. Bergdahl A et al (2005) Plasticity of TRPC expression in arterial smooth muscle: correlation with store-operated Ca^{2+} entry. Am J Physiol Cell Physiol 288:C872–880
121. Dietrich A et al (2007) Pressure-induced and store-operated cation influx in vascular smooth muscle cells is independent of TRPC1. Pflugers Arch 455:465–477
122. Takahashi S et al (2008) Nitric oxide-cGMP-protein kinase G pathway negatively regulates vascular transient receptor potential channel TRPC6. J Physiol 586:4209–4223
123. Sweeney M et al (2002) Inhibition of endogenous TRP1 decreases capacitative Ca^{2+} entry and attenuates pulmonary artery smooth muscle cell proliferation. Am J Physiol Lung Cell Mol Physiol 283:L144–155
124. Li J et al (2008) Interactions, functions, and independence of plasma membrane STIM1 and TRPC1 in vascular smooth muscle cells. Circ Res 103:e97–104
125. Gudermann T, Hofmann T, Mederos y Schnitzler M, Dietrich A (2004) Activation, subunit composition and physiological relevance of DAG-sensitive TRPC proteins. Novartis Found Symp 258:103–118, discussion 118-122, 155-109, 263-106
126. Hofmann T et al (1999) Direct activation of human TRPC6 and TRPC3 channels by diacylglycerol. Nature 397:259–263
127. Chen J, Crossland RF, Noorani MM, Marrelli SP (2009) Inhibition of TRPC1/TRPC3 by PKG contributes to NO-mediated vasorelaxation. Am J Physiol Heart Circ Physiol 297:H417–424
128. Liu D et al (2009) Increased transient receptor potential canonical type 3 channels in vasculature from hypertensive rats. Hypertension 53:70–76
129. Xie A et al (2007) Novel mechanism of endothelin-1-induced vasospasm after subarachnoid hemorrhage. J Cereb Blood Flow Metab 27:1692–1701, Official journal of the International Society of Cerebral Blood Flow and Metabolism
130. Lindsey SH, Tribe RM, Songu-Mize E (2008) Cyclic stretch decreases TRPC4 protein and capacitative calcium entry in rat vascular smooth muscle cells. Life Sci 83:29–34

131. Xu SZ, Boulay G, Flemming R, Beech DJ (2006) E3-targeted anti-TRPC5 antibody inhibits store-operated calcium entry in freshly isolated pial arterioles. Am J Physiol Heart Circ Physiol 291:H2653–2659
132. Xu SZ et al (2006) A sphingosine-1-phosphate-activated calcium channel controlling vascular smooth muscle cell motility. Circ Res 98:1381–1389
133. Flemming PK et al (2006) Sensing of lysophospholipids by TRPC5 calcium channel. J Biol Chem 281:4977–4982
134. Shi J, Ju M, Saleh SN, Albert AP, Large WA (2010) TRPC6 channels stimulated by angiotensin II are inhibited by TRPC1/C5 channel activity through a Ca^{2+}- and PKC-dependent mechanism in native vascular myocytes. J Physiol 588:3671–3682
135. Zulian A et al (2010) Upregulation of Na+/Ca^{2+} exchanger and TRPC6 contributes to abnormal Ca^{2+} homeostasis in arterial smooth muscle cells from Milan hypertensive rats. Am J Physiol Heart Circ Physiol 299:H624–633
136. Graham S et al (2010) Canonical transient receptor potential 6 (TRPC6), a redox-regulated cation channel. J Biol Chem 285: 23466–23476
137. Erac Y, Selli C, Kosova B, Akcali KC, Tosun M (2010) Expression levels of TRPC1 and TRPC6 ion channels are reciprocally altered in aging rat aorta: implications for age-related vasospastic disorders. Age 32:223–230
138. Ju M, Shi J, Saleh SN, Albert AP, Large WA (2010) Ins(1,4,5)P3 interacts with PIP2 to regulate activation of TRPC6/C7 channels by diacylglycerol in native vascular myocytes. J Physiol 588:1419–1433
139. Albert AP, Saleh SN, Large WA (2008) Inhibition of native TRPC6 channel activity by phosphatidylinositol 4,5-bisphosphate in mesenteric artery myocytes. J Physiol 586: 3087–3095
140. Bae YM et al (2007) Enhancement of receptor-operated cation current and TRPC6 expression in arterial smooth muscle cells of deoxycorticosterone acetate-salt hypertensive rats. J Hypertens 25:809–817
141. Jung S, Strotmann R, Schultz G, Plant TD (2002) TRPC6 is a candidate channel involved in receptor-stimulated cation currents in A7r5 smooth muscle cells. Am J Physiol Cell Physiol 282:C347–359
142. Inoue R et al (2001) The transient receptor potential protein homologue TRP6 is the essential component of vascular alpha(1)-adrenoceptor-activated Ca^{2+}-permeable cation channel. Circ Res 88:325–332
143. Soboloff J et al (2005) Role of endogenous TRPC6 channels in Ca^{2+} signal generation in A7r5 smooth muscle cells. J Biol Chem 280:39786–39794
144. Poburko D, Fameli N, Kuo KH, van Breemen C (2008) Ca^{2+} signaling in smooth muscle: TRPC6. NCX and LNats in nanodomains. Channels (Austin) 2:10–12
145. Fellner SK, Arendshorst WJ (2008) Angiotensin II-stimulated Ca2+ entry mechanisms in afferent arterioles: role of transient receptor potential canonical channels and reverse Na+/Ca^{2+} exchange. Am J Physiol Renal Physiol 294:F212–219
146. Welsh DG, Morielli AD, Nelson MT, Brayden JE (2002) Transient receptor potential channels regulate myogenic tone of resistance arteries. Circ Res 90:248–250
147. Yu Y et al (2004) Enhanced expression of transient receptor potential channels in idiopathic pulmonary arterial hypertension. Proc Natl Acad Sci USA 101:13861–13866
148. Yu Y et al (2003) PDGF stimulates pulmonary vascular smooth muscle cell proliferation by upregulating TRPC6 expression. Am J Physiol Cell Physiol 284:C316–330
149. Weissmann N et al (2006) Classical transient receptor potential channel 6 (TRPC6) is essential for hypoxic pulmonary vasoconstriction and alveolar gas exchange. Proc Natl Acad Sci USA 103:19093–19098
150. Walker RL, Hume JR, Horowitz B (2001) Differential expression and alternative splicing of TRP channel genes in smooth muscles. Am J Physiol Cell Physiol 280:C1184–1192
151. George SJ, Dwivedi A (2004) MMPs, cadherins, and cell proliferation. Trends Cardiovasc Med 14:100–105
152. Hedin U, Roy J, Tran PK (2004) Control of smooth muscle cell proliferation in vascular disease. Curr Opin Lipidol 15:559–565
153. Rivard A, Andres V (2000) Vascular smooth muscle cell proliferation in the pathogenesis of atherosclerotic cardiovascular diseases. Histol Histopathol 15:557–571
154. Sanz-Gonzalez SM et al (2000) Control of vascular smooth muscle cell growth by cyclin-dependent kinase inhibitory proteins and its implication in cardiovascular disease. Front Biosci 5:D619–628, A journal and virtual library
155. Schachter M (1997) Vascular smooth muscle cell migration, atherosclerosis, and calcium channel blockers. Int J Cardiol 62(Suppl 2): S85–90
156. Sriram V, Patterson C (2001) Cell cycle in vasculoproliferative diseases: potential

interventions and routes of delivery. Circulation 103:2414–2419

157. Afroze T, Husain M (2001) Cell cycle dependent regulation of intracellular calcium concentration in vascular smooth muscle cells: a potential target for drug therapy. Curr Drug Targets Cardiovasc Haematol Disord 1:23–40
158. Mason RP, Marche P, Hintze TH (2003) Novel vascular biology of third-generation L-type calcium channel antagonists: ancillary actions of amlodipine. Arterioscler Thromb Vasc Biol 23:2155–2163
159. Schachter M (1997) Calcium antagonists and atherosclerosis. Int J Cardiol 62(Suppl 2): S9–15
160. Banner KH, Igney F, Poll C (2011) TRP channels: emerging targets for respiratory disease. Pharmacol Ther 130:371–384
161. Li S, Westwick J, Cox B, Poll CT (2004) TRP channels as drug targets. Novartis Found Symp 258:204–213, discussion 213–221, 263–206
162. Li S, Westwick J, Poll C (2003) Transient receptor potential (TRP) channels as potential drug targets in respiratory disease. Cell Calcium 33:551–558
163. Watanabe H (2009) Pathological role of TRP channels in cardiovascular and respiratory diseases. Nihon Yakurigaku Zasshi 134:127–130, Folia pharmacologica Japonica
164. Corteling RL et al (2004) Expression of transient receptor potential C6 and related transient receptor potential family members in human airway smooth muscle and lung tissue. Am J Respir Cell Mol Biol 30:145–154
165. Antigny F et al (2011) Transient receptor potential canonical channel 6 links Ca2+ mishandling to cystic fibrosis transmembrane conductance regulator channel dysfunction in cystic fibrosis. Am J Respir Cell Mol Biol 44:83–90
166. Gosling M, Poll C, Li S (2005) TRP channels in airway smooth muscle as therapeutic targets. Naunyn Schmiedebergs Arch Pharmacol 371:277–284
167. Nassini R, Materazzi S, De Siena G, De Cesaris F, Geppetti P (2010) Transient receptor potential channels as novel drug targets in respiratory diseases. Curr Opin Investig Drugs 11:535–542
168. Nilius B (2007) TRP channels in disease. Biochim Biophys Acta 1772:805–812
169. Wang YX, Zheng YM (2011) Molecular expression and functional role of canonical transient receptor potential channels in airway smooth muscle cells. Adv Exp Med Biol 704:731–747
170. Ong HL, Brereton HM, Harland ML, Barritt GJ (2003) Evidence for the expression of transient receptor potential proteins in guinea pig airway smooth muscle cells. Respirology 8:23–32
171. White TA et al (2006) Role of transient receptor potential C3 in TNF-alpha-enhanced calcium influx in human airway myocytes. Am J Respir Cell Mol Biol 35:243–251
172. Ong HL et al (2002) Specific detection of the endogenous transient receptor potential (TRP)-1 protein in liver and airway smooth muscle cells using immunoprecipitation and Western-blot analysis. Biochem J 364: 641–648
173. Godin N, Rousseau E (2007) TRPC6 silencing in primary airway smooth muscle cells inhibits protein expression without affecting OAG-induced calcium entry. Mol Cell Biochem 296:193–201
174. Ay B, Prakash YS, Pabelick CM, Sieck GC (2004) Store-operated Ca2+ entry in porcine airway smooth muscle. Am J Physiol Lung Cell Mol Physiol 286:L909–917
175. Xiao JH, Zheng YM, Liao B, Wang YX (2010) Functional role of canonical transient receptor potential 1 and canonical transient receptor potential 3 in normal and asthmatic airway smooth muscle cells. Am J Respir Cell Mol Biol 43:17–25
176. Rosker C et al (2004) Ca^{2+} signaling by TRPC3 involves Na^{+} entry and local coupling to the Na^{+}/Ca^{2+} exchanger. J Biol Chem 279:13696–13704
177. Liu XS, Xu YJ (2005) Potassium channels in airway smooth muscle and airway hyperreactivity in asthma. Chin Med J (Engl) 118: 574–580
178. Damann N, Owsianik G, Li S, Poll C, Nilius B (2009) The calcium-conducting ion channel transient receptor potential canonical 6 is involved in macrophage inflammatory protein-2-induced migration of mouse neutrophils. Acta Physiol (Oxf) 195:3–11
179. Finney-Hayward TK et al (2010) Expression of transient receptor potential c6 channels in human lung macrophages. Am J Respir Cell Mol Biol 43:296–304
180. Sel S et al (2008) Loss of classical transient receptor potential 6 channel reduces allergic airway response. Clin Exp Allergy 38:1548–1558, Journal of the British Society for Allergy and Clinical Immunology

Chapter 5

TRPM2 Function and Potential as a Drug Target

Barbara A. Miller

Abstract

The TRPM subfamily of transient receptor potential channels includes a number of members which are involved in cell proliferation or cell survival. TRPM2, the second member to be cloned, has a key role in the response to oxidative stress. After exposure to oxidant stress, TNFα, concanavalin A, or amyloid β-peptide, ADP-ribose is produced, which binds to a NUDT9-H domain in the C terminus and is a key regulator of channel opening and calcium influx. The important roles of TRPM2 isoforms in cell proliferation and oxidant-induced cell death have been well established in divergent cell types using a broad range of techniques including overexpression, channel knockout, depletion, or inhibition, and calcium chelation. Related to its central role in oxidative stress-induced cell injury, TRPM2 has been shown to mediate a number of disease processes. TRPM2 is involved in regulation of metabolism in diabetes (insulin secretion, β-cell injury, and insulin resistance), and obesity. It has a role in inflammation, immunity, and cancer. Modulation of TRPM2 may be able to ameliorate a number of these disease processes as well as those resulting from ischemia–reperfusion injury, vascular endothelial damage, and traumatic brain injury. A review of the involvement of TRPM2 in these physiological and pathophysiological processes, and its tremendous potential as a drug target is presented below.

Key words: TRP channels, TRPM2, Oxidative stress, Inflammation, Diabetes, Cardiomyocytes, Endothelial cells

1. Introduction

TRPM subfamily members are among the first TRP channels shown to have roles in cell proliferation and survival, including TRPM2 (1, 2), TRPM4 (3), TRPM5 (4), TRPM7 (5, 6), and TRPM8 (7, 8). The TRPM subfamily was named after the first described member, TRPM1 or melastatin. TRPM1 is expressed on melanocytes and expression levels correlate inversely with melanoma aggressiveness and metastatic potential, suggesting that it functions as a tumor suppressor (9–11). TRPM2, the second member of the TRPM2 channel to be cloned, is a 1,503 amino acid protein (12). It is expressed in many cell types including brain,

Arpad Szallasi and Tamás Bíró (eds.), *TRP Channels in Drug Discovery: Volume I*, Methods in Pharmacology and Toxicology, DOI 10.1007/978-1-62703-077-9_5, © Springer Science+Business Media, LLC 2012

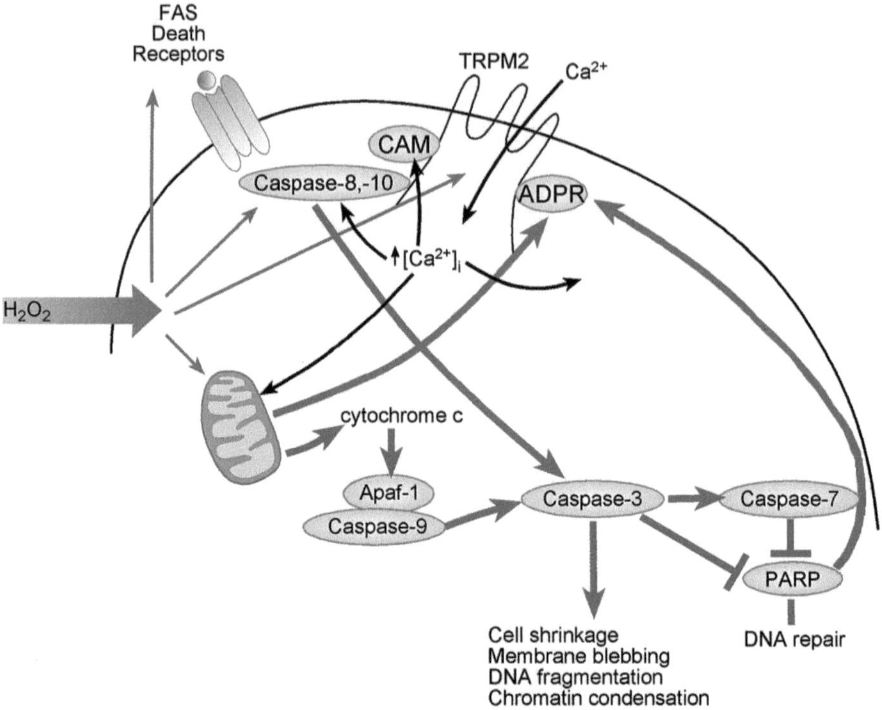

Fig. 1. Mechanisms of TRPM2 activation and induction of cell death by H_2O_2. H_2O_2 activates production of ADP-ribose (ADPR) in the mitochondria and through activation of PARP/PARG. The increase in ADPR activates TRPM2 by binding to the C-terminal NUDT9-H domain. Ca^{2+} influx ensues, which enhances calmodulin (CAM) binding to TRPM2 and further channel opening. $[Ca^{2+}]_i$ rises, and in association with other oxidative stress-induced signals results in activation of extrinsic and intrinsic cell death pathways, leading to caspase-3 activation and PARP cleavage and inactivation (Printed with permission from Transient Receptor Potential Channels, 2011, pg 535, Chapter 29 TRP Channels as Mediators of Oxidative Stress, BA Miller and W Zhang, Fig. 29.1).

hematopoietic cells, heart, vascular smooth muscle, endothelial cells, lung, and the endocrine system (2, 12–16). TRPM2 channels are permeable to sodium, potassium, and calcium. Extracellular signals which activate TRPM2 include oxidative stress, TNFα, amyloid β-peptide, and concanavalin A (1, 15, 17, 18).

Stimulation with these extracellular signals results in a common activation pathway, production of ADP-ribose (ADPR), which activates TRPM2 by binding to the TRPM2 C-terminal NUDT9-H domain (Fig. 1) (2, 18–21). This NUDT9-H domain has a very low level of ADPR pyrophosphatase activity. Cyclic adenosine diphosphoribose (cADPR) potentiates the effects of ADPR on TRPM2 at low concentrations and also can gate TRPM2 by itself (19). Nicotinamide adenine dinucleotide (NAD) may also induce TRPM2 opening, but the predominance of evidence suggests that activation by NAD may not be direct, but instead through conversion to or contamination by ADPR. ADPR may arise from mitochondria (20) or through activation of poly (ADPR) polymerase (PARP) (22, 23). PARP-1 attaches ADPR polymers to produce long and branched poly (ADPR) molecules, which are then degraded into free ADPR by poly (ADR-ribose) glycohydrolase (PARG) (24, 25). TRPM2 is also positively regulated by intracellular calcium (26–28). Interaction of ADPR with TRPM2 supports limited calcium entry through TRPM2. The subsequent increase in Ca^{2+}-bound calmodulin enhances calmodulin binding to an IQ-motif in the N terminus of TRPM2, providing positive feedback for TRPM2 activation and increased Ca^{2+} influx (28). TRPM2 with mutant ADPR binding sites can be activated by intracellular calcium concentration $[Ca^{2+}]_i$, and it may be activated in a wide range of physiological situations through this mechanism (27). Recently, strong inhibition of TRPM2 currents by extra- or intracellular acidification has been reported, and this suggests a mechanism for limitation of $[Ca^{2+}]_i$, under conditions of lower pH, which often occur in inflammation or ischemia (29–31).

Naturally occurring TRPM2 isoforms resulting from alternative splicing have been described (Fig. 2). These include TRPM2-L (full length or wild type), TRPM2-S (short) (32), TRPM2-ΔN (17), TRPM2-ΔC (17), and TRPM2-TE (33). Because TRP channels including TRPM2 function as tetramers, isoforms can coassociate and levels of isoform expression can influence TRPM2-L function. TRPM2-S has a deletion of four of six C-terminal transmembrane domains, the putative calcium pore, and the entire C terminus (32). TRPM2-S suppresses Ca^{2+} influx through TRPM2-L

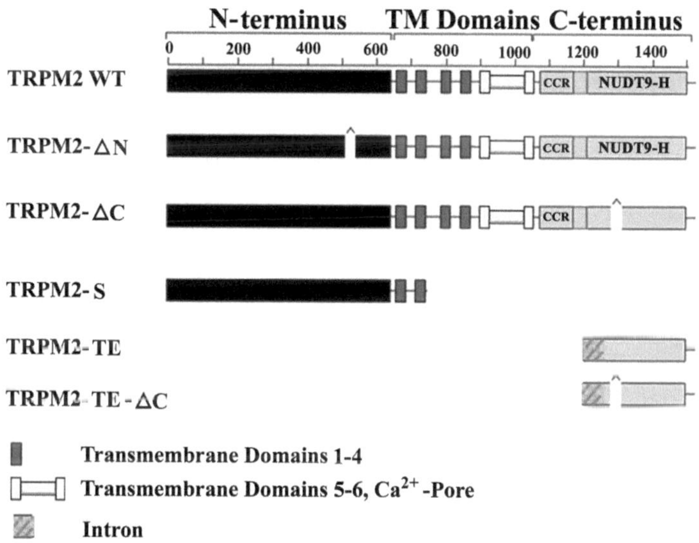

Fig. 2. Schematic representation of TRPM2 isoforms. Membrane spanning domains 1–4 and the putative pore region including transmembrane domains 5–6 are indicated. CCR represents the coiled coil region which may mediate protein/protein interactions. NUDT9-H represents the NUDT9 ADP-ribose hydrolase domain (Printed with permission from Transient Receptor Potential Channels, 2011, pg 537, Chapter 29 TRP Channels as Mediators of Oxidative Stress, BA Miller and W Zhang, Fig. 29.2, Springer).

and inhibits cell death induced by oxidative stress when cells are exposed to high concentrations of H_2O_2 (32). The mechanism of inhibition is not known, but one possibility is that TRPM2-S associates TRPM2-L, altering the tertiary structure of the tetramer and reducing ion permeability. New data suggest that TRPM2-S expression may reduce cell viability compared to TRPM2-L when cells are exposed to low, physiological concentrations of H_2O_2 (Chen et al. manuscript in preparation). TRPM2-ΔN has deletion of amino acids 538–557 from the N terminus, and does not respond to H_2O_2 or ADPR, suggesting that this deletion disrupts channel gaiting or assembly (17). TRPM2-ΔC results from deletion of nucleotides encoding amino acids 1,292–1,325 in the C terminus of TRPM2-L (17). TRPM2-ΔC has decreased affinity for ADPR, and cells expressing this deletion do not respond to ADPR. However, they do respond to H_2O_2, suggesting that oxidative stress may activate TRPM2 through mechanisms independently of ADPR. TRPM2-TE was identified by Orfanelli et al. (33) when they were searching for antisense transcripts in melanoma in order to identify new tumor suppressor genes, using a software program called AntiHunter. TRPM2-TE transcripts encode either a 218 amino acid protein or a 184 amino acid protein including the ΔC deletion (TRPM2-TE-ΔC). Expression of TRPM2-TE is thought to result following hypomethylation of a CpG island in the TRPM2 C terminus, and TRPM2-TE is highly expressed in tumor cells including melanoma and lung compared to normal tissue. Coexpression of TRPM2-TE and TRPM2-L protected cells from apoptosis. At this point, little is known about the mechanisms thorough which alternative splicing is regulated in vivo, or the physiological functions and interactions of these isoforms.

A number of reports have examined the mechanisms through which TRPM2 isoforms are involved is modulation of cell death. In oxidative stress, increased reactive oxygen species (ROS) enhance ADPR production, activating TRPM2-L, increasing $[Ca^{2+}]_i$ and susceptibility to cell death. Downstream signaling pathways were explored in the human monocytic cell line U937 (34). Procaspases 8, 9, 3, and 7 were cleaved, and apoptosis inducing factor (AIF) was translocated from the inner mitochondrial membrane to the nucleus (25, 34). PARP, an important enzyme in DNA repair, was inactivated. These data demonstrate that reduced cell viability involves both the extrinsic (caspase 8) and intrinsic (caspase 9) signaling cascades. Data suggest a feedback loop in which although TRPM2 is activated by PARP involvement in ADPR production, TRPM2 activation in turn results in PARP cleavage and inactivation. The rise in $[Ca^{2+}]_i$ is an important part of the cell death cascade, because inhibition of the rise in $[Ca^{2+}]_i$ with the intracellular Ca^{2+} chelator BAPTA blocked caspase and PARP cleavage and enhanced cell viability (34). Oxidative stress results from a disturbance in the balance between oxidants and antioxidants, and can lead to tissue injury depending on severity and duration (2, 35).

Oxidative stress plays an important role in tissue damage in a large number of physiological processes including inflammation and ischemia–reperfusion injury, and disease processes including diabetes mellitus, autoimmune disorders, cardiovascular disease, neurodegenerative disorders, cancer, and aging (34–40). The mechanisms through which TRPM2 mediates most of these processes need to be better understood. Nevertheless, these data suggest that because of the important role of TRPM2 in mediating cell survival in oxidative stress, targeting TRPM2 will have significant implications for therapy of a number of diseases. At present, TRPM2 inhibitors lack specificity of action, and knowledge of therapeutic potential of TRPM2 relies on knockout mice and in vitro coexpression with dominant negative isoforms or depletion with siRNAs. Current knowledge of TRPM2 function in specific tissues and pathological conditions and its potential as a drug target is presented in this review.

2. TRPM2 Function in Diabetes

Type 2 diabetes is one of the most common metabolic diseases in the world, and cardiovascular disease is the leading cause of mortality in diabetes. Elevations in glucose and/or free fatty acids contribute to production of ROS, and the resultant oxidative stress is thought to play an important role in β-cell death and development of diabetes types 1 and 2 (41, 42). Because of low expression of antioxidant enzymes, particularly glutathione peroxidase and catalase (43), β cells have reduced capacity for protection from ROS. TRPM2 is expressed in pancreatic β cells (44, 45). Bari et al. demonstrated a requirement for calcium entry inhibitable by *N*-(p-amylcinnamoyl) anthranilic acid (ACA) in hydrogen-peroxide-induced β-cell death (44). The currents exhibited biophysical properties characteristic of TRPM2. Using short interfering RNA and a knockout mouse, TRPM2 was also shown to have a dual role in β cells as a plasma membrane Ca^{2+} influx channel and as an intracellular lysosomal Ca^{2+}-release channel, both activated by intracellular ADPR (45). Release of Ca^{2+} from lysosomes is important for phosphatidylserine redistribution from the inner plasma membrane to the cell surface (46), where it serves as a recognition ligand for elimination of apoptotic cells. These reports strongly link TRPM2 to H_2O_2-induced β-cell death (47). A key role for TRPM2 in insulin secretion in pancreatic β cells has also been demonstrated. Using isolated β cells, TRPM2 knockout mice showed impaired insulin secretion in response to high glucose and incretin hormones (48).

Insulin resistance is a characteristic of type 2 diabetes and is casually associated with obesity. Because oxidative stress and inflammation play a major role in insulin resistance, we examined the role of TRPM2 in glucose metabolism. Work from our and collaborating

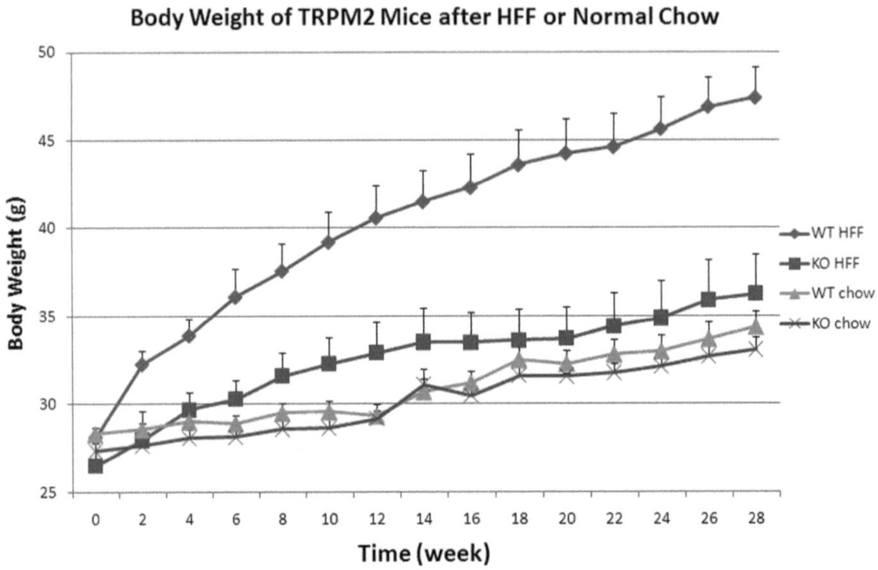

Fig. 3. Body weights of TRPM2 knockout (KO, from GlaxoSmithKline) mice and their wild type (WT) littermates during 28 weeks of feeding either normal chow (10 mice/group) or a high fat diet of 55 % fat by calories (HFF, 18 mice/group).

laboratories showed that TRPM2 deletion protected mice from developing diet-induced obesity and insulin resistance (manuscript submitted). TRPM2 knockout mice demonstrated significantly reduced weight gain on a high fat diet compared to their wild type littermates (Fig. 3), through a mechanism involving increased energy expenditure and increased insulin sensitivity. The roles of TRPM2 in enhancement of oxidative stress induced β-cell injury, diet induced obesity, and insulin resistance suggest depletion of TRPM2 may be a novel therapy to treat type 2 diabetes. Although reduced TRPM2 function may impair insulin secretion, this may also be beneficial because chronic hyperinsulinemia may exacerbate obesity and insulin resistance (49).

3. TRPM2 Function in Inflammation and Immunity

An important function of granulocytes and monocytes is the generation of ROS, including H_2O_2, known as the respiratory burst, to eliminate invading pathogens, which is accompanied by an increase in $[Ca^{2+}]_i$. Monocytes and macrophages interact with lipopolysaccharide (LPS), the major outer membrane component of bacteria, resulting in production of ROS, proinflammatory cytokines, and chemokines which are part of the inflammatory response. When produced in excessively large quantities, this can lead to lethal septic shock. A role for TRPM2 in neutrophil activation was proposed nearly 10 years ago (17, 50, 51). In granulocytes, transmembrane CD38 converts extracellular NAD^+ to ADPR, and both ADPR and H_2O_2 participate in TRPM2 activation, mediating increased $[Ca^{2+}]_i$ through Ca^{2+} influx. Recently, a requirement for TRPM2 in LPS-induced cytokine production was demonstrated in human monocytes (52). This was dependent on Ca^{2+} influx. Incubation of human monocytes with LPS resulted in upregulation of TRPM2 mRNA and ADPR-induced membrane currents. Downregulation of TRPM2 in monocytes with shRNA resulted in reduced rise in $[Ca^{2+}]_i$, and reduced production of IL-6, IL-8, IL-10, and TNF-α in response to LPS. H_2O_2 also activated TRPM2 in microglia, the host macrophages found in the central nervous system (53). In TRPM2 knockout mice, Ca^{2+} influx through TRPM2 was shown to control the ROS-induced signaling cascade responsible for chemokine production (40). Chemokines are chemotactic cytokines which mediate recruitment of inflammatory cells to inflamed sites. In the U937 human monocytic cell line, H_2O_2 evoked Ca^{2+} influx to activate the tyrosine kinase Pyk2 and amplify Erk signaling through Ras GTPase. This resulted in nuclear translocation of NF κB and production of the chemokine interleukin-8 (CXCL8) (40, 54). In monocytes from TRPM2-deficient mice, Ca^{2+} influx and production of macrophage inflammatory protein-2 (CXCL2), the mouse homolog of CXCL8, in response to H_2O_2 were impaired, leading to reduced inflammation (40). TRPM2 depletion attenuated neutrophil infiltration and ulceration in colitis induced by dextran sulfate sodium (DSS), and TRPM2 knockout mice were largely protected from DSS-mediated colitis. This model has many similarities to human ulcerative colitis. These results suggest that TRPM2-dependent release of CXCL2 from macrophages contributes to neutrophil accumulation in the colon of DSS-treated mice. Of note, DSS-induced macrophage recruitment, and T-cell, NK-cell, and dendritic-cell recruitment were not impaired in TRPM2 knockout mice. These experiments suggest that inhibition of TRPM2 function may be a therapeutic approach to treat a number of inflammatory diseases, particularly those triggered by CXCL8. Other inflammatory diseases in which CXCL8 is involved, but the role of TRPM2 has not yet been explored, include rheumatoid arthritis, inflammatory liver disease (55), and chronic obstructive pulmonary disease (54, 56).

However, ROS also are involved in host defense by promoting bacterial killing and production of cytokines. TRPM2 knockout mice were extremely susceptible to Listeria monocytogenes infection, which was lethal (57). TRPM2 knockout mice had reduced levels of IL-12, IFNγ, and NOS-inducible monocytes, but retained IFNγ responsiveness, which when given supplementally reversed the lethal phenotype. These data suggested that complete suppression of TRPM2-L may result in susceptibility to specific severe infections including Listeria due to weakened immunity.

4. TRPM2 Function in Cardiovascular Disease

Production of ROS is a major cause of cell death in ischemic–reperfusion injury (38). This type of injury may affect a number of organs, including cardiac myocytes in both animal models and patients post ischemia/reperfusion. TRPM2 mRNA and protein are expressed in cardiac myocytes (13, 38, 58). In rats, H_2O_2 activated TRPM2 through increased ADPR/NAD^+ formation, leading to sodium and calcium overload in mitochondria and myocyte apoptosis through mitochondrial membrane disruption, cytochrome c release, and caspase 3-dependent chromatin condensation and fragmentation (38). H_2O_2 exposure also resulted in PARP-induced $(ATP)_i$/NAD^+ depletion, increased membrane permeabilization, and necrotic cell death. Whereas inhibition of either TRPM2 or PARP partially inhibited death, inhibition of both completely prevented ultrastructural changes and myocyte death in vitro.

The vascular endothelium regulates the passage of cells from blood to tissues, and is a target of oxidative stress. ROS are important mediators of vascular barrier dysfunction at sites of inflammation and injury, and increase endothelial barrier permeability and local edema formation. TRPM2 is expressed in human pulmonary artery endothelial cells, and mediates the H_2O_2-induced increase in endothelial permeability through enhanced Ca^{2+} entry (16). H_2O_2 decreased trans-monolayer transendothelial electrical resistance, indicating opening of interendothelial junctions. Overexpression of TRPM2-L enhanced H_2O_2-induced Ca^{2+} entry and the transendothelial electrical resistance decrease. In contrast, Ca^{2+} entry and the transendothelial electrical resistance decrease in response to H_2O_2 were inhibited by TRPM2 depletion or overexpression of the dominant negative splice variant TRPM2-S. Of note, suppression of TRPM2 resulted in only a 50 % reduction in transendothelial electrical resistance response to H_2O_2, suggesting that up to half of the increased permeability induced by H_2O_2 may be due to other mediators. These data suggest that TRPM2 is a potential therapeutic target to protect against endothelial barrier disruption in oxidant injury, which would reduce subsequent edema, as well as cardiac myocyte death in ischemic–reperfusion injury (16, 39, 59).

5. TRPM2 Function in Brain

TRPM2 is expressed widely in brain and although many of its roles in neuronal tissue remain to be clarified, it has been shown to modulate oxidative stress-induced tissue damage. The striatum is highly vulnerable to ischemia/reperfusion injury (60). Rat striatal cells express TRPM2, and a novel short TRPM2 transcript that is expressed only in the striatum of human brain (61). Differences in function of this short isoform compared to TRPM2-L are not known. In primary striatal cultures, H_2O_2 induced an increase in $[Ca^{2+}]_i$ and cell death, which was inhibited by dominant negative TRPM2-S or by depletion of endogenous TRPM2 with siRNA (15). Amyloid β-peptide is the main component of senile plaques in Alzheimer's disease, and may also mediate neuronal injury through generation of oxidative stress (62). Striatal cells were also protected from amyloid β-peptide-induced $[Ca^{2+}]_i$ rise and cell death by TRPM2 depletion or inhibition (15). TRPM2-S did not reduce ROS in H_2O_2 or amyloid β-peptide treated cells, suggesting that neuroprotective effects of TRPM2-S are mediated by direct inhibition of ROS-induced TRPM2 activation (15, 63). Recently, a report showed increased TRPM2 mRNA and protein in cortical and hippocampal neurons in rat brains after 3–5 days following traumatic brain injury (64). Previously, rat TRPM2 transcripts were also found to be elevated in a time-dependent manner following transient middle cerebral artery occlusion (65). These data suggest that TRPM2 may contribute to neuronal cell death post traumatic brain injury and ischemic injury. TRPM2-L and TRPM2-S isoforms were both expressed in pyramidal neurons of the hippocampus but not in astrocytes (66). Astrocytes expressed TRPV4; they are thought to express TRPM2 under oxidative stress. Exposure to low doses of H_2O_2 resulted in preferential damage to pyramidal neurons, which was reduced by exposure to antioxidants. However, in these experiments, blockers of TRPM2 channels clotrimazole, *N*-(p-amylcinnomoyl) anthranilic acid, or flufenamic acid failed to protect pyramidal neurons from H_2O_2-induced damage and actually increased damage caused by mercaptosuccinate or buthionine sulfoximine, which enhance endogenous ROS production (66). These data suggest that either these inhibitors are ineffective in blocking TRPM2 at the concentrations used, that H_2O_2-induced Ca^{2+} influx is not the sole determinant of cell viability, or their lack of specificity plays a role in an unknown manner. In a recent report, blockade of channel opening inhibited Ca^{2+} influx, but did not correlate with protection from cell death (67). While TRPM2 isoforms are involved in H_2O_2-induced cell death and are important and potentially widely applicable drug targets, more will need to be learned about the mechanisms through which they cause cell death to develop a rational therapeutic approach. This is particularly relevant since genetic variants of the TRPM2 gene have recently been reported which may confer susceptibility to developing Western Pacific amyotrophic lateral sclerosis and parkinsonism-dementia (68) or bipolar disorder (69).

6. TRPM2 Function in Cancer

Oxidative stress is a major molecular mechanism for carcinogenesis. Excess reactive oxygen and nitrogen species can cause single- and double-strand DNA breaks and other DNA modifications as well as epigenetic alterations, followed by selection of proliferation of successfully adapted cells in specific environments (70). These events may contribute to oncogene activation or tumor suppressor inactivation. In addition, roles for modulation of $[Ca^{2+}]_i$ in cancer cell proliferation (71) and in tumor metastasis (72) have been proposed. Although TRPM2 is likely to have a role in carcinogenesis because of its functions in oxidative stress and calcium modulation, little is known about its roles at this time. Two TRPM2 isoforms, TRPM2-AS (antisense) and TRPM2-TE (tumor enriched), were identified in a search to find antisense transcripts to novel tumor suppressor genes in melanoma (33). TRPM2-TE and TRPM2-AS were highly expressed in tumor cells including melanoma, lung cancer, and breast cancer cell lines compared to normal tissue. Overexpression of TRPM2-L or functional knockout of TRPM2-TE increased melanoma susceptibility to cell death. In contrast to the above observation, knockdown of TRPM2 with siRNA was found to inhibit the growth of prostate cancer but not of noncancerous cells (73). In prostate cancer, TRPM2 was also found to be localized to the nucleus in a clustered pattern, which was also not observed in normal cells. Evidence supports the conclusion that TRPM2 isoforms have important roles in cell proliferation and survival, but more information is needed about their function in different cancers to understand how inhibitors or enhancers can be best used for effective tumor treatments.

7. Future Perspective

TRPM2 is an ion channel with critical roles in regulating cellular responses to oxidative stress, TNFα, and several other activators. Factors which modulate endogenous activation in addition to extracellular signals include the quantity of different TRPM2 isoforms expressed, subcellular localization, and interaction with second messengers and other proteins, many of which still need to be identified. As the understanding of TRPM2 activation mechanisms has grown, so has the understanding of the complexity of its regulation and the number of physiological and pathological pathways in which it is involved. Inhibition of TRPM2 activation has tremendous potential as a future drug target to ameliorate a number of disease processes including obesity, diabetes, autoimmune disorders, ischemia–reperfusion injury, vascular endothelial damage, traumatic brain injury, and cancer. However, while potential therapeutic applications are huge, potential issues with depletion, such as reduction in innate immunity in mice, are also recognized. Application of the full therapeutic potential of TRPM2 requires a more complete understanding of TRPM2 in vivo function, activation mechanisms, and the downstream pathways it modulates.

References

1. Hara Y, Wakamori M, Ishii M, Maeno E, Nishida M, Yoshida T, Yamada H, Shimizu S, Mori E, Kudoh J, Shimizu N, Kurose H, Okada Y, Imoto K, Mori Y (2002) LTRPC2 Ca^{2+}-permeable channel activated by changes in redox status confers susceptibility to cell death. Mol Cell 9:163–173
2. Miller BA, Zhang W (2011) TRP channels as mediators of oxidative stress. Adv Exp Med Biol 704:531–544
3. Suguro M, Tagawa H, Kagami Y, Okamoto M, Ohshima K, Shiku H, Morishima Y, Nakamura S, Seto M (2006) Expression profiling analysis of the CD5+ diffuse large B-cell lymphoma subgroup: development of a CD5 signature. Cancer Sci 97:868–874
4. Prawitt D, Enklaar T, Klemm G, Gartner B, Spangenberg C, Winterpacht A, Higgins M, Pelletier J, Zabel B (2000) Identification and characterization of MTR1, a novel gene with homology to melastatin (MLSN1) and the trp gene family located in the BWS-WT2 critical region on chromosome 11p15.5 and showing allele-specific expression. Hum Mol Genet 9:203–216
5. Aarts M, Iihara K, Wei WL, Xiong ZG, Arundine M, Cerwinski W, MacDonald JF, Tymianski M (2003) A key role for TRPM7 channels in anoxic neuronal death. Cell 115:863–877
6. Guilbert A, Gautier M, Dhennin-Duthille I, Haren N, Sevestre H, Ouadid-Ahidouch H (2009) Evidence that TRPM7 is required for breast cancer cell proliferation. Am J Physiol Cell Physiol 297:C493–C502
7. Tsavaler L, Shapero MH, Morkowski S, Laus R (2001) Trp-p8, a novel prostate-specific gene, is up-regulated in prostate cancer and other malignancies and shares high homology with transient receptor potential calcium channel proteins. Cancer Res 61:3760–3769
8. Lehen'kyi V, Prevarskaya N (2011) Oncogenic TRP channels. Adv Exp Med Biol 704: 929–945
9. Duncan LM, Deeds J, Hunter J, Shao J, Holmgren LM, Woolf EA, Tepper RI, Shyjan AW (1998) Down-regulation of the novel gene melastatin correlates with potential for melanoma metastasis. Cancer Res 58:1515–1520
10. Fang D, Setaluri V (2000) Expression and up-regulation of alternatively spliced transcripts of melastatin, a melanoma metastasis-related gene, in human melanoma cells. Biochem Biophys Res Commun 279:53–61
11. Deeds J, Cronin F, Duncan LM (2000) Patterns of melastatin mRNA expression in melanocytic tumors. Hum Pathol 31:1346–1356
12. Nagamine K, Kudoh J, Minoshima S, Kawasaki K, Asakawa S, Ito F, Shimizu N (1998) Molecular cloning of a novel putative Ca^{2+} channel protein (TRPC7) highly expressed in brain. Genomics 54:124–131
13. Sano Y, Inamura K, Miyake A, Mochizuki S, Yokoi H, Matsushime H, Furuichi K (2001) Immunocyte Ca^{2+} influx system mediated by LTRPC2. Science 293:1327–1330
14. Togashi K, Hara Y, Tominaga T, Higashi T, Konishi Y, Mori Y, Tominaga M (2006) TRPM2 activation by cyclic ADP-ribose at body temperature is involved in insulin secretion. EMBO J 25:1804–1815
15. Fonfria E, Marshall IC, Boyfield I, Skaper SD, Hughes JP, Owen DE, Zhang W, Miller BA, Benham CD, McNulty S (2005) Amyloid beta-peptide(1-42) and hydrogen peroxide-induced toxicity are mediated by TRPM2 in rat primary striatal cultures. J Neurochem 95:715–723
16. Hecquet CM, Ahmmed GU, Vogel SM, Malik AB (2008) Role of TRPM2 channel in mediating H2O2 induced Ca^{2+} entry and endothelial hyperpermeability. Circ Res 102:347–355
17. Wehage E, Eisfeld J, Heiner I, Jungling E, Zitt C, Luckhoff A (2002) Activation of the cation channel long transient receptor potential channel 2 (LTRPC2) by hydrogen peroxide. A splice variant reveals a mode of activation independent of ADP-ribose. J Biol Chem 277:23150–23156
18. Gasser A, Glassmeier G, Fliegert R, Langhorst MF, Meinke S, Hein D, Kruger S, Weber K, Heiner I, Oppenheimer N, Schwarz JR, Guse AH (2006) Activation of T cell calcium influx by the second messenger adp-ribose. J Biol Chem 281:2489–2496

19. Kolisek M, Beck A, Fleig A, Penner R (2005) Cyclic ADP-ribose and hydrogen peroxide synergize with ADP-ribose in the activation of TRPM2 channels. Mol Cell 18:61–69
20. Perraud AL, Takanishi CL, Shen B, Kang S, Smith MK, Schmitz C, Knowles HM, Ferraris D, Li W, Zhang J, Stoddard BL, Scharenberg AM (2005) Accumulation of free ADP-ribose from mitochondria mediates oxidative stress-induced gating of TRPM2 cation channels. J Biol Chem 280:6138–6148
21. Toth B, Csanady L (2010) Identification of direct and indirect effectors of the transient receptor potential melastatin 2 (TRPM2) cation channel. J Biol Chem 285:30091–30102
22. Fonfria E, Marshall ICB, Benham CD, Boyfield I, Brown JD, Hill K, Hughes JP, Skaper SD, Scharenberg AM, McNulty S (2004) TRPM2 channel opening in response to oxidative stress is dependent on activation of poly(ADP-ribose) polymerase. Br J Pharmacol 143:186–192
23. Buelow B, Song Y, Scharenberg AM (2008) The poly(ADP-ribose) polymerase PARP-1 is required for oxidative stress-induced TRPM2 activation in lymphocytes. J Biol Chem 283: 24571–24583
24. Kuhn FJ, Heiner I, Luckhoff A (2005) TRPM2: a calcium influx pathway regulated by oxidative stress and the novel second messenger ADP-ribose. Pflugers Arch 451:212–219
25. Blenn C, Wyrsch P, Bader J, Bollhalder M, Althaus FR (2011) Poly(ADP-ribose)glycohydrolase is an upstream regulator of Ca^{2+} fluxes in oxidative cell death. Cell Mol Life Sci 68: 1455–1466
26. McHugh D, Flemming R, Xu SZ, Perraud AL, Beech DJ (2003) Critical intracellular Ca^{2+} dependence of transient receptor potential melastatin 2 (TRPM2) cation channel activation. J Biol Chem 278:11002–11006
27. Du J, Xie J, Yue L (2009) Intracellular calcium activates TRPM2 and its alternative spliced isoforms. Proc Natl Acad Sci USA 106: 7239–7244
28. Tong Q, Zhang W, Conrad K, Mostoller K, Cheung JY, Peterson BZ, Miller BA (2006) Regulation of the TRP channel TRPM2 by the Ca^{2+} sensor calmodulin. J Biol Chem 281: 9076–9085
29. Du J, Xie J, Yue L (2009) Modulation of TRPM2 by acidic pH and the underlying mechanisms for pH sensitivity. J Gen Physiol 134:471–488
30. Starkus JG, Fleig A, Penner R (2010) The calcium-permeable non-selective cation channel TRPM2 is modulated by cellular acidification. J Physiol 588:1227–1240
31. Csanady L (2010) Permeating proton found guilty in compromising TRPM2 channel activity. J Physiol 588:1661–1662
32. Zhang W, Chu X, Tong Q, Cheung JY, Conrad K, Masker K, Miller BA (2003) A novel TRPM2 isoform inhibits calcium influx and susceptibility to cell death. J Biol Chem 278: 16222–16229
33. Orfanelli U, Wenke AK, Doglioni C, Russo V, Bosserhoff AK, Lavorgna G (2008) Identification of novel sense and antisense transcription at the TRPM2 locus in cancer. Cell Res 18:1128–1140
34. Zhang W, Hirschler-Laszkiewicz I, Tong Q, Conrad K, Sun SC, Penn L, Barber DL, Stahl R, Carey DJ, Cheung JY, Miller BA (2006) TRPM2 is an ion channel that modulates hematopoietic cell death through activation of caspases and PARP cleavage. Am J Physiol Cell Physiol 290:C1146–C1159
35. Langley B, Ratan RR (2004) Oxidative stress-induced death in the nervous system: cell cycle dependent or independent? J Neurosci Res 77:621–629
36. Waring P (2005) Redox active calcium ion channels and cell death. Arch Biochem Biophys 434:33–42
37. Duchen MR, Verkhratsky A, Muallem S (2008) Mitochondria and calcium in health and disease. Cell Calcium 44:1–5
38. Yang KT, Chang WL, Yang PC, Chien CL, Lai MS, Su MJ, Wu ML (2006) Activation of the transient receptor potential M2 channel and poly(ADP-ribose) polymerase is involved in oxidative stress-induced cardiomyocyte death. Cell Death Differ 13:1815–1826
39. Hecquet CM, Malik AB (2009) Role of H(2) O(2)-activated TRPM2 calcium channel in oxidant-induced endothelial injury. Thromb Haemost 101:619–625
40. Yamamoto S, Shimizu S, Kiyonaka S, Takahashi N, Wajima T, Hara Y, Negoro T, Hiroi T, Kiuchi Y, Okada T, Kaneko S, Lange I, Fleig A, Penner R, Nishi M, Takeshima H, Mori Y (2008) TRPM2-mediated Ca^{2+}influx induces chemokine production in monocytes that aggravates inflammatory neutrophil infiltration. Nat Med 14:738–747
41. Evans JL, Goldfine ID, Maddux BA, Grodsky GM (2002) Oxidative stress and stress-activated signaling pathways: a unifying hypothesis of type 2 diabetes. Endocr Rev 23:599–622
42. Rosen P, Nawroth PP, King G, Moller W, Tritschler HJ, Packer L (2001) The role of oxidative stress in the onset and progression of diabetes and its complications: a summary of a Congress Series sponsored by UNESCO-MCBN,

the American Diabetes Association and the German Diabetes Society. Diabetes Metab Res Rev 17:189–212

43. Lenzen S, Drinkgern J, Tiedge M (1996) Low antioxidant enzyme gene expression in pancreatic islets compared with various other mouse tissues. Free Radic Biol Med 20:463–466

44. Bari MR, Akbar S, Eweida M, Kuhn FJ, Gustafsson AJ, Luckhoff A, Islam MS (2009) H2O2-induced Ca^{2+} influx and its inhibition by N-(p-amylcinnamoyl) anthranilic acid in the beta-cells: involvement of TRPM2 channels. J Cell Mol Med 13:3260–3267

45. Lange I, Yamamoto S, Partida Sanchez S, Mori Y, Fleig A, Penner R (2009) TRPM2 functions as a lysosomal Ca^{2+}-release channel in beta cells. Sci Signal 2:ra23

46. Mirnikjoo B, Balasubramanian K, Schroit AJ (2009) Mobilization of lysosomal calcium regulates the externalization of phosphatidylserine during apoptosis. J Biol Chem 284: 6918–6923

47. Scharenberg AM (2009) TRPM2 and pancreatic beta-cell responses to oxidative stress. Islets 1:165–166

48. Uchida K, Dezaki K, Damdindorj B, Inada H, Shiuchi T, Mori Y, Yada T, Minokoshi Y, Tominaga M (2011) Lack of TRPM2 impaired insulin secretion and glucose metabolisms in mice. Diabetes 60:119–126

49. Park SY, Cho YR, Kim HJ, Hong EG, Higashimori T, Lee SJ, Goldberg IJ, Shulman GI, Najjar SM, Kim JK (2006) Mechanism of glucose intolerance in mice with dominant negative mutation of CEACAM1. Am J Physiol Endocrinol Metab 291:E517–E524

50. Heiner I, Eisfeld J, Luckhoff A (2003) Role and regulation of TRP channels in neutrophil granulocytes. Cell Calcium 33:533–540

51. Heiner I, Radukina N, Eisfeld J, Kuhn F, Luckhoff A (2005) Regulation of TRPM2 channels in neutrophil granulocytes by ADP-ribose: a promising pharmacological target. Naunyn Schmiedebergs Arch Pharmacol 371: 325–333

52. Wehrhahn J, Kraft R, Harteneck C, Hauschildt S (2010) Transient receptor potential melastatin 2 is required for lipopolysaccharide-induced cytokine production in human monocytes. J Immunol 184:2386–2393

53. Kraft R, Grimm C, Grosse K, Hoffmann A, Sauerbruch S, Kettenmann H, Schultz G, Harteneck C (2004) Hydrogen peroxide and ADP-ribose induce TRPM2-mediated calcium influx and cation currents in microglia. Am J Physiol Cell Physiol 286:C129–C137

54. Yamamoto S, Takahashi N, Mori Y (2010) Chemical physiology of oxidative stress-activated TRPM2 and TRPC5 channels. Prog Biophys Mol Biol 103:18–27

55. Videla LA, Rodrigo R, Orellana M, Fernandez V, Tapia G, Quinones L, Varela N, Contreras J, Lazarte R, Csendes A, Rojas J, Maluenda F, Burdiles P, Diaz JC, Smok G, Thielemann L, Poniachik J (2004) Oxidative stress-related parameters in the liver of non-alcoholic fatty liver disease patients. Clin Sci (Lond) 106:261–268

56. Barnes PJ (2004) COPD: is there light at the end of the tunnel? Curr Opin Pharmacol 4:263–272

57. Knowles H, Heizer JW, Li Y, Chapman K, Ogden CA, Andreasen K, Shapland E, Kucera G, Mogan J, Humann J, Lenz LL, Morrison AD, Perraud AL (2010) Transient receptor potential melastatin 2 (TRPM2) ion channel is required for innate immunity against Listeria monocytogenes. Proc Natl Acad Sci USA 108:11578–11583

58. Perraud AL, Fleig A, Dunn CA, Bagley LA, Launay P, Schmitz C, Stokes AJ, Zhu Q, Bessman MJ, Penner R, Kinet JP, Scharenberg AM (2001) ADP-ribose gating of the calcium-permeable LTRPC2 channel revealed by Nudix motif homology. Nature 411:595–599

59. Hecquet CM, Ahmmed GU, Malik AB (2010) TRPM2 channel regulates endothelial barrier function. Adv Exp Med Biol 661:155–167

60. Lipton P (1999) Ischemic cell death in brain neurons. Physiol Rev 79:1431–1568

61. Uemura T, Kudoh J, Noda S, Kanba S, Shimizu N (2005) Characterization of human and mouse TRPM2 genes: identification of a novel N-terminal truncated protein specifically expressed in human striatum. Biochem Biophys Res Commun 328:1232–1243

62. Butterfield DA (2003) Amyloid beta-peptide [1-42]-associated free radical-induced oxidative stress and neurodegeneration in Alzheimer's disease brain: mechanisms and consequences. Curr Med Chem 10:2651–2659

63. Aarts MM, Tymianski M (2005) TRPMs and neuronal cell death. Pflugers Arch 451: 243–249

64. Cook NL, Vink R, Helps SC, Manavis J, van den Heuvel C (2010) Transient receptor potential melastatin 2 expression is increased following experimental traumatic brain injury in rats. J Mol Neurosci 42:192–199

65. Fonfria E, Mattei C, Hill K, Brown JT, Randall A, Benham CD, Skaper SD, Campbell CA, Crook B, Murdock PR, Wilson JM, Maurio FP,

Owen DE, Tilling PL, McNulty S (2006) TRPM2 is elevated in the tMCAO stroke model, transcriptionally regulated, and functionally expressed in C13 microglia. J Recept Signal Transduct Res 26:179–198

66. Bai JZ, Lipski J (2010) Differential expression of TRPM2 and TRPV4 channels and their potential role in oxidative stress-induced cell death in organotypic hippocampal culture. Neurotoxicology 31:204–214
67. Wilkinson JA, Scragg JL, Boyle JP, Nilius B, Peers C (2008) H_2O 2-stimulated Ca^{2+} influx via TRPM2 is not the sole determinant of subsequent cell death. Pflugers Arch 455:1141–1151
68. Hermosura MC, Cui AM, Go RC, Davenport B, Shetler CM, Heizer JW, Schmitz C, Mocz G, Garruto RM, Perraud AL (2008) Altered functional properties of a TRPM2 variant in Guamanian ALS and PD. Proc Natl Acad Sci USA 105:18029–18034
69. Xu C, Li PP, Cooke RG, Parikh SV, Wang K, Kennedy JL, Warsh JJ (2009) TRPM2 variants and bipolar disorder risk: confirmation in a family-based association study. Bipolar Disord 11:1–10
70. Toyokuni S (2008) Molecular mechanisms of oxidative stress-induced carcinogenesis: from epidemiology to oxygenomics. IUBMB Life 60:441–447
71. Parkash J, Asotra K (2010) Calcium wave signaling in cancer cells. Life Sci 87:587–595
72. Prevarskaya N, Skryma R, Shuba Y (2011) Calcium in tumour metastasis: new roles for known actors. Nat Rev Cancer 11:609–618
73. Zeng X, Sikka SC, Huang L, Sun C, Xu C, Jia D, Abdel-Mageed AB, Pottle JE, Taylor JT, Li M (2010) Novel role for the transient receptor potential channel TRPM2 in prostate cancer cell proliferation. Prostate Cancer Prostatic Dis 13:195–201

Chapter 6

The Ca^{2+}-Activated Monovalent Cation-Selective Channels TRPM4 and TRPM5

Barbara Colsoul, Miklos Kecskes, Koenraad Philippaert, Aurelie Menigoz, and Rudi Vennekens

Abstract

In this review we summarize the current data on the ion channels TRPM4 and TRPM5. These proteins are to date the only molecular candidates for the class of Ca^{2+}-activated cation channels (CAN channels). They form monovalent cation-selective and Ca-impermeable ion channels, which are activated by a rise of the cytosolic Ca^{2+} concentration. They have been described in a plethora of tissues, and their physiological importance is only now being explored since knockout mice have become available.

Key words: TRPM4, TRPM5, Ca^{2+} activated non-selective cation channels, TRP ion channels, Membrane depolarization

1. Introduction

TRP ion channels are important signaling proteins involved in many physiological processes. In this review we will focus on the role of TRPM4 and TRPM5. They are two closely related members of the TRPM subfamily, named after its first characterized member melastatin, a putative tumor suppressor protein (1). TRPM4 and TRPM5 are structurally much related to each other showing ~40% amino acid homology. Based on homology they are both thought to contain six transmembrane regions and assemble in tetramers. In contrast to the other proteins in the TRP family, TRPM4 and TRPM5 are the only two known TRP channels that are activated by increase in intracellular calcium and are impermeable for calcium and only permeable for monovalent cations. This

Arpad Szallasi and Tamás Bíró (eds.), *TRP Channels in Drug Discovery: Volume I*, Methods in Pharmacology and Toxicology, DOI 10.1007/978-1-62703-077-9_6, © Springer Science+Business Media, LLC 2012

makes them to date the only molecular candidates for the class of calcium-activated calcium-impermeable monovalent cation channels (CAN channels).

CAN channels with properties reminiscent of TRPM4 or TRPM5 were first described in cardiac myocytes (2), but have since been described in a wide range of excitable cells including several neuron types (3–7), vascular and smooth muscle cells (8–10), and in endocrine pancreatic tissue (11). They have also been found in non-excitable cells such as red blood cells (12, 13), exocrine cells from the pancreas (14–16), brown adipocytes (17, 18), kidney epithelial cells (19), cochlear hair cells (20), and vascular endothelial cells (21–23). These CAN channels have not yet been proven to be TRPM4 or TRPM5. However, they have been identified and characterized in several cases.

1.1. Gene Expression

The cloning of TRPM4 was first described in 2001 (1) after screening human cDNA libraries with an expressed sequence tag. A 4.0-kb long cDNA fragment was found to encode a predicted protein of 1,040 aa. This was mapped to the 19q13.32 region spanning 54 kb in the human genome. Later, additional TRPM4 encoding sequences were found that were different in size compared to the originally described variant. A longer sequence with an additional 174 N-terminal residues (24) as well as a shorter variant of 678 aa was found (25). These three splice variants are, respectively, named TRPM4a for the 1,040 aa protein first described, TRPM4b for the 1,214 aa full length clone, and TRPM4c for the shortest splice variant encoding for 678 aa (26). In subsequent screening also the mouse ortholog for TRPM4a and TRPM4b was found on chromosome 7 (27) together with additional splice variants (25). There is no clear functional difference between the identified splice variants and for most functional analysis of the TRPM4 protein, either the human or murine full-length clone of TRPM4b is used. TRPM4 is a widely expressed protein present in most adult tissues but abundantly expressed in the heart, prostate, colon, and kidneys. TRPM4 was also found to be endogenously expressed in HEK-293 cells (24) and CHO-cells (28) and was characterized as a Ca^{2+} activated nonselective cation channel mediating cell membrane depolarization.

The TRPM5 gene was first identified in a functional analysis of chromosomal region 11p15.5 (29) and consists of a 4.5 kb mRNA transcript that was identified and expressed in several tissues. Two alternative splice variants were found: the longest one has 1,165 aa and six predicted transmembrane domains whereas the second variant lacks exon 18, leading to a premature stop codon after 872 aa. The murine variant of TRPM5 has also been identified on chromosome 7 with a highly conserved exon–intron structure compared to the human gene. Unlike the human TRPM5, however, the mouse clone does not have an alternative splice variant.

Expression has been proven in a broad range of tissues including brain, spleen, lung, kidney, and heart (30). A restricted expression of TRPM5 was found in taste tissue, stomach, and small intestine (31). In taste cells, TRPM5 is acting as a modulator of bitter, sweet, and amino acid (unami) taste (32). TRPM5 was reported to be a voltage-modulated and Ca^{2+}-activated monovalent selective cation channel and is activated by PLCβ2 in a Ca^{2+}-independent manner in taste cells (33).

1.2. Voltage Dependency and Ca^{2+} Activation

Non-selective cation channels activated by intracellular Ca^{2+} are important in Ca^{2+} signaling and membrane excitability. TRPM4 and TRPM5 are molecular candidates for this kind of channels: they are expressed in many tissues, behave in a similar manner, and have similar unitary conductances of 25 and 23 pS, respectively (24, 33). There are, however, several differences between the two channels such as the different Ca^{2+} sensitivity and time course of activation (34). The Ca^{2+} concentration for half-maximal current activation in whole cell patch clamping for TRPM4 is 20.2 ± 4.0 μM whereas TRPM5 has an EC_{50} of 0.7 ± 0.12 μM. Also, the current amplitude of TRPM4 was largely reduced by $[Ca^{2+}]_i$ concentrations above 300 μM and no such observation was made with $[Ca^{2+}]_i$ concentrations up to 1mM on TRPM5. The Ca^{2+} sensitivity is greatly reduced in inside-out patches where the EC_{50} concentration is 167 ± 11 μM for TRPM4 and 2.3 ± 0.3 μM for TRPM5 (34) although there is a large variety in the reported values for Ca^{2+} sensitivity (35). This observation indicates that the loss of a cytosolic factor or changes in the phosphorylation state of the channels can lead to a strongly reduced Ca^{2+} sensitivity. The currents of both TRPM4 and TRPM5 in response to voltage ramps show outward rectification but is more pronounced in TRPM4 (25, 33, 36). This voltage dependency might be a common feature in TRPM channels as it has been reported not only for (37) and TRPM4 but also for TRPM8 (37). It has been hypothesized that gating and modulation of TRPM4 and TRPM5 are essentially dependent on the weak voltage dependency (38), as has been described for other TRP channels (39). In an overexpression system, there is a fast decrease of the inward currents observed after reaching the initial peak amplitude. The desensitization of TRPM5 after Ca^{2+}-induced activation is a consistent observation and may play a role in sensory adaptation (33, 35, 36). The rundown of the current is observed following intracellular Ca^{2+} dialysis not only in whole cell recording mode but also in perforated patch recordings after activation by a bath-applied Ca^{2+}-ionophore, arguing that the rundown is not exclusively due to the washout of cytosolic signaling molecules (35).

Activation of TRPM4 and TRPM5 is also temperature dependent (40): warm temperatures promote the activation of the channel. This is in line with the observed temperature-dependent taste sensation mediated by TRPM5 (41) (see below).

1.3. Ion Selectivity

Both TRPM4 and TRPM5 are activated by an increase in intracellular Ca^{2+} but are impermeable to Ca^{2+}. The pore of both channels is thus selective for monovalent cations but poorly discriminates between K^+, Na^+, Cs^+, or Li^+. Both ion channels are virtually impermeable to divalent cations such as Ca^{2+} or Mg^{2+}. The selectivity filter for TRPM4 has been identified in a study using chimeric ion channels with the pore sequence of TRPV6, a distantly related Ca^{2+} permeable ion channel (42). The chimeric channel provides the same gating properties as TRPM4 but is Ca^{2+} permeable. The region between the fifth and the sixth transmembrane domain of TRPM4 and the homologous region of TRPM5 contains a stretch of 5–6 acidic amino acid residues responsible for the ion selectivity (43).

1.4. Cellular Modulation of TRPM4 and TRPM5

$PI(4,5)P_2$ has been indicated as an important cofactor in the activation of TRPM5 (44) and TRPM4 (45) and depletion of PIP_2 could be the reason for the rundown of the current amplitude over time. There is a rescue of the TRPM4 current decay observed by the application of PIP_2 to the cell, indicating that PIP_2 hydrolysis is indeed the reason for the current rundown in both TRPM4 and TRPM5 (46). However, TRPM4 but not TRPM5 shows a second phase of activation after a variable delay of 250 ± 40 s even without loading the cell with PIP_2 (34). Ca^{2+} is an important factor in the TRPM5 activation and also a brief elevation of intracellular IP_3 can gate TRPM5 channels in intact taste cells but not in inside-out patches (47), suggesting that the IP_3 receptor could be the missing link in the signaling transduction between PLC and TRPM5 (48). Ca^{2+} release from the SR through the IP_3 receptor can activate TRPM4 without Ca^{2+} in the extracellular solution (49, 50). TRPM4 cannot be directly activated by PIP_2, but application of PIP_2 greatly increases the Ca^{2+} sensitivity and gives a drastic shift of the voltage dependency of activation towards negative potentials, strongly increasing the open probability at physiological membrane potential (45).

In addition ATP can help to restore the Ca^{2+} sensitivity of TRPM4 in inside-out patches when MgATP is applied to the cytosolic side of the membrane (51). ATP plays an important role in maintaining the Ca^{2+} sensitivity of TRPM4 channels, although ATP has been found to block TRPM4 channel activity (52). ATP blocking the channel could explain not only these apparent contradictory findings but also priming it for reactivation by Ca^{2+}, whereby the Ca^{2+} removes ATP and allows for channel opening. It is, however, unclear whether the inhibitory ATP binding site and the facilitating binding site on TRPM4 are identical (53). Furthermore, TRPM5 was found to be blocked by extracellular acidification. A fast reversible block of the channel and a slower irreversible inactivation of TRPM4 were seen under the same circumstances. The amino acid residues responsible for the blocking by lower pH were identified and are conserved between human and mouse TRPM5, indicating an evolutionary conserved regulating role, but are not found in TRPM4 (54).

1.5. Pharmacology

Selective blocking agents of these ion channels can be used for specific current identification in primary cells. Most identified compounds show, however, a poor selectivity between TRPM4 and TRPM5 (and other ion channels). Spermine is identified as an intracellular blocking agent of TRPM4 (52) but is equally potent to block TRPM5 currents (34). Furthermore, Flufenamic acid (FFA) blocks both TRPM4 (EC_{50} of 2.5 μM) and TRPM5 (EC_{50} of 24.5 μM) and also clotrimazole is shown to inhibit both channels. Some compounds have been shown to inhibit TRPM4 but not TRPM5. Decavanadate has been found to modulate TRPM4 by introducing a dramatic shift in voltage dependency, which results in an increased open probability at negative potentials. It also interferes with the ATP gating of TRPM4 but has no effect on TRPM5 (53). MBP-104 not only was found to block TRPM4 but is also an activator of CFTR, making it hard to identify TRPM4 specific currents using MBP-104 in primary tissue. 9-Phenanthrol, another hydroxytricyclic compound, provides a reversible pharmacological block of TRPM4 without effects on CFTR or TRPM5, in this way giving a clear tool for identifying TRPM4 currents in native cells (55). The specificity of 9-phenanthrol has also been shown in smooth muscle cells, where the compound has no effect on TRPC3, TRPC6, or other channels (56). Triphenylphosphine oxide has been identified as an inhibitor of TRPM5 without blocking TRPM4, TRPV1, or TRPA1 and can therefore be considered as a relative selective blocker for TRPM5 (57). Activation of TRPM5 is possible with arachidonic acid and in this way taste signaling is modulated (58).

1.6. Endogenous TRPM4 and TRPM5 Currents

The expression of both TRPM4 and TRPM5 was shown in several different tissues, but there is much less known about the functional role of these channels in native tissues. These data are summarized in Tables 1 (TRPM4) and 2 (TRPM5). There is evidence for a TRPM4-like current in cardiac myocytes (69): a Ca^{2+}-activated nonselective cation channel (CAN channel) was functionally identified with a unitary conductance of 25 ± 0.6 pS and a voltage-dependent activation. These observations, together with expression data in ventricular cells showing expression of TRPM4 and TRPM5, suggest strongly one of these channels as the carrier of that current. As a current decrease was observed upon addition of intracellular ATP and extracellular FFA was found to have an EC_{50} of 5.5 μM on the observed current (which is far below the EC_{50} value for TRPM5 but close to the TRPM4 EC_{50}), the identified current might be TRPM4.

It has been shown in mast cells that TRPM4 is functionally active by comparing the results of wild type and *Trpm4*$^{-/-}$ animals (61). The detected current had a single-channel conductance of 25.3 ± 0.7 pS and a linear voltage dependency. There was clear Ca^{2+} dependency and ion selectivity similar to TRPM4 as assessed in TRPM4-overexpressing HEK cells. There were no

Table 1
TRPM4

Tissue	Species/cell lines	Method	Proposed function	References
Pituitary gland	Human	RT-PCR		(59)
Lung	Human	RT-PCR		(59)
Liver	Human fetal liver	RT-PCR Northern blot		(59) (24)
Kidney	Human	RT-PCR		(59)
Chromaffin cells	Mouse	Functional KO study Immunostaining	Regulation of blood pressure	(60)
Skeletal muscle	Human	Northern blot RT-PCR		(24)
Spleen	Human	RT-PCR Northern blot		(59) (24)
Adipose tissue	Human	RT-PCR		(59)
Macrophages				
Mast Cells	Mouse	RT-PCR Functional KO study Immunostaining	Degranulation, migration	(61)
Bone	Human	RT-PCR		(62)
Prostate	Human	RT-PCR Northern blot		(25) (24)
Bladder	Urothelium cells	RTP-PCR Western blot		(63)
Taste cells	Mouse taste cells	RT-PCR	Unknown	(64)
Pancreas	Several beta cell lines; alpha cell line INR1G9; mouse and human islets	RT-PCR, immunoprecipitation; western blot; immunofluorescence	Controversy	(36, 61, 65, 66)
HEK-293 cells		Northern blot		(24)
CHO cells		Immunoblot RT-PCR Functional data		(28)
Cardiac myocytes	Human Rat Mouse	Functional data(patch clamp) Genetic analysis RT-PCR PCR, Western blot	Ca(2+)-activated non-selective cation channel Electrical conductance Contributor in transient inward current Same as in rat	(67) (68) (69) (70)
Vascular smooth muscle	Rat	RT-PCR	Myogenic tone, Bayliss effect	(71)
Peripheral blood mononuclear cells	Human	RT-PCR		(36, 59)

Table 2
TRPM5

Tissue	Species/cell lines	Method	Proposed function	References
Pituitary (fetal) brain	Human Mice	RT-PCR Northern blot		(29, 59) (30)
Fetal kidney	Human Mice	Northern blot RT-PCR		(29) (30, 59)
Fetal liver	Human	RT-PCR Northern blot		(59) (29)
Testes	Human Mice	Northern blot		(29) (30)
Enteroendocrine cells Duodenum Ilium Jejunum Stomach	Human Mice Mice Mice Mice	RT-PCR Immunostaining Northern blot Northern blot	Sensing nutrients	(72) (73, 74) (29) (73)
Intestine	Human	RT-PCR		(59)
Prostate	Human	RT-PCR Northern blot		(59) (29))
Olfactory tissue	Mice	Immunostaining		(75)
Taste cells	Mouse taste cells	Northern blot; in situ hybridization; immunostaining	Transduction of sweet, bitter and umami taste responses	(31, 32, 73, 76)
Pancreas	INS-1; MIN6; mouse and human islets	RT-PCR; immunofluorescence	Positive regulation of glucose-stimulated insulin secretion	(36, 77, 78)
Heart	Human Mice	RT-PCR Northern blot		(29, 59) (30)
Peripheral blood mononuclear cells	Human	RT-PCR		(59)

such currents measured in cells isolated from *Trpm4*$^{-/-}$ mice under the same experimental conditions. Similar results were obtained in dendritic cells (79) where an outwardly rectified, Ca^{2+}-activated current was carried by TRPM4. The conductance was measured to be 22.1 ± 0.6 pS and there was no selectivity for K^+ compared to Na^+. These currents were also absent in

Trpm4$^{-/-}$ animals. Perforated patch recordings and pharmacological block were used to identify TRPM4 channel activity in smooth muscle cells (49, 56).

In taste cells, an outward rectifying current was observed upon release of intracellular Ca^{2+} with IP_3. This current was absent in *Trpm5*$^{-/-}$ mice (47). The ionic selectivity was in agreement with the selectivity for TRPM5 in overexpression systems and the current was activated by raising the intracellular Ca^{2+} level. The TRPM5 channels in taste cells also show time-dependent activation and are voltage dependent.

Finally, in β cells from pancreatic islets, a Ca^{2+} activated cation current was identified. This current shares many of the properties of TRPM5 as assessed in overexpression systems, including a bell-shaped dependency on intracellular Ca^{2+} levels, weak voltage dependence, and selectivity properties. Moreover, in cells isolated from *Trpm5*$^{-/-}$ mice, this current was largely absent, indicating that it is at least partly dependent on the presence of a functional TRPM5 protein (77).

2. Functional Role of TRPM4 and TRPM5 in the Cardiovascular System

Using molecular techniques, TRPM4 and TRPM5 have been detected in large numbers of tissues including those of the cardiovascular system. In humans and rodents, TRPM4 has been detected in every cell type of the heart (80), and also in vascular endothelium and smooth muscle. The expression of TRPM5 is more restricted but the channel is also shown to be expressed in heart tissue, albeit at a low level (59). Since the cloning of TRPM4 by Launay et al. (24), the function of the channel has been described in several physiological and pathological processes in the cardiovascular system. In short, TRPM4 is involved in progressive familiar heart block type 1 (81), in hypertension and hypertrophy (60, 70), and recently found as a possible candidate causing arrhythmia (82). It is also expressed in vascular smooth muscle (VSM) and has a critical role in pressure-induced membrane depolarization (71). The role of TRPM5 in cardiovascular system is poorly investigated. In this chapter, we focus on the results dealing with the role of TRPM4 in the cardiovascular system.

2.1. Progressive Familiar Heart Block I

Progressive familiar heart block type I (PFHBI) is a cardiac bundle branch disease in the His–Purkinje system that shows autosomal-dominant inheritance. The first paper, published by Kruse et al., shows connection between TRPM4 and PFHBI. They identified the mutation c19. G→A in the *Trpm4* gene at chromosomal locus 19q13.3 in three branches of a large South African family (68). Cellular expression studies showed that the c.19G→A missense mutation reduced the deSUMOylation of the TRPM4 channel. The resulting SUMOylatied mutant TRPM4 channel caused impaired endocytosis and led to elevated TRPM4 channel density at the cell surface. One year later, Liu et al. investigated one Lebanese and two French families with autosomal-dominant isolated cardiac conduction blocks (83). They again found missense mutations in the *Trpm4* gene in each family and these mutations again caused an increased TRPM4 channel density at the cell surface. However, it has to be noted that this result was the outcome of the investigation of 12 candidate genes from the linked genomic interval which contains about 300 genes, therefore mutations in other genes cannot be excluded. However, these results clearly suggest the participation of the TRPM4 channel in heart block disease, where the increased channel density is the result of the SUMOylation, which blocks the ubiquitylation or the outcome of trafficking defects or probably both, but the molecular mechanism remains unclear. They showed that TRPM4 was prominently expressed in Purkinje fibers and play an important role in heart conductance. It is hypothesized that the increased TRPM4 channel density would inhibit conduction by increasing the membrane leak conductance. This would depolarize the cellular resting membrane potential, and thereby limit the driving force for Ca^{2+} entry in Purkinje cells, but the precise mechanism remains to be determined (68).

2.2. Hypertension and Hypertrophy

A role of TRPM4 in blood pressure regulation and cardiac hypertrophy has recently been suggested in several studies. *Trpm4* is expressed in several organs involved in blood pressure regulation such as kidney, heart, adrenal glands, and VSM. Mathar et al. showed a mild, but stable blood pressure increase in *Trpm4*$^{-/-}$ mice (60): an average increase of about 10 mmHg is present during both the resting and active period and is not due to changes in locomotor activity. Furthermore, heart rate, cardiac output, ejection fraction, and cardiac contractility are not changed in *Trpm4*$^{-/-}$ mice. However, those mice show increased plasma epinephrine levels and increased urinary excretion of catecholamine products. The source of this epinephrine was then shown to be the chromaffin cells in the medulla of the adrenal gland.

Cardiac hypertrophy is an adaptive process that occurs in response to increased physical stress on the heart. A common reason for cardiac hypertrophy is high blood pressure or heart valve stenosis. Recently, several TRP channels including TRPC1, TRPC3, and TRPC6 have been shown to be related to various aspects of cardiac hypertrophy (70). The first report about functionally active TRPM4 in sino-atrial node (SAN) cells was from Demion et al. They identified a CAN which is permeable for Na^+ and K^+ but does not conduct Ca^{2+}, could be activated by intracellular Ca^{2+}, and inhibited by intracellular ATP, FFA, and glibenclamide. Finally, they identified this channel as the TRPM4 molecule (84). As a follow up, Guinamard et al. showed a higher expression of this channel in cardiomyocytes of spontaneously hypertensive rats (SHR), a model of hypertension and cardiac hypertrophy. They also found an increase in the TRPM4 current activity in myocytes from SHR rats (69). While mechanisms underlying the onset of hypertrophy are still not clear, intracellular Ca^{2+} levels appear to be a major component of the process. Thus, the contribution of TRPM4 is likely supported, but the mechanism how TRPM4 may modify this Ca^{2+} signaling or modify the known signaling cascades (NFAT, JNK, ERK) involved in triggering hypertrophy still remains to be determined. In addition, the higher blood pressure in *Trpm4*$^{-/-}$ mice and the increased expression of *Trpm4* in hypertensive rats show a possible side effect which needs to be clarified.

2.3. TRPM4 and Myogenic Tone

Local control of cerebral blood flow is regulated by myogenic constriction of resistance arteries; this regulation was described originally as the Bayliss effect. A stretch of the muscle membrane opens a stretch-activated ion channel. The cells become depolarized and this results in a Ca^{2+} signal. It is generally hypothesized that TRPC6 and TRPM4 are involved in the VSM depolarization as mechano-sensitive ion channels (71, 85, 86). Earley and coworkers presented a detailed analysis of the role of TRPM4 in VSM in several studies. In detail, they found that pressure-induced VSM cell depolarization was attenuated by TRPM4 antisense oligodeoxynucleotides application downregulating the channel subunit expression. The same result has been found with intact vessels, measured in vessel chamber (49, 56, 71, 87–90). Moreover, 9-phenanthrol hyperpolarized the membrane from −40 to −70 mV, leading to the abolishment of the myogenic tone (56). Furthermore, in vivo suppression of TRPM4 using antisense technology decreases myogenic constriction in cerebral arteries and leads to a higher cerebral blood flow both at resting and at elevated mean arterial pressure, confirming a role for TRPM4 in myogenic constriction and cerebral blood flow regulation (91). Later, it was shown that TRPM4 activation during pressure-induced depolarization is protein kinase C (PKC) dependent. Indeed, stimulation of PKC activity increased the intracellular Ca^{2+} sensitivity of TRPM4 (90), as it was found previously by others in HEK293 cells overexpressing TRPM4 (51). A recent paper shows the translocation of TRPM4 to the plasma membrane during PKC activation in live-cell confocal imaging experiments. Using a green fluorescein protein (GFP) tagged TRPM4, they showed that PKC activation with phorbol 12-myristate 13-acetate (PMA) increased about threefold the levels of TRPM4-GFP protein on the cell surface. Furthermore, they proved that this translocation was independent of PKC-alpha and PKC-beta activity but was inhibited by blockade of PKC-delta with rottlerin (88). Taken together, it has become clear that TRPM4 has a critical role in VSM depolarization as a mechano-sensitive ion channel together with PKC-delta activation. However, it should be mentioned that the Bayliss effect was not changed in *Trpm4*$^{-/-}$ mice, (60) and the mechano-sensitivity of TRPM4 was not reported by other authors.

2.4. TRPM4 and Arrhythmias

Cardiac arrhythmia can occur in a wide variety of conditions when the intracellular Ca^{2+} concentration is increased, which is attributed to the activation of a transient inward current. This inward current could be three different intracellular Ca^{2+}-sensitive currents: the Na^{+}–Ca^{2+} exchange current, a Ca^{2+} activated chloride current, or a Ca^{2+}-activated non-selective cation current. Functional characterization of CAN channels in human atrial cardiomyocytes using the patch-clamp technique showed that this channel is likely TRPM4 (67) The exact role of the channel in cardiac function remains unclear, but it may be involved in an arrhythmogenic mechanism in situations of intracellular Ca^{2+} overload or cytoplasmic ATP depletion, generating a delayed after-depolarization. Interestingly, a recent study shows that hypoxia and reoxygenation-induced early after-depolarizations could be abolished by the TRPM4 inhibitor 9-phenanthrol, suggesting involvement of TRPM4 in early after-depolarizations and arrhythmias (92). Recently, another study showed evidence that TRPM4 could participate in triggering arrhythmia together with purinergic signaling (82). They reported that during early-ischemic events, ATP/UTP binding to P2Y2 purinergic receptors activates TRPC3/7 channels, and the release of Ca^{2+} could also activate TRPM4. In these papers, it is likely confirmed that the TRPM4 may somehow participate in the generation of cardiac arrhythmia, but we are far away from understanding its precise role.

In conclusion, the role of TRPM4 in the cardiovascular system has been extensively investigated in the past decade, such as in cardiac conduction block, hypertension, cardiac hypertrophy, myogenic tone regulation, and arrhythmia. The importance of TRPM4 channel in the treatment of several cardiac diseases is clear. However, its function in different diseases and different cell types remains largely to be characterized in detail and needs to be elucidated further for potential pharmaceutical targeting.

3. TRPM4, TRPM5, and Glucose Sensing in Taste Tissue and the Pancreas

3.1. Functional Role for TRPM5 in Taste

Taste transduction, via the recognition of the five basic tastes salty, sour, bitter, sweet, and umami, plays an important role in the evaluation of the nutritious value, toxicity, sodium content, and acidity of food (93). For example, sweet, bitter, and umami taste compounds each activate taste GPCRs that are expressed in discrete sets of taste receptor (type II) cells. These GPCRs activate heterotrimeric GTP-binding proteins (α-gustducin), resulting in functional interaction with PLCβ2, formation of IP_3, and release of Ca^{2+} from intracellular stores. This finally leads to the release of ATP by ATP-permeable gap junction hemichannels composed of pannexin 1 (94, 95). TRPM5 has been shown to play an essential role in this process. TRPM5 is expressed selectively in a subset of taste receptor cells, along with the taste signaling molecules α-gustducin, $G_{\gamma 13}$, phoshpolipase C-β_2 (PLCβ_2), and IP_3R3 (31, 73). Deletion of TRPM5 results in an impaired sensitivity to bitter, sweet, and umami tastes, although there is some debate whether some additional pathways, next to the TRPM5-dependent pathway, exist (32, 96, 97). Indeed, whereas one study shows complete abolishment of behavioral responses to sweet, bitter, and umami tastes in *Trpm5*$^{-/-}$ mice (32), other studies detect some residual activity in response to these components (96, 97). The bitter tastant nicotine has been shown to activate both TRPM5-dependent and TRPM5-independent pathways (98). This last pathway, which involves the nicotinic acetylcholine receptor, is however not a common one for bitter compounds as the bitter tastant quinine acts only through the TRPM5-dependent pathway (98). Interestingly, *Trpm5*$^{-/-}$ mice have been shown to develop preferences for glucose over amino acid solutions independent of taste quality and caloric load, an effect associated with the ability of a given nutrient to regulate glucose metabolism and stimulate brain dopamine centers (99).

Activation of TRPM5 in taste receptor cells occurs downstream of IP_3-mediated Ca^{2+} release (47). Indeed, brief elevation of either intracellular IP_3 or Ca^{2+} is sufficient to gate TRPM5-dependent currents in intact taste cells, but only intracellular Ca^{2+} is able to activate TRPM5-dependent currents in excised patches (47). The activation of TRPM5 allows Na^+ influx and depolarization of the membrane potential, a feature necessary for the consequent release of ATP. In fact, elimination of TRPM5 current prevents the secretion of ATP from taste cells and this release can be restored when the membrane is depolarized by other means such as elevated K^+ (100, 101). Since the gap junction hemichannel pannexin 1, responsible for the ATP release, can be opened by membrane depolarization or by elevation of intracellular Ca^{2+} (102), it can be concluded that both the IP_3-mediated Ca^{2+} release and the TRPM5-mediated membrane depolarization are responsible for the ATP release in response to stimulation of taste receptor cells (100). Finally, it is shown that TRPM5 can be activated by arachidonic acid, a fatty acid that is released from diacylglycerol (DAG) and 2-arachidonicglycerol, molecules generated downstream of PLCβ_2 after stimulation by taste components (58). The physiological role of this phenomenon remains, however, unclear.

As TRPM5 is an essential component of taste signaling, it can be expected that modulation of TRPM5 influences taste perception.

TRPM5 channel activity can be modulated by heat: the channel is activated by increasing temperatures between 15 and 35 °C and this heat activation is due to a temperature-dependent shift of the voltage-activation curve to negative potentials (40). In line with this, TRPM5-dependent sweet taste responses in mouse gustatory nerves are strongly enhanced by heating, a feature lost in *Trpm5*$^{-/-}$ mice (40). This temperature sensitivity of TRPM5 might underlie certain effects of taste, such as enhanced sweetness perception at high temperatures and thermal taste (a phenomenon whereby heating or cooling of the tongue evokes sensations of taste in the absence of tastants) (40). Furthermore, the bitter tastant quinine dose-dependently inhibits TRPM5 currents and abolishes the gustatory response of sweet-sensitive gustatory nerves in WT but not in *Trpm5*$^{-/-}$ mice (103). This inhibition of TRPM5 by bitter compounds constitutes the molecular basis of bitter–sweet taste interactions, whereby bitter tastants directly inhibit the sweet transduction pathway.

There is converging evidence that, in addition to its well-recognized texture and odor components, the flavor of fat includes a taste component and that fat, as a consequence, may also be considered as a basic taste quality (93, 104). For example, it has been shown that mice show a preference for linoleic acid emulsions (105). A recent study reports that the polyunsaturated long-chain free fatty acid linoleic acid depolarizes taste cells (64). This occurs via an increase in intracellular Ca^{2+} that is dependent on the PLC pathway. Furthermore, linoleic acid activates a monovalent cation-selective current in taste cells that is most likely mediated by TRPM5 as the current can be abolished by the TRPM5 antagonist TPPO and is largely reduced in *Trpm5*$^{-/-}$ cells (64). These data suggest that fatty acids can activate taste cells in a manner similar to sweet, bitter, and umami tastes. Interestingly, the preferences for linoleic acid and soybean oil emulsions are lost in *Trpm5*$^{-/-}$ mice (64, 106), although the soybean oil preference could be partially recovered by extensive experience (106).

Finally, very little is known about a possible role for TRPM4 in taste transduction. Interestingly, the channel is expressed in mouse taste cells (64). Furthermore, taste cells seem to have a second type of CAN channels. This current shows many features (including low sensitivity to Ca^{2+}, block by free ATP, and a conductance of ~30 pS) reminiscent of heterologously expressed TRPM4 (47). However, as this current could not be measured in the whole-cell mode by uncaging IP_3 or Ca^{2+} (47), it remains unclear what function TRPM4 might have in taste receptor cells.

3.2. Functional Role for TRPM4 and TRPM5 in Pancreas

Both TRPM4 and TRPM5 have been suggested to influence insulin release from pancreatic beta cells. The secretion of insulin by the pancreatic β cell is a complex process driven by electrical activity and oscillations of the intracellular Ca^{2+} concentration, $[Ca^{2+}]_i$ (107). Indeed, ATP produced by the glucose metabolism closes

ATP-sensitive K^+ channels, allowing depolarization of the membrane potential and activation of L-type Ca^{2+} channels. This leads to Ca^{2+} increase and secretion of insulin-containing vesicles (108–111). As insulin is the only glucose-lowering hormone, its secretion is essential for glucose homeostasis. Although Ca^{2+}-activated non-selective cation channels had been described in beta cells from the pancreas, their contribution to the electrical activity and insulin release was unclear (112–114). With *Trpm4*$^{-/-}$ and *Trpm5*$^{-/-}$ mice available, their possible role in pancreatic beta cells could be analyzed in a more direct way.

The role of TRPM4 in insulin release is quite controversial. Protein expression and channel activity are detected in several beta cell lines such as INS-1, HIT-T15, RINm5F, β-TC3, and MIN-6 (65, 66). Furthermore, expression of TRPM4 protein was detected in mouse pancreatic islets and human beta cells (66). Ca^{2+} increase and insulin secretion of INS-1 cells after stimulation by glucose, AVP (arginine–vasopressin, a Gq-coupled receptor agonist in β-cells), or glyburide (glibenclamide) were decreased when TRPM4 was inhibited by the dominant-negative construct ΔN TRPM4 (65, 66), suggesting that depolarizing currents generated by TRPM4 are an important component in the control of intracellular Ca^{2+} signals necessary for insulin secretion. Furthermore, Ca^{2+}-dependent exocytosis might cause dynamic translocation of TRPM4-containing vesicles to the plasma membrane, a regulatory mechanism that would allow beta cells to regulate the electrical activity in response to glucose and other nutrients (65). However, *Trpm4*$^{-/-}$ mice show no defect in glucose-induced insulin secretion from pancreatic islets or in glucose tolerance after an intraperitoneal injection of glucose (61), suggesting that TRPM4 is not involved in glucose-induced Ca^{2+} signaling and insulin release. Further studies are required to elucidate whether TRPM4 might be involved in G_q- or G_s-receptor-coupled signaling pathways, for example, during stimulation with, e.g., glucagon or GLP-1.

In contrast to TRPM4, the function of TRPM5 in insulin release is more established. The channel is expressed in beta cells from mouse and human pancreatic islets (36, 77). Moreover, TRPM5-channel activity is an important constituent of the Ca^{2+}-activated non-selective monovalent cation channel measured in murine β cells (77). The channel plays an important role in glucose-induced electrical activity: whereas normal WT islets display three types of V_m and Ca^{2+} oscillations (slow, mixed, or fast) in response to glucose, deletion of TRPM5 leads to a specific lack of fast glucose-induced V_m and Ca^{2+} oscillations. TRPM5 contributes to the slow depolarization in the slow interburst interval of the glucose-induced electrical activity, in this way shortening the interburst interval and leading to faster glucose-induced oscillations in V_m and Ca^{2+} (77). Why TRPM5 is only functionally relevant in a (fast-oscillating) subpopulation of the islets remains unclear. Fast Ca^{2+} oscillations are shown to be more efficient than slow oscillations in triggering exocytosis of secretory vesicles and insulin release (115). In agreement with this, two independent laboratories show that glucose-induced insulin release is impaired in isolated pancreatic islets from *Trpm5*$^{-/-}$ mice (77, 78). This beta cell dysfunction leads to lower plasma insulin levels and impaired glucose tolerance after both oral and intraperitoneal glucose challenges (77, 78). The relevance of this prediabetic phenotype during conditions of higher insulin demand (such as pregnancy, obesitas, aging, etc.) remains to be shown. However, it is interesting to note that *Trpm5* expression is negatively correlated with blood glucose concentrations in the small intestine of diabetic patients (116). Moreover, TRPM5 variants are associated with prediabetic phenotypes in subjects at risk for type 2 diabetes. TRPM5 SNPs were related to not only insulin secretion, plasma glucose, and GLP-1 levels but also to insulin sensitivity (117). How these TRPM5 variants might affect insulin sensitivity remains elusive, as *Trpm5*$^{-/-}$ mice show a normal insulin tolerance test (77). Furthermore, the functional impact of these mutations on TRPM5 channel activity has not been clarified yet. However, these data indicate a possible link between TRPM5 and type 2 diabetes mellitus.

Less is known about a possible role for TRPM4 and TRPM5 in the secretion of other hormones by the pancreatic islet. Indeed, next to the insulin-secreting beta cells, the islet of Langerhans comprises several other cell types such as glucagon-releasing α cells and somatostatin-producing δ cells. TRPM4 is expressed in the alpha cell line INR1G9 (66) and TRPM5 is shown to be expressed, albeit to a low level, in alpha cells from murine islets (77). TRPM4 inhibition decreased the magnitude of intracellular Ca^{2+} signals and glucagon secretion in response to several agonists such as AVP and high K^+ in the alpha cell line INR1G9, suggesting a function for this channel in glucagon secretion (118). However, more experiments are required to elucidate the exact function of TRPM4 (and TRPM5) in this process.

4. TRPM4 and TRPM5 In the Central Nervous System

Trpm4 and *Trpm5* mRNA are detected in the central nervous system. RT PCR experiments showed *Trpm4* and *Trpm5* expression in brain and spinal cord extracts from mouse and rat (24, 119, 120). The Allen Brain Atlas confirmed these results in mouse by In Situ *Hybridization* and showed different levels of expression according to the structures: *Trpm4* is expressed at a quite high level in hip pocampus and hypothalamus, and at a reduced level in the olfactory bulb, cerebellum, pons, and cortex. *Trpm5* is highly detectable by *ISH* and using reporter mice in the olfactory bulb and to a lesser extent in the thalamus (75). So far, it seems that *Trpm4* and *Trpm5* expression in brain is mainly in neurons although the protein expression has not been described. This expression pattern might suggest a profound physiological and pathophysiological role for both channels in the CNS. However, the physiological roles of these two channels in neuronal cells remain quite unclear. For instance, although TRPM4 and TRPM5 can be detected by *ISH* in the spinal cord of young and adult mice, until now nothing is known about these channels in this structure. In this section, we will focus on the role of TRPM4 and TRPM5 only in brain neurons.

4.1. TRPM4, TRPM5, and Burst Firing

Some brain neurons present a specific firing behavior called burst firing. This spiking behavior is characterized by a sustained firing activity. Such a burst firing activity is involved in different brain processes like reward circuit, short-term memory in an emotional- and experience-dependent learning context, and respiratory rhythms regulation. CAN currents play a role in brain as a key player of sustained firing activity mechanisms (121). TRPM4 and TRPM5 are considered as the candidate channels underlying CAN currents and might in this way contribute to many brain processes (24, 33).

So far, the most investigated process of burst firing behavior in which TRPM4 plays a role is in pre-Bötzinger Complex (preBötC) neurons (122). The preBötC is involved in the respiratory rythmogenesis (123). These neurons are characterized by an oscillating activity and by the synchronization of their burst firing. Only 20% of these neurons present a pacemaker activity, meaning that most of the neurons generate inspiratory drive potentials by evoking post-synaptic currents that depend on intrinsic membrane properties (124). $I_{(CAN)}$ has been proposed to be the candidate to amplify glutamatergic synaptic drive by transforming the glutamatergic synaptic inputs to membrane depolarization (122, 125). Pace et al. showed that Ca^{2+} influx was able to induce some plateau potentials, and external Na^{+} substitution and FFA exposure attenuated those plateau potentials. $I_{(CAN)}$ activation by glutamatergic inputs could be direct (via NMDAR Ca^{2+} influx) or indirect (via mGluR induced IP_3-dependent Ca^{2+}-release or AMPAR activation of voltage-gated Ca^{2+} channels). Crowder et al. detected *Trpm4* and *Trpm5* expression by RT-PCR in preBötC neurons and showed that excess PIP_2 augmented the inspiratory drive potential. This effect was modulated by FFA application (120). Although in these neurons both *Trpm4* and *Trpm5* are expressed, the electrophysiological characteristics of the $I_{(CAN)}$ are more reminiscent of TRPM4. Indeed, the current can be blocked by Gd^{3+} at 30 μM and the application of a negative pressure in the patch pipette can activate the current, which resembles the behavior of TRPM4 in myocytes. TRPM4 current could be activated by Ca^{2+} waves in the soma and generate inspiratory bursts by boosting glutamatergic synaptic inputs.

More recently, a novel pathway of activation of TRPM4 has been described: the Epac/cAMP pathway. Epac agonist application on preBötC neurons sensitized Ca^{2+} mobilization from IP_3 internal Ca^{2+} stores that stimulated TRPM4 and potentiated bursts of action potentials (50). It remains unclear, however, whether TRPM4 activity itself is regulated by this mechanism.

This mechanism of activation via glutamatergic synaptic inputs and the role of TRPM4/5 in burst firing activity might be conserved in other brain structures. Mrejeru et al. have described a similar mechanism in dopaminergic (DA) neurons of substantia nigra (126). Those neurons present two different behaviors: tonic firing and bursts of action potentials. They showed by electrophysiology that NMDA currents recruit an $I_{(CAN)}$ current capable of generating a plateau potential. This $I_{(CAN)}$ can be blocked by FFA and 9-phenanthrol application. Since mRNA expression of *Trpm2* and *Trpm4* has been detected by RT-PCR (*Trpm5* could not be detected), they hypothesized TRPM4 to be the channel involved in the burst firing behavior. Although TRPM4 current has not been directly recorded in dopaminergic neurons, and the specificity of FFA and 9-phenanthrol on brain slices has not been determined yet, Mrejeru et al. provide the first evidences that TRPM channels (TRPM2 and TRPM4) are expressed in substancia nigra neurons and could be part of the reward circuit by boosting NMDA currents during burst firing.

The neurons of the lateral nucleus of amygdala also display a sustained firing activity. The graded increase in firing is linked to an $I_{(CAN)}$ and is blocked by FFA application (127). The Allen Brain Atlas shows *Trpm4/5* mRNA expression in amygdala, leading to the conclusion that either one or two channels are involved in burst firing activity in the lateral nucleus and then are part of the mechanism for sustaining information about novel items in a short-term memory in the context of emotional- and experience-dependent learning.

Although no direct evidence of TRPM4 or TRPM5 currents are now available, similar process of sustained firing activity dependent on $I_{(CAN)}$ exists in diverse structures such as the motoneurons of the nucleus ambiguous, the layer II neurons of the entorhinal cortex (128), the blanes cells of the olfactory bulbs (75), indicating possible new roles for TRPM4 and TRPM5 in firing behavior in brain physiology.

4.2. TRPM4, TRPM5, and Pathophysiology

TRPM4 is thought to underly the boosting of NMDA current in DA neurons (126). Since those neurons are vulnerable to neurodegeneration, this $I_{(CAN)}$ boost mechanism may also explain the high sensitivity of DA neurons for excitotoxicity. In this case, TRPM4 could be considered as a potential drug target in Parkinson disease.

TRPM4 and TRPM5 might also be key players in epilepsy. An epileptic seizure is composed of recurrent bursts of intense firing. Schiller recorded $I_{(CAN)}$ on neocortex slices treated with bicuculline to induce seizure (129). This current was unaffected by changing chloride concentrations but was sensitive to intracellular Ca^{2+} changes and was blocked by FFA application. This $I_{(CAN)}$ is activated by Ca^{2+} influx through NDMA receptors and voltage-gated Ca^{2+} channels. This is the first direct evidence that $I_{(CAN)}$ is involved in a pathological process. Indeed this current could support sustained seizure-like events (129). Interestingly, the mechanism seems to be similar to what has been described in the preBötC and substantia nigra neurons. Further investigation in *Trpm4*$^{-/-}$ and *Trpm5*$^{-/-}$ mice could improve the understanding of the pathophysiological process leading to epileptic seizure.

Acknowledgments

The authors are supported by the FWO-Vlaanderen en het Bijzonder Onderzoeksfonds (KU Leuven). We thank Prof. Em. Bernd Nilius and all the members of the Laboratory of Ion Channel Research for inspiring discussions.

References

1. Xu XZ, Moebius F, Gill DL, Montell C (2001) Regulation of melastatin, a TRP-related protein, through interaction with a cytoplasmic isoform. Proc Natl Acad Sci USA 98:10692–10697
2. Colquhoun D, Neher E, Reuter H, Stevens CF (1981) Inward current channels activated by intracellular Ca in cultured cardiac cells. Nature 294:752–754
3. Magistretti J, Alonso A (2002) Fine gating properties of channels responsible for persistent sodium current generation in entorhinal cortex neurons. J Gen Physiol 120: 855–873
4. Shalinsky MH, Magistretti J, Ma L, Alonso AA (2002) Muscarinic activation of a cation current and associated current noise in entorhinal-cortex layer-II neurons. J Neurophysiol 88:1197–1211
5. Liman ER (2003) Regulation by voltage and adenine nucleotides of a Ca^{2+}-activated cation channel from hamster vomeronasal sensory neurons. J Physiol 548:777–787
6. El-Sherif Y, Wieraszko A, Banerjee P, Penington NJ (2001) ATP modulates Na^{+} channel gating and induces a non-selective cation current in a neuronal hippocampal cell line. Brain Res 904:307–317
7. Simard JM, Chen M (2004) Regulation by sulfanylurea receptor type 1 of a non-selective cation channel involved in cytotoxic edema of reactive astrocytes. J Neurosurg Anesthesiol 16:98–99
8. Kim CJ, Weir BK, Macdonald RL, Zhang H (1998) Erythrocyte lysate releases Ca^{2+} from IP3-sensitive stores and activates Ca(2+)-dependent K^{+} channels in rat basilar smooth muscle cells. Neurol Res 20:23–30
9. Eto W et al (2003) Intracellular alkalinization induces Ca^{2+} influx via non-voltage-operated Ca^{2+} channels in rat aortic smooth muscle cells. Cell Calcium 34:477–484
10. Miyoshi H, Yamaoka K, Garfield RE, Ohama K (2004) Identification of a non-selective cation channel current in myometrial cells isolated from pregnant rats. Pflugers Arch 447:457–464
11. Sturgess NC, Carrington CA, Hales CN, Ashford ML (1987) Nucleotide-sensitive ion channels in human insulin producing tumour cells. Pflugers Arch 410:169–172
12. Rodighiero S, De Simoni A, Formenti A (2004) The voltage-dependent nonselective cation current in human red blood cells studied by means of whole-cell and nystatin-perforated patch-clamp techniques. Biochim Biophys Acta 1660:164–170

13. Kaestner L, Bernhardt I (2002) Ion channels in the human red blood cell membrane: their further investigation and physiological relevance. Bioelectrochemistry 55:71–74
14. Gogelein H, Pfannmuller B (1989) The nonselective cation channel in the basolateral membrane of rat exocrine pancreas. Inhibition by 3′,5-dichlorodiphenylamine-2-carboxylic acid (DCDPC) and activation by stilbene disulfonates. Pflugers Arch 413:287–298
15. Gray MA, Argent BE (1990) Non-selective cation channel on pancreatic duct cells. Biochim Biophys Acta 1029:33–42
16. Suzuki K, Petersen OH (1988) Patch-clamp study of single-channel and whole-cell K^+ currents in guinea pig pancreatic acinar cells. Am J Physiol 255:G275–G285
17. Ringer E, Russ U, Siemen D (2000) Beta(3)-adrenergic stimulation and insulin inhibition of non-selective cation channels in white adipocytes of the rat. Biochim Biophys Acta 1463:241–253
18. Halonen J, Nedergaard J (2002) Adenosine 5 -monophosphate is a selective inhibitor of the brown adipocyte nonselective cation channel. J Membr Biol 188:183–197
19. Hurwitz CG, Hu VY, Segal AS (2002) A mechanogated nonselective cation channel in proximal tubule that is ATP sensitive. Am J Physiol Renal Physiol 283:F93–F104
20. Van den Abbeele T, Tran Ba Huy P, Teulon J (1996) Modulation by purines of calcium-activated non-selective cation channels in the outer hair cells of the guinea-pig cochlea. J Physiol 494(Pt 1):77–89
21. Suh SH, Watanabe H, Droogmans G, Nilius B (2002) ATP and nitric oxide modulate a Ca(2+)-activated non-selective cation current in macrovascular endothelial cells. Pflugers Arch 444:438–445
22. Watanabe H et al (2002) Heat-evoked activation of TRPV4 channels in a HEK293 cell expression system and in native mouse aorta endothelial cells. J Biol Chem 277: 47044–47051
23. Csanady L, Adam-Vizi V (2003) Ca(2+)- and voltage-dependent gating of Ca(2+)- and ATP-sensitive cationic channels in brain capillary endothelium. Biophys J 85:313–327
24. Launay P et al (2002) TRPM4 is a Ca^{2+}-activated nonselective cation channel mediating cell membrane depolarization. Cell 109: 397–407
25. Nilius B et al (2003) Voltage dependence of the Ca^{2+}-activated cation channel TRPM4. J Biol Chem 278:30813–30820
26. Montell C et al (2002) A unified nomenclature for the superfamily of TRP cation channels. Mol Cell 9:229–231
27. Murakami M et al (2003) Identification and characterization of the murine TRPM4 channel. Biochem Biophys Res Commun 307: 522–528
28. Yarishkin OV et al (2008) Endogenous TRPM4-like channel in Chinese hamster ovary (CHO) cells. Biochem Biophys Res Commun 369:712–717
29. Prawitt D et al (2000) Identification and characterization of MTR1, a novel gene with homology to melastatin (MLSN1) and the trp gene family located in the BWS-WT2 critical region on chromosome 11p15.5 and showing allele-specific expression. Hum Mol Genet 9:203–216
30. Enklaar T et al (2000) Mtr1, a novel biallelically expressed gene in the center of the mouse distal chromosome 7 imprinting cluster, is a member of the Trp gene family. Genomics 67:179–187
31. Perez CA et al (2002) A transient receptor potential channel expressed in taste receptor cells. Nat Neurosci 5:1169–1176
32. Zhang Y et al (2003) Coding of sweet, bitter, and umami tastes: different receptor cells sharing similar signaling pathways. Cell 112:293–301
33. Hofmann T, Chubanov V, Gudermann T, Montell C (2003) TRPM5 is a voltage-modulated and Ca(2+)-activated monovalent selective cation channel. Curr Biol 13:1153–1158
34. Ullrich ND et al (2005) Comparison of functional properties of the Ca^{2+}-activated cation channels TRPM4 and TRPM5 from mice. Cell Calcium 37:267–278
35. Liu D, Liman ER (2003) Intracellular Ca^{2+} and the phospholipid PIP2 regulate the taste transduction ion channel TRPM5. Proc Natl Acad Sci USA 100:15160–15165
36. Prawitt D et al (2003) TRPM5 is a transient Ca^{2+}-activated cation channel responding to rapid changes in $[Ca^{2+}]_i$. Proc Natl Acad Sci USA 100:15166–15171
37. Rohacs T, Lopes CM, Michailidis I, Logothetis DE (2005) PI(4,5)P2 regulates the activation and desensitization of TRPM8 channels through the TRP domain. Nat Neurosci 8:626–634
38. Nilius B et al (2005) Gating of TRP channels: a voltage connection? J Physiol 567:35–44
39. Voets T et al (2004) The principle of temperature-dependent gating in cold- and heat-sensitive TRP channels. Nature 430: 748–754
40. Talavera K et al (2005) Heat activation of TRPM5 underlies thermal sensitivity of sweet taste. Nature 438:1022–1025

41. Talavera K et al (2007) Influence of temperature on taste perception. Cell Mol Life Sci 64:377–381
42. Nilius B et al (2005) The selectivity filter of the cation channel TRPM4. J Biol Chem 280:22899–22906
43. Owsianik G, Talavera K, Voets T, Nilius B (2006) Permeation and selectivity of TRP channels. Annu Rev Physiol 68:685–717
44. Suh BC, Hille B (2005) Regulation of ion channels by phosphatidylinositol 4,5-bisphosphate. Curr Opin Neurobiol 15:370–378
45. Nilius B et al (2006) The Ca^{2+}-activated cation channel TRPM4 is regulated by phosphatidylinositol 4,5-biphosphate. EMBO J 25:467–478
46. Zhang Z, Okawa H, Wang Y, Liman ER (2005) Phosphatidylinositol 4,5-bisphosphate rescues TRPM4 channels from desensitization. J Biol Chem 280:39185–39192
47. Zhang Z, Zhao Z, Margolskee R, Liman E (2007) The transduction channel TRPM5 is gated by intracellular calcium in taste cells. J Neurosci 27:5777–5786
48. Hisatsune C et al (2007) Abnormal taste perception in mice lacking the type 3 inositol 1,4,5-trisphosphate receptor. J Biol Chem 282:37225–37231
49. Gonzales AL, Amberg GC, Earley S (2010) Ca^{2+} release from the sarcoplasmic reticulum is required for sustained TRPM4 activity in cerebral artery smooth muscle cells. Am J Physiol Cell Physiol 299:C279–C288
50. Mironov SL, Skorova EY (2011) Stimulation of bursting in pre-Botzinger neurons by Epac through calcium release and modulation of TRPM4 and K-ATP channels. J Neurochem 117:295–308
51. Nilius B et al (2005) Regulation of the Ca^{2+} sensitivity of the nonselective cation channel TRPM4. J Biol Chem 280:6423–6433
52. Nilius B et al (2004) Decavanadate modulates gating of TRPM4 cation channels. J Physiol 560:753–765
53. Nilius B, Prenen J, Voets T, Droogmans G (2004) Intracellular nucleotides and polyamines inhibit the Ca^{2+}-activated cation channel TRPM4b. Pflugers Arch 448: 70–75
54. Liu D, Zhang Z, Liman ER (2005) Extracellular acid block and acid-enhanced inactivation of the Ca^{2+}-activated cation channel TRPM5 involve residues in the S3-S4 and S5-S6 extracellular domains. J Biol Chem 280:20691–20699
55. Grand T et al (2008) 9-Phenanthrol inhibits human TRPM4 but not TRPM5 cationic channels. Br J Pharmacol 153:1697–1705
56. Gonzales AL, Garcia ZI, Amberg GC, Earley S (2010) Pharmacological inhibition of TRPM4 hyperpolarizes vascular smooth muscle. Am J Physiol Cell Physiol 299: C1195–C1202
57. Palmer RK et al (2010) Triphenylphosphine oxide is a potent and selective inhibitor of the transient receptor potential melastatin-5 ion channel. Assay Drug Dev Technol 8:703–713
58. Oike H et al (2006) Arachidonic acid can function as a signaling modulator by activating the TRPM5 cation channel in taste receptor cells. Biochim Biophys Acta 1761: 1078–1084
59. Fonfria E et al (2006) Tissue distribution profiles of the human TRPM cation channel family. J Recept Signal Transduct Res 26: 159–178
60. Mathar I et al (2010) Increased catecholamine secretion contributes to hypertension in TRPM4-deficient mice. J Clin Invest 120: 3267–3279
61. Vennekens R et al (2007) Increased IgE-dependent mast cell activation and anaphylactic responses in mice lacking the calcium-activated nonselective cation channel TRPM4. Nat Immunol 8:312–320
62. Abed E et al (2009) Expression of transient receptor potential (TRP) channels in human and murine osteoblast-like cells. Mol Membr Biol 26:146–158
63. Yu W, Hill WG, Apodaca G, Zeidel ML (2011) Expression and distribution of transient receptor potential (TRP) channels in bladder epithelium. Am J Physiol Renal Physiol 300:F49–F59
64. Liu P, Shah BP, Croasdell S, Gilbertson TA (2011) Transient receptor potential channel type M5 is essential for fat taste. J Neurosci 31:8634–8642
65. Cheng H et al (2007) TRPM4 controls insulin secretion in pancreatic beta-cells. Cell Calcium 41:51–61
66. Marigo V et al (2009) TRPM4 impacts Ca(2+) signals during agonist-induced insulin secretion in pancreatic beta-cells. Mol Cell Endocrinol 299(2):194–203
67. Guinamard R et al (2004) Functional characterization of a Ca(2+)-activated non-selective cation channel in human atrial cardiomyocytes. J Physiol 558:75–83
68. Kruse M et al (2009) Impaired endocytosis of the ion channel TRPM4 is associated with human progressive familial heart block type I. J Clin Invest 119:2737–2744
69. Guinamard R et al (2006) Functional expression of the TRPM4 cationic current in ventricular cardiomyocytes from spontaneously hypertensive rats. Hypertension 48:587–594

70. Guinamard R, Bois P (2007) Involvement of transient receptor potential proteins in cardiac hypertrophy. Biochim Biophys Acta 1772: 885–894
71. Earley S, Waldron BJ, Brayden JE (2004) Critical role for transient receptor potential channel TRPM4 in myogenic constriction of cerebral arteries. Circ Res 95:922–929
72. Nakamura E et al (2010) New frontiers in gut nutrient sensor research: luminal glutamate-sensing cells in rat gastric mucosa. J Pharmacol Sci 112:13–18
73. Perez CA, Margolskee RF, Kinnamon SC, Ogura T (2003) Making sense with TRP channels: store-operated calcium entry and the ion channel Trpm5 in taste receptor cells. Cell Calcium 33:541–549
74. Kokrashvili Z et al (2009) Release of endogenous opioids from duodenal enteroendocrine cells requires Trpm5. Gastroenterology 137:598–606, 606 e1-2
75. Lin W et al (2007) Olfactory neurons expressing transient receptor potential channel M5 (TRPM5) are involved in sensing semiochemicals. Proc Natl Acad Sci USA 104:2471–2476
76. Kaske S et al (2007) TRPM5, a taste-signaling transient receptor potential ion-channel, is a ubiquitous signaling component in chemosensory cells. BMC Neurosci 8:49
77. Colsoul B et al (2010) Loss of high-frequency glucose-induced Ca^{2+} oscillations in pancreatic islets correlates with impaired glucose tolerance in $Trpm5^{-/-}$ mice. Proc Natl Acad Sci USA 107:5208–5213
78. Brixel LR et al (2010) TRPM5 regulates glucose-stimulated insulin secretion. Pflugers Arch 460:69–76
79. Barbet G et al (2008) The calcium-activated nonselective cation channel TRPM4 is essential for the migration but not the maturation of dendritic cells. Nat Immunol 9:1148–1156
80. Watanabe H et al (2008) TRP channel and cardiovascular disease. Pharmacol Ther 118: 337–351
81. Mills RW, Milan DJ (2010) TRPM4-linked isolated cardiac conduction defects: bad trafficking causes electrical gridlock. Circ Cardiovasc Genet 3:311–313
82. Vassort G, Bideaux P, Alvarez J (2010) Could early ischemic arrhythmia triggered by purinergic activation of the transient receptor potential channels be prevented by creatine? Exp Clin Cardiol 15:e104–e108
83. Liu H et al (2011) Gain-of-function mutations in TRPM4 cause autosomal dominant isolated cardiac conduction disease. Circ Cardiovasc Genet 3:374–385
84. Demion M, Bois P, Launay P, Guinamard R (2007) TRPM4, a Ca^{2+}-activated nonselective cation channel in mouse sino-atrial node cells. Cardiovasc Res 73:531–538
85. Morita H et al (2007) Membrane stretch-induced activation of a TRPM4-like nonselective cation channel in cerebral artery myocytes. J Pharmacol Sci 103:417–426
86. Brayden JE, Earley S, Nelson MT, Reading S (2008) Transient receptor potential (TRP) channels, vascular tone and autoregulation of cerebral blood flow. Clin Exp Pharmacol Physiol 35:1116–1120
87. Gonzales AL, Earley S (2012) Endogenous cytosolic Ca(2+) buffering is necessary for TRPM4 activity in cerebral artery smooth muscle cells. Cell Calcium 51(1):82–93
88. Garcia ZI, Bruhl A, Gonzales AL, Earley S (2011) Basal protein kinase Cdelta activity is brequired for membrane localization and activity of TRPM4 channels in cerebral artery smooth muscle cells. Channels (Austin) 5: 210–214
89. Crnich R et al (2010) Vasoconstriction resulting from dynamic membrane trafficking of TRPM4 in vascular smooth muscle cells. Am J Physiol Cell Physiol 299:C682–C694
90. Earley S, Straub SV, Brayden JE (2007) Protein kinase C regulates vascular myogenic tone through activation of TRPM4. Am J Physiol Heart Circ Physiol 292: H2613–H2622
91. Reading SA, Brayden JE (2007) Central role of TRPM4 channels in cerebral blood flow regulation. Stroke 38:2322–2328
92. Simard C, Salle L, Rouet R, Guinamard R (2012) TRPM4 inhibitor 9-phenanthrol abolishes arrhythmias induced by hypoxia and reoxygenation in mouse ventricle. Br J Pharmacol 165(7):2354–2364
93. Chaudhari N, Roper SD (2010) The cell biology of taste. J Cell Biol 190:285–296
94. Dando R, Roper SD (2009) Cell-to-cell communication in intact taste buds through ATP signalling from pannexin 1 gap junction hemichannels. J Physiol 587:5899–5906
95. Huang YJ et al (2007) The role of pannexin 1 hemichannels in ATP release and cell-cell communication in mouse taste buds. Proc Natl Acad Sci USA 104:6436–6441
96. Damak S et al (2006) Trpm5 null mice respond to bitter, sweet, and umami compounds. Chem Senses 31:253–264
97. Eddy MC et al (2011) A conditioned aversion study of sucrose and SC45647 taste in TRPM5 knockout mice. Chem Senses 37(5): 391–401
98. Oliveira-Maia AJ et al (2009) Nicotine activates TRPM5-dependent and independent

taste pathways. Proc Natl Acad Sci USA 106: 1596–1601
99. Ren X et al (2010) Nutrient selection in the absence of taste receptor signaling. J Neurosci 30:8012–8023
100. Huang YA, Roper SD (2011) Intracellular Ca(2+) and TRPM5-mediated membrane depolarization produce ATP secretion from taste receptor cells. J Physiol 588: 2343–2350
101. Romanov RA, Rogachevskaja OA, Khokhlov AA, Kolesnikov SS (2008) Voltage dependence of ATP secretion in mammalian taste cells. J Gen Physiol 132:731–744
102. Locovei S, Wang J, Dahl G (2006) Activation of pannexin 1 channels by ATP through P2Y receptors and by cytoplasmic calcium. FEBS Lett 580:239–244
103. Talavera K et al (2008) The taste transduction channel TRPM5 is a locus for bitter-sweet taste interactions. FASEB J 22:1343–1355
104. Laugerette F et al (2007) Do we taste fat? Biochimie 89:265–269
105. Laugerette F et al (2005) CD36 involvement in orosensory detection of dietary lipids, spontaneous fat preference, and digestive secretions. J Clin Invest 115:3177–3184
106. Sclafani A, Zukerman S, Glendinning JI, Margolskee RF (2007) Fat and carbohydrate preferences in mice: the contribution of alpha-gustducin and Trpm5 taste-signaling proteins. Am J Physiol Regul Integr Comp Physiol 293:R1504–R1513
107. MacDonald PE, Rorsman P (2006) Oscillations, intercellular coupling, and insulin secretion in pancreatic beta cells. PLoS Biol 4:e49
108. Drews G, Krippeit-Drews P, Dufer M (2010) Electrophysiology of islet cells. Adv Exp Med Biol 654:115–163
109. Henquin JC (2009) Regulation of insulin secretion: a matter of phase control and amplitude modulation. Diabetologia 52:739–751
110. Henquin JC, Nenquin M, Ravier MA, Szollosi A (2009) Shortcomings of current models of glucose-induced insulin secretion. Diabetes Obes Metab 11(Suppl 4):168–179
111. MacDonald PE, Rorsman P (2007) The ins and outs of secretion from pancreatic beta-cells: control of single-vesicle exo- and endocytosis. Physiology (Bethesda) 22:113–121
112. Leech CA, Habener JF (1998) A role for Ca^{2+}-sensitive nonselective cation channels in regulating the membrane potential of pancreatic beta-cells. Diabetes 47:1066–1073
113. Roe MW et al (1998) Characterization of a Ca^{2+} release-activated nonselective cation current regulating membrane potential and $[Ca^{2+}]_i$ oscillations in transgenically derived beta-cells. J Biol Chem 273:10402–10410
114. Sturgess NC, Hales CN, Ashford ML (1987) Calcium and ATP regulate the activity of a non-selective cation channel in a rat insulinoma cell line. Pflugers Arch 409:607–615
115. Berggren PO et al (2004) Removal of Ca^{2+} channel beta3 subunit enhances Ca^{2+} oscillation frequency and insulin exocytosis. Cell 119:273–284
116. Young RL et al (2009) Expression of taste molecules in the upper gastrointestinal tract in humans with and without type 2 diabetes. Gut 58:337–346
117. Ketterer C et al (2011) Genetic variation within the TRPM5 locus associates with prediabetic phenotypes in subjects at increased risk for type 2 diabetes. Metabolism 60(9):1325–1333
118. Nelson PL et al (2011) Regulation of Ca(2+)-entry in pancreatic alpha-cell line by transient receptor potential melastatin 4 plays a vital role in glucagon release. Mol Cell Endocrinol 335:126–134
119. Yoo JC et al (2010) Cloning and characterization of rat transient receptor potential-melastatin 4 (TRPM4). Biochem Biophys Res Commun 391:806–811
120. Crowder EA et al (2007) Phosphatidylinositol 4,5-bisphosphate regulates inspiratory burst activity in the neonatal mouse preBotzinger complex. J Physiol 582:1047–1058
121. Rubin JE, Hayes JA, Mendenhall JL, Del Negro CA (2009) Calcium-activated nonspecific cation current and synaptic depression promote network-dependent burst oscillations. Proc Natl Acad Sci USA 106: 2939–2944
122. Pace RW, Mackay DD, Feldman JL, Del Negro CA (2007) Inspiratory bursts in the preBotzinger complex depend on a calcium-activated non-specific cation current linked to glutamate receptors in neonatal mice. J Physiol 582:113–125
123. Feldman JL, Del Negro CA (2006) Looking for inspiration: new perspectives on respiratory rhythm. Nat Rev Neurosci 7:232–242
124. Del Negro CA et al (2005) Sodium and calcium current-mediated pacemaker neurons and respiratory rhythm generation. J Neurosci 25:446–453
125. Mironov SL (2008) Metabotropic glutamate receptors activate dendritic calcium waves and TRPM channels which drive rhythmic respiratory patterns in mice. J Physiol 586: 2277–2291
126. Mrejeru A, Wei A, Ramirez JM (2011) Calcium-activated non-selective cation currents are involved in generation of tonic and

bursting activity in dopamine neurons of the substantia nigra pars compacta. J Physiol 589:2497–2514

127. Egorov AV, Unsicker K, von Bohlen und Halbach O (2006) Muscarinic control of graded persistent activity in lateral amygdala neurons. Eur J Neurosci 24:3183–3194
128. Zhu ZT, Munhall A, Shen KZ, Johnson SW (2004) Calcium-dependent subthreshold oscillations determine bursting activity induced by N-methyl-D-aspartate in rat subthalamic neurons in vitro. Eur J Neurosci 19:1296–1304
129. Schiller Y (2004) Activation of a calcium-activated cation current during epileptiform discharges and its possible role in sustaining seizure-like events in neocortical slices. J Neurophysiol 92:862–872

Chapter 7

The Emerging Role of TRPM7 in the Regulation of Magnesium Homeostasis

Vladimir Chubanov, Jonathan T. Eggenschwiler, Lillia V. Ryazanova, Thomas Gudermann, and Alexey G. Ryazanov

Abstract

TRPM7 and its closely related protein TRPM6 are unique bifunctional molecules consisting of a kinase domain fused to an ion channel. The ion channel belongs to the transient receptor potential (TRP) family of ion channels, TRPM subfamily, while the kinase domain belongs to a family of atypical protein kinases called alpha kinases that are known to phosphorylate substrates within alpha helices. Recent evidence suggests that TRPM6 and TRPM7 are involved in the regulation of cellular and whole body Mg^{2+} balance. Mutations in human TRPM6 were found to cause hypomagnesemia. Deficiency in TRPM7 inhibits cell proliferation, and this growth defect can be rescued by supplementation with high concentrations of Mg^{2+}. In this review, we discuss the role of TRPM6 and TRPM7 in the regulation of Mg^{2+} homeostasis. We present evidence that TRPM7 is involved in the regulation of Mg^{2+} influx into proliferating cells, while the TRPM7–TRPM6 complex is involved in the regulation of whole body Mg^{2+} balance.

Key words: TRPM7, TRPM6, Channel kinases, TRP channels, Alpha kinases, Mg^{2+} homeostasis, Cell proliferation

1. Introduction

Until recently, little was known about the mechanisms that control magnesium homeostasis. An advance in our understanding of the regulation of cellular and whole body Mg^{2+} balance came from the discovery of the channel kinase TRPM7 and its genetically related protein, TRPM6 (Fig. 1). TRPM7 and TRPM6 are unusual bifunctional molecules that have a unique structure; they are ion channels fused to a protein kinase. Among known ion channels, only TRPM6 and TRPM7 are covalently coupled to kinase

Arpad Szallasi and Tamás Bíró (eds.), *TRP Channels in Drug Discovery: Volume I*, Methods in Pharmacology and Toxicology, DOI 10.1007/978-1-62703-077-9_7,

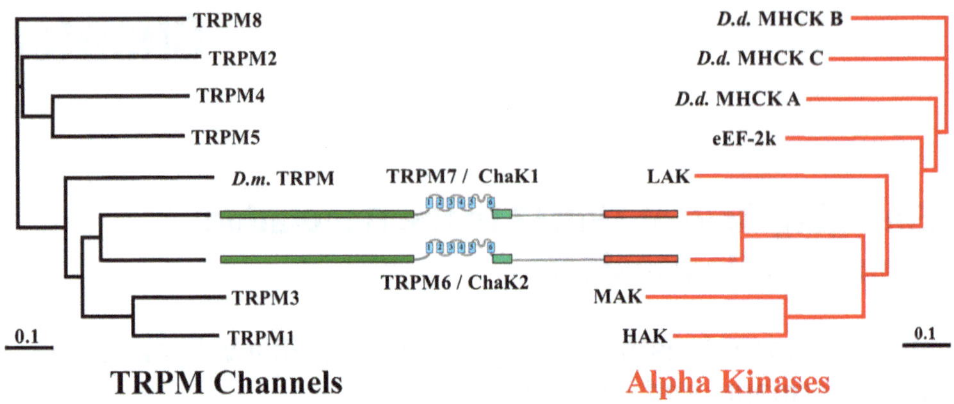

Fig. 1. Phylogenetic trees of TRPM channels and alpha kinases, including TRPM7 and TRPM6.

domains (1–6). The ion channel portion shows high homology to other melastatin-related TRP (TRPM) channels (reviewed in (2, 7–9)). The kinase domains of TRPM6 and TRPM7 are classified as alpha kinases (Fig. 1), a family of atypical protein kinases that do not display homology to conventional eukaryotic protein kinases (6, 10).

Recent evidence suggests that TRPM6 and TRPM7 channels regulate Mg^{2+} homeostasis in vertebrates. The first indication that channel kinases were involved in the regulation of Mg^{2+} homeostasis came from the observation that TRPM7 can function as a Mg^{2+}-permeable channel regulated by intracellular Mg^{2+} and Mg-ATP (1). It was also found that TRPM7-deficient lymphoid cells cannot grow, but can be rescued by supplementation with Mg^{2+} (11). Further evidence came from the identification of mutations in TRPM6 that are linked to hypomagnesemia with secondary hypocalcemia (HSH), a disorder characterized by very low levels of Mg^{2+} in serum (3, 4). HSH manifests itself in early infancy with generalized convulsions and increased neuromuscular excitability leading to muscle spasms. Another important finding was the demonstration that TRPM6 cannot efficiently form functional channels by itself and assembles with TRPM7 to form heteromeric channel complexes on the cell surface (5). These observations suggest that TRPM7, which is ubiquitously expressed, is involved in the regulation of cellular Mg^{2+} homeostasis, while TRPM6, expressed predominantly in the intestine and kidney epithelium, is involved in the regulation of systemic Mg^{2+} balance.

The role of TRPM6 was extensively overviewed in a number of recent articles (8, 9, 12, 13). The scope of this chapter is to discuss recent findings supporting a view that TRPM7 plays a crucial role in the regulation of cellular and systemic Mg^{2+} balance.

2. The Importance of Intracellular and Whole Body Magnesium Balance

Mg^{2+} is the second-most abundant intracellular cation and is essential for a normal cellular physiology. In healthy subjects, serum Mg^{2+} concentrations are maintained in a narrow range (0.7–1.1 mM). Less than 1% of total body Mg^{2+} is in circulation, 50–60% is stored in the bones and the rest is found in intracellular compartments (14). Mg^{2+} homeostasis is tightly balanced by absorption in the intestine and renal excretion. Diminished dietary Mg^{2+} uptake or excessive renal excretion leads to a depletion of Mg^{2+} from the bones and soft tissues, reduction of serum Mg^{2+} levels (hypomagnesemia), and secondary changes in metabolism of Ca^{2+}, K^{+} and inorganic phosphate (14). The clinical importance of Mg^{2+} was underscored by numerous studies which have shown that hypomagnesemia is a frequent finding in patients subjected to drug therapy (diuretics, aminoglycosides, immuno-suppressive, and antitumor drugs) (15–19). In addition, a growing number of common disorders (e.g., diabetes, osteoporosis, chronic kidney disease, asthma, several tumor types) are associated with a misbalance in Mg^{2+} (15, 20–24). Severe forms of acquired or inherited hypomagnesaemia induce secondary defects in the cardiovascular system, motor activity and behavior, and can be fatal if not properly treated (12, 14, 16, 25, 26).

3. Functional Characterization of the TRPM7 Channel

TRPM7 has been functionally characterized in heterologous expression systems where it behaves as a constitutively active channel that is highly permeable to divalent cations such as Mg^{2+} and Ca^{2+} (1). Remarkably, endogenous TRPM7-like currents have been detected in all cell types examined. The TRPM7 channel pore is believed to be composed of a homo-oligomeric tetramer (27, 28). TRPM7 elicits very small (few pA) inward currents at physiological negative membrane potentials and large outward currents (up to several nA) at positive potentials. It has been shown that the channel activity of TRPM7 is regulated by Mg^{2+}, Mg-ATP, PIP2, pH, cAMP, polyamines, osmolarity and mechanical stress (1, 11, 29–38). While the physiological relevance for these modulations is not yet understood, current evidence indicates that the sensitivity of TRPM7 to Mg^{2+} is central for its role in vivo (39, 40). External Mg^{2+} was found to be a permanent blocker of the TRPM7 channel pore (1, 7, 41). Several recent studies (7, 41) showed that a short amino acid stretch of the predicted pore-loop forms an ion "selectivity filter" in TRPM7. In particular, E1047 and Y1049 residues of mouse TRPM7 were found to be essential for the high divalent cation permeability of the channel. Of note, neutralization of the negatively charged E1047 (E1047Q mutation) is sufficient to eliminate the permeation Mg^{2+} block suggesting that E1047 interacts directly with ionized Mg^{2+} (7, 41).

The TRPM7 channel is inhibited by intracellular Mg^{2+} and Mg-ATP in a negative feedback loop (1). Internal Mg^{2+} and Mg-ATP suppress TRPM7 channel activity through a nucleotide-binding site located in the kinase catalytic domain synergistically with another Mg^{2+} binding site(s) (11, 32). Recent studies (39) with TRPM7 gene-modified mice suggest that the kinase/Mg^{2+}/Mg-ATP-interdependent modulations of TRPM7 currents play an important role in vivo (discussed below). However, exact molecular mechanisms underlying the sensitivity of TRPM7 to Mg^{2+} and Mg-ATP remain to be determined.

4. Pharmacological Inhibition of TRPM7 Channel

A temporal and reversible pharmacological targeting of either kinase or channel domains of TRPM7 may be indispensable in overcoming of the current experimental limitations. However, chemical compounds inhibiting the TRPM7 kinase are not yet produced. Until recently, only a few synthetic modulators of TRPM7 channel were available (Table 1), including a subset of unspecific ion channel blockers like 2-aminoethoxydiphenyl borate (2-APB) (42–48). Recently, efforts of independent groups resulted in a subset of TRPM7 inhibitors such as carvacrol, NDGA, AA861, MK886, Nafamostat, Waixenicin A, Quinine, CyPPA, Dequalinium, NS8593, SKA31, UCL 1684 and NS8593 (Table 1). Interestingly, unlike other TRPM7 inhibitors (Table 1), NS8593 and Waixenicin A act on TRPM7 in a Mg^{2+}-dependent manner (42–48) which may be instrumental in studying mechanisms of TRPM7 sensitivity to intracellular Mg^{2+}. Thus, a pharmacological targeting of TRPM7 channels is feasible, and future studies testing the experimental and therapeutic potential of drugs targeting TRPM7 are awaited.

5. Functional Characterization of the TRPM7 Kinase

TRPM7, and its close relative TRPM6, were initially discovered and cloned while searching for proteins containing kinase domains similar to eukaryotic elongation factor 2 kinase, or eEF2K (Fig. 1) (6, 49, 50). Five mammalian proteins containing eEF2K-like kinase domains were identified (6). These kinases were unique because they displayed no homology to conventional protein kinases (6, 51). Furthermore, unlike conventional protein kinases that phosphorylate

Table 1
Compounds inhibiting TRPM7

Inhibitor	IC_{50} (μM)[a]	Description of effects	References
2-APB	178	Reversible	Prakriya et al. (43), Li et al. (42)
Carvacrol	306	Reversible	Parnas et al. (44)
NDGA	n.d.	Tested only at 10 and 20 μM	Chen et al. (45)
AA861	n.d.	Tested only at 10 and 40 μM	Chen et al. (45)
MK886	n.d.	Tested only at 10 μM	Chen et al. (45)
Quinine	n.d.	Reversible, tested only at 30 μM	Chubanov et al. (46)
CyPPA	n.d.	Reversible, tested only at 30 μM	Chubanov et al. (46)
Dequalinium	n.d.	Reversible, tested only at 30 μM	Chubanov et al. (46)
SKA31	n.d.	Reversible, tested only at 30 μM	Chubanov et al. (46)
UCL 1684	n.d.	Reversible, tested only at 30 μM	Chubanov et al. (46)
Nafamostat	617	Reversible, voltage and extracellular divalents dependent	Chen et al. (47)
Waixenicin A	7.0	Irreversible, $[Mg^{2+}]_i$ dependent	Zierler et al. (48)
NS8593	1.6	Reversible, $[Mg^{2+}]_i$ dependent	Chubanov et al. (46)

[a]IC_{50} values were determined for recombinant TRPM7 in patch-clamp measurements in the absence of internal Mg^{2+}

substrates within β-turns, loops or irregular structures (52), the existing evidence suggests that these enzymes phosphorylate amino acids located within alpha-helices; therefore, they were named "Alpha kinases" (10, 53). Since their initial discovery, a large number of alpha kinases have been identified in various phylogenetic groups and are particularly prevalent in unicellular protozoans (54).

A comparison of the X-ray crystal structures of murine TRPM7 and *Dictyostelium* Myosin Heavy Chain Kinase A (MHCKA) alpha kinase domains demonstrated that the architecture of alpha kinase domains is highly conserved (55, 56). Most of the invariant residues identified in this domain are either directly involved in catalysis or associate with catalytic sites (54). Several conserved residues in TRPM7's active site, K1646, D1775, D1765, and Q1767, are important for Mg-ATP binding (56). One equivalent residue in the active site of MHCKA, D765, was found to be phosphorylated suggesting that the alpha kinase catalytic mechanism differs from that of conventional eukaryotic protein kinases (55). This residue may serve as a phosphotransfer intermediate in a novel two-step catalytic mechanism that is in contrast to the direct phosphate transfer used by conventional protein kinases.

TRPM7 kinase activity may also regulate downstream pathways by phosphorylating substrate proteins. Annexin 1 was the first known substrate of TRPM7 (57). TRPM7 phosphorylates annexin A1 at Ser5 (57), a modification identified by several groups in different cell lines (58–60). Recent data indicate that the phosphorylation of annexin A1 at Ser5 disrupts the ability of its N-terminal alpha-helix to interact with membranes and the S100A11 protein (61). Since there are many studies suggesting that annexin A1 has anti-proliferative activity (62–66) and is downregulated in many cancers (67–71), the phosphorylation of annexin A1 by TRPM7 may promote cell proliferation.

The amino acid residues preferred at each position surrounding the TRPM7 phosphorylation site were determined using a comprehensive peptide library screening approach (72). The TRPM7 phosphorylation motif is characterized by preference for hydrophobic residues at +1 position, aromatic residues or Met at the −2 and +2 positions, as well as selectivity for basic residues at the −1, +3, and +4 positions. This information was utilized to predict other TRPM7 protein substrates. The predicted TRPM7 substrates include membrane proteins and G-protein coupled receptors such as TRPM2, opsin 4, G-protein coupled receptor GPR31, type-1 angiotensin II receptor (AT1), and α-2C adrenergic receptor.

In addition, the kinase domains of TRPM7 and TRPM6 have been shown to cross-phosphorylate subunits within the homo/hetero-tetramer (5, 9). The functional consequences of substrate phosphorylation and cross-phosphorylation have yet to be determined.

Accumulating evidence suggests that the TRPM7 kinase domain may play a role in Mg^{2+}-sensitivity of the TRPM7 channel. While the kinase domain of TRPM7 is dispensable for channel function, loss of kinase activity decreases TRPM7's sensitivity to inhibition by Mg^{2+} and Mg-ATP (11). Deletion of the kinase domain enhances the sensitivity of the channel to inhibition by Mg^{2+}. It is still unclear how the kinase domain directs TRPM7's capacity to be regulated by intracellular Mg^{2+}.

6. Role of TRPM7 in the Regulation of Cell Cycle Progression

Early studies by Rubin et al. set out to determine which components of conventional tissue culture media were indispensable for cell proliferation. Aside from serum, the only constituent found to be essential was Mg^{2+} (73, 74). Most intracellular Mg^{2+} is bound to proteins, phospholipids, nucleic acids, and nucleotides like ATP and GTP. As a cell grows and divides, it must double these components as well as Mg^{2+} content (75, 76). This explains why the levels of intracellular Mg^{2+} are is so closely tied to the ability of cells to proliferate. Further studies investigating specific requirements for Mg^{2+}

during the cell cycle revealed that Mg^{2+} is vital throughout G1, a period where all cellular constituents are doubled (77). While Mg^{2+} restriction inhibits cell proliferation, supplementation with high levels of Mg^{2+} has been shown to promote growth (78).

Recent findings suggest that TRPM7 plays an important role in regulating magnesium homeostasis during cell cycle progression (2, 39). B lymphoma cells lacking TRPM7 are incapable of proliferating in the presence of physiological concentrations of Mg^{2+} (1 mM) in the culture medium. Supplementation of growth media with high Mg^{2+} (10–15 mM) can restore their proliferative capacity (11). Follow-up studies revealed that TRPM7 is also required for proliferation of osteoblasts, epithelial, and endothelial cells (39, 79–81).

TRPM7 was found to be upregulated in several cancer types, including breast and pancreatic adenocarcinoma (82, 83) suggesting that TRPM7 may be essential for tumor growth. Consistent with this, depletion of TRPM7 halted the growth of breast and pancreatic adenocarcinoma cells and induced senescence (82, 83).

One important target of Mg^{2+} in cell growth is thought to be the PI3K/AKT/mTOR pathway (40, 84, 85). Andrew Scharenberg's laboratory showed that B lymphocytes lacking TRPM7 arrest in the G0 phase of the cell cycle and cease to accumulate cell mass when cultured with normal levels of Mg^{2+} (11, 86). This is paralleled by a down-regulation of PI3K–AKT signaling and accumulation of the CDK inhibitor p27kip1 (86, 87). Reconstitution of TRPM7-deficient B lymphocytes with wild-type TRPM7 restored their proliferative capacity, as did exogenous expression of the Mg^{2+} transporter SLC41A1/A2, MagT1, or Mg^{2+} supplementation (88).

Deletion of the TRPM7 gene or the kinase domain of TRPM7 both result in embryonic lethality in mice (39). Embryonic stem cells isolated from $TRPM7^{\Delta kinase/\Delta kinase}$ or $TRPM7^{-/-}$ blastocysts fail to proliferate at physiological concentrations of Mg^{2+} (39). The growth arrest phenotype of $TRPM7^{-/-}$ ES cells can be completely rescued with high levels of Mg^{2+}. These data suggest that the regulation of intracellular Mg^{2+} concentrations is dependent upon TRPM7's channel domain. However, it remains unclear whether the alpha kinase domain of TRPM7 is involved in the regulation of cell cycle progression.

7. In Vivo Role of TRPM7 in Animal Models

The roles of TRPM7 and TRPM7-related invertebrate channels in the living organism have been investigated by a number of laboratories using mouse, zebrafish, *Xenopus*, *C. elegans*, and *Drosophila*.

Trpm7 null mutant mice die during early development, arresting during gastrulation stages (embryonic day 6.5–7.5; (89)) and conditional mutagenesis indicated that TRPM7 is required within the embryo proper. Mouse mutants lacking the kinase region of TRPM7 (*Trpm7*$^{\Delta kinase/\Delta kinase}$) also die during embryonic development (39). Such mutants arrest shortly after e7.5 and, although they execute early gastrulation events, they are noticeably smaller than control embryos by e7.5. The reason for the arrest of *Trpm7*$^{-/-}$ and *Trpm7*$^{\Delta kinase/\Delta kinase}$ mutants remains to be investigated.

We found that Mg^{2+} levels are reduced in the plasma, bone, and urine of *Trpm7*$^{\Delta kinase/+}$ heterozygote mice. Importantly, a substantial fraction of *Trpm7*$^{\Delta kinase/+}$ animals died shortly after placing them on a Mg^{2+}-deficient diet, whereas wild-type control mice were unaffected. These Trpm7$^{\Delta kinase/+}$ mice showed behavioral defects indicative of Mg^{2+}-deficiency (clasping, tremors, and seizures) on a Mg^{2+}-deficient diet. Collectively, these data indicate that an important aspect of mammalian TRPM7 function is a regulation of Mg^{2+} homeostasis.

Using conditional mutagenesis, the authors found that TRPM7 is differentially required depending on tissue developmental stage. Jin et al. (89) reported that a conditional disruption of TRPM7 in developing thymocytes and mature T cells suppressed thymocyte differentiation and reduced the number of T cells without affecting intracellular Mg^{2+} levels (89). Jin et al. suggested that TRPM7 is not involved in Mg^{2+} homeostasis in T cells. However, it is unclear whether these experiments were performed on proliferating or non-cycling T cells. As discussed above, numerous studies have established that proliferating cells are critically dependent from Mg^{2+} uptake unlike to non-proliferating cells capable to function with excising reserves of intracellular Mg^{2+} which was estimated up to 17–20 mM (90). Accordingly, TRPM7 may be functionally redundant in non-cycling cells such as resting T cells.

TRPM7 also appears to regulate Mg^{2+} homeostasis in zebrafish and *Xenopus*. Different mutations in zebrafish *Trpm7* have been identified (91–93). Such mutants undergo early morphogenesis normally but exhibit a number of defects including loss of touch responsiveness, defective melanin synthesis and apoptotic death of melanophores, defective cell cycle progression in the exocrine pancreas epithelium, and lethality in late larval life (91–93). Zebrafish *Trpm7* mutants exhibit reduced total levels of both Mg^{2+} and Ca^{2+} (94), and addition of supplemental Mg^{2+}, but not Ca^{2+}, partially rescued melanophore survival and proliferation in the exocrine pancreas (93, 95). Knock-down of Xenopus TRPM7 transcripts revealed that the channel, together with the non-canonical Wnt pathway, regulates Rac-mediated cell polarity and migration during gastrulation (96). Importantly, the TRPM7-deficient phenotypes of Xenopus can be suppressed by increasing the levels of Mg^{2+} (but not Ca^{2+}) or by overexpressing the Mg^{2+} transporter SLC41A2. This work indicates that TRPM7-mediated influx of Mg^{2+} plays an important role in directing polarized cell migration in Xenopus. This work suggests that TRPM7-mediated influx of Mg^{2+} plays an important role in the development of a wide range of vertebrates.

The roles of TRPM7-related proteins have also been studies in invertebrate species. *Drosophila* harbors a single TRPM family member which, in contrast to mammalian TRPM6 or TRPM7, does not comprise an enzyme domain. *Drosophila trpm* is expressed in the equivalent of mammalian kidneys, the Malpighian tubules. *Trpm* mutants develop slowly as larvae and arrest as prepupae with morphological defects in the Malpighian tubules (97). Unexpectedly, *trpm*-deficient larvae show increased Mg^{2+} levels in the body when raised on Mg^{2+}-enriched diets, suggesting that *Drosophila* TRPM controls homeostatic removal Mg^{2+} from the hemolymph (blood) by the Malpighian tubules.

In *C. elegans*, there are three TRPM channels. Two of these, GON-2 and GTL-1, are expressed in the intestine and regulate Mg^{2+} homeostasis, while the third, GTL-2, acts within the excretory cell to mediate the excretion of excess Mg^{2+} (98, 99). Interestingly, *gon-2/gtl-1* double mutant worms show diminished body Mg^{2+} and display a growth defect, which is suppressed by increasing dietary Mg^{2+}, but not Ca^{2+} (98).

Collectively, the data support an idea that TRPM7 plays a role in Mg^{2+} uptake and this function is evolutionarily conserved.

8. Current Model for TRPM7 and the Regulation of Mg^{2+} Homeostasis

Based on the aforementioned findings, the following model for a cellular role of TRPM7 can be suggested. During cell growth and division a substantial amount of additional Mg^{2+} is required for synthesis of new macromolecules and metabolites resulting in reduced levels of free Mg^{2+}. This, in turn, will cause activation of TRPM7 channel and influx of Mg^{2+} from the extracellular space. Once Mg^{2+} levels are re-adjusted, TRPM7 will be inactivated. This Mg^{2+}-dependent feedback loop ensures that the optimal level of intracellular Mg^{2+} is maintained by TRPM7.

9. Concluding Remarks

Recent experiments with cultured cells and studies with genetically modified mice have revealed that the channel kinase TRPM7 plays essential role in both cellular and systemic Mg^{2+} homeostasis.

A role of the kinase in Mg^{2+} balance is less understood. Several TRPM7 substrates have been identified in vitro. Whether TRPM7 targets these substrates in vivo has not yet been determined.

References

1. Nadler MJ et al (2001) LTRPC7 is a Mg.ATP-regulated divalent cation channel required for cell viability. Nature 411:590–595
2. Schlingmann KP et al (2007) TRPM6 and TRPM7–Gatekeepers of human magnesium metabolism. Biochim Biophys Acta 1772: 813–821
3. Schlingmann KP et al (2002) Hypomagnesemia with secondary hypocalcemia is caused by mutations in TRPM6, a new member of the TRPM gene family. Nat Genet 31:166–170
4. Walder RY et al (2002) Mutation of TRPM6 causes familial hypomagnesemia with secondary hypocalcemia. Nat Genet 31:171–174
5. Chubanov V et al (2004) Disruption of TRPM6/TRPM7 complex formation by a mutation in the TRPM6 gene causes hypomagnesemia with secondary hypocalcemia. Proc Natl Acad Sci USA 101:2894–2899
6. Ryazanov AG (2002) Elongation factor-2 kinase and its newly discovered relatives. FEBS Lett 514:26–29
7. Mederos y Schnitzler M et al (2008) Evolutionary determinants of divergent calcium selectivity of TRPM channels. FASEB J 22:1540–1551
8. Chubanov V, Gudermann T, Schlingmann KP (2005) Essential role for TRPM6 in epithelial magnesium transport and body magnesium homeostasis. Pflugers Arch 451:228–234
9. Chubanov V et al (2005) Emerging roles of TRPM6/TRPM7 channel kinase signal transduction complexes. Naunyn Schmiedebergs Arch Pharmacol 371:334–341
10. Ryazanov AG, Pavur KS, Dorovkov MV (1999) Alpha-kinases: a new class of protein kinases with a novel catalytic domain. Curr Biol 9:R43–R45
11. Schmitz C et al (2003) Regulation of vertebrate cellular Mg^{2+} homeostasis by TRPM7. Cell 114:191–200
12. Quamme GA (1997) Renal magnesium handling: new insights in understanding old problems. Kidney Int 52:1180–1195
13. Hsu YJ, Hoenderop JG, Bindels RJ (2007) TRP channels in kidney disease. Biochim Biophys Acta 1772:928–936
14. Konrad M, Schlingmann KP, Gudermann T (2004) Insights into the molecular nature of magnesium homeostasis. Am J Physiol Renal Physiol 286:F599–F605
15. Alexander RT, Hoenderop JG, Bindels RJ (2008) Molecular determinants of magnesium homeostasis: insights from human disease. J Am Soc Nephrol 19:1451–1458
16. Regolisti G et al (2010) Severe hypomagnesemia during long-term treatment with a proton pump inhibitor. Am J Kidney Dis 56:168–174
17. Cundy T, Mackay J (2011) Proton pump inhibitors and severe hypomagnesaemia. Curr Opin Gastroenterol 27:180–185
18. Buckley JE et al (1984) Hypomagnesemia after cisplatin combination chemotherapy. Arch Intern Med 144:2347–2348
19. Petrelli F et al (2011) Risk of anti-EGFR monoclonal antibody-related hypomagnesemia: systematic review and pooled analysis of randomized studies. Expert Opin Drug Saf 11(Suppl 1):S9–S19
20. Spiegel DM (2011) Magnesium in chronic kidney disease: unanswered questions. Blood Purif 31:172–176
21. Barbagallo M, Belvedere M, Dominguez LJ (2009) Magnesium homeostasis and aging. Magnes Res 22:235–246
22. Wolf FI et al (2009) Hypomagnesaemia in oncologic patients: to treat or not to treat? Magnes Res 22:5–9
23. Rodriguez-Moran M et al (2011) The role of magnesium in type 2 diabetes: a brief based-clinical review. Magnes Res 24:156–162
24. Rude RK, Singer FR, Gruber HE (2009) Skeletal and hormonal effects of magnesium deficiency. J Am Coll Nutr 28:131–141
25. Touyz RM (2004) Magnesium in clinical medicine. Front Biosci 9:1278–1293
26. Kramer JH et al (2009) Neurogenic inflammation and cardiac dysfunction due to hypomagnesemia. Am J Med Sci 338:22–27
27. Fujiwara Y, Minor DL Jr (2008) X-ray crystal structure of a TRPM assembly domain reveals an antiparallel four-stranded coiled-coil. J Mol Biol 383:854–870
28. Schmitz C et al (2005) The channel kinases TRPM6 and TRPM7 are functionally nonredundant. J Biol Chem 280:37763–37771
29. Runnels LW, Yue L, Clapham DE (2002) The TRPM7 channel is inactivated by PIP(2) hydrolysis. Nat Cell Biol 4:329–336
30. Jiang J, Li M, Yue L (2005) Potentiation of TRPM7 inward currents by protons. J Gen Physiol 126:137–150
31. Bessac BF, Fleig A (2007) TRPM7 channel is sensitive to osmotic gradients in human kidney cells. J Physiol 582:1073–1086
32. Demeuse P, Penner R, Fleig A (2006) TRPM7 channel is regulated by magnesium nucleotides via its kinase domain. J Gen Physiol 127:421–434

33. Takezawa R et al (2004) Receptor-mediated regulation of the TRPM7 channel through its endogenous protein kinase domain. Proc Natl Acad Sci USA 101:6009–6014
34. Numata T, Shimizu T, Okada Y (2007) Direct mechano-stress sensitivity of TRPM7 channel. Cell Physiol Biochem 19:1–8
35. Numata T, Shimizu T, Okada Y (2007) TRPM7 is a stretch- and swelling-activated cation channel involved in volume regulation in human epithelial cells. Am J Physiol Cell Physiol 292:C460–C467
36. Kozak JA, Kerschbaum HH, Cahalan MD (2002) Distinct properties of CRAC and MIC channels in RBL cells. J Gen Physiol 120:221–235
37. Oancea E, Wolfe JT, Clapham DE (2006) Functional TRPM7 channels accumulate at the plasma membrane in response to fluid flow. Circ Res 98:245–253
38. Wei C et al (2009) Calcium flickers steer cell migration. Nature 457:901–905
39. Ryazanova LV et al (2010) TRPM7 is essential for Mg($^{2+}$) homeostasis in mammals. Nat Commun 1:109
40. Rubin H (2005) The membrane, magnesium, mitosis (MMM) model of cell proliferation control. Magnes Res 18:268–274
41. Li M et al (2007) Molecular determinants of Mg^{2+} and Ca^{2+} permeability and pH sensitivity in TRPM6 and TRPM7. J Biol Chem 282: 25817–25830
42. Li M, Jiang J, Yue L (2006) Functional characterization of homo- and heteromeric channel kinases TRPM6 and TRPM7. J Gen Physiol 127:525–537
43. Prakriya M, Lewis RS (2002) Separation and characterization of currents through store-operated CRAC channels and Mg^{2+}-inhibited cation (MIC) channels. J Gen Physiol 119 :487–507
44. Parnas M et al (2009) Membrane lipid modulations remove divalent open channel block from TRP-like and NMDA channels. J Neurosci 29:2371–2383
45. Chen X et al (2010) The modulation of TRPM7 currents by nafamostat mesilate depends directly upon extracellular concentrations of divalent cations. Mol Brain 3:38
46. Chubanov V et al (2012) Natural and synthetic modulators of SK (K(ca)2) potassium channels inhibit magnesium-dependent activity of the kinase-coupled cation channel TRPM7. Br J Pharmacol 166(4):1357–1376
47. Chen HC et al (2010) Blockade of TRPM7 channel activity and cell death by inhibitors of 5-lipoxygenase. PLoS One 5:e11161
48. Zierler S et al (2011) Waixenicin A inhibits cell proliferation through magnesium-dependent block of transient receptor potential melastatin 7 (TRPM7) channels. J Biol Chem 286: 39328–39335
49. Riazanova LV et al (2001) Novel type of signaling molecules: protein kinases covalently linked to ion channels. Mol Biol (Mosk) 35: 321–332
50. Ryazanova LV et al (2004) Characterization of the protein kinase activity of TRPM7/ChaK1, a protein kinase fused to the transient receptor potential ion channel. J Biol Chem 279: 3708–3716
51. Ryazanov AG et al (1997) Identification of a new class of protein kinases represented by eukaryotic elongation factor-2 kinase. Proc Natl Acad Sci USA 94:4884–4889
52. Pinna LA, Ruzzene M (1996) How do protein kinases recognize their substrates? Biochim Biophys Acta 1314:191–225
53. Enkemann SA et al (1999) Does prothymosin alpha affect the phosphorylation of elongation factor 2? J Biol Chem 274:18644–18650
54. Drennan D, Ryazanov AG (2004) Alpha-kinases: analysis of the family and comparison with conventional protein kinases. Prog Biophys Mol Biol 85:1–32
55. Ye Q et al (2010) Crystal structure of the alpha-kinase domain of Dictyostelium myosin heavy chain kinase A. Sci Signal 3:ra17
56. Yamaguchi H et al (2001) Crystal structure of the atypical protein kinase domain of a TRP channel with phosphotransferase activity. Mol Cell 7:1047–1057
57. Dorovkov MV, Ryazanov AG (2004) Phosphorylation of annexin I by TRPM7 channel-kinase. J Biol Chem 279:50643–50646
58. Wang W, Creutz CE (1992) Regulation of the chromaffin granule aggregating activity of annexin I by phosphorylation. Biochemistry 31:9934–9939
59. Olsen JV et al (2010) Quantitative phosphoproteomics reveals widespread full phosphorylation site occupancy during mitosis. Sci Signal 3:ra3
60. Rigbolt KT et al (2011) System-wide temporal characterization of the proteome and phosphoproteome of human embryonic stem cell differentiation. Sci Signal 4:rs3
61. Dorovkov MV, Kostyukova AS, Ryazanov AG (2011) Phosphorylation of annexin A1 by TRPM7 kinase: a switch regulating the induction of an alpha-helix. Biochemistry 50: 2187–2193
62. Croxtall J, Flower R, Perretti M (1993) The role of lipocortin 1 in the regulation of A549

cell proliferation and leukocyte migration. J Lipid Mediat 6:295–302
63. Alldridge LC et al (1999) The annexin protein lipocortin 1 regulates the MAPK/ERK pathway. J Biol Chem 274:37620–37628
64. Croxtall JD et al (1993) N-terminal peptide fragments of lipocortin-1 inhibit A549 cell growth and block EGF-induced stimulation of proliferation. Int J Cancer 54:153–158
65. Alldridge LC, Bryant CE (2003) Annexin 1 regulates cell proliferation by disruption of cell morphology and inhibition of cyclin D1 expression through sustained activation of the ERK1/2 MAPK signal. Exp Cell Res 290: 93–107
66. Lim LH, Pervaiz S (2007) Annexin 1: the new face of an old molecule. FASEB J 21: 968–975
67. Zhang L et al (2009) Decreased expression of Annexin A1 correlates with pathologic differentiation grade in oral squamous cell carcinoma. J Oral Pathol Med 38:362–370
68. Petrella et al (2006) Annexin-1 downregulation in thyroid cancer correlates to the degree of tumor differentiation. Cancer Biol Ther 5:643–647
69. Luthra R et al (2008) MicroRNA-196a targets annexin A1: a microRNA-mediated mechanism of annexin A1 downregulation in cancers. Oncogene 27:6667–6678
70. Wang D et al (2010) Annexin-1 downregulation is associated with clinical outcome in Chinese patients with hilar cholangiocarcinoma. Eur Surg Res 45:151–157
71. Wang LP et al (2010) Annexin A1 expression and its prognostic significance in human breast cancer. Neoplasma 57:253–259
72. Hutti JE et al (2004) A rapid method for determining protein kinase phosphorylation specificity. Nat Methods 1:27–29
73. Rubin H, Koide T (1976) Mutual potentiation by magnesium and calcium of growth in animal cells. Proc Natl Acad Sci USA 73:168–172
74. Rubin H (1975) Central role for magnesium in coordinate control of metabolism and growth in animal cells. Proc Natl Acad Sci USA 72:3551–3555
75. Rubin H (2005) Magnesium: the missing element in molecular views of cell proliferation control. Bioessays 27:311–320
76. Wolf FI (2004) TRPM7: channeling the future of cellular magnesium homeostasis? Sci STKE 2004:pe23
77. Hazelton B, Mitchell B, Tupper J (1979) Calcium, magnesium, and growth control in the WI-38 human fibroblast cell. J Cell Biol 83:487–498
78. Zhang A, Altura BT, Altura BM (1997) Elevation of extracellular magnesium rapidly raises intracellular free Mg^{2+} in human aortic endothelial cells: is extracellular Mg^{2+} a regulatory cation? Front Biosci 2:a13–a17
79. Abed E, Moreau R (2007) Importance of melastatin-like transient receptor potential 7 and cations (magnesium, calcium) in human osteoblast-like cell proliferation. Cell Prolif 40:849–865
80. Abed E, Moreau R (2009) Importance of melastatin-like transient receptor potential 7 and magnesium in the stimulation of osteoblast proliferation and migration by platelet-derived growth factor. Am J Physiol Cell Physiol 297:C360–C368
81. Baldoli E, Maier JA (2011) Silencing TRPM7 mimics the effects of magnesium deficiency in human microvascular endothelial cells. Angiogenesis 15(1):47–57
82. Guilbert A et al (2009) Evidence that TRPM7 is required for breast cancer cell proliferation. Am J Physiol Cell Physiol 297:C493–C502
83. Yee NS, Zhou W, Lee M et al (2012) Transient receptor potential ion channel TRPM7 is required for preventing replicative senescence in pancreatic adenocarcinoma cells. Cancer Lett 318(1):99–105
84. Krueger A et al (2001) Cellular FLICE-inhibitory protein splice variants inhibit different steps of caspase-8 activation at the CD95 death-inducing signaling complex. J Biol Chem 276:20633–20640
85. Rubin H (2005) Central roles of Mg^{2+} and MgATP2- in the regulation of protein synthesis and cell proliferation: significance for neoplastic transformation. Adv Cancer Res 93:1–58
86. Sahni J et al (2010) TRPM7 regulates quiescent/proliferative metabolic transitions in lymphocytes. Cell Cycle 9:3565–3574
87. Sahni J, Scharenberg AM (2008) TRPM7 ion channels are required for sustained phosphoinositide 3-kinase signaling in lymphocytes. Cell Metab 8:84–93
88. Deason-Towne F, Perraud AL, Schmitz C (2011) The Mg^{2+} transporter MagT1 partially rescues cell growth and Mg^{2+} uptake in cells lacking the channel-kinase TRPM7. FEBS Lett 585:2275–2278
89. Jin J et al (2008) Deletion of Trpm7 disrupts embryonic development and thymopoiesis without altering Mg^{2+} homeostasis. Science 322:756–760
90. Romani AM (2011) Cellular magnesium homeostasis. Arch Biochem Biophys 512:1–23
91. Elizondo MR et al (2005) Defective skeletogenesis with kidney stone formation in dwarf

zebrafish mutant for trpm7. Curr Biol 15: 667–671

92. Low SE et al (2011) TRPM7 is required within zebrafish sensory neurons for the activation of touch-evoked escape behaviors. J Neurosci 31:11633–11644
93. Yee NS, Zhou W, Liang IC (2011) Transient receptor potential ion channel Trpm7 regulates exocrine pancreatic epithelial proliferation by Mg^{2+}-sensitive Socs3a signaling in development and cancer. Dis Model Mech 4:240–254
94. Elizondo MR, Budi EH, Parichy DM (2010) trpm7 regulation of in vivo cation homeostasis and kidney function involves stanniocalcin 1 and fgf23. Endocrinology 151:5700–5709
95. McNeill MS et al (2007) Cell death of melanophores in zebrafish trpm7 mutant embryos depends on melanin synthesis. J Invest Dermatol 127:2020–2030
96. Liu W et al (2011) TRPM7 regulates gastrulation during vertebrate embryogenesis. Dev Biol 350:348–357
97. Hoffmann AA (2010) Physiological climatic limits in Drosophila: patterns and implications. J Exp Biol 213:870–880
98. Teramoto T, Lambie EJ, Iwasaki K (2005) Differential regulation of TRPM channels governs electrolyte homeostasis in the *C. elegans* intestine. Cell Metab 1:343–354
99. Teramoto T et al (2010) Magnesium excretion in *C. elegans* requires the activity of the GTL-2 TRPM channel. PLoS One 5:e9589

Chapter 8

TRPM8 Channels as Potential Therapeutic Targets for Pain, Analgesia, and Thermoregulation

David D. McKemy

Abstract

The perception of temperature is critical in not only acute responses to changes in the environment but also fundamental in regulating homeostatic mechanisms like core body temperature. The somatosensory system detects changes in ambient temperature via the coordinated efforts of thermosensory nerves expressing temperature-sensitive members of the TRP family of ion channels. These channels respond over a wide range of temperatures that covers the entire perceived spectrum from comfortably warm to painfully hot and from pleasingly cool to excruciatingly cold. Many of these channels are receptors for natural products that elicit distinct psychophysical sensations, such as the heat associated with capsaicin and the cold felt with menthol, the latter influential in the discovery of TRPM8, the first TRP channel shown to be responsive to cold temperatures. TRPM8, a member of the melastatin TRP channel subfamily, is a receptor for a number of cold mimetics such as menthol and icilin, and is activated by temperatures that range from innocuous cool (26–15°C) to noxious cold (<15°C). In vivo, TRPM8 is the predominant mammalian cold sensor and is involved in most, if not all, aspects of cold thermal transduction, including the transduction of innocuous cool and noxious cold, hypersensitivity to cold caused by inflammation, nerve injury or chemotherapeutic toxicity, regulation of core body temperature, and provides the analgesic effect produced by cold or chemical cooling compounds. This chapter describes the physiological roles attributed to TRPM8 and highlights some potential uses of both TRPM8 antagonists and agonists in the treatment of pain and metabolic homeostasis.

Key words: TRPM8, Cold, Pain, Temperature, Allodynia, Menthol, Analgesia, Ion channel, Somatosensation, Nociception

1. Introduction

Our perception of cold is initiated when the skin is cooled by as little as 1°C from the customary skin temperature of 32°C. This is generally a pleasant sensation, but once temperatures approach 10°C, the perception of cold pain is felt, with qualities richly

Arpad Szallasi and Tamás Bíró (eds.), *TRP Channels in Drug Discovery: Volume I*, Methods in Pharmacology and Toxicology, DOI 10.1007/978-1-62703-077-9_8,

described as sharp, stinging, throbbing, aching, and pricking. Indeed, when subjects are asked to provide such verbal descriptors for the sensation of a noxious cold stimulus versus that of noxious heat, a wider range of words are used for the former (1). For example, only "burning" was used to describe heat pain by more than half of the subjects tested, whereas cold pain prompted at least half of the subjects to use at least four descriptor words to describe the sensation (1). Thus, the perception of cold generates a diverse, and perhaps more ambiguous, range of sensations compared to heat.

In addition, the temporal aspect of cold pain is remarkably distinct from that of heat. Subjects exposed to either noxious heat or cold report that the pain is felt even after the stimulus is removed, presumably due to the time needed for subcutaneous skin to return to basal temperatures. However, this delay in pain cessation is significantly different between the two stimuli with the perception of heat pain ending after about 5 s, whereas cold pain lasts for up to 14 s after stimulus removal (1). Temporal cessation of innocuous warmth and cool is similar to that of noxious heat, observations that has led to the postulate that nociceptive cold nerve endings are dispersed deeper in the skin compared to those of other thermosensitive nerve fibers.

At the level of the afferent nerve, a subset of both Aδ- and C-fibers from either the dorsal root ganglion (DRG) or trigeminal ganglia (TG) respond to cold temperatures with thermal thresholds for activation ranging from below 30°C to near freezing. Cold fibers account for only 15% of all somatosensory afferents, but within this population there is significant diversity in the types of neurons that respond to cold as well as an expansive range of cold activation thresholds. Most cold-sensitive neurons are also sensitive to the cooling compound menthol, a cyclic terpene alcohol found in mint leaves (2). It has been well appreciated since antiquity that moderate concentrations of menthol induce a pleasant cool sensation, such as that felt when using menthol-containing products. Indeed, menthol is intertwined in mythological lore and named after the naiad Menthe, a fresh water nymph who was beloved by Hades yet was transformed into a plant by Persephone, Hades wife, in a fit of jealous rage (3). Hades was still attracted to Menthe so gave the plant a delightful smell that becomes more robust when crumbled or ground, much in the manner Persephone was to have treated Menthe. This of course is the story behind the plant we know as mint or peppermint and the nymph's name gives the origin of the word menthol.

This lore highlights the pleasant aspect of menthol on our senses, but when present at higher doses menthol can be noxious, causing burning, irritation, and pain (4). In influential studies conducted by Hensel and Zotterman in the middle of the twentieth century, it was shown for the first time that menthol likely elicits a sensation of "cool" by shifting the threshold temperature for activation of lingual cold fibers towards warmer temperatures (5). Furthermore, if the temperature of the tongue was maintained at supraphysiological temperatures near 40°C, menthol-induced impulses were not observed, thereby suggesting that menthol stimulation was temperature dependent. Indeed, they hypothesized that menthol exerted its actions on "an enzyme" that was involved in the activation of these nerves by cold (5). Consistent with this postulate, topical application of menthol to the skin of healthy human subjects evokes cold hypersensitivity by sensitizing cold-sensitive C-fibers in a manner that mimics cold allodynia (pain to stimuli that are normally felt as pleasant or innocuous) observed in patients with neuropathies of various etiologies, and this approach has become a suitable model for human studies into cold hypersensitivity (6, 7). Indeed, in a patient with a small-fiber neuropathy whose principal complaint was cold allodynia, responsiveness of C-fibers to menthol and cold was significantly enhanced (8).

At the cellular level, a small fraction of neurons from either the DRG or TG respond to cold temperatures with thresholds for activation initiated below 30°C. Such stimuli evoke a robust calcium influx (9) similar to that caused by menthol (10). Indeed there is a strong correlation between menthol and cold sensitivity in vitro (10–12). Electrophysiological recordings first showed that cooling generates an inward cation current when DRG neurons are held at negative membrane potentials, with an average temperature threshold near 29°C (13). This threshold shifted to warmer temperatures when the recordings were conducted in the presence of menthol, as was predicted in Hensel and Zotterman's original hypothesis (5). Similar responses are observed in cultured TG neurons and both menthol and cold evoke rapidly activating, nonselective cation conductances that are characterized by strong outward rectification (10). Thus, menthol has been most useful in identifying and describing mechanisms of cold sensing in both acute and pathological settings. Moreover, this ubiquitous cold mimetic was critical in identifying the predominant cold sensor in mammals, TRPM8 (10, 14).

2. The Cloning of TRPM8

Two independent groups using different experimental approaches concurrently cloned TRPM8 from sensory afferents, identifying it as a cold- and menthol-sensitive ion channel (10, 14). The first strategy used menthol-induced Ca^{2+} influx in heterologous cells as a functional readout to expression clone a complementary DNA (cDNA; a synthesized copy of an RNA transcript) for a cold receptor from rat TG neurons (10). The second strategy searched for

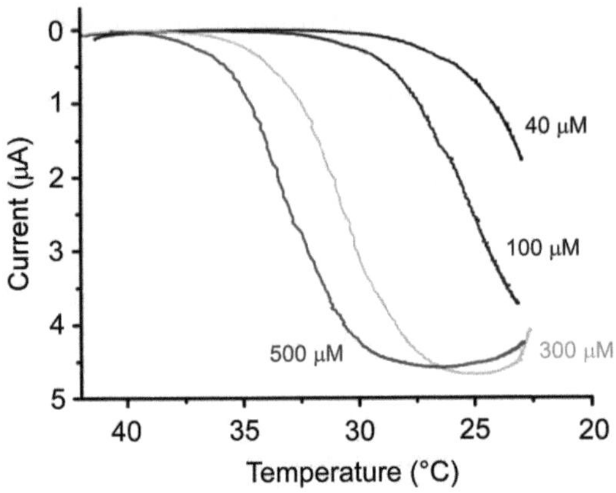

Fig. 1. Menthol shifts the temperature sensitivity of TRPM8. In a heterologous cell (*Xenopus oocyte*) expressing recombinant rat TRPM8 channels, inclusion of menthol in the bath solutions causes a leftward shift in the temperature–response profile of cold-evoked TRPM8 currents.

TRP channel-like transcripts expressed in mouse DRG neurons and tested these recombinant channels for temperature sensitivity (14). With these divergent approaches, both groups simultaneously identified TRPM8 (also referred to as trp-p8 or CMR1), a member of the melastatin channel subfamily (15). Of note, TRPM8 had previously been identified as a transcriptional marker of prostate epithelia cancer cells (16).

In addition to menthol, a number of natural and synthetic chemicals have been shown to activate TRPM8 (17), including the natural cooling product in eucalyptus, eucalyptol, as well as the synthetic chemical icilin (AG-3-5), the latter a much more potent TRPM8 agonist that was originally described as a "super-cooling" compound (10, 18). A number of menthol-derived compounds also activate TRPM8, with WS-12 being the most potent known agonist to date (19). However, very little is known of endogenous TRPM8 activators with only lysophospholipids produced by the activity of calcium-independent phospholipases (iPLA2) shown to potentiate TRPM8-mediated responses (20, 21). Regardless of the ligand, the effect of chemical activators of TRPM8 is to shift the temperature gating curve of the channel so that it is more likely to open at warmer temperatures (see Fig. 1). Thus, TRPM8 ligands can be considered as channel modulators, shifting temperature sensitivity of the channel with the end result of channel activation still dependent upon the as yet unknown temperature-dependent gating mechanism of this and other thermosensitive TRP ion channels.

Biophysically, cold- and menthol-evoked TRPM8 currents have surprisingly similar properties to those recorded in both cultured DRG and TG neurons using similar experimental paradigms.

These include ion selectivity, menthol potency, and voltage dependence of membrane currents induced by either cold or menthol (10). Like almost all TRP channels, recombinant TRPM8 is a nonselective cation channel that displays strong outward rectification and has relatively high selectivity for calcium and little selectivity among monovalent cations (10, 14). More remarkably, TRPM8 currents are also evoked by temperature decreases with an activation temperature threshold of ~26°C in heterologous cells, with activity increasing somewhat linearly in magnitude down to 8°C. The physical mechanism of temperature sensing by TRPM8, as well as other thermosensitive TRP channels, is unclear. These channels are weakly voltage dependent (10, 22, 23), and temperature has been shown to affect channel open probability in response to voltage changes, whereas a change in the channel's ability to sense voltage affects its thermal gating (24–26). For instance, when positively charged residues in the fourth transmembrane domain (S4) and the S4–S5 linker domain of the channel are neutralized, the number of gating charges was reduced, suggesting localization of a voltage sensor (26). However, several studies have shown that temperature-, agonist-, and voltage-dependent gating work allosterically as distinct activation domains for each have been identified (24, 27–30). For example, phosphatidylinositol 4,5-bisphosphate (PIP_2)- and phospholipase C (PLC)-mediated adaptation leads to a change in the voltage dependence of the channel but does not alter thermal sensitivity or menthol dose–response relationships (30). This, as well as recent evidence of the dissociation of thermal and voltage gating in TRPV1, suggests that the mechanisms of activation of TRPM8 by cold and by voltage, although related, are separate processes (25, 30, 31).

Interestingly, the broad range of temperatures that activate TRPM8 spans what are considered both innocuous cool (~30–15°C) and noxious cold (<15°C) in animal behavioral studies and human psychophysics (1, 32). Nonetheless, the relatively warm activation threshold of recombinant TRPM8 channels was suggested to preclude the channel from mediating responses to noxious cold temperatures. However, care must be taken in such assumptions when the context of channel expression in vivo is considered. As with all neuronal environmental sensors, the channel's active site is predominantly the peripheral nerve terminal, in this case in cutaneous structures such as the skin, and surface skin temperatures are considerably colder than those recorded subcutaneously. For example, Morin and Bushnell correlated subcutaneous skin temperatures (recorded at a site 1 mm below the skin surface near the epidermal–dermal boundary) in humans and found that when a subfreezing temperature of −5°C was applied to the skin for 30 s, subcutaneous temperatures were reduced to only ~17°C (1). Thus, in order for temperatures near the nerve terminal to reach the activation threshold for TRPM8 (~26°C), surface skin temperatures need to be near that considered noxious cold. When considered in this context, the ranges of temperatures that activate TRPM8 in vitro are more reflective of a noxious cold sensor than one detecting innocuous cool.

In both TG and DRG, TRPM8 is expressed in <15% of small-diameter sensory neurons, consistent with the proportion of neurons shown to be cold and menthol sensitive in neuronal cultures (10, 14, 33–35). TRPM8 transcripts are more abundant in trigeminal versus dorsal root ganglia, and those expressing this channel have a diverse neurochemical phenotype, suggesting that distinct subsets of TRPM8 neurons may mediate different aspects of cold sensing (10, 14, 33, 34). A cohort of mouse TRPM8 neurons are both Aδ and C-fibers based on co-expression with neurofilament 200 (NF200) and peripherin, respectively, as well as in the location of their peripheral terminals in structures such as the tooth (34). Thus, first and second cold pain sensations are likely dependent on TRPM8 activation (36). Moreover, a fraction of TRPM8 neurons also express the nociceptive markers TRPV1 and calcitonin gene-related peptide (CGRP), further supporting a role for this channel in cold nociception (33, 34). Thus, TRPM8 defines a small and discrete population of sensory afferents that innervate tissues known to be highly sensitive to cold and nociceptive stimuli.

3. Physiological Roles of TRPM8

3.1. Innocuous or Noxious Cold

As stated above, TRPM8 becomes active at innocuous temperatures below 26°C in vitro, and the steep temperature dependence of TRPM8 currents extends its activity into the noxious range, reported to begin at temperatures lower than 15°C in psychophysical and animal studies (1, 32, 37). Thus, it was conjectured whether TRPM8 mediates detection of innocuous cool, noxious cold, or both. Insights into these and other questions regarding the role of this channel in vivo came to light in 2007 when three independent groups reported on the phenotype of genetically modified mice lacking functional TRPM8 channels (12, 38, 39). These elegant studies not only provided the first evidence that TRPM8 is the principle sensor of cold temperatures in mammals but also highlighted some of the difficulties in evaluating animal behavior to cold thermal stimuli (40).

One of the characteristic difficulties in analyzing cold-evoked animal behaviors using traditional methods, such as the thermal plate, is that rodents seem little bothered by cold in comparison to the robust responses generated by noxious heat. In the standard paw withdrawal assay, where measurements such as number of paw lifts or latency to lifts are recorded, previous reports have found rodents to be poorly responsive. For example, in rats placed on a surface chilled to 5°C, latencies of longer than 1 min have been observed (32, 41). Moreover, there is significant variability in these behaviors between research studies. In the analyses of control mice in the three TRPM8 knockout studies, the time to paw withdrawal at near freezing temperatures ranged from 5 to 50 s (5, 20, and 50 s) between the three studies (12, 38, 39). Moreover, in cold plate assays, Colburn et al. found that TRPM8-nulls exhibit longer paw withdrawal latencies at 0°C than their wild-type littermates (38), whereas no differences in behavioral responses were observed in the two other mouse lines under similar experimental conditions (12, 39). These significant differences in animal behavior highlight the difficulty of these assays and demonstrate the need for additional experimental paradigms.

One informative behavioral assay used to assess cold behaviors is acetone evaporative cooling in which nocifensive licking and flinching behaviors are recorded in response to a droplet of acetone applied to the hindpaw (12, 38, 39, 42). Skin surface temperature drops to approximately 17°C and evokes a robust, yet transient, nocifensive response that is diminished in TRPM8$^{-/-}$ mice. Another novel paradigm is the use of the classical cold plate assay with lightly restrained mice, a method allowing for easier measurements of paw latency that removes the burden of weight distribution which seems to complicate traditional cold plate assays (43, 44). Additionally, this assay reduces whole body exposure to cold and subsequent reduction in mobility as seen in the cold plate assay. However, the acts of habituating the animals, particularly mice, to being restrained can be problematic. Using this assay, Gentry et al. found that TRPM8$^{-/-}$ animals had significantly longer withdrawal latencies than wild-type mice when their hindpaws were placed on a 10°C plate (WT = ~15 s, TRPM8-KO = ~29 s) (43).

Perhaps the most robust and commonly used assay for cold sensitivity is the two-temperature plate test, an experimental system used to analyze all three TRPM8-null lines (40). In these assays, animals are presented with a choice between two surfaces that are independently thermally controlled with one held typically near 30°C, the preferred thermal clime of wild-type mice (45, 46) and the other test surface set to a range of temperatures. When both surfaces are set to the same temperature, mice will explore and spend equal time on both surfaces. However, once the temperature of one surface is shifted to colder temperatures (or hot), wild-type mice show a preference for the 30°C side. In the case of mice lacking functional TRPM8 channels, there are robust deficits in their ability to show preference for the warmer side, yet discrepancies emerged as to the exact temperature ranges affected (40). Bautista et al. showed that TRPM8-nulls display no preference until temperatures near 10°C, where they do increase their time on the 30°C side, but not to the levels exhibited by wild-type mice (12). However, Colburn et al. found that their line of TRPM8-nulls were incapable of discriminating temperatures below 5°C (38).

It has yet to be determined if these differences are due to methodology or genetic background, but this highlighted one important caveat for the preference assay. Although two mouse lines lacking functional TRPM8 were shown to spend more time on the plate held at 30°C than the one held at 10°C or lower (12, 39, 47), it is not clear if this was due to a drive to avoid an unpleasant stimulus, or an equally compelling attraction towards a comfortably warm environment. The latter input is likely vital for thermoregulatory mechanisms involved in the maintenance of body temperature and the absence of this input from warm fibers when the animals are standing on the cold surface would promote them to actively seek a warmer environment. Therefore, while TRPM8$^{-/-}$ mice are unable to detect the noxious cold signal, they are still attracted to the warm surface, thereby showing a preference for this surface in the likely absence of any cold inputs. To support this hypothesis, avoidance behaviors were assessed in the two-temperature plate test, finding that when the number of times an animal crosses from the 30°C surface to the test surface and back again is quantified, wild-type mice show a precipitous drop in the number of crossing events as one plate is cooled (47). At noxious cold temperatures (<15°C), wild-type animals will essentially sample the cold surface once and then largely never return, whereas TRPM8-nulls continue to sample the cold surface without aversion, even at a highly noxious temperature of 5°C (47). Thus, even when TRPM8$^{-/-}$ mice are showing a strong preference for the warmer surface over one held at a noxious cold temperature, they display no aversion to the cold and continue to sample it without reservation. These and other data strongly suggest that TRPM8 has the unique property amongst thermally gated TRP channels in that it is necessary for the perception of both innocuous and noxious stimuli.

3.2. Injury-Induced Cold Hypersensitivity

Injury greatly impacts thermosensation, resulting in heightened sensitivity to temperatures already perceived as painful (hyperalgesia) and the novel perception of pain to stimuli that are normally felt as pleasant or innocuous (allodynia). Indeed, menthol is commonly used as a mechanism to mimic cold allodynia in human subjects (6, 48). Thus, in light of the established role of TRPM8 in acute cold pain, one outstanding question regarding TRPM8 is whether or not the channels serve a role in cold-evoked hypersensitivity under different pathological conditions (37). To address this, researchers have used several models to assess the involvement of TRPM8 in injury-evoked hypersensitivity to cold, including peripheral inflammation, neuropathic nerve injury, and toxicity associated with chemotherapeutics (49). Under each experimental setting, animals (rodents and humans) become increasingly sensitive to cold, and analyses of TRPM8-null mice have clearly shown that these behaviors are dependent on TRPM8. Peripheral inflammation is classically studied in rodents after unilateral intraplantar injection of complete Freund's adjuvant (CFA) into the hindpaw and an assessment of any resulting change in sensitivity that usually occurs within the first 24-h postinjection. Using the acetone evaporative cooling assay, several groups have shown that mice become increasingly sensitive to cold within the first day of the injection, a condition that persists for several days but is dependent on expression of TRPM8 channels (32, 38, 42). Similarly, cold neuropathies induced by irritation of the sciatic nerve using the chronic constriction injury model (CCI) (50), or after systemic injection of the platinum-based chemotherapeutic oxaliplatin (51), a key drug in the treatment of colorectal cancer, have also been shown to be TRPM8 dependent (38, 42, 52, 53). Lastly, pharmacological antagonism of TRPM8 channels can alleviate cold allodynia in both inflammatory and neuropathic pain models (42). Taken together, these results suggest that TRPM8 may mediate the majority of cold allodynia and hyperalgesia. However, how injury alters cold sensitivity at either the cellular or molecular level, via TRPM8, is currently unclear.

3.3. Analgesia by Cooling

In light of the role of TRPM8 in cold hypersensitivity, an interesting paradox has emerged which further implicates the channel in a diverse range of physiological functions: TRPM8 is also involved in cooling-mediated analgesia. Cold packs and cooling compounds, such as menthol, have long been used for their analgesic properties in treatment of both acute and chronic pain symptoms (4, 54). Under most conditions, cold clearly is helpful in reducing pain by limiting the extent of inflammation after acute injury as well as quelling the hyperactivity of nociceptive nerve fibers. However, there is intriguing data suggesting that activation of TRPM8 may lead to pain relief under certain established acute and chronic pain conditions. The first indication that TRPM8 may serve this role arose when Proudfoot et al. demonstrated that topical application of cold or cooling compounds produces a temporary analgesic effect mediated by TRPM8-expressing afferents in the CCI and CFA injury models (55). It was found that treating injured rats with cool temperatures or either topical or intrathecal menthol or icilin largely abolished both thermal (heat) and mechanical hypersensitivity in both the inflammatory and neuropathic pain models. The analgesia persisted for over 30 min after which the animals regained their hypersensitivity at levels similar to before the cool stimuli and cold mimetics were applied. It is important to note that only modest cooling (≥16°C) and low doses of cooling compounds produced analgesia, and when more robust stimuli were applied to the injured hindpaws, the animals showed cold sensitization. More strikingly, when TRPM8 expression was reduced by molecular methods, this form of analgesia was abolished. An independent study in TRPM8$^{-/-}$ mice further confirmed a role for TRPM8 in cooling-mediated analgesia, in this case in the formalin pain test which evokes acute pain followed by inflammation (39). When formalin was injected into the hindpaws of wild-type mice, cooling to 17°C produced a marked decrease in pain behaviors (licking and lifting hindpaws) during the acute pain phase (39). However, TRPM8$^{-/-}$ mice did not behave similarly in that they continued to show nocifensive responses even after exposed to the cool surface and were indistinguishable from wild-type animals that were not exposed to cool temperatures. Together, these data indicate that TRPM8 is mediating the analgesia provided by cool temperatures and cooling compounds, suggesting that modest activation of TRPM8 afferent nerves can serve as an endogenous mechanism to promote pain relief.

3.4. Regulation of Core Body Temperature

In addition to the roles of TRPM8 in somatosensation and nociception, the channel is also involved in thermoregulation and homeostatic mechanisms. The channel's involvement in thermoregulation would not be entirely unexpected in light of evidence that other temperature-sensitive ion channels, particularly TRPV1, play a role in regulating body temperature as well (56–59). Moreover, intravenous administration of icilin induces a robust thermogenic type behavior characterized as "wet dog shakes" (18) as well as induces a significant increase in core body temperature, mimicking exposure to cold, effects that are absent in TRPM8$^{-/-}$ animals (39, 42, 60, 61). The TRPM8 dependence of the icilin-induced changes in core temperature is not due to a general deficiency in the ability of these animals to respond to chemical thermal-mimetics, as TRPM8-nulls do mount an appropriate hypothermic response to capsaicin injections (42).

In addition to TRP channel agonists' leading to changes in thermoregulatory behaviors, antagonists also produce reciprocal alterations in core temperatures. Blockade of TRPV1 channels induces a robust and reproducible hyperthermic response, suggesting that the absence of input from heat fibers induced thermoregulatory mechanisms to generate heat (58, 59). Similarly, pharmacological blockade of TRPM8 channels leads to a dramatic hypothermic response in wild-type mice but has no effect in TRPM8-nulls (42). Thus, emerging evidence strongly supports a fundamental role of thermosensory afferent neurons in controlling thermoregulatory mechanisms via the activity of temperature-sensitive TRP ion channels.

4. Regulation of TRPM8 Activity

4.1. Lipid Regulation of TRPM8 Activity

Temperature sensation is dynamic and like other sensory systems can easily adapt to different stimuli, a process observed in both psychophysical and cellular assays. Similarly, cold- or menthol-induced TRPM8 currents adapt or desensitize in a calcium-dependent manner during prolonged stimulation (10, 30). In addition to Ca^{2+} sensitivity, TRPM8 adaptation is also temperature dependent, meaning less adaptation is observed at colder temperatures, and that channel activity can be recovered, but only when the cell is returned to temperatures near that of skin (~32°C) (10, 30). These observations suggest that a Ca^{2+}- and temperature-dependent process mediates adaptation at the cellular and channel level. TRPM8 activity is also highly sensitive to plasmalemmal levels of the PIP_2 (30, 62, 63). These observations led to the hypothesis that adaptation occurs as a result of TRPM8-mediated Ca^{2+} entry leading to activation of PLC, thereby promoting breakdown of PIP_2, a hypothesis confirmed when it was shown that Ca^{2+}- and receptor-independent activation of PLC, in addition to PLC-independent reductions in PIP_2 levels, leads to TRPM8 adaptation (30, 63). Adaption does not alter either menthol or cold sensitivity of the channel but shifts the voltage dependence of TRPM8 currents indicative of a change in channel gating (30). At the molecular level, a number of amino acid residues have been identified in the carboxy-terminal domain of the channel, adjacent to the sixth transmembrane domain, that appear to be involved in PIP_2's effects on TRPM8 (63). Interestingly, these residues are near the highly conserved TRP box of the channel and are found in other PIP_2-sensitive TRPM channels including TRPM4 and TRPM5 (64).

In addition to phosphoinositides, lysophospholipids, and polyunsaturated fatty acids, products of phospholipase A2 (PLA2) have also been found to modulate TRPM8 channel activity (20, 21). Inhibition of the Ca^{2+}-independent isoform of PLA2 (iPLA2) largely abolished cold and icilin responses in cultured DRG neurons but surprisingly only slightly diminished channel activation by menthol (65). PLA2 enzymes cleave glycerophospholipids to generate equimolar amounts of lysophospholipids and polyunsaturated fatty acids. When the former is added to TRPM8 channels, the temperature threshold for activation was shifted towards warmer temperatures, much in the way menthol acts, whereas the latter products, such as arachidonic acid, inhibited activation by temperature, icilin, and menthol (65). Even though equimolar amounts of each of the counter-acting products are released by PLA2 activity, the greater potency of the various lysophospholipids leads overall to positive modulation of channel function (21, 65). Moreover, in vivo, lysophospholipids were shown to increase cold sensitivity in a TRPM8-dependent manner (43). While it is not clear how increased PLA2 activity leads to modulation of TRPM8 function physiologically, these results are intriguing in that they are the first insights into endogenous pathways that may regulate channel function in tissues that express TRPM8 but are not exposed to changes in temperature, such as the prostate, or in tissues like muscle and bladder which are innervated by TRPM8-expressing afferents, but where thermal stimuli are unlikely to be relevant.

4.2. Other Regulators

In addition to direct effects of PIP_2 breakdown, increased kinase activity has also been suggested to modulate TRPM8 channel function (66, 67). A consequence of PIP_2 breakdown is the generation of diacylglycerol (DAG) and inositol trisphosphate (IP_3), which are second messengers that promote protein kinase C (PKC) activity. Phosphorylation is a common mechanism whereby channel activity is modulated, and increased PKC activity causes decreased TRPM8 membrane currents. Interestingly, PKC activation does not lead to increased incorporation of phosphate on TRPM8 but rather a decrease in phosphorylation. Thus, the PKC-mediated effects are not due to direct phosphorylation of TRPM8, but that PKC plays a role upstream of channel phosphorylation.

When intracellular pH increases to supraphysiological levels, TRPM8 activity is inhibited. These effects of pH are thought to be mediated intracellularly, but there is disagreement on the effects of pH on cold-, menthol-, and icilin-evoked currents. One study reported that menthol's ability to activate TRPM8 is unaffected by pH, but that cold and icilin responses are inhibited (68), whereas another also found that icilin was less effective in activating TRPM8 at high pH, but in contrast, menthol-evoked responses were also suppressed (69).

Lastly, polyphosphates, an inorganic polymer comprised of tens to hundreds of phosphates linked by phosphoanhydride bonds, has been shown to form a complex with TRPM8 and modulate channel activity. These long phosphate chains are known to associate with a number of membrane proteins and modulate their activity. For TRPM8 it was observed that channel activity was inhibited in heterologous cells expressing TRPM8 after enzymatic breakdown of polyphosphates by the exopolyphosphatase scPPX1 (70). In an intriguing set of studies, it was shown that scPPX1-mediated reduction in polyphosphates blocked menthol and cold-evoked channel activity of purified TRPM8 channels reconstituted into planar lipid bilayers even in the presence of PIP_2. Thus, it seems likely that either cell-to-cell variation in temperature thresholds for cold, or altered sensitivity due to experience and pathological state of the neuron, may be a result of TRPM8 regulation via cellular levels of PIP_2, protons, or kinase activity. This level of channel modulation may also account for the complexity and variability in cold-evoked temperature responses observed both in vivo and in vitro.

5. The Pharmacology of TRPM8

A growing list of cooling compounds has been shown to activate TRPM8 with various potencies and efficacies. Indeed one of the weakest TRPM8 agonists is menthol with a half effective concentration in the mid micromolar range (10). Insights into the mechanism of menthol activation came out of a large study which used high-throughput screening of several thousand random TRPM8 mutants to search for clones that affected menthol sensitivity without altering thermal gating (71). Important amino acid residues that are determinants of menthol sensitivity were found in the S2 and the C-terminal TRP domain of the channel, and analysis of these mutants showed that gating by cold and cold mimetics could be separated, as well as the effects of voltage and PIP_2 on channel function. The latter was of interest as basic residues in the TRP domain had been implicated in the effects of PIP_2 (63), yet mutations in this region affected menthol sensitivity without alter PIP_2 (71).

The most potent TRPM8 agonist is WS-12 (19), which is one of many cooling compounds that came out of an extensive synthesis and evaluation program by Wilkinson Sword Ltd. intent on developing confectionary compounds that did not have the minty sensations associated with menthol (2, 72). In addition, another cooling compound icilin (also known as AG-3-5) was identified in the early 1980s which bears little resemblance to menthol structurally but is more potent and effective in activating TRPM8 (10, 18). Of note, when given intravenously, icilin will induce the characteristic shivering or "wet dog" shakes as well as an increase in core body temperature as noted above. Interestingly, the mechanism whereby icilin activates TRPM8 is different than that of menthol or cold in that it requires a coincident rise in cytoplasmic calcium, either via permeation through the channel or by release from intracellular stores, in order to evoke TRPM8 currents (73). This requirement of a calcium rise for TRPM8 activity is not needed for cold- or menthol-induced responses, suggesting the channel can be activated by multiple mechanisms. Additionally, a critical amino acid was identified, which when mutated renders icilin incapable of activating TRPM8. This residue was located between the second and third *trans*-membrane domains of the channel, a region known to be important for capsaicin sensitivity of TRPV1 (74). Moreover, this region is also near sites identified as important for menthol activation of TRPM8 channels (71), suggesting a conserved mechanism for ligand activation of these thermosensitive TRP channels.

While various compounds activate TRPM8, a more relevant class of molecules that may be of use clinically is those that antagonize or block the channel. Unfortunately, many TRPM8 antagonists that have been identified to date also antagonize the heat-gated channel TRPV1, suggesting a conserved mechanisms amongst thermosensitive channels (75). Capsazepine, *N*-(4-tert-butylphenyl)-4-(3-chloropyridin-2-yl)piperazine-1-carboxamide (BCTC), and Thio-BCTC, known TRPV1 antagonists, were found to inhibit TRPM8 activity using a fluorometric imaging plate reader (FLIPR) assay (69). This is in addition to capsazepine's known roles as a TRPV1 antagonist with nonspecific activity on voltage-gated calcium channels and nicotinic acetylcholine receptors (76–79). BCTC has also been shown to be a TRPA1 agonist (80). *N*-(3-aminopropyl)-2-{[(3-methylphenyl) methyl)oxy}-*N*-(2-thienylmethyl)benzamide hydrochloride salt (AMTB) is a TRPM8 blocker in cellular assays (81), and application of it intravenously was able to diminish the frequency of volume-induced bladder contractions, without any effects on contraction. These results suggest that TRPM8 afferents are good targets for treating overactive and painful bladder conditions. Finally, clotrimazole, an anti-fungal medication, was shown in a recent study to activate TRPV1 and TRPA1 (consistent with its commonly reported side effects of irritation and burning), while also serving as a potent TRPM8 antagonist (82). Together, these antagonists present a range of pharmacological tools to regulate TRPM8 function, yet each of these compounds has off-target effects and, while of interest pharmacologically, these results complicate the search for selective agents for these channels that are such good targets for drug discovery.

However, a robust and selective TRPM8 antagonist, PBMC (1-phenylethyl-4-(benzyloxy)-3-methoxybenzyl(2-aminoethyl)carbamate), was recently described (42). In vitro, PBMC is the most potent TRPM8 antagonist reported to date and inhibits both chemical and thermal channel activation. Indeed, the IC_{50} was less than 1 nM, a dosage approximately 100-fold lower than the most potent TRPM8 antagonist reported to date, CTPC (44, 75). Unlike other TRPM8 antagonists, no cross-reactivity with both TRPV1 and TRPA1 was observed, and the two-orders-of-magnitude higher affinity of PBMC makes this compound a more amenable reagent in the study of TRPM8 channel function. This latter point was validated when it was shown that this antagonist inhibited acute cold sensitivity, attenuated injury-induced cold hypersensitivity, as well as induced a striking hypothermic response that contrasts the hyperthermia evoked by TRPM8 activation with icilin (42).

6. Conclusions

The elucidation of TRP channels as molecular detectors of thermal stimuli addressed a fundamental issue in sensory transduction: how are thermal stimuli converted into neuronal activity? These proteins have now become interesting and potentially valuable drug targets for a range of conditions from pain, visceral organ function, and thermal homeostasis. Currently, we are at the beginning in our understanding of how modulation of TRPM8 can be beneficial clinically. Both direct genetic evidence and indirect data from the effects of menthol and cooling clearly show that the channel serves a fundamental role in sensory physiology. Indeed the breadth of the in vivo properties of TRPM8 are surprising, raising the fundamental question of how can a single sensor of environmental stimuli lead to such a diverse assortment of functions, ranging from acute cold detection, injury-evoked hypersensitivity, and analgesia. One likely hypothesis is that while TRPM8 enables neurons to respond to cold, the cellular context of channel expression is the final determinant of our perception, and that there may be distinct and nonoverlapping TRPM8-mediated neural pathways that are segregated into transducing temperature, pain, and analgesia (83). Nonetheless, the identification of TRPM8 established the first molecular detector of cold stimuli and confirmed Hensel and Zotterman's half-century-old hypothesis for how menthol makes us feel cool.

Acknowledgments

This work was supported by a National Institutes of Health Grants NS054069 and NS071364 (D.D.M).

References

1. Morin C, Bushnell MC (1998) Temporal and qualitative properties of cold pain and heat pain: a psychophysical study. Pain 74(1):67–73
2. Eccles R (1994) Menthol and related cooling compounds. J Pharm Pharmacol 46(8): 618–630
3. Graves R (1955) The Greek myths. Penguin, Baltimore, MD, 2 vols
4. Green BG (1992) The sensory effects of l-menthol on human skin. Somatosens Mot Res 9(3):235–244
5. Hensel H, Zotterman Y (1951) The effect of menthol on the thermoreceptors. Acta Physiol Scand 24:27–34
6. Wasner G et al (2004) Topical menthol – a human model for cold pain by activation and sensitization of C nociceptors. Brain 127(Pt 5):1159–1171
7. Binder A et al (2011) Topical high-concentration (40%) menthol-somatosensory profile of a human surrogate pain model. J Pain 12(7): 764–773
8. Serra J et al (2009) C-nociceptors sensitized to cold in a patient with small-fiber neuropathy and cold allodynia. Pain 147:46–53
9. Suto K, Gotoh H (1999) Calcium signaling in cold cells studied in cultured dorsal root ganglion neurons. Neuroscience 92(3):1131–1135
10. McKemy DD, Neuhausser WM, Julius D (2002) Identification of a cold receptor reveals a general role for TRP channels in thermosensation. Nature 416(6876):52–58
11. Jordt SE et al (2004) Mustard oils and cannabinoids excite sensory nerve fibres through the TRP channel ANKTM1. Nature 427(6971):260–265
12. Bautista DM et al (2007) The menthol receptor TRPM8 is the principal detector of environmental cold. Nature 448(7150):204–208
13. Reid G, Flonta ML (2001) Physiology. Cold current in thermoreceptive neurons. Nature 413(6855):480
14. Peier AM et al (2002) A TRP channel that senses cold stimuli and menthol. Cell 108(5):705–715

15. Montell C (2005) The TRP superfamily of cation channels. Sci STKE 2005(272):re3, 1–24
16. Tsavaler L et al (2001) Trp-p8, a novel prostate-specific gene, is up-regulated in prostate cancer and other malignancies and shares high homology with transient receptor potential calcium channel proteins. Cancer Res 61(9):3760–3769
17. Eid SR, Cortright DN (2009) Transient receptor potential channels on sensory nerves. Handb Exp Pharmacol 194:261–281
18. Wei ET, Seid DA (1983) AG-3-5: a chemical producing sensations of cold. J Pharm Pharmacol 35(2):110–112
19. Bodding M, Wissenbach U, Flockerzi V (2007) Characterisation of TRPM8 as a pharmacophore receptor. Cell Calcium 42(6):618–628
20. Andersson DA, Nash M, Bevan S (2007) Modulation of the cold-activated channel TRPM8 by lysophospholipids and polyunsaturated fatty acids. J Neurosci 27(12):3347–3355
21. Vanden Abeele F et al (2006) Ca^{2+}-independent phospholipase A2-dependent gating of TRPM8 by lysophospholipids. J Biol Chem 281(52):40174–40182
22. Caterina MJ et al (1997) The capsaicin receptor: a heat-activated ion channel in the pain pathway. Nature 389(6653):816–824
23. Tominaga M et al (1998) The cloned capsaicin receptor integrates multiple pain-producing stimuli. Neuron 21(3):531–543
24. Brauchi S, Orio P, Latorre R (2004) Clues to understanding cold sensation: thermodynamics and electrophysiological analysis of the cold receptor TRPM8. Proc Natl Acad Sci USA 101(43):15494–15499
25. Voets T et al (2004) The principle of temperature-dependent gating in cold- and heat-sensitive TRP channels. Nature 430(7001):748–754
26. Voets T et al (2007) TRPM8 voltage sensor mutants reveal a mechanism for integrating thermal and chemical stimuli. Nat Chem Biol 3(3):174–182
27. Brauchi S et al (2007) Dissection of the components for PIP2 activation and thermosensation in TRP channels. Proc Natl Acad Sci USA 104(24):10246–10251
28. Matta JA, Ahern GP (2007) Voltage is a partial activator of rat thermosensitive TRP channels. J Physiol 585(2):469–482
29. Brauchi S et al (2006) A hot-sensing cold receptor: C-terminal domain determines thermosensation in transient receptor potential channels. J Neurosci 26(18):4835–4840
30. Daniels RL, Takashima Y, McKemy DD (2009) Activity of the neuronal cold sensor TRPM8 Is regulated by phospholipase C via the phospholipid phosphoinositol 4,5-bisphosphate. J Biol Chem 284(3):1570–1582
31. Grandl J et al (2010) Temperature-induced opening of TRPV1 ion channel is stabilized by the pore domain. Nat Neurosci 13(6):708–714
32. Allchorne AJ, Broom DC, Woolf CJ (2005) Detection of cold pain, cold allodynia and cold hyperalgesia in freely behaving rats. Mol Pain 1:36
33. Dhaka A et al (2008) Visualizing cold spots: TRPM8-expressing sensory neurons and their projections. J Neurosci 28(3):566–575
34. Takashima Y et al (2007) Diversity in the neural circuitry of cold sensing revealed by genetic axonal labeling of transient receptor potential melastatin 8 neurons. J Neurosci 27(51):14147–14157
35. Viana F, de la Pena E, Belmonte C (2002) Specificity of cold thermotransduction is determined by differential ionic channel expression. Nat Neurosci 5(3):254–260
36. Jyvasjarvi E, Kniffki KD (1987) Cold stimulation of teeth: a comparison between the responses of cat intradental A delta and C fibres and human sensation. J Physiol 391:193–207
37. McKemy DD (2005) How cold is it? TRPM8 and TRPA1 in the molecular logic of cold sensation. Mol Pain 1(1):16
38. Colburn RW et al (2007) Attenuated cold sensitivity in TRPM8 null mice. Neuron 54(3):379–386
39. Dhaka A et al (2007) TRPM8 is required for cold sensation in mice. Neuron 54(3): 371–378
40. Daniels RL, McKemy DD (2007) Mice left out in the cold: commentary on the phenotype of TRPM8-nulls. Mol Pain 3:23
41. Jasmin L et al (1998) The cold plate as a test of nociceptive behaviors: description and application to the study of chronic neuropathic and inflammatory pain models. Pain 75(2–3):367–382
42. Knowlton WM et al (2011) Pharmacological blockade of TRPM8 ion channels alters cold and cold pain responses in mice. PLoS One 6(9):e25894
43. Gentry C et al (2010) The roles of iPLA2, TRPM8 and TRPA1 in chemically induced cold hypersensitivity. Mol Pain 6:4
44. Knowlton WM, McKemy DD (2011) TRPM8: from cold to cancer, peppermint to pain. Curr Pharm Biotechnol 12(1):68–77
45. Moqrich A et al (2005) Impaired thermosensation in mice lacking TRPV3, a heat and camphor sensor in the skin. Science 307(5714):1468–1472

46. Lee H et al (2005) Altered thermal selection behavior in mice lacking transient receptor potential vanilloid 4. J Neurosci 25(5): 1304–1310
47. Knowlton WM et al (2010) TRPM8, but not TRPA1, is required for neural and behavioral responses to acute noxious cold temperatures and cold-mimetics in vivo. Pain 150(2): 340–350
48. Altis K et al (2009) Analgesic efficacy of tramadol, pregabalin and ibuprofen in menthol-evoked cold hyperalgesia. Pain 147:116–121
49. McCoy DD, Knowlton WM, McKemy DD (2011) Scraping through the ice: uncovering the role of TRPM8 in cold transduction. Am J Physiol Regul Integr Comp Physiol 300(6): R1278–R1287
50. Bennett GJ, Xie YK (1988) A peripheral mononeuropathy in rat that produces disorders of pain sensation like those seen in man. Pain 33(1):87–107
51. Attal N et al (2009) Thermal hyperalgesia as a marker of oxaliplatin neurotoxicity: a prospective quantified sensory assessment study. Pain 144(3).245–252
52. Descoeur J et al (2011) Oxaliplatin-induced cold hypersensitivity is due to remodelling of ion channel expression in nociceptors. EMBO Mol Med 3(5):266–278
53. Gauchan P et al (2009) Involvement of increased expression of transient receptor potential melastatin 8 in oxaliplatin-induced cold allodynia in mice. Neurosci Lett 458(2): 93–95
54. Galeotti N et al (2002) Menthol: a natural analgesic compound. Neurosci Lett 322(3): 145–148
55. Proudfoot CJ et al (2006) Analgesia mediated by the TRPM8 cold receptor in chronic neuropathic pain. Curr Biol 16(16):1591–1605
56. Iida T et al (2005) Attenuated fever response in mice lacking TRPV1. Neurosci Lett 378(1): 28–33
57. Montell C, Caterina MJ (2007) Thermoregulation: channels that are cool to the core. Curr Biol 17(20):R885–R887
58. Gavva NR (2008) Body-temperature maintenance as the predominant function of the vanilloid receptor TRPV1. Trends Pharmacol Sci 29(11):550–557
59. Gavva NR et al (2008) Pharmacological blockade of the vanilloid receptor TRPV1 elicits marked hyperthermia in humans. Pain 136(1–2):2 02–210
60. Ding Z et al (2008) Icilin induces a hyperthermia in rats that is dependent on nitric oxide production and NMDA receptor activation. Eur J Pharmacol 578(2–3):201–208
61. Tajino K et al (2011) Cooling-sensitive TRPM8 is thermostat of skin temperature against cooling. PLoS One 6(3):e17504
62. Liu B, Qin F (2005) Functional control of cold- and menthol-sensitive TRPM8 ion channels by phosphatidylinositol 4,5-bisphosphate. J Neurosci 25(7):1674–1681
63. Rohacs T et al (2005) PI(4,5)P2 regulates the activation and desensitization of TRPM8 channels through the TRP domain. Nat Neurosci 8(5):626–634
64. Rohacs T, Nilius B (2007) Regulation of transient receptor potential (TRP) channels by phosphoinositides. Pflugers Arch 455(1): 157–168
65. Andersson DA et al (2008) Transient receptor potential A1 is a sensory receptor for multiple products of oxidative stress. J Neurosci 28(10): 2485–2494
66. Abe J et al (2006) Ca^{2+}-dependent PKC activation mediates menthol-induced desensitization of transient receptor potential M8. Neurosci Lett 397(1–2):140–144
67. Premkumar LS et al (2005) Downregulation of transient receptor potential melastatin 8 by protein kinase C-mediated dephosphorylation. J Neurosci 25(49):11322–11329
68. Andersson DA, Chase HW, Bevan S (2004) TRPM8 activation by menthol, icilin, and cold is differentially modulated by intracellular pH. J Neurosci 24(23):5364–5369
69. Behrendt HJ et al (2004) Characterization of the mouse cold-menthol receptor TRPM8 and vanilloid receptor type-1 VR1 using a fluorometric imaging plate reader (FLIPR) assay. Br J Pharmacol 141(4):737–745
70. Zakharian E et al (2009) Inorganic polyphosphate modulates TRPM8 channels. PLoS One 4(4):e5404
71. Bandell M et al (2006) High-throughput random mutagenesis screen reveals TRPM8 residues specifically required for activation by menthol. Nat Neurosci 9(4):493–500
72. Watson HR et al (1978) New compounds with the menthol cooling effect. J Soc Cosmet Chem 29:185–200
73. Chuang HH, Neuhausser WM, Julius D (2004) The super-cooling agent icilin reveals a mechanism of coincidence detection by a temperature sensitive TRP channel. Neuron 43(6): 859–869
74. Jordt SE, Julius D (2002) Molecular basis for species-specific sensitivity to "hot" chili peppers. Cell 108(3):421–430
75. Weil A et al (2005) Conservation of functional and pharmacological properties in the distantly related temperature sensors TRPV1 and TRPM8. Mol Pharmacol 68(2):518–527

76. Walpole CS et al (1994) The discovery of capsazepine, the first competitive antagonist of the sensory neuron excitants capsaicin and resiniferatoxin. J Med Chem 37(13): 1942–1954
77. Docherty RJ, Yeats JC, Piper AS (1997) Capsazepine block of voltage-activated calcium channels in adult rat dorsal root ganglion neurones in culture. Br J Pharmacol 121(7): 1461–1467
78. Liu L, Simon SA (1997) Capsazepine, a vanilloid receptor antagonist, inhibits nicotinic acetylcholine receptors in rat trigeminal ganglia. Neurosci Lett 228(1):29–32
79. Valenzano KJ, Sun Q (2004) Current perspectives on the therapeutic utility of VR1 antagonists. Curr Med Chem 11(24):3185–3202
80. Madrid R et al (2006) Contribution of TRPM8 channels to cold transduction in primary sensory neurons and peripheral nerve terminals. J Neurosci 26(48):12512–12525
81. Lashinger ES et al (2008) AMTB, a TRPM8 channel blocker: evidence in rats for activity in overactive bladder and painful bladder syndrome. Am J Physiol Renal Physiol 295(3):F803–F810
82. Meseguer V et al (2008) Transient receptor potential channels in sensory neurons are targets of the antimycotic agent clotrimazole. J Neurosci 28(3):576–586
83. Madrid R et al (2009) Variable threshold of trigeminal cold-thermosensitive neurons is determined by a balance between TRPM8 and Kv1 potassium channels. J Neurosci 29(10): 3120–3131

Chapter 9

TRPML Channels in Function, Disease, and Prospective Therapies

David A. Zeevi

Abstract

The transient receptor potential mucolipin (TRPML) subfamily of transient receptor potential cation channels consists of three members (TRPML1-3) that function at various stages of endocytosis. Conventional research in the TRPML field suggests that dysfunction along these endocytic stages underlies the severe psychomotor impairment in mucolipidosis type IV (MLIV). However, recent studies intimate that TRPMLs may be implicated in other neuropathological disorders as well. This review follows the historical development of TRPML research from the clinical description of the first MLIV patient until present-day characterization of TRPML1-based defects in MLIV on the molecular and cell biological levels. In addition, the aberrant role of TRPML3 in *varitint-waddler* mouse pathology is elucidated and the normal function of the protein and its paralogs are described. TRPML electrophysiology, pharmacology, and animal models are discussed and TRPML-associated systemic and neurological disorders, not including MLIV, are also addressed. Recently, a number of prospective TRPML-based therapies have been proposed in the treatment of these disorders. These prospects are carefully considered here as well. Altogether, the aforementioned descriptions aim to highlight the transformation of TRPML research from a single discipline, single gene, and single disorder field into a multidisciplinary, multigene endeavor with wide application to therapeutic treatment of several neurobiological disorders.

Key words: TRP channels, TRPML, Mucolipidosis type IV, Varitint-waddler, Membrane trafficking, Endocytosis, Lysosomal exocytosis

1. Introduction: Historical and Clinical Background

Arguably, the TRPML story began in the year 1945 with the publication of a genetic study describing the discovery of the *varitint-waddler* mouse. The *varitint-waddler* is, as the authors of the original study put it, "just what [its] name implies" (1). Variable coat color and a "duck-like walk" comprised the easily identifiable external features of this haphazard by-product of

Arpad Szallasi and Tamás Bíró (eds.), *TRP Channels in Drug Discovery: Volume I*, Methods in Pharmacology and Toxicology, DOI 10.1007/978-1-62703-077-9_9,

genetic crosses. However, further investigation revealed that the *varitint-waddler* (*Va*) mouse is also deaf, demonstrates acute circling behavior, and is generally not viable when homozygous for a dominant *Va* allele (1). Some years later, another genetic study identified a milder phenotypic version of the *Va* mouse, termed *varitint-waddler-J (VaJ)*, that presented with deafness but also featured less exaggerated pigmentation defects and no abnormal circling behavior (2). These seminal studies correctly predicted that, despite the unusual heterogeneity of features in the *Va* or *VaJ* phenotypes, all heritable traits are linked to a single causative monogenic allele. However, this single genetic determinant would not be identified until close to six decades after the original description of the *varitint-waddler* mouse.

In the meantime, the TRPML narrative developed further in 1974, with the publication of a case report describing a "hitherto undescribed form of mucolipid storage disease" (3). The report described an infant, born to parents of European Jewish descent, displaying congenital corneal opacity and progressive motor impairment together with unique biochemical findings that led the authors to classify the patient as the first identified case of mucolipidosis type IV (MLIV; MIM 252650). Subsequent studies identified MLIV as a genetic disorder of autosomal recessive inheritance with increased incidence in the Ashkenazi Jewish population. Patients were identified, clinically, by corneal opacification, retinal degeneration, and psychomotor impairment that appeared, typically, within the first 1.5 years of life (4).

Eventually, as awareness of MLIV increased worldwide, the clinical spectrum of MLIV symptoms grew as well. It was found that although psychomotor impairment is severe and it appears in most MLIV patients early in life, the disease progression is very slow and reaches an apparent steady-state level that allows patients to live with a lengthened and presently undefined life expectancy (5). In addition, although eye, muscle, and cognitive dysfunction are considered primary indications of MLIV onset, some milder cases of the disease, lacking in one or two of these abnormalities, have been identified (6). Moreover, many but not all MLIV patients also suffer from iron deficiency anemia (7). Thus, from a diagnostic standpoint, identification of MLIV has relied upon other more common features of the disorder. For example, elevated blood gastrin levels, resulting from constitutive achlorhydria, were identified in all severely affected MLIV patients together with a characteristically thin corpus collosum in MRIs (7). Also, in the era prior to the identification of the genetic determinant of MLIV, patient diagnosis relied heavily upon MLIV-specific ultrastructural abnormalities of affected patients' cells (5, 8) (see later for a more detailed description of these abnormalities). Hence, from a clinical perspective, the need to identify a genetic determinant for MLIV pathogenesis grew enormously during the ensuing decades since the disorder was first described.

2. Molecular Genetics

With the development of molecular genetic tools, towards the end of the twentieth century, came the necessary breakthroughs in the search for the MLIV-associated gene. In 1999 a linkage study, involving DNA derived from Ashkenazi Jewish families affected by MLIV, narrowed the search down to a 1.5-cM region on chromosome 19p. This same study identified two MLIV-associated founder haplotypes in the Ashkenazi Jewish population that together comprised 95% of MLIV-associated alleles in Ashkenazi Jews. The more frequent major haplotype was found in 73% of Ashkenazi chromosomes, whereas the less frequent minor founder haplotype was identified in 23% of the same pool of chromosomes (9). One year later and 26 years after the description of the first MLIV patient, positional candidate gene sequencing identified deleterious mutations in the *MCOLN1* gene. The major haplotype was linked to a splice site mutation in the acceptor site of *MCOLN1* intron 3 and the minor haplotype was linked to a partial gene deletion removing exons 1–6 and part of exon 7 (out of 14 *MCOLN1* exons) (10–12). Both of these mutations led to a significant reduction in mRNA expression (12) and, more importantly, to complete absence of coding protein (13, 14). Additional loss-of-function mutations were also identified in *MCOLN1* and these mutations were generally associated with non-Jewish haplotypes. Indeed, only 70–80% of described MLIV patients are from Ashkenazi Jewish descent (15). Other MLIV patients possess various missense, nonsense, and short insertion/deletion mutations that have been summarized elsewhere (6). Overall, the heterozygote carrier frequency, in the Ashkenazi Jewish population, for both the major and minor MLIV mutations is ~1:100 (16). The identification of this relatively high carrier frequency has led to the successful establishment of carrier screening programs, in countries around the world with high concentrations of Ashkenazi Jews, to prevent the incidence of MLIV in at-risk couples (5).

The *MCOLN1* gene encodes a protein called mucolipin 1 or transient receptor potential mucolipin 1 (TRPML1). The TRPML1 protein is an integral membrane protein, 580 amino acids long, and features six predicted transmembrane domains (TM) with striking similarity to the transient receptor potential (TRP) superfamily of cation channels in the pore-forming TM5–TM6 regions. Upon identification of the TRPML1-encoding gene, sequence-based searches immediately identified TRPML1 as a founding member of a novel subfamily of TRP channels because two very similar TRPML1 paralogs are also present in humans and other vertebrates. These proteins were designated mucolipin-2/TRPML2 (566 a.a.) and mucolipin-3/TRPML3 (553 a.a.) and their encoding genes were designated *MCOLN2* and *MCOLN3*, respectively (10). All three proteins share a similar 6 TM structure and a highly conserved TRP-like channel pore in their TM5 and TM6 regions as well.

Two years after the identification of *MCOLN1* via positional candidate gene methodology, a similar approach led to the identification of the *varitint-waddler*-associated gene. A linkage study narrowed the *VaJ* causative allele to a 0.14-cM region on mouse chromosome 3 (17). Subsequently, all three genes located in this region were sequenced and novel missense mutations were identified in the *mcoln3* gene, of *VaJ* and *Va* mice, that did not appear in any parental controls or other inbred or wild-derived mice (18). *Mcoln3* encodes a 553 amino acid murine homologue of human TRPML3. The severely affected *Va* mice were shown to have an A419P substitution in their predicted TM5 of TRPML3 and the mildly affected *VaJ* mice were shown to possess an additional I362T missense mutation that segregated in *cis* with A419P. These initial findings led the authors to speculate that the A419P substitution in murine TRPML3 is either a gain-of-function or dominant-negative loss-of-function mutation that is lethal to ear inner hair cells unless combined in *cis* with the I362T intragenic suppressor mutation (18).

3. TRPML3 Gain of Function Leads to *Varitint-Waddler* Pathology

Following its identification, the elucidation of TRPML3's role in *varitint-waddler* pathogenesis has been fairly straightforward. *Va* and *VaJ* mice present with readily identifiable ear hair cell, organ of Corti, and skin melanocyte degeneration (18–20). Electrophysiological recordings with heterologously expressed TRPML3-A419P indicated that the mutant *Va*-implicated channel is constitutively active, cell surface expressed, and conducts Ca^{2+} with robust inward rectifying currents (20–23). These robust inward rectifying currents are identified in outer hair cells of *Va* mice (19) and, in general, they are thought to overload TRPML3-A419P-expressing cells with Ca^{2+} to effectively cause apoptosis (24). In contrast, the wild-type TRPML3 channel was found not to be constitutively active, but rather carefully regulated by extracellular Na^+ and H^+ (22, 25). Moreover, the constitutively active TRPML3-A419P channel was shown to possess significant changes in biophysical conductance properties that are not present in the wild-type channel (25, 26). Altogether, these findings argue that the A419P substitution, in TM5 of the TRPML3 channel pore, is a gain-of-function mutation that transforms TRPML3 into a cytotoxic protein that induces apoptosis in all mutant channel expressing cells (namely hair, cochlear, and melanoycte cells) (18, 20, 23). This tissue-selective degeneration leads to the characteristic vestibular, hearing, and pigmentation defects in *varitint-waddler* mice. It has also been suggested that the *VaJ* mice feature a milder phenotype due to the reduced cell surface expression of the constitutively active I362T-A419P *cis* TRPML3 mutant therein (22).

With the initial identification of TRPML3 as the cause of deafness in *varitint-waddler* mice came the possibility that the first hair cell mechanotransductory channel had been discovered (27). However, as seen in the previous paragraph, the *Va*-associated mutation in TRPML3 is a gain-of-function toxic variant. Moreover, a recent study describing a TRPML3 loss-of-function mouse, featuring a missing channel pore, did not present with any auditory deficits (28). Thus, until further evaluation of TRPML3 loss-of-function animal models is performed, it is yet unknown whether TRPML3 loss of function leads to any gross pathological abnormalities in mammalian organisms.

4. TRPML1 Loss of Function Leads to Lysosomal Pathology

Ultrastructural pathology is found in cells from all MLIV patients together with loss-of-function mutations in TRPML1 (29). The characteristic cytological abnormalities in MLIV consist of heterogeneous lysosomal bodies that appear as laminar electron-dense inclusions and/or granulated amorphic vacuoles (15). These characteristic lysosomal inclusions are found in every analyzed tissue and organ from MLIV patients and they principally indicate MLIV as a lysosomal storage disorder. Early biochemical studies with cultured fibroblasts from MLIV patients identified storage of various gangliosides, neutral lipids, phospholipids, and mucopolysaccharides (30–35). However, unlike in other "classical" lysosomal disorders (such as Tay–Sachs or Gaucher disease), no deficiency in any particular hydrolase activity is detected in TRPML1-deficient cells (36). Rather, MLIV cells feature a profound membrane trafficking defect whereby endocytosed membrane markers are excessively transported from late endosomes towards lysosomes and delayed in their efflux from these organelles to the trans-Golgi network (34). Lysosomal degradation of these endocytosed components takes place, but at an altered rate and pattern (37). These features of MLIV pathology correlate well with the clinical presentation of the disease. During the months that ensue from birth, as intracellular biochemicals slowly accumulate in nerve and brain tissue (the organs most severely affected by lysosomal storage), MLIV patients slowly begin to suffer from physical and mental disabilities that manifest acutely towards the end of their first year of life. This impaired psychomotor condition continues to deteriorate until lysosomal accumulation/degradation of heterogeneous hydrophobic and hydrophilic compounds reaches a steady-state level.

At this stage, residual slow-acting lysosomal hydrolyzing machinery takes hold in patient brain and nerve cells to allow survival of even the most severely affected MLIV patients for decades (15).

Recently, it has also been proposed that MLIV cells store various cations in late endosomal/lysosomal compartments. These stored ions include Fe^{2+} (38), Zn^{2+} (39), and Ca^{2+} (40). In addition, protons are also suspected to be stored in TRPML1-deficient cells (41), although this assertion has been debated extensively (see below). On the other hand, given that MLIV storage is also autofluorescent (42), suggesting lipofuscin accumulation; heavy metal storage in lysosomal compartments is not a remote possibility (43–45). This assertion also correlates with aberrant autophagic machinery that is apparent in MLIV, as evident by the appearance of immature autophagosomes in TRPML1-deficient cells and enhanced accumulation of polyubiquitin and insoluble p62 (46, 47).

Although clinical disorders, stemming from TRPML2 and/or TRPML3 loss of function have yet to be described in humans, it is interesting to note that shRNA knockdown of these proteins leads to pathological phenotypes reminiscent of MLIV. Indeed, ultrastructural analysis of TRPML2 and TRPML3-depleted cells revealed lysosomal inclusions, similar to that in TRPML1-deficient cells, which suggest a common function for these proteins in the regulation of membrane trafficking (14). This observation may ultimately lead to successful treatment of TRPML1-associated disorders (see below).

5. TRPML Animal Models of MLIV

During the past decade, the search for treatments of MLIV has focused on gaining a better understanding of TRPML1 function. At present, four different animal models (two knockout mice, one *Drosophila*, and one *C. elegans* model) have been described to aid in this endeavor and provide a "testing ground" for potential therapeutic strategies.

TRPML1 knockout mice ($TRPML1^{-/-}$) effectively recapitulate most of MLIV pathology. At ~6 months of age, $TRPML1^{-/-}$ mice present with delayed motor milestones and reduced gating that deteriorates to hind-limb paralysis by ~8.5 months postpartum. Retinal degeneration is observed but not corneal opacities. Characteristic MLIV-like ultrastructural storage pathology is also detected in $TRPML1^{-/-}$ nervous tissue (48). A different $TRPML1^{-/-}$ mouse model features elevated plasma gastrin levels and constitutive hypochlorhydria that is detected together with ultrastructural parietal cell pathology similar to that in MLIV patients (49, 50). Biochemically, $TRPML1^{-/-}$ mice were shown to store gangliosides throughout their central nervous system and the storage material in their brains was autofluorescent (51). In addition, macroautophagy appears to be aberrantly affected in TRPML1$^{-/-}$ mice just as in MLIV patients. Autophagic markers, LC3-II and polyubiquitin, accumulate in TRPML1$^{-/-}$ mouse neuronal cell culture together with p62 aggregates in insoluble protein fractions. This autophagosomal accumulation in TRPML1$^{-/-}$ cells is proposed to result from decreased degradation in the autophagic pathway rather than upregulated autophagic machinery (46). Altogether, these findings suggest that TRPML1$^{-/-}$ mice will serve as invaluable tools, in the future, for testing MLIV-indicated therapies.

Other models for TRPML loss of function have been described in *Drosophila* and *C. elegans.* These invertebrate organisms express a singular TRPML unlike vertebrates which, as described earlier, express three paralogous TRPML proteins. Nevertheless, these models recapitulate many of the features of MLIV pathology. The fly *trpml* mutant presents with retinal degeneration and motor deficits that result from massive neuronal cell death. Furthermore, it was found that neurotoxic lipofuscin accumulates in mutant fly brains together with signs of oxidative stress and aberrant degradation in the autophagic pathway. Importantly, this TRPML-deficient model also demonstrated that the motor deficit could be rescued by clearance of excess apoptotic nerve cells with an hsp70-homologue (52).

Cup-5, the TRPML homologue in *C. elegans,* plays a critical role in worm development given that cup-5 null mutants are embryonic lethal (53). However, this model has proven invaluable towards the elucidation of TRPML-deficiency pathogenesis. Indeed, the enlarged vacuoles that accumulate in cup-5 mutant cells (54) can be rescued by expression of either TRPML1 or TRPML3 human homologs (55). Moreover, it has recently been demonstrated that the vacuoles accumulating in cup-5 mutant cells represent late endosomal–lysosomal hybrid (56) and/or autolysosome (57) organelles that fail to reform lysosomes after fusion. These findings indicate that autophagic degradation is also impaired in the *C. elegans* model of MLIV, just as in the mouse and fly models described earlier. In addition, these discoveries have contributed key points to the debate over where the true defect lies in TRPML1-deficient cells (see below).

6. Elucidating TRPML1 Function: Lysosomal Biogenesis Versus Lysosomal Metabolism

With the discovery of TRPML1 deficiency as the cause of MLIV, a searing debate has ensued over the primary cellular defect that underlies disease pathogenesis. One model, termed the "lysosomal biogenesis" model, proposes that storage molecules accumulate in MLIV cells due to a buildup of prelysosomal compartments that do not reform endpoint lysosomes. This primary defect, in turn, causes a bottleneck that interferes with or delays intracellular anterograde

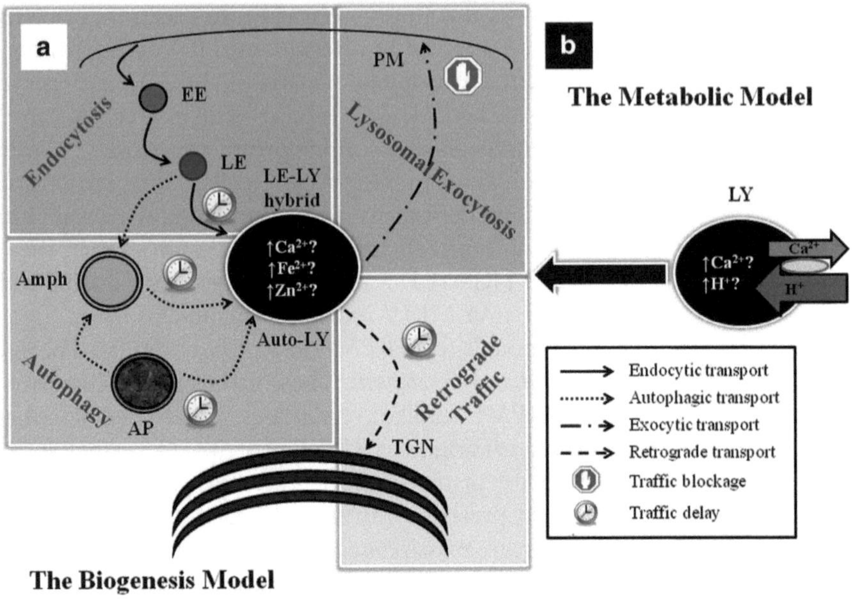

Fig. 1. Models of MLIV disease pathogenesis. (**a**) The lysosomal biogenesis model argues that loss of TRPML1 function leads to reduced juxta-organellar Ca^{2+} release from late endosomal–lysosomal hybrid (LE–LY hybrid) organelles that fail to reform endpoint lysosomes. This defect causes a bottleneck that delays or blocks various membrane trafficking events. Thus, although early endosomes (EE) successfully mature into late endosomes (LE) during early stage endocytosis (*top left panel*), late endosomes fail to mature into lysosomes during late stage endocytosis. Under these circumstances, endocytosed materials are still processed in LE–LY hybrid organelles, but only in a delayed manner. A similar bottleneck is also found in autophagic machinery (*bottom left panel*). Immature lysosomes successfully merge with double membrane-enclosed autophagosomes (AP), carrying damaged intracellular organelles; and amphisomes (Amph), carrying endocytic waste from late endosomes; to form auto-lysosome (Auto-LY) hybrid organelles. However, these Auto-LY hybrids fail to mature into endpoint lysosomes thereby delaying degradation of macromolecules in autophagy. Hence, it has been suggested that, in addition to heterogeneous lipid and mucopolysaccharide storage, these trafficking delays also lead to the storage of heavy metals, like Fe^{2+} and Zn^{2+}, and ultimately neurotoxic lipofuscin in immature LE–LY or Auto-LY vesicles. Other trafficking defects resulting from the lysosomal biogenesis bottleneck in MLIV include delayed retrograde traffic of lysosome-processed materials to the trans-Golgi network (TGN; *bottom right panel*) and blocked lysosomal membrane fusion with the plasma membrane (PM) in lysosomal exocytosis (*top right panel*). (**b**) The metabolic model of MLIV pathogenesis argues that all of the trafficking defects in the biogenesis model result from a primary defect in the regulation of lysosomal pH. Specifically, the heterogeneous lysosomal storage in MLIV results from increased [H^+] in lysosomal compartments (LY) that reduces the activity of a variety of pH-sensitive lysosomal hydrolases. This primary metabolic defect leads to accumulation of undigested lipids in lysosomes and, ultimately, effects a terminal inhibition of intracellular membrane trafficking, as depicted in (**a**). Presently, the mechanistic link between TRPML1 loss of function and lysosomal hyperacidification is unclear. However, it has been suggested that reduced Ca^{2+} release from TRPML1-deficient lysosomes may possibly stimulate excess H^+ influx through a yet to be identified lysosomal Ca^{2+}–H^+ exchanger.

and retrograde membrane trafficking (Fig. 1a). A second model, termed the "metabolic" model, proposes that storage molecules accumulate in MLIV cells due to overacidification of lysosomal compartments. This primary defect interferes with the proper functioning of a heterogeneous assortment of lysosomal hydrolases and ultimately leads to aberrant degradation of various intracellular biochemicals (Fig. 1b) (43). The debate over TRPML1 function is of fundamental importance as it bears heavily on the development of effective therapeutic strategies for the treatment of MLIV and related disorders. Hence, the following sections will be devoted to summarizing the evidence that support and detract from the current mainstream theories.

6.1. Lysosomal Biogenesis

The following discoveries provide compelling evidence to support the "lysosomal biogenesis" theory. First, it was found that hydrolytic activity of lysosomal enzymes in MLIV fibroblasts exists but the activity proceeds at an altered rate from normal (36, 37). Thus, although lysosomal hydrolases are aberrantly affected by TRPML1 dysfunction, their activity is not completely compromised, as the metabolic model purports. Second, the *C. elegans* cup-5 mutants, described earlier, clearly depict an accumulation of late endosome–lysosome hybrid organelles consistent with the biogenesis model assertion that TRPML deficiency leads to defects in prelysosomal compartments (55, 56). Third, the membrane trafficking bottleneck that forms from accumulation of prelysosomal compartments in TRPML1-deficient cells has been found to inhibit retrograde traffic from lysosomes to the trans-Golgi network (34, 58) and delay anterograde traffic of endocytosed molecules towards lysosomes (59) (see Fig. 1a for a summary of these findings).

Over the years, a mechanistic approach to explain the role/s of TRPML1 in membrane trafficking and lysosomal biogenesis has also been put forward. A series of experiments, involving Ca^{2+} chelators, have indicated that juxta-organellar Ca^{2+} release is a critical step in the homotypic (i.e., endosome–endosome) and heterotypic (i.e., endosome–lysosome) fusion of endosomal vesicles (60–62). In addition, vesicle fission events have also been shown to undergo regulation by juxta-organellar Ca^{2+} release (61, 63, 64). Collectively, these findings correlate well with biochemical and electrophysiological experiments describing the TRPML1 channel. Indeed, TRPML1 has been characterized as a late endosomal/lysosomal (LEL)-localized, inwardly rectifying (lumen to cytosol), Ca^{2+} permeable channel that is activated by an LEL-specific phosphoinositide (38, 65–68). Moreover, a recent report identified Ca^{2+} storage in lysosomes of TRPML1-depleted cells to further cement a role for the protein in intracellular Ca^{2+} homeostasis (40). Altogether, these findings support the notion that TRPML1 serves as an essential LEL Ca^{2+} release channel that functions in various membrane fusion and fission events in the endocytic pathway.

Interestingly, recent studies have also identified TRPML1 function in lysosomal exocytosis, another pathway that is dependent on LEL Ca^{2+} release (69–71). Initially, Ca^{2+} ionophore-induced lysosomal exocytosis was found to be impaired in MLIV fibroblasts (72). Subsequently, a constitutively active gain-of-function TRPML1 channel pore mutant was shown to enhance lysosomal exocytosis in HEK 293 cells (73). Finally, a very recent investigation demonstrated that a transcription factor (TFEB), responsible for upregulating expression of *MCOLN1* and other genes involved in lysosomal biogenesis (74), also upregulates lysosomal exocytosis. TRPML1 was found to perform a key role in this process because lysosomal membrane fusion with the plasma membrane was only achieved through TRPML1-dependent intracellular Ca^{2+} release (Fig. 1a) (75).

Thus in aggregate, all of the aforementioned findings provide very persuasive evidence to support the lysosomal biogenesis theory of TRPML1 function. However, conflicting evidence to detract from this theory has also been described in the literature and this evidence is discussed in the next section.

6.2. Lysosomal Metabolism

The metabolic model of TRPML1 function emerged from inconsistencies that were observed in the biogenesis model. Of particular importance is the fact that the metabolic model points to a fundamental issue with the biogenesis model's investigation of cells derived from MLIV patients. Given that MLIV pathology likely results from a slow and steady accumulation of intracellular biochemicals to levels that eventually become toxic to nerve and brain tissue; any investigation based on cells from clinically ill MLIV patients solely describes the endpoint or near-endpoint of TRPML1-implicated pathology. To be clear, the metabolic model of MLIV pathogenesis does not dispute the fact that membrane trafficking defects exist in endpoint cells from MLIV patients or TRPML1-deficient animal models. Rather, the metabolic model argues that these defects emerged secondarily to the primary metabolic defect; a defect that can only be identified when TRPML1 expression is acutely reduced by RNAi, for a short time period, prior to the emergence of endpoint storage inclusions (76).

What is the primary metabolic defect? The proponents of the metabolic model argue that lysosomes in cells, with freshly depleted TRPML1 expression levels, become hyperacidified before any trafficking defects can be identified (Fig. 1b) (77). Indeed, siRNA depletion of TRPML1 in HeLa cells led to reduced lysosomal pH while anterograde and retrograde endocytic trafficking proceeded as normal. Likewise, the degradation of protein and lipid molecules along these pathways was largely unaffected (77). In another study, it was found that reduced lysosomal pH ultimately led to the detection of inactive lysosomal hydrolase activities in lysosomal preparations from endpoint MLIV fibroblasts (41). Hence this model claims that MLIV pathogenesis results from a metabolic defect in pH-sensitive lysosomal hydrolases. This defect leads to accumulation of undigested lipids in lysosomes and, ultimately, effects a terminal inhibition of intracellular membrane trafficking (76).

To further support the metabolic model, one study identified TRPML1 as a proton-conducting channel (41). However, the same study portrays TRPML1 as an outwardly rectifying (cytosol to lumen) channel that prefers to conduct protons into lysosomes (to acidify them when the channel is active) rather than out. The authors of the study explain this inconsistency by claiming that the channel is inactivated by proteolytic cleavage upon reaching lysosomes (13) but the proposed mechanism of TRPML1-mediated lysosomal pH regulation is not clear. Nevertheless, some years after their original study, the authors acknowledged that their observation of TRPML1-mediated proton conductance may not be correct (78). This debate over TRPML1 conductance is discussed later.

Although the metabolic model is lacking in a mechanistic explanation that relates TRPML1 dysfunction to lysosomal hyperacidification, the Ca^{2+} conductance of TRPML1 may provide a link. Indeed, if a Ca^{2+}–H^{+} exchanger exists in lysosomes, it is possible that a lacking in TRPML1-mediated LEL Ca^{2+} release leads to aberrant lysosomal pH regulation in MLIV cells. This possibility is discussed further in another recent review (43) (see also Fig. 1b).

Regardless of which theory ultimately proves to be correct, the very existence of a debate over the true model of MLIV pathogenesis highlights the need to resolve the inherent properties of native TRPML1 and that of its closely related sister channels, TRPML2 and TRPML3. These topics are discussed in the following section.

7. TRPML Subcellular Localization, Channel Properties, and Interacting Proteins

7.1. The Subcellular Localization of TRPML1

TRPML1 is an LEL-localized protein. This point has been firmly established in numerous studies describing the trafficking of recombinant TRPML1 (13, 79–81). In addition, two separate groups have convincingly identified native TRPML1 in LEL compartments of human cells (14, 82). The targeting of TRPML1 to LEL is mediated by two different acidic dileucine clathrin adaptor protein targeting motifs on the amino- and carboxyl-terminal tails of the protein. The amino-terminal motif interacts with the adaptor proteins AP-1 and AP-3 to facilitate direct delivery from the trans-Golgi network to early endosomes and late endosomes, respectively (76, 80, 81). The carboxyl-terminal motif interacts with the AP-2 adaptor protein to facilitate endocytosis of TRPML1, should it reach the plasma membrane (81), perhaps during lysosomal exocytosis (5).

7.2. TRPML1 Channel Properties

The electrophysiological properties of TRPML1 have been difficult to characterize due to the strict intracellular localization of the native channel. This challenge likely accounts for the conflicting descriptions of TRPML1 conductance and selectivity that have surfaced in the literature since the protein was first identified. These conflicting reports have been thoroughly summarized in a previous review (76). Therefore, in order to provide a fresh angle to this topic, the current discussion will focus on the historical development of methodology in the characterization of TRPML1 channel properties.

Shortly after TRPML1 was first implicated in MLIV pathogenesis, pilot electrophysiological studies described TRPML1 channel activity in *Xenopus* oocytes (66, 67). These reports provided the first evidence of TRPML1 functionality as a nonselective cation channel. However, the channel was ectopically expressed on the plasma membrane and LEL membranes of *Xenopus* oocytes in order to facilitate current recordings. Thus, from a technical perspective, the observed channel properties in these studies may have been distorted by the presence of endogenous *Xenopus* channels and modulators.

A second set of experiments described in vitro-translated TRPML1 in reconstituted lipid bilayers (83, 84). This setup effectively demonstrated that TRPML1 is an inherently functional cation channel in a controlled environment. It should be noted that lipid reconstitution of any functional TRP channel is quite a feat given that *Drosophila* TRP, the founding member of the TRP superfamily, has yet to be functionally expressed in such a system (85). Nevertheless, beyond confirmation of the inherent functionality of TRPML1 as a nonselective cation channel, these reports did not characterize the protein in a physiologically relevant environment.

The next series of studies addressed the physiological role of TRPML1 by heterologously expressing the channel in HEK 293 cells (13, 41). Although these experiments were the first to characterize the human TRPML1 channel in a human cell line, they also featured whole-cell current recordings that required mistargeting of TRPML1 (an LEL-localized protein, as described earlier) to the plasma membrane via extreme protein overexpression (13). This departure from physiological context may have contributed to the controversial finding, in these reports, that TRPML1 is a proton-conducting channel (41).

The controversy regarding TRPML1 proton permeability officially took hold when another group described a constitutively active variant of TRPML1 that was not proton permeable but was activated by low pH (20). The same study identified the *varitint-waddler*-associated A419P mutation in TRPML3 as a channel-activating pore substitution that, when adapted to the homologous amino acid position in TRPML1 (V432P), effectively activates TRPML1 channel activity as well. In parallel, a separate group confirmed the effectiveness of this genetic activation mechanism on TRPML1 and other TRP channels (21). Subsequently, these studies necessitated the development of channel deactivating mechanisms that offset the constitutive activity of TRPML1-V432P. Thus, a separate genetic deactivating mutation was also identified in the TRPML1 channel pore (DD471/472KK) that overrides the V432P activating mutation and transforms the protein into a dominant-negative channel (58, 86). Overall, these genetic activating/deactivating mechanisms have facilitated some basic investigation into the physiological role of TRPML1 conductivity in cell biological function (38, 58, 73, 86). However, the main issue with these descriptions of TRPML1 conductance is that the mutations described earlier, which facilitated current recordings and cell biological analysis, do not exist in nature (although at least one MLIV-associated TRPML1 pore mutant (F465L) does feature dominant-negative properties in vitro (86)). In addition, it is also a concern that the V432P activating mutation, in TRPML1, drives cell surface expression of an otherwise intracellular channel (73). Thus, the search for TRPML1 channel activators later intensified in order to address these issues.

During the same period that the aforementioned experiments were performed, yet another group took a very different approach to characterizing the electrophysiological properties of TRPML1. Zhang et al. isolated lysosomal membranes by subcellular fractionation on a percoll gradient and subsequently reconstituted the membrane fractions into lipid bilayers for single channel current recordings. Interestingly, the investigators found that nicotinic acid adenine dinucleotide phosphate (NAADP)-mediated Ca^{2+} release from these membrane preparations was highly TRPML1 dependent. Indeed, they showed that pretreatment of lysosomal preparations with anti-TRPML1 antibodies, specifically, inhibited NAADP-induced currents (87). Moreover, gene silencing of TRPML1, immunodepletion of TRPML1, and TRPML1-deficient MLIV fibroblasts all featured significantly reduced NAADP-mediated lysosomal Ca^{2+} release (68, 88).

On the physiological level, these studies accomplished more than any of the aforementioned electrophysiological studies had accomplished beforehand. Rather than assaying ectopically expressed or cell surface mislocalized channels, the authors of these reports assayed the actual organelle where native TRPML1 is localized, namely lysosomes, and they did so with a potent activator in NAADP. However, there is one major caveat to the lessons gleaned from these experiments. The authors provide no concrete evidence that NAADP directly activates the TRPML1 channel. Indeed, a recent investigation has shown that NAADP is a very potent activator of intracellular Ca^{2+} release via another population of lysosomal channels called two-pore channels (TPCs) (89). In addition, yet another report has recently shown that despite the fact that recombinant TRPML1 interacts with recombinant TPCs, NAADP-induced Ca^{2+} release was found to be TPC dependent and TRPML1 independent (90). Of course, the protein–protein interaction between TPCs and TRPML1 may explain the TRPML1 dependence for NAADP-mediated Ca^{2+} release, as observed by Zhang et al., given that downregulation of TRPML1 may affect the functionality of TPCs. Nevertheless, no such effect was observed in pancreatic acinar cells from TRPML1 knockout mice (90). Therefore, unless NAADP proves to activate an isolated recombinant form of TRPML1, these studies may teach the importance of assaying channel properties in both biological and artificial reconstituted systems.

It is assumed that the identification of TRPML1 channel activators will help elucidate the physiological function of the protein in LEL membranes. Towards this end, a high throughput study recently identified synthetic small molecules that activate recombinant TRPML1 (91). However, the authors of this report did not assay TRPML1 currents in intracellular membranes. Rather, in order to assay TRPML1 response to small molecule activators, the authors mutated the lysosomal targeting signals on the amino- and carboxyl-tails of the protein so as to facilitate plasma membrane delivery of the channel for whole-cell patch-clamp recordings in HEK 293 cells (91). Given that TRPML1 was genetically modified and ectopically expressed on the cell surface, it remains to be seen whether these recently identified synthetic channel activators are capable of activating the wild type and/or native channel in a more physiologically relevant environment.

Towards the ultimate goal of characterizing TRPML1 conductance in native LEL membranes, a key breakthrough has recently been reported in patch-clamp methodology. Using vacuolin-1 to enlarge LEL vesicles for current recordings, Dong et al. developed the "whole-lysosome patch-clamp" to characterize wild-type TRPML1 in its native environment (38). After further experimentation with this technique, the same group identified an intracellular phospho-inositide, PI(3,5)P_2, as a specific and potent activator of recombinant as well as endogenous TRPML1. The authors also found that PI(3,5)P_2 is a physiological TRPML1 channel activator, whose specific depletion in LEL compartments correlates with a marked decrease in basal TRPML1-derived currents (65). These studies constitute significant forward steps in the fields of TRPML1 electrophysiology and MLIV research, in general. First, the authors have managed to develop an assay to measure TRPML1 conductance in its native environment. Second, they have done so with a wild-type TRPML1 channel. Third, they have discovered a potent and physiologically meaningful channel activator in PI(3,5)P_2. Fourth, they have used that activator to activate heterologously expressed TRPML1. Finally, they have used their activator to activate the native channel in its natural setting. The only drawback to their experimental setup is that it requires pretreating cells with an LEL membrane-modulating compound in vacuolin-1 (92). The downstream limitations of this caveat, if any, should become evident if in vivo experiments fail to recapitulate in vitro electrical recordings. Nevertheless, given all the challenges that have obstructed progress in the field of TRPML1 electrophysiology over the past decade, the "whole-lysosome patch-clamp" assay should play heavily into the elucidation of TRPML1 channel function going forward.

What are the inherent properties of the TRPML1 channel? Presently, it appears that TRPML1 is an inwardly rectifying (lumen to cytosol) nonselective cation channel that is permeable to Na^+, K^+, Ca^{2+}, Fe^{2+}, Mn^{2+}, Zn^{2+}, and other divalent trace metals. TRPML1 is not permeable to Fe^{3+} or H^+ although low pH (~4.6) does potentiate the basal currents of the channel (38, 65, 73). TRPML1 currents are inhibited by verapamil, Gd^{3+}, and La^{3+} (20) and the channel is activated by $PI(3,5)P_2$ (65). Altogether, these findings have led to the suggestion that TRPML1 is an LEL Ca^{2+} (66, 67, 72, 73), Fe^{2+} (38), and possibly Zn^{2+} (39) release channel that is responsible for regulating a subset of LEL membrane fusion/fission events (93). In addition, the channel may also participate in the regulation of lysosomal Ca^{2+}, Fe^{2+}, and Zn^{2+} homeostasis (40).

7.3. TRPML1 Interacting Proteins

A number of TRPML1 binding partners have been identified to help elucidate protein function. These interacting proteins are discussed below.

A yeast two-hybrid screen using the TRPML1 luminal domain between TM1 and TM2 identified two binding partners, Hsc70 and Hsp40, that are components of the chaperone-mediated autophagy (CMA) complex (94). Subsequently, a series of experiments were conducted to determine if CMA is aberrantly affected in TRPML1-deficient MLIV cells. These experiments indicated that oxidized proteins accumulate in MLIV cells together with downregulated CMA-associated proteolysis and LAMP-2A lysosomal uptake (94). The latter findings correlate well with the general trend of defective macroautophagy in MLIV (47) but it has not yet been clearly established whether CMA and macroautophagic dysfunction are primary effects of TRPML1 loss of function or secondary effects of gross lysosomal biogenesis/pH regulation dysfunction. Acute RNAi-mediated knockdown of TRPML1 expression should help to answer this question. Another issue that is unresolved is how TRPML1 purportedly participates in CMA via interaction with Hsc70 and Hsp40. Does TRPML1 facilitate lysosomal membrane docking of the CMA complex? This possibility is intriguing given the recent discovery that the same domain on TRPML1 that interacts with Hsc70 and Hsp40 also features inherent membrane re-shaping function (95). Future experiments will be needed to explore this possibility further in addition to assaying TRPML1 channel function involvement, if any, in CMA.

LAPTM4a and LAPTM4b are two proteins of unknown function that were identified as TRPML1 interacting proteins in a split-ubiquitin yeast two-hybrid screen (96). These LAPTMs traffic to

LEL membranes, like TRPML1, but they are not dependent on TRPML1 for LEL localization. However, there is perhaps some sort of functional inter-relationship between these proteins because lysosomal swelling resulting from overexpression of LAPTM4b is rescued by co-expression with TRPML1. Moreover, siRNA knock-down of both LAPTM4s leads to the formation of lysosomal inclusions reminiscent of MLIV ultrastructural pathology (96). Previous studies have speculated that LAPTMs serve as lysosomal transporters responsible for translocation of substances across lysosomal membranes (97, 98), but this possibility still awaits empirical evidence. Therefore, until the physiological roles of LAPTMs are better clarified, their interaction with TRPML1 provides limited information to assess protein function.

Interestingly, another investigation identified ALG-2, a putative Ca^{2+} sensor protein, as a Ca^{2+}-dependent interactor with the amino terminal of TRPML1 (99). Aside from the intriguing Ca^{2+} dependence of this interaction, a TRPML1 mutant that fails to interact with ALG-2 also traffics to LELs with reduced efficiency (99). Furthermore, it has been found that ALG-2 interacts with many of the residues, on the N-terminus tail of TRPML1, that also bind $PI(3,5)P_2$ to activate the channel (100). Altogether, these observations have led to speculation that ALG-2 may serve as a Ca^{2+} sensor that regulates vesicle fusion via $PI(3,5)P_2$-mediated spatiotemporal activation of TRPML1 in LEL compartments (93). This possibility, if proven true, would provide substantial support to the lysosomal biogenesis model of MLIV pathogenesis.

TRPML1 and two-pore channels, TPC1 and TPC2, have each been implicated in NAADP-mediated Ca^{2+} release from acidic intracellular vesicles (68, 88, 89, 101). This led to speculation that these channels might share functional association (43). Indeed, Yamaguchi et al. have recently described physical interactions between recombinant TRPML1 and both TPC channel isoforms (90). However, despite their physical interaction, a number of electrophysiological experiments argue against TRPML1–TPC functional co-regulation of NAADP-mediated LEL Ca^{2+} release. Indeed, NAADP-mediated intracellular Ca^{2+} release was found to be uniquely regulated by TPCs even in the presence of a dominant-negative TRPML1 mutant and even in the absence of TRPML1 in $TRPML1^{-/-}$ mouse pancreatic acinar cells (90). Nevertheless, the physical interaction between TPCs and TRPML1 should not be overlooked given that TPC modulation of $PI(3,5)P_2$-mediated activation of TRPML1 has not yet been assayed.

The heteromerization of all three TRPML channels constitute the first described sets of protein–protein interactions involving members of the TRPML subfamily (102). These interactions, while quite robust in heterologous protein expression systems (91), are actually quite limited in native tissue (14, 82). Indeed, endogenous TRPML1 co-immunoprecipitates and co-localizes with only a small subset of native TRPML2 and TRPML3, each in separate LEL vesicles (14). Nonetheless, TRPML1 does form functional heteromeric cation channel complexes with TRPML2 and TRPML3 (103) and these complex assemblies modulate several cellular processes, including intracellular Ca^{2+} overload prevention and starvation-induced autophagy (86). Interestingly, overexpression of an MLIV-associated TRPML1 channel pore mutant resulted in dominant-negative phenotypes when assaying effects on either TRPML2 or TRPML3 function (86). Altogether, these studies indicate that all three TRPMLs share some physiological functions even though they are functionally distinct, in general. This avenue of possibility should encourage further research into TRPML2 and TRPML3 channel activation mechanisms because new discoveries in this field may ultimately help to treat MLIV patients, who lack TRPML1, with either one of the channel's closely related paralogs (see below).

7.4. The Subcellular Localization of TRPML2

TRPML2 is the least characterized member of the TRPML subfamily because the protein is yet to be implicated in any genetic disorder. However, as will be seen later, the elucidation of TRPML2 function has contributed to increasing speculation that the protein may serve a therapeutic role in the treatment of TRPML-associated disorders.

The first clues in the characterization of TRPML2 function came from a subcellular localization study involving the recombinant protein. TRPML2 was identified at the cell surface and in tubular intracellular vesicles that correspond with Arf6-regulated recycling endosomes (104). In addition, a fraction of recombinant TRPML2 was also identified in CD63-positive LEL vesicles (104) and this was further corroborated in experiments describing the native protein (14). Indeed, endogenous TRPML2 distributes primarily to tubulo-vesicular structures while only a limited fraction associates with endogenous TRPML1 in LAMP1-positive LEL vesicles (14). The implications of these findings are discussed later, in the context of TRPML2 current recordings and protein interactions.

7.5. TRPML2 Channel Properties

The first current recordings involving TRPML2 applied *varitint-waddler*-based genetic activation. Sequence alignment of TRPML2 with TRPML3 identified the A424P activating mutation (and other corresponding mutations in murine channel isoforms) in the TM5 pore region of TRPML2. Like TRPML3-A419P, the TRPML2-A424P mutant is a constitutively active, inwardly rectifying, Ca^{2+}-permeable channel that presents with measurable currents in the whole-cell patch clamp configuration (38, 105, 106). However, unlike the Fe^{2+}-impermeable constitutively active TRPML3 mutant, the Na^{+}, K^{+}, Ca^{2+}, and Fe^{2+} permeability of TRPML2-A424P more closely resembles that of its paralogous TRPML1-V432P mutant (38, 43).

Genetic deactivating mutants of TRPML2 have also been described. Like TRPML1-V432P-DD471/472KK, the paralogous TRPML2-A424P-DD463/464KK pore mutant is also an inactive channel with dominant-negative properties (86). In general, the DD-KK substitution in all three TRPML channel pores has resulted in inactive dominant-negative channels that have been utilized to explore the physiological roles of TRPML channel function (20, 58, 82, 86, 104). Regarding TRPML2, this pore mutant inhibited the recycling of CD59 (a GPI-anchored protein) from tubulo-vesicular recycling endosomes to the plasma membrane. This suggests that TRPML2 channel activity is crucial to recycling endosomal function (104). Given that internalized CD59 is dependent upon Arf6-mediated recycling (107) and given that TRPML2 is an in vitro activator of Arf6 that localizes to recycling endosomes, these data suggest that TRPML2 currents facilitate Arf6-mediated recycling in vivo (104). It remains to be seen, however, whether TRPML2 truly regulates this and other processes in membrane trafficking and turnover.

Of utmost importance, when considering the physiological function of an ion channel, is the verification that the channel features inherent functionality even in the absence of genetic manipulation. This has recently been demonstrated with wild-type TRPML2 expression constructs. Single channel conductance was identified in lipid bilayers containing in vitro-translated TRPML2 (103) and ectopic expression of the channel in *Drosophila* S2 cells also elicited robust TRPML2-A424P-like inwardly rectifying whole-cell currents (105). The wild-type TRPML2 channel is Ca^{2+} permeable and inhibited by moderately low extracellular (or luminal) pH (~6.0) (105), although much lower pH (~4.7) seems to facilitate the currents mediated by the mutant TRPML2-A424P isoform (38). The physiological explanation for these pH-dependent properties of TRPML2 is not known. However, one may speculate that pH regulation of TRPML2-mediated currents would correlate with the channel's function in the various acidic intracellular vesicles where it has been detected. Indeed, the appearance of lysosomal inclusions in TRPML2-depleted cells (14) suggests that TRPML2 participates in key elements of membrane trafficking and turnover in these acidified vesicles. Hopefully, the newly identified phospho-inositide and small molecule activators of TRPML2 (65, 91) will aid in drawing such connections between the channel's intracellular distribution, its localized pH-dependent currents, and its associated physiological functions.

7.6. TRPML2 Interacting Proteins

Knowledge of TRPML2 interacting proteins is rather limited. At present, TRPML2 is known only to directly interact with TRPML1 and TRPML3 to form functional heteromeric channel complexes (14, 86, 102, 103). Even so, these interactions comprise just a small subset of the intracellular distribution of native TRPML2 (14). Nevertheless, TRPML2 expression levels in certain tissues seem to be regulated by TRPML1 function (106) and perhaps this provides some physiological relevance to the complex assembly.

Interestingly, although physical interactions were described between endogenous TRPML2 and TRPML3, the two channels do not overlap in LAMP1-positive LEL vesicles. This is in contrast to the only overlap that is detected between native TRPML2 and TRPML1 and between native TRPML3 and TRPML1 (14). These observations suggest that it is highly unlikely for all three native TRPMLs to co-assemble into the same heteromeric complex, and more importantly, these observations suggest that each particular TRPML channel assembly may serve to regulate differing intracellular functions. This latter possibility is intriguing given that loss of function of each distinctive heteromeric TRPML complex could theoretically lead to the pathogenesis of TRPML-associated disorders. Nevertheless, TRPML2 function must be further clarified before substantiating such salient claims.

7.7. The Subcellular Localization of TRPML3

With the increasing realization that TRPML3 causes deafness and pigmentation defects in the *varitint-waddler* mouse due to mutational gain of function, the focus has shifted in recent years towards elucidation of the natural physiological role of the wild-type protein. The first step in this endeavor has been the characterization of the intracellular distribution of recombinant TRPML3.

At first, it was suggested that TRPML3 is retained in the endoplasmic reticulum of HEK 293 cells unless co-expressed with either TRPML2 or TRPML1. These TRPML3-interacting proteins purportedly facilitated the trafficking of TRPML3 to acidic intracellular vesicles (102). However, this observation is inconsistent with more recent reports describing autonomous TRPML3 localization along the endocytic pathways of HeLa and ARPE-19 cells (82, 108). These studies identified a substantial portion of recombinant TRPML3 on the cell surface and in early endosomal vesicles. In addition, a smaller fraction of TRPML3 was also identified in LEL vesicles by both immunofluorescence and subcellular fractionation-based assays (82, 108). Importantly, these latter findings are largely consistent with the observed distribution of the native protein as well (14, 82).

Just as the recycling endosomal localization of TRPML2 hinted towards recycling endosomal function of the endogenous protein (104), the early endosomal localization of TRPML3 has provided similar clues in the investigation of native TRPML3 function (Fig. 2). Indeed, RNAi-mediated knockdown of TRPML3 was shown to enhance the kinetics of epidermal growth factor receptor (EGFR) degradation in the endocytic pathway (82, 108) in addition to the endocytic uptake of transferrin and epidermal growth factor (EGF) (82). This aberrant upregulation of endocytosis is attributed to misregulated Ca^{2+} concentrations in the endosomal lumen that hinder the natural acidification of endocytic vesicles. Specifically, TRPML3 depletion has been associated with a

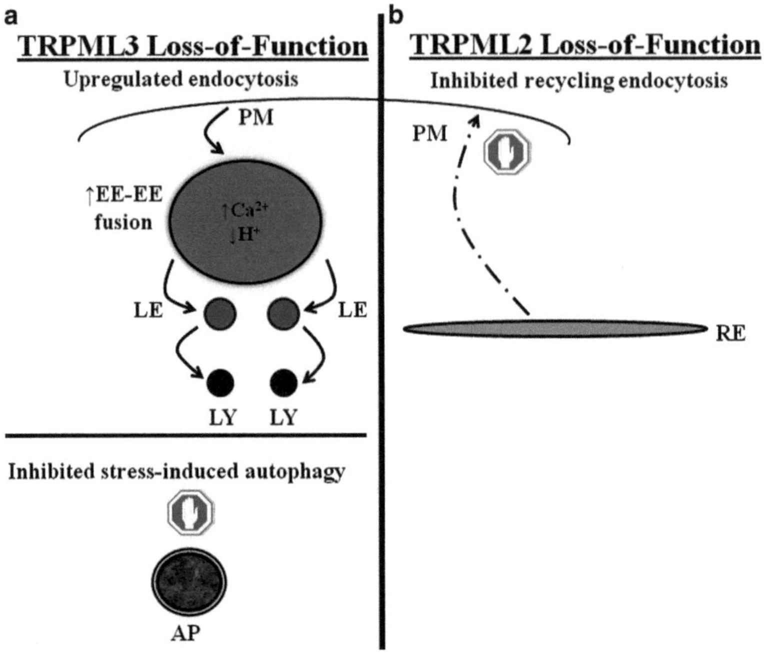

Fig. 2. Membrane trafficking defects in TRPML2/3 loss-of-function models. (**a**) *Top*, TRPML3 loss of function leads to elevated early endosomal [Ca^{2+}] and reduced [H^{+}]. This, in turn, stimulates increased homotypic fusion of early endosomes (↑EE–EE fusion) and accelerated processing of endocytosed proteins in late endosomes (LE) and lysosomes (LY). (**a**) *Bottom*, TRPML3 loss of function also blocks autophagosome (AP) formation under cell stress conditions. (**b**) TRPML2 loss of function leads to reduced recycling of Arf6-dependent recycling-endosomal proteins to the plasma membrane (PM).

significant increase in measurable endosomal [Ca^{2+}] and this has been correlated with elevated pH measurements inside these compartments (109). Remarkably, the defective hypo-acidified endocytic vesicles in TRPML3-deficient cells undergo enhanced homotypic endosomal fusion in an in vitro assay (109) and present as MLIV-like vesicular inclusions in electron micrographs (14). Not surprisingly, each of the aforementioned membrane trafficking defects, in TRPML3 knockdown cells, strongly resembles those observed in TRPML1-deficient MLIV cells. The major difference between TRPML3- and TRPML1-associated intracellular pathologies is that the former defect presumably takes place at the earlier stages of endocytosis (i.e., homotypic [endosome–endosome] fusion) while the latter defect takes place at the later endocytic stages (i.e., heterotypic late endosomal–lysosomal fusion).

Another interesting discovery, pertaining to wild-type TRPML3, is that its subcellular distribution appears to be quite dynamic. Although the majority of TRPML3 is found in endosomal vesicles, some of the protein is also identified on the cell surface (14, 82, 108) and this cell surface fraction is enhanced when a very early stage of endocytosis is inhibited with a dominant-negative dynamin mutant (82). This suggests that TRPML3 traffics dynamically along the endocytic pathway by internalizing from the cell surface into dynamin-dependent endocytic vesicles. However, it is not yet known whether TRPML3 is capable of recycling back to the plasma membrane from intracellular vesicles.

An additional dynamic process that appears to be regulated by TRPML3 is macroautophagy. Indeed, cell starvation or application of other cellular stressors has clearly shown a marked recruitment of recombinant TRPML3 into LC3-positive autophagosomal compartments (82, 108). Moreover, the very formation of autophagosomes seems to be highly dependent upon native TRPML3 function given that RNAi knockdown of TRPML3 hinders the appearance of LC3-positive vesicles upon autophagy induction (Fig. 2a) (82, 86). Thus, much insight has been gained in the elucidation of TRPML3 function just from peering into cells and assaying the protein's subcellular localization under challenging and nonchallenging growth conditions.

7.8. TRPML3 Channel Properties

Assaying the channel properties of TRPML3 has been crucial in the assessment of the protein's true physiological function. As mentioned earlier, the *varitint-waddler*-associated TRPML3-A419P variant is constitutively active. From a biophysical point of view, both this gain-of-function mutant and the wild-type channel are similar in current–voltage relationship and permeability profile (20, 22, 25). However, the conductance of the mutant for each respective cation (Ca^{2+}, in particular) and its physical channel pore size are quite different from wild type (25, 26). Hence, proper examination of the genuine electrophysiological properties of TRPML3 has focused on the wild-type channel and more natural activation mechanisms.

The innate cell surface localization of TRPML3 has greatly facilitated current recordings of the wild-type channel. Initially, Kim et al. described an unusual mechanism by which wild-type recombinant TRPML3-mediated currents can be facilitated in HEK 293 cells. They found that removal of Na^+ from the extracellular solution, transiently activated TRPML3 (22). Subsequently, they identified a stretch of histidines on the extracellular/luminal face of the channel that mediate inhibition of wild-type TRPML3 currents upon exposure to low extracellular/luminal pH (~6.0) (25). While low luminal pH inhibition seems logical for an early endosome-localized channel, the physiological relevance of this inhibition and low Na^+-mediated activation of TRPML3 were not clarified in these studies. Nonetheless, a recent report proposes that TRPML3 currents, which are facilitated at neutral pH by low Na^+, correspond well with endolymph salt concentrations in developing postnatal mouse cochlear hair cell stereocilia (91). Remarkably, at stages where TRPML3 was previously shown to be highly expressed, the Na^+ concentration in extracellular endolymph of cochlear hair cells is low, but as ear tissue further develops, TRPML3 expression diminishes, reciprocally, with increasing TRPML3 current-inhibiting Na^+ concentration in the endolymph (19, 91, 110). Nevertheless, the applicability of this putative Na^+-mediated regulation to endogenous TRPML3 channel activity, in postnatal cochlear hair cells, still remains to be seen.

Using a high-throughput approach, Grimm et al. identified various TRPML3-specific small molecule channel activators (91). These activators stimulated robust Ca^{2+} influx from surface expressed, inwardly rectifying, recombinant TRPML3 channels. Moreover, the activation of TRPML3 was synergized with low Na^+ extracellular solutions to generate even more robust inwardly rectifying currents (91). Theoretically, these newly identified TRPML3 activators should help to elucidate native TRPML3 function and help develop TRPML3-based therapies in the future. However, they have yet to activate native TRPML3 channels in their natural environment (91).

The prospect of activating endogenous TRPML3 channels may emerge from burgeoning research in the field of phosphoinositide-dependent membrane trafficking. Recently, the LEL-localized $PI(3,5)P_2$ was shown to activate recombinant TRPML3 by whole-LEL patch clamp (as described for TRPML1 earlier), but it is not yet known whether $PI(3,5)P_2$ facilitates endogenous channel currents (65). Given that $PI(3,5)P_2$ is an indispensable component of the endocytic machinery (111), the possibility that it might activate native TRPML3 in endocytic vesicles is highly worth investigating in the near future.

As discussed earlier, a series of knockdown experiments have indicated that TRPML3 is a functioning component of endocytic machinery whose downregulation leads to Ca^{2+} "storage" in, and hypo-acidification of, early endosomes (82, 108, 109). In order to assay whether TRPML3 channel function participates in these regulatory processes, each of the aforementioned experiments was repeated in the presence of the dominant-negative TRPML3-DD463/464KK pore mutant (a mutant that is homologous to the TRPML1- and TRPML2-inactivating DD-KK mutants, as described earlier). Not surprisingly, overexpression of the dominant-negative TRPML3 channel completely recapitulated the effects of TRPML3 knockdown in each case (82, 108, 109) strongly suggesting that high intra-endosomal $[Ca^{2+}]$ in TRPML3-depleted cells arises from loss of TRPML3 channel function. Given that TRPML3 is an inwardly rectifying (lumen to cytosol) Ca^{2+}-conducting channel that is activated (at least in recombinant form) by LEL-localized $PI(3,5)P_2$ (26, 65), the abovementioned findings indicate that native TRPML3 functions as an endosomal Ca^{2+}-release channel that facilitates a subset of vesicle fusion/fission events during endocytosis. This tempting prospect must be further verified, however, because there is no direct evidence that native

TRPML3 channels are functional in early or late endosomes and there is no clear association between TRPML3-mediated endosomal Ca^{2+} release and in vivo vesicle fusion. Nevertheless, further research in this area may provide a new avenue for treating TRPML-associated disorders in the very near future.

7.9. TRPML3 Interacting Proteins

Knowledge of TRPML3 interacting proteins is still rather sparse. At present, TRPML3 is known only to physically interact with TPCs (90) and TRPMLs.

Recently, recombinant TRPML3 was shown to co-immunoprecipitate in part with recombinant TPC1 and, more so with TPC2. These physical interactions correlate with an appreciable overlap between TRPML3 and TPCs in intracellular vesicular compartments (90). TPC2 has been identified as a primarily LEL-localized Ca^{2+}-release channel (89) that shares a highly similar intracellular distribution with TRPML1 (90). TPC1 also co-localizes with LEL markers, but to a lesser extent than TPC2, as the protein has also been identified in early endosomes and other unidentified vesicular compartments (89, 101). Both human TPCs have shown to mediate robust NAADP-induced Ca^{2+} release from acidic intracellular stores (89, 101), but the functional interpretation of their interactions with TRPML3 (and TRPML1, as described earlier) has not been clarified. Given the physical interactions involving TPCs and TRPMLs, it is tempting to speculate that they co-regulate NAADP-mediated Ca^{2+} release from LEL vesicles. However, this hypothesis was tested and could not be verified, at least with regard to TRPML1 functionality in complex with TPC1 or TPC2 (90). Further investigation into this matter, including verification of the physical interactions between endogenous TRPML1/3 and TPC1/2, should clarify these issues.

The studies describing the physical, functional, and physiological interactions between TRPML3 and its paralogous TRPML subfamily members have been discussed earlier at length. Only the therapeutic strategies that have emerged from these investigations remain for consideration. This topic is discussed in the following section.

8. Prospective Therapies for TRPML-Associated Disorders

In light of the aforementioned data regarding TRPML function, a number of therapeutic models are currently being developed towards the treatment of an assortment of human diseases. These diseases include mucolipidosis type IV (MLIV) and other lysosomal storage disorders; in addition to Charcot–Marie–Tooth (CMT) and amyotrophic lateral sclerosis (ALS).

8.1. Prospective Therapies for MLIV

In the search for treatments of mucolipidosis type IV, the TRPML1-associated channelopathy, a core debate has emerged over which element/s of MLIV pathogenesis to target. The metabolic model of disease pathogenesis argues that excess lysosomal H^+ is the target (41, 77) while the lysosomal biogenesis model argues that lysosomal Ca^{2+} is the objective (43, 93). Excess lysosomal Fe^{2+} and Zn^{2+} have also been put forward as potential MLIV-treating targets (38, 39, 43). Of course, it could be that each of these accumulating cations, identified in MLIV patients and animal models alike, must be downregulated concomitantly in order to effectively treat the disease, but this matter is surely more complex. Lysosomal inclusion bodies accumulate in just about every tissue and organ of MLIV patients (112) yet, aside from gastric parietal cells (50), only brain and nerve cells are noticeably incapacitated by this pathology (7). Hence, any intracellular targeted drug must also be carefully designed to act on degenerating nerve tissue without affecting other "healthy" systemic tissues. Moreover, given the eye and brain pathology in MLIV patients (7), many prospective drugs that effectively target gastric parietal cells may not be able to overcome the highly selective blood–brain barrier (BBB) surrounding the brain and central nervous system (113). Nevertheless, these obstacles are not necessarily insurmountable. Quite a few compounds are currently known to cross the BBB and the burgeoning field of nanotechnology has already offered various solutions to afford drug delivery across the BBB as well (114, 115).

The metabolic model of MLIV pathogenesis indicates that hyperacidified lysosomes must be neutralized in order to reverse the storage defect (41). Thus, an appropriate therapeutic approach was assessed on cultured MLIV fibroblasts. Two compounds that were previously shown to elevate lysosomal pH in culture (116, 117) were used to rescue the ultrastructural pathology in MLIV cells. However, neither nigericin (a proton antiporter) nor chloroquine (a weak base) effectively reduced the incidence of lysosomal inclusions in these cells (118). Nevertheless, assuming TRPML1 is an LEL Ca^{2+} release channel, it could be that an upregulated lysosomal Ca^{2+}–H^+ exchanger (yet to be identified) is causing excess acidification of lysosomes in the absence of functional TRPML1 (Fig. 1b) (43). If this is so, then this unidentified Ca^{2+}–H^+ exchanger may serve as a more effective target for reversing the hyperacification of MLIV lysosomes.

The lysosomal biogenesis model of MLIV pathogenesis assumes that lysosomal inclusions accumulate in diseased cells due to reduced juxta-endosomal Ca^{2+} release transients in the absence of functional TRPML1 (Fig. 1a) (63, 93, 119). Unfortunately, this model does not easily lend itself to drug target development owing to a general lacking in pathology-inducing hyperactive proteins or other intracellular compounds. Indeed, drugs are much more effectively developed by targeting and neutralizing hyperactive substances, but much less effectively developed when designed to replace hypoactive compounds or ion channels, as in the case of MLIV (120). Nevertheless, current research is still pursuing the possibility that agonist activation of either TRPML2 or TRPML3 in TRPML1-deficient MLIV cells might rescue some of the trafficking defects therein. This prospect is supported by recent studies identifying common biophysical characteristics among the three TRPMLs (86, 105) and even common physiological functions in vesicular Ca^{2+} release (65, 75, 109). However, even though some progress has been made in the identification of TRPML2- and TRPML3-specific agonists (65, 91), the field of TRPML channel activation is still in its earlier stages of development. In addition, further experimentation, with currently available TRPML2/3 agonists, is still required in order to evaluate the effectiveness of this pharmacological strategy in treating MLIV.

Another avenue for treatment of MLIV pathology relates to the increased lysosomal storage of Fe^{2+} and Zn^{2+} in MLIV and TRPML1 knockout cells (Fig. 1a) (38, 39). It is suggested that accumulation of these metals in lysosomes leads to neurotoxic lipofuscin accumulation and/or free radical excess (44, 121). Hence, it may be possible to prevent this metal-induced neurotoxicity by treating MLIV patients with lysosome-targeted Fe^{2+} and/or Zn^{2+} chelators. This mode of treatment for the disease seems quite straightforward but experimental evidence to support the claims that Fe^{2+} and Zn^{2+} are mediating a significant portion of the neurodegeneration in MLIV is still lacking. Future experimentation with brain and nervous tissue from available MLIV animal models should focus on this question going forward.

8.2. Prospective TRPML-Based Therapies for Other Lysosomal Storage Disorders

A very recent report described a crucial role for TRPML1 in TFEB-mediated lysosomal exocytosis. As mentioned earlier, experiments involving overexpression of the TFEB transcription factor showed that TFEB stimulates TRPML1-mediated vesicular Ca^{2+} release to effect fusion of lysosomes with the plasma membrane and exocytosis of intra-vesicular contents (Fig. 1a) (75). The authors of this study then manipulated this process to facilitate treatment of various lysosomal storage disorders (LSDs) in vitro and in vivo. They found that TFEB induced lysosomal exocytosis effectively cleared the storage material in cultured cells from multiple sulfatase deficiency (MSD; MIM 272200), mucopolysaccharidosis IIIA (MPS-IIIA; Sanfilippo A; MIM 252900), Batten (CLN3; MIM 204200), and Pompe (Glycogen Storage Disease; MIM 232300) disease patients without rescuing the storage in TRPML1-deficient MLIV cultured cells. In addition, gene therapy-mediated transduction of TFEB effectively cleared stored glycose amino-glycans and pathological phenotypes (macrophage infiltration and apoptosis) in a mouse model of MSD (75). Hence, this study offers new hope for treating many different kinds of devastating lysosomal storage disorders by converting TRPML1, a ubiquitously expressed protein (12, 91, 106), into a makeshift "garbage truck" that deposits accumulating intracellular toxic waste materials into extracellular space for immune system clearance.

However, as promising as this prospective therapy appears from preliminary experiments, there is still much to be clarified and tested before TRPML1-dependent lysosomal exocytosis is utilized for widespread treatment of LSDs. First, the authors utilized a gene therapy-based method in order to rescue MSD mice. This strategy must be adapted to humans since mouse-effective vectors are not necessarily as effective on humans (122). Second, gene therapy is generally more effective and longer lasting in treating systemic diseases such as Pompe disease (123) but neurological disorders of the central nervous system, such as metachromatic leukodystrophy (MIM 250100), are best treated with multiple injections with only proximal localized and transient effectiveness (122). Third, while sucrose and trehalose disaccharides have already been identified as pharmacological agents that stimulate TFEB (and presumably TRPML1) expression (74, 124), they have not yet been tested in the context of lysosomal exocytosis and LSD pathology rescue. This avenue is certainly worth investigating in the future, but the packaging of easily digestible disaccharide sugars for tissue-specific delivery (especially brain and CNS tissues on the other end of the BBB) is a concern for prospective drug development. Finally, any therapeutic method that relies on enhanced TRPML1 stimulation, for extended periods of time, runs the risk that new pathologies may surface from channel hyperstimulation. More comprehensive experimentation with animal models from various classes of LSDs will hopefully provide positive resolutions to these crucial issues.

8.3. Prospective TRPML-Based Therapies for Charcot–Marie–Tooth, Amyotrophic Lateral Sclerosis, and Francois-Neetens Corneal Fleck Dystrophy

Charcot–Marie–Tooth (CMT) diseases are a heterogeneous class of genetic peripheral neuropathies that are characterized by progressive distal muscle weakness and atrophy. CMT type 2 is considered the axonal form of the disease, featuring higher nerve conduction velocities than the demyelinating CMT types 1, 3, and 4. The demyelinating CMT types that feature autosomal recessive inheritance are designated CMT type 4 (125). Within this subclass of genetic neuropathies is CMT4J (MIM 611228), a severe form of CMT that results from one null mutation and one hypomorphic missense mutation in the *FIG4* gene (126, 127).

Amyotrophic lateral sclerosis (ALS) is a severe late-onset neurological disorder that is characterized by selective neurodegeneration of motor neurons in the central nervous system and brain that ultimately results in paralysis and death (128). The genetic determinants of the disorder are many, but one report found that ~1–2% of ALS patients of European descent were heterozygous for a deleterious *FIG4* allele (129). Interestingly, the *FIG4* gene codes for a $PI(3,5)P_2$ 5-phospatase that is critical for regulating $PI(3,5)P_2$ levels in vivo (126) and, as mentioned earlier, $PI(3,5)P_2$ has recently been identified as a potent TRPML channel activator (65).

As a $PI(3,5)P_2$ 5-phosphatase, the primary role of the Fig4 enzyme would appear to be conversion of $PI(3,5)P_2$ into PI(3)P (130). Thus, one would predict that loss-of-function mutations in Fig4 would lead to enhanced $PI(3,5)P_2$ levels. However, genetic studies in yeast and mice revealed that ablation of Fig4 homologues actually led to reduced $PI(3,5)P_2$ levels (126, 131, 132). This complicated phenotype is explained by biochemical and genetic evidence that Fig4 is an essential component of a core protein complex that is responsible for regulating both $PI(3,5)P_2$ synthesis and turnover in LEL membranes (111). Indeed, Fig4 has been shown to co-associate with an essential scaffold protein, Vac14, and PIKfyve/Fab1, the only known kinase with the ability to synthesize $PI(3,5)P_2$ from PI(3)P (111). When any member of this complex is downregulated, either by complete removal or by missense mutation at a site of protein interaction, the whole complex fails to assemble and $PI(3,5)P_2$ is not synthesized (127, 133). Interestingly, the $PI(3,5)P_2$ deficiency that results from genetic ablation of either PIKfyve, Vac14, or Fig4 leads to a readily identified enlarged vacuolar/endosomal phenotype (125) that is similar to the granulated vacuolar inclusions in MLIV cells (although MLIV-like lamellar inclusions are not observed in these models). Moreover, corneal pathology is also described for heterozygous carriers of loss-of-function mutations in PIKfyve. This pathology, termed autosomal dominant Francois-Neetens corneal fleck dystrophy (CFD, MIM 121850), is not nearly as severe as the corneal opacity in MLIV patients, but enlarged vesicles of unknown origin are observed in the cornea of affected individuals (134).

Grippingly, current research on TRPML channels is offering new avenues for treating these $PI(3,5)P_2$-dependent disorders. A recent report showed that overexpression of TRPML1 was sufficient to rescue the vacuolar phenotype in $PI(3,5)P_2$-deficient mouse Vac14 knockout cells. This rescue was shown to be $PI(3,5)P_2$ dependent because a TRPML1 mutant that failed to bind $PI(3,5)P_2$ also failed to rescue the $Vac14^{-/-}$ cells (65). Importantly, a subsequent report indicated that these experiments also warrant comparison to Fig4 downregulated disorders like CMT4J and ALS because even Fig4 protein levels are significantly reduced in $Vac14^{-/-}$ mice (127). This indicates that TRPML1 overexpression rescued the vacuolar phenotype in the absence of functional Fig4 as well as Vac14. Therefore, in order to compensate for $PI(3,5)P_2$ deficiency in CMT4J, ALS, and CFD patients, investigators are currently searching for ways to activate endogenous TRPML1 in target tissues of these disorders. Possible strategies include pharmacological induction of TFEB-mediated TRPML1 expression activation or small molecule targeted activation of TRPML1 currents. However, these prospective therapies will require the development of activators that can somehow reach target tissue while still maintaining effectiveness; a tall order for newly identified channel activators.

9. Concluding Remarks

A story that began with a waddling, hearing-impaired by-product of genetic crosses (the *varitint-waddler* mouse) and the unfortunate discovery of a devastating psychomotor disorder (MLIV), has flourished into an exciting narrative with multiple twists and turns. The story is not over yet, but TRPML research has already gained much momentum. The cell biology of TRPML channels has taught invaluable lessons regarding membrane trafficking defects in neuropathology. Indeed, these lessons are crucial because one of the greatest challenges in modern medicine is identification of the proverbial "problem." Once the problem is known, a physician is better able to treat his patient. Similarly, now that the issues with TRPML channel dysfunction along endocytic/exocytic pathways are coming to light, the opportunities for generating appropriate treatments for TRPML-associated disorders are proceeding apace. This search for therapeutic solutions is now well underway.

Acknowledgements

The author is grateful to Dr. Shaya Lev and Maximilian Peters for careful reading of the manuscript. This review was made possible by the generous support of Rabbi David and Mrs. Anita Fuld.

References

1. Cloudman AM, Bunker LE Jr (1945) The varitint-waddler mouse: a dominant mutation in Mus musculus. J Hered 36:259–263
2. Lane PW (1972) Two new mutations in linkage group XVI of the house mouse. Flaky tail and varitint-waddler-J. J Hered 63:135–140
3. Berman ER et al (1974) Congenital corneal clouding with abnormal systemic storage bodies: a new variant of mucolipidosis. J Pediatr 84:519–526
4. Amir N, Zlotogora J, Bach G (1987) Mucolipidosis type IV: clinical spectrum and natural history. Pediatrics 79:953–959
5. Zeevi DA, Frumkin A, Bach G (2007) TRPML and lysosomal function. Biochim Biophys Acta 1772:851–858
6. Wakabayashi K et al (2011) Mucolipidosis type IV: an update. Mol Genet Metab 104(3): 206–213
7. Altarescu G et al (2002) The neurogenetics of mucolipidosis type IV. Neurology 59: 306–313
8. Alroy J, Ucci AA (2006) Skin biopsy: a useful tool in the diagnosis of lysosomal storage diseases. Ultrastruct Pathol 30:489–503
9. Slaugenhaupt SA et al (1999) Mapping of the mucolipidosis type IV gene to chromosome 19p and definition of founder haplotypes. Am J Hum Genet 65:773–778
10. Bargal R et al (2000) Identification of the gene causing mucolipidosis type IV. Nat Genet 26:118–123

11. Bassi MT et al (2000) Cloning of the gene encoding a novel integral membrane protein, mucolipidin-and identification of the two major founder mutations causing mucolipidosis type IV. Am J Hum Genet 67:1110–1120
12. Sun M et al (2000) Mucolipidosis type IV is caused by mutations in a gene encoding a novel transient receptor potential channel. Hum Mol Genet 9:2471–2478
13. Kiselyov K et al (2005) TRP-ML1 is a lysosomal monovalent cation channel that undergoes proteolytic cleavage. J Biol Chem 280: 43218–43223
14. Zeevi DA et al (2009) A potentially dynamic lysosomal role for the endogenous TRPML proteins. J Pathol 219:153–162
15. Bach G (2001) Mucolipidosis type IV. Mol Genet Metab 73:197–203
16. Bargal R et al (2001) Mucolipidosis type IV: novel MCOLN1 mutations in Jewish and non-Jewish patients and the frequency of the disease in the Ashkenazi Jewish population. Hum Mutat 17:397–402
17. Kim HJ, Jackson T, Noben-Trauth K (2003) Genetic analyses of the mouse deafness mutations varitint-waddler (Va) and jerker (Espnje). J Assoc Res Otolaryngol 4:83–90
18. Di PF et al (2002) Mutations in Mcoln3 associated with deafness and pigmentation defects in varitint-waddler (Va) mice. Proc Natl Acad Sci USA 99:14994–14999
19. van Aken AF et al (2008) TRPML3 mutations cause impaired mechano-electrical transduction and depolarization by an inward-rectifier cation current in auditory hair cells of varitint-waddler mice. J Physiol 586:5403–5418
20. Xu H et al (2007) Activating mutation in a mucolipin transient receptor potential channel leads to melanocyte loss in varitint-waddler mice. Proc Natl Acad Sci USA 104: 18321–18326
21. Grimm C et al (2007) A helix breaking mutation in TRPML3 leads to constitutive activity underlying deafness in the varitint waddler mouse. Proc Natl Acad Sci USA 104: 19583–19588
22. Kim HJ et al (2007) Gain-of-function mutation in TRPML3 causes the mouse varitint-waddler phenotype. J Biol Chem 282: 36138–36142
23. Nagata K et al (2008) The varitint-waddler (Va) deafness mutation in TRPML3 generates constitutive, inward rectifying currents and causes cell degeneration. Proc Natl Acad Sci USA 105:353–358
24. Grimm C, Jors S, Heller S (2009) Life and death of sensory hair cells expressing constitutively active TRPML3. J Biol Chem 284: 13823–13831
25. Kim HJ et al (2008) A novel mode of TRPML3 regulation by extracytosolic PH absent in the varitint-waddler phenotype. EMBO J 27:1197–1205
26. Kim HJ et al (2010) Properties of the TRPML3 channel pore and its stable expansion by the varitint-waddler-causing mutation. J Biol Chem 285:16513–16520
27. Steel KP (2002) Varitint-waddler: a double whammy for hearing. Proc Natl Acad Sci USA 99:14613–14615
28. Jors S et al (2010) Genetic inactivation of Trpml3 does not lead to hearing and vestibular impairment in mice. PLoS One 5:e14317
29. Slaugenhaupt SA (2002) The molecular basis of mucolipidosis type IV. Curr Mol Med 2:445–450
30. Bach G, Cohen MM, Kohn G (1975) Abnormal ganglioside accumulation in cultured fibroblasts from patients with mucolipidosis IV. Biochem Biophys Res Commun 66:1483–1490
31. Bach G et al (1977) Mucopolysaccharide accumulation in cultured skin fibroblasts derived from patients with mucolipidosis IV. Am J Hum Genet 29:610–618
32. Bargal R, Bach G (1988) Phospholipids accumulation in mucolipidosis IV cultured fibroblasts. J Inherit Metab Dis 11:144–150
33. Caimi L et al (1982) Mucolipidosis IV, a sialolipidosis due to ganglioside sialidase deficiency. J Inherit Metab Dis 5:218–224
34. Chen CS, Bach G, Pagano RE (1998) Abnormal transport along the lysosomal pathway in mucolipidosis, type IV disease. Proc Natl Acad Sci USA 95:6373–6378
35. Crandall BF et al (1982) Review article: mucolipidosis IV. Am J Med Genet 12:301–308
36. Bargal R, Bach G (1997) Mucolipidosis type IV: abnormal transport of lipids to lysosomes. J Inherit Metab Dis 20:625–632
37. Bach G, Chen CS, Pagano RE (1999) Elevated lysosomal pH in mucolipidosis type IV cells. Clin Chim Acta 280:173–179
38. Dong XP et al (2008) The type IV mucolipidosis-associated protein TRPML1 is an endolysosomal iron release channel. Nature 455: 992–996
39. Eichelsdoerfer JL et al (2010) Zinc dyshomeostasis is linked with the loss of mucolipidosis IV-associated TRPML1 ion channel. J Biol Chem 285:34304–34308
40. Kiselyov K et al (2011) TRPML: transporters of metals in lysosomes essential for cell survival? Cell Calcium 50:288–294

41. Soyombo AA et al (2006) TRP-ML1 regulates lysosomal pH and acidic lysosomal lipid hydrolytic activity. J Biol Chem 281: 7294–7301
42. Goldin E et al (1999) Mucolipidosis IV consists of one complementation group. Proc Natl Acad Sci USA 96:8562–8566
43. Cheng X et al (2010) Mucolipins: intracellular TRPML1-3 channels. FEBS Lett 584: 2013–2021
44. Kurz T et al (2008) Lysosomes in iron metabolism, ageing and apoptosis. Histochem Cell Biol 129:389–406
45. Kurz T, Eaton JW, Brunk UT (2011) The role of lysosomes in iron metabolism and recycling. Int J Biochem Cell Biol 43(12): 1686–1697
46. Curcio-Morelli C et al (2010) Macroautophagy is defective in mucolipin-1-deficient mouse neurons. Neurobiol Dis 40:370–377
47. Vergarajauregui S et al (2008) Autophagic dysfunction in mucolipidosis type IV patients. Hum Mol Genet 17:2723–2737
48. Venugopal B et al (2007) Neurologic, gastric, and opthalmologic pathologies in a murine model of mucolipidosis type IV. Am J Hum Genet 81:1070–1083
49. Chandra M et al (2011) A role for the Ca2+ channel TRPML1 in gastric acid secretion, based on analysis of knockout mice. Gastroenterology 140:857–867
50. Schiffmann R et al (1998) Constitutive achlorhydria in mucolipidosis type IV. Proc Natl Acad Sci USA 95:1207–1212
51. Micsenyi MC et al (2009) Neuropathology of the Mcoln1(-/-) knockout mouse model of mucolipidosis type IV. J Neuropathol Exp Neurol 68:125–135
52. Venkatachalam K et al (2008) Motor deficit in a Drosophila model of mucolipidosis type IV due to defective clearance of apoptotic cells. Cell 135:838–851
53. Schaheen L, Dang H, Fares H (2006) Basis of lethality in *C. elegans* lacking CUP-5, the mucolipidosis type IV orthologue. Dev Biol 293:382–391
54. Fares H, Greenwald I (2001) Regulation of endocytosis by CUP-5, the *Caenorhabditis elegans* mucolipin-1 homolog. Nat Genet 28:64–68
55. Treusch S et al (2004) *Caenorhabditis elegans* functional orthologue of human protein H-mucolipin-1 is required for lysosome biogenesis. Proc Natl Acad Sci USA 101: 4483–4488
56. Campbell EM, Fares H (2010) Roles of CUP-5, the *Caenorhabditis elegans* orthologue of human TRPML1, in lysosome and gut granule biogenesis. BMC Cell Biol 11:40
57. Sun T et al (2011) CUP-5, the *C. elegans* ortholog of the mammalian lysosomal channel protein MLN1/TRPML1, is required for proteolytic degradation in autolysosomes. Autophagy 7:1308–1315
58. Pryor PR et al (2006) Mucolipin-1 is a lysosomal membrane protein required for intracellular lactosylceramide traffic. Traffic 7:1388–1398
59. Thompson EG et al (2007) Lysosomal trafficking functions of mucolipin-1 in murine macrophages. BMC Cell Biol 8:54
60. Chen JL, Ahluwalia JP, Stamnes M (2002) Selective effects of calcium chelators on anterograde and retrograde protein transport in the cell. J Biol Chem 277:35682–35687
61. Hay JC (2007) Calcium: a fundamental regulator of intracellular membrane fusion? EMBO Rep 8:236–240
62. Pryor PR et al (2000) The role of intraorganellar Ca(2+) in late endosome-lysosome heterotypic fusion and in the reformation of lysosomes from hybrid organelles. J Cell Biol 149:1053–1062
63. Luzio JP, Bright NA, Pryor PR (2007) The role of calcium and other ions in sorting and delivery in the late endocytic pathway. Biochem Soc Trans 35:1088–1091
64. Luzio JP, Pryor PR, Bright NA (2007) Lysosomes: fusion and function. Nat Rev Mol Cell Biol 8:622–632
65. Dong XP et al (2010) PI(3,5)P(2) controls membrane trafficking by direct activation of mucolipin Ca($^{2+}$) release channels in the endolysosome. Nat Commun 1:38
66. LaPlante JM et al (2002) Identification and characterization of the single channel function of human mucolipin-1 implicated in mucolipidosis type IV, a disorder affecting the lysosomal pathway. FEBS Lett 532:183–187
67. LaPlante JM et al (2004) Functional links between mucolipin-1 and Ca^{2+}-dependent membrane trafficking in mucolipidosis IV. Biochem Biophys Res Commun 322: 1384–1391
68. Zhang F et al (2009) TRP-ML1 functions as a lysosomal NAADP-sensitive Ca^{2+} release channel in coronary arterial myocytes. J Cell Mol Med 13:3174–3185
69. Andrews NW (2000) Regulated secretion of conventional lysosomes. Trends Cell Biol 10: 316–321
70. Rodriguez A et al (1997) Lysosomes behave as Ca^{2+}-regulated exocytic vesicles in fibroblasts and epithelial cells. J Cell Biol 137:93–104

71. Rodriguez A et al (1999) CAMP regulates Ca^{2+}-dependent exocytosis of lysosomes and lysosome-mediated cell invasion by trypanosomes. J Biol Chem 274:16754–16759
72. LaPlante JM et al (2006) Lysosomal exocytosis is impaired in mucolipidosis type IV. Mol Genet Metab 89:339–348
73. Dong XP et al (2009) Activating mutations of the TRPML1 channel revealed by proline-scanning mutagenesis. J Biol Chem 284:32040–32052
74. Sardiello M et al (2009) A gene network regulating lysosomal biogenesis and function. Science 325:473–477
75. Medina DL et al (2011) Transcriptional activation of lysosomal exocytosis promotes cellular clearance. Dev Cell 21:421–430
76. Puertollano R, Kiselyov K (2009) TRPMLs: in sickness and in health. Am J Physiol Renal Physiol 296:F1245–F1254
77. Miedel MT et al (2008) Membrane traffic and turnover in TRP-ML1-deficient cells: a revised model for mucolipidosis type IV pathogenesis. J Exp Med 205:1477–1490
78. Kiselyov K, Muallem S (2008) Mitochondrial Ca^{2+} homeostasis in lysosomal storage diseases. Cell Calcium 44:103–111
79. Manzoni M et al (2004) Overexpression of wild-type and mutant mucolipin proteins in mammalian cells: effects on the late endocytic compartment organization. FEBS Lett 567: 219–224
80. Miedel MT et al (2006) Posttranslational cleavage and adaptor protein complex-dependent trafficking of mucolipin-1. J Biol Chem 281:12751–12759
81. Vergarajauregui S, Puertollano R (2006) Two di-leucine motifs regulate trafficking of mucolipin-1 to lysosomes. Traffic 7:337–353
82. Kim HJ et al (2009) The Ca(2+) channel TRPML3 regulates membrane trafficking and autophagy. Traffic 10:1157–1167
83. Cantiello HF et al (2005) Cation channel activity of mucolipin-1: the effect of calcium. Pflugers Arch 451:304–312
84. Raychowdhury MK et al (2004) Molecular pathophysiology of mucolipidosis type IV: pH dysregulation of the mucolipin-1 cation channel. Hum Mol Genet 13:617–627
85. Katz B, Minke B (2009) Drosophila photoreceptors and signaling mechanisms. Front Cell Neurosci 3:2
86. Zeevi DA et al (2010) Heteromultimeric TRPML channel assemblies play a crucial role in the regulation of cell viability models and starvation-induced autophagy. J Cell Sci 123: 3112–3124
87. Zhang F, Li PL (2007) Reconstitution and characterization of a nicotinic acid adenine dinucleotide phosphate (NAADP)-sensitive Ca^{2+} release channel from liver lysosomes of rats. J Biol Chem 282:25259–25269
88. Zhang F et al (2011) Reconstitution of lysosomal NAADP-TRP-ML1 signaling pathway and its function in TRP-ML1(-/-) cells. Am J Physiol Cell Physiol 301:C421–C430
89. Calcraft PJ et al (2009) NAADP mobilizes calcium from acidic organelles through two-pore channels. Nature 459:596–600
90. Yamaguchi S et al (2011) Transient receptor potential mucolipin 1 (TRPML1) and two-pore channels are functionally independent organellar ion channels. J Biol Chem 286:22934–22942
91. Grimm C et al (2010) Small molecule activators of TRPML3. Chem Biol 17:135–148
92. Cerny J et al (2004) The small chemical vacuolin-1 inhibits Ca(2+)-dependent lysosomal exocytosis but not cell resealing. EMBO Rep 5:883–888
93. Shen D, Wang X, Xu H (2011) Pairing phosphoinositides with calcium ions in endolysosomal dynamics: phosphoinositides control the direction and specificity of membrane trafficking by regulating the activity of calcium channels in the endolysosomes. Bioessays 33:448–457
94. Venugopal B et al (2009) Chaperone-mediated autophagy is defective in mucolipidosis type IV. J Cell Physiol 219:344–353
95. LaPlante JM et al (2011) The cation channel mucolipin-1 is a bifunctional protein that facilitates membrane remodeling via its serine lipase domain. Exp Cell Res 317:691–705
96. Vergarajauregui S, Martina JA, Puertollano R (2011) LAPTMs regulate lysosomal function and interact with mucolipin 1: new clues for understanding mucolipidosis type IV. J Cell Sci 124:459–468
97. Cabrita MA et al (1999) Mouse transporter protein, a membrane protein that regulates cellular multidrug resistance, is localized to lysosomes. Cancer Res 59:4890–4897
98. Pak Y et al (2006) Transport of LAPTM5 to lysosomes requires association with the ubiquitin ligase Nedd4, but not LAPTM5 ubiquitination. J Cell Biol 175:631–645
99. Vergarajauregui S, Martina JA, Puertollano R (2009) Identification of the penta-EF-hand protein ALG-2 As a Ca^{2+}-dependent interactor of mucolipin-1. J Biol Chem 284:36357–36366
100. Abe K, Puertollano R (2011) Role of TRP channels in the regulation of the endosomal pathway. Physiology (Bethesda) 26:14–22

101. Brailoiu E et al (2009) Essential requirement for two-pore channel 1 in NAADP-mediated calcium signaling. J Cell Biol 186:201–209
102. Venkatachalam K, Hofmann T, Montell C (2006) Lysosomal localization of TRPML3 depends on TRPML2 and the mucolipidosis-associated protein TRPML1. J Biol Chem 281:17517–17527
103. Curcio-Morelli C et al (2010) Functional multimerization of mucolipin channel proteins. J Cell Physiol 222:328–335
104. Karacsonyi C, Miguel AS, Puertollano R (2007) Mucolipin-2 localizes to the Arf6-associated pathway and regulates recycling of GPI-APs. Traffic 8:1404–1414
105. Lev S et al (2010) Constitutive activity of the human TRPML2 channel induces cell degeneration. J Biol Chem 285:2771–2782
106. Samie MA et al (2009) The tissue-specific expression of TRPML2 (MCOLN-2) gene is influenced by the presence of TRPML1. Pflugers Arch 459:79–91
107. Caplan S et al (2002) A tubular EHD1-containing compartment involved in the recycling of major histocompatibility complex class I molecules to the plasma membrane. EMBO J 21:2557–2567
108. Martina JA, Lelouvier B, Puertollano R (2009) The calcium channel mucolipin-3 is a novel regulator of trafficking along the endosomal pathway. Traffic 10:1143–1156
109. Lelouvier B, Puertollano R (2011) Mucolipin-3 regulates luminal calcium, acidification, and membrane fusion in the endosomal pathway. J Biol Chem 286:9826–9832
110. Atiba-Davies M, Noben-Trauth K (2007) TRPML3 and hearing loss in the varitint-waddler mouse. Biochim Biophys Acta 1772:1028–1031
111. Ho CY, Alghamdi TA, Botelho RJ (2011) Phosphatidylinositol-3,5-bisphosphate: no longer the poor PIP(2). Traffic 13(1):1–8
112. Folkerth RD et al (1995) Mucolipidosis IV: morphology and histochemistry of an autopsy case. J Neuropathol Exp Neurol 54:154–164
113. Silva GA (2006) Neuroscience nanotechnology: progress, opportunities and challenges. Nat Rev Neurosci 7:65–74
114. Silva GA (2007) Nanotechnology approaches for drug and small molecule delivery across the blood brain barrier. Surg Neurol 67:113–116
115. Silva GA (2010) Nanotechnology applications and approaches for neuroregeneration and drug delivery to the central nervous system. Ann N Y Acad Sci 1199:221–230
116. Poole B, Ohkuma S (1981) Effect of weak bases on the intralysosomal pH in mouse peritoneal macrophages. J Cell Biol 90:665–669
117. Reijngoud DJ, Oud PS, Tager JM (1976) Effect of ionophores on intralysosomal pH. Biochim Biophys Acta 448:303–313
118. Kogot-Levin A et al (2009) Mucolipidosis type IV: the effect of increased lysosomal pH on the abnormal lysosomal storage. Pediatr Res 65:686–690
119. Piper RC, Luzio JP (2004) CUPpling calcium to lysosomal biogenesis. Trends Cell Biol 14:471–473
120. Krogsgaard-Larsen P, StrÃmgaard K, Madsen U (2010) Textbook of drug design and discovery, 4th edn. CRC, Boca Raton, FL
121. Lee SJ, Cho KS, Koh JY (2009) Oxidative injury triggers autophagy in astrocytes: the role of endogenous zinc. Glia 57:1351–1361
122. Sands MS, Davidson BL (2006) Gene therapy for lysosomal storage diseases. Mol Ther 13:839–849
123. Sun B et al (2005) Correction of glycogen storage disease type II by an adeno-associated virus vector containing a muscle-specific promoter. Mol Ther 11:889–898
124. Dehay B et al (2010) Pathogenic lysosomal depletion in parkinson's disease. J Neurosci 30:12535–12544
125. Nicot AS, Laporte J (2008) Endosomal phosphoinositides and human diseases. Traffic 9:1240–1249
126. Chow CY et al (2007) Mutation of FIG4 causes neurodegeneration in the pale tremor mouse and patients with CMT4J. Nature 448:68–72
127. Lenk GM et al (2011) Pathogenic mechanism of the FIG4 mutation responsible for Charcot-Marie-Tooth disease CMT4J. PLoS Genet 7:1002104
128. Pasinelli P, Brown RH (2006) Molecular biology of amyotrophic lateral sclerosis: insights from genetics. Nat Rev Neurosci 7:710–723
129. Chow CY et al (2009) Deleterious variants of FIG4, a phosphoinositide phosphatase, in patients with ALS. Am J Hum Genet 84:85–88
130. Sbrissa D et al (2007) Core protein machinery for mammalian phosphatidylinositol 3,5-bisphosphate synthesis and turnover that regulates the progression of endosomal transport. Novel Sac phosphatase joins the ArPIKfyve-PIKfyve complex. J Biol Chem 282:23878–23891

131. Duex JE et al (2006) Phosphoinositide 5-phosphatase Fig 4p is required for both acute rise and subsequent fall in stress-induced phosphatidylinositol 3,5-bisphosphate levels. Eukaryot Cell 5:723–731
132. Duex JE, Tang F, Weisman LS (2006) The Vac14p-Fig4p complex acts independently of Vac7p and couples PI3,5P2 synthesis and turnover. J Cell Biol 172:693–704
133. Ikonomov OC et al (2010) ArPIKfyve regulates Sac3 protein abundance and turnover: disruption of the mechanism by Sac3I41T mutation causing Charcot-Marie-Tooth 4J disorder. J Biol Chem 285: 26760–26764
134. Li S et al (2005) Mutations in PIP5K3 are associated with Francois-Neetens mouchetee fleck corneal dystrophy. Am J Hum Genet 77:54–63

Chapter 10

The TRPP Signaling Module: TRPP2/Polycystin-1 and TRPP2/PKD1L1

Alexis Hofherr

Abstract

TRPP ion channels assemble with polycystin-1 family proteins into receptor-channel complexes. TRPP2–Polycystin-1 is required to coordinate renal three-dimensional tissue organization, whereas TRPP2–PKD1L1 is essential for establishment of left–right asymmetry during early embryonic development. The availability of orthologous and heterologous animal models made it feasible not only to characterize the function of single genes, but also to combine gene deficiencies and/or gene overexpression to establish molecular interdependencies. Canonical and noncanonical Wnt signaling have been associated with polycystic kidney disease. Furthermore, PRKCSH and SEC63 have been implicated in TRPP/polycystin protein maturation. Putative targets for therapeutic agents have set the stage for clinical intervention studies. The purpose of this review is to critically summarize recent in vivo evidence and to highlight inconsistencies that should be resolved if an accurate understanding of TRPP channels and polycystins is to be achieved. It is hoped that a deeper understanding of associated phenotypes will help to design rational therapies for autosomal dominant polycystic kidney disease.

Key words: TRPP2, PKD1, PKD1L1, TRPP heteromerization, Autosomal dominant polycystic kidney disease, Situs inversus

1. Introduction

Transient receptor potential (TRP) channels serve as intrinsic sensors of the cellular environment (1, 2). A characteristic of these channels is their responsiveness to multiple distinct physical and chemical stimuli (3). Therefore, the physiological activity of any given TRP channel will be governed by the specifics of its cellular context, e.g., ligand concentrations, phosphorylation status, lipid environment, and interacting proteins (3). The TRP-Polycystin (TRPP) subfamily and their polycystin 1 interaction partners have become of particular interest because of their relevance to renal

Arpad Szallasi and Tamás Bíró (eds.), *TRP Channels in Drug Discovery: Volume I*, Methods in Pharmacology and Toxicology, DOI 10.1007/978-1-62703-077-9_10, © Springer Science+Business Media, LLC 2012

ailment and early embryonic development. Much effort has been spent to elucidate the joint physiological role of these evolutionary conserved proteins. Excellent experimental and observational studies have broadened our understanding of the involved proteins and their relation to cellular signaling pathways.

2. Autosomal Dominant Polycystic Kidney Disease

In 2009, more than 26,800 United States citizens were suffering from kidney failure due to PKD (4). Two etiologically distinct subgroups of PKD have been established: autosomal dominant polycystic kidney disease (ADPKD) and rare autosomal recessive PKD (5). The hallmark of ADPKD is the age-dependent massive enlargement of the kidney characterized by dilated or cystic renal tubular segments. The defective three-dimensional tissue organization in ADPKD begins in utero and is slowly progressive; signs of the disorder may not be detected until late adulthood. Approximately 50% of affected individuals ultimately develop end-stage renal disease (ESRD) by the sixth decade of life (6–9). With only dialysis and kidney transplant as treatment options available, ESRD remains a severe illness with poor life expectancy. The 5-year probability of survival among 2004 incident PKD ESRD patients was 38.8% (4). Clearly, specific and efficacious treatment options are in urgent need for ESRD and ADPKD in particular.

3. ADPKD Is a Multisystem Disorder

A wide spectrum of mostly nonlife-threatening abnormalities accompanies ADPKD. Pain is a recurrent symptom in adult patients, often caused by cyst hemorrhage, cyst infection, or kidney stones (10, 11). Gross and microscopic hematuria are frequently present (12). A major contributor to renal disease progression in addition to cardiovascular morbidity and mortality is arterial hypertension (13). Hepatic cysts are the primary extra-renal lesion (14). Cysts might develop, furthermore, in the arachnoid membrane, pancreas, and seminal tract (15, 16). Cardiac and vascular manifestations include an increased frequency of valvular heart disease, coronary artery, and intracranial aneurysms, together with chronic subdural hematoma (16–21).

4. Genetics of ADPKD: PKD1 and PKD2

ADPKD is regardless of ethnicity the most common potentially lethal single-gene disorder in humans affecting one in 400–1,000 live births (22). As its name implies, ADPKD is an autosomal dominant trait. Patients carry one copy of the mutated gene and there is on average a 50% chance to inherit the disorder, which is completely penetrant. Most individuals with ADPKD have a positive family history of PKD; the incidence of cases due to spontaneous mutation is approximately 5–10% (23, 24).

Genetic studies have indicated that almost all cases of ADPKD can be attributed to mutations in either one of two genes, polycystic kidney disease 1 (PKD1) and polycystic kidney disease 2 (PKD2), respectively (#173900 and #613095 in Online Mendelian Inheritance of Man, http://www.ncbi.nlm.nih.gov/omim/) (25–28). Mutations of PKD1 on chromosome 16p13.3 are more prevalent (accounting for ~70% of cases) than mutations of PKD2 on chromosome 4q21 (accounting for ~30% of cases) (29–32). An existence of a third gene causing rare forms of ADPKD has been hypothesized (33–36). However, progressive assignment of families to PKD1 and PKD2 has left few with no linkage to either of the known gene loci. Errors in genotyping, DNA sample mix-up, nonpaternity, and misdiagnosis (including phenocopies and nonpenetrance) have been reported to lead to false exclusion of linkage (37). Thus, the existence of a third gene causing ADPKD is rather unlikely.

The clinical course of ADPKD may vary from person to person, from neonatal death to adequate renal function into old age. Recently, mutations in PKD1 and PKD2 have also been associated with perinatal death, which might be clinically indistinguishable from autosomal recessive PKD (31, 38–40). Although the clinical manifestations overlap completely between PKD1 and PKD2, there is a strong locus effect on renal pathology. On a population scale, PKD1 mutations cause earlier PKD with ESRD occurring at 54.3 years versus 74 years in PKD2 (7, 41). Gender correlates with phenotypic penetrance. Men with PKD2 affection experience a faster worsening symptomatology than women, with ESRD at 68.1 years and 76 years, respectively (7, 42). Whether this link holds true in PKD1 remains to be validated (43–45).

Genotypes homozygous and compound heterozygous for PKD1 or PKD2 have been reported to be lethal, resulting in embryonic death with PKD, cardiac failure, and gross edema (46–51). Individuals heterozygous for both PKD1 and PKD2 mutations survive to adulthood but show a harsh progression of disease (52, 53).

The specific nature of PKD1 and PKD2 sequence alterations is rather heterogeneous. Hundreds of unique mutations have been associated with ADPKD and very limited information is available on genotype–phenotype correlation (The ADPKD Mutation Database, http://pkdb.mayo.edu). A weak allelic effect on renal pathology may exist for PKD1, with mutations in the 5′ region of the gene presenting with a significantly more severe course of disease (44). No such effect has been described for PKD2 (42).

A significant inter- and intra-familiar phenotypic variability points to an involvement of genetic and environmental modifiers to the ADPKD phenotype (54). Subtle genetic determinants including the level of PKD1 protein expression and the penetrance of pathogenic alleles might modify the disorder's slow progression over the course of decades (44, 55). In the short-term, environmental factors play an important part in renal pathology, e.g., ADPKD cyst enlargement is significantly accelerated by acute kidney injury events (56–58).

5. TRPP and Polycystin Proteins Form a Receptor-Channel Complex

PKD1 and PKD2 encode for Polycystin-1 and TRPP2, respectively (25–28). Together the two proteins are thought to form the core of a signaling cascade that uses Ca^{2+} as a second messenger (59–64).

Polycystin-1 is a 4,302 amino acids long integral membrane protein with 11 transmembrane segments (S1–S11) (65, 66). It resembles structurally a receptor or adhesion molecule. However, the ligand or activating stimulus of Polycystin-1 is unknown. Four additional PKD1-like genes were identified through homology cloning that are now members of the Polycystin-1 family: (1) polycystic kidney disease 1 like 1 (PKD1L1), (2) polycystic kidney disease 1 like 2 (PKD1L2), (3) polycystic kidney disease 1 like 3 (PKD1L3), and (4) polycystic kidney disease receptor for egg jelly (PKDREJ, also referred to as polycystin receptor for egg jelly (PC-REJ)) (67–71).

TRPP2 is the archetypal TRPP ion channel (1). The 968 amino acids long protein features the familiar TRP composition with six transmembrane segments (S1–S6) and a pore loop (between S5 and S6) (25). It functions as a Ca^{2+}- permeable nonselective cation channel (62, 72, 73). Two closely related proteins were identified through homology cloning: TRPP3 (formerly named Polycystin L, encoded by polycystic kidney disease 2 like 1 (PKD2L1)) and TRPP5 (encoded by polycystic kidney disease 2 like 2 (PKD2L2)) (71, 74–77). Together with TRPP2 they now constitute the TRPP family of ion channels (2).

Polycystin-1 and TRPP2 have been extensively investigated at all levels of complexity, from the single molecule to the whole organism. The other members of the TRPP and polycystin-1 families are less well understood, but first discoveries have been made. Heteromerization of polycystin receptors and TRPP ion channels has added another layer of complexity to these sensory modules and might have an important contribution to the diversity of their physiological function (71). This review will draw attention to recent studies from various model systems.

6. A "Two-Hit" Mechanism Explains the Focal Nature of ADPKD

A human kidney is composed of approximately 1,000,000 nephrons (78). Nevertheless, an ADPKD patient will develop only about a thousand cysts which will eventually destroy the kidney in a process associated with a decline in glomerular filtration and ESRD (23). The quantity of overall cysts in relation to nephrons emphasizes that only a very small number of renal tubules undergo cystic transformation. The focal nature of ADPKD is consistent with a "two-hit" model. According to this model, each cyst arises as a consequence of a distinct somatic mutation event ("second hit") on the background of an inherited germ line mutation ("first hit") (79, 80). The acquired homozygous cellular inactivation of PKD1 or PKD2 is a necessary pathogenic step leading to clonal expansion of cysts (81–83). It has been demonstrated that the "second hit" instant of time is of particular importance to cyst growth (84). Inactivation of *Pkd1* in mice before postnatal day 13 provokes severely cystic kidneys within 3 weeks, whereas inactivation at day 14 or later results in cysts only after 5 months (84). Thus, it is most likely that the kidney's developmental status defines the pathologic consequences of *Pkd1* deletion.

7. Cyst Lining Cells Have Converted from Absorptive to Secretory

Kidney cysts in ADPKD originate from focal dilatations along all nephron segments, including occasional glomeruli (85). Over the space of decades adjacent renal parenchyma is displaced and destroyed (85). Extracellular matrix remodeling, interstitial inflammation, and fibrosis are concomitant pathogenetic factors (85–87). Furthermore, in cyst-neighboring tubular and glomerular cells an increased rate of apoptosis has been observed (88–90).

Nascent cysts still preserve continuity with the parental nephron and glomerular filtrate may enter (85, 91). Later on, in most cysts beyond two millimeters in diameter, no evidence for continuity with contiguous tubule segments can be demonstrated, leaving an isolated sac of fluid (91). Cysts are lined by a single layer of epithelium that is characterized by an increased cellular proliferation and decreased differentiation (71, 91–93). Nevertheless, ADPKD is not associated with an increased malignant potential (94). Cysts are, in fact, fluid filled benign renal tubular neoplasms. The phenotype of cyst lining cells has converted from absorptive to secretory; this promotes the accumulation of fluid within the cyst and drives cyst growth (95). Most likely, active net transepithelial chloride secretion is the motive force for the subsequent movement of sodium and water into the cysts (95–97). However, although progress has been made in understanding the pathophysiology of cysts, the initial event defining cyst formation, i.e., giving rise to the change from tubular to spherical morphology, is still unknown.

8. Cyst Growth and Renal Function: Linked or Unlinked Outcomes?

Cyst burden is an intuitive therapeutic target in ADPKD. It is assumed that effective therapies should slow or arrest cyst expansion and hence preserve renal function. A slowed decline in glomerular filtration rate could delay ESRD for decades and, although not cure PKD, alleviate suffering considerably.

Experimental evidence concerning the serine-threonine kinase mammalian target of rapamycin (mTOR) offered the rationale for clinical trials of mTOR inhibitors in ADPKD. On the one hand, aberrant mTOR activation is linked to tubular cell proliferation in animal models and humans with ADPKD (98). On the other hand, pharmacological inhibition of mTOR signaling by Sirolimus and Everolimus was effective in suppressing cyst growth and preserving renal function in rodent models of PKD (98, 99). However, the negative findings from two clinical trials contrasted the positive preclinical results (100, 101). Sirolimus for 18 months did not slow kidney growth and there was no difference in glomerular filtration rate (100). Everolimus appeared to transiently slow kidney enlargement during the 2-year trial, but glomerular filtration rate declined more rapidly in the Everolimus group than in the placebo group (101).

Two questions have to be considered for these discordant results. First, was the failure to detect any health benefit due to the challenges inherent in executing definitive clinical trials involving patients with ADPKD? The gradual course of disease over decades and the stage of chronic kidney disease may have been confounding factors in these short-term intervention studies. Second, do the preclinical models provide a close enough approximation of the human clinical condition to predict human outcomes? Improved surrogate indexes and the evaluation of hard endpoints in animal models will be essential to legitimate future clinical trials.

The observation that renal volume did not correlate with functional kidney impairment in the Everolimus trial raises the concern that a reduction in cyst growth might not necessarily have a beneficial effect on glomerular filtration rate (101). Certainly, a longer follow-up period would have been required to determine additional clinical endpoints. In summary, the use of mTOR inhibitors in patients with ADPKD may be more complicated than anticipated (102).

Hopes for an ADPKD intervention are now pinned on the TEMPO study (Tolvaptan Efficacy and Safety in Management of Autosomal Dominant Polycystic Kidney Disease and Its Outcomes) (103, 104).

9. Not All Cysts Are Created Equal

In addition to ADPKD, several rare cystic disorders have been described, including autosomal recessive PKD, nephronophthisis, and Bardet–Biedl syndrome (105–107). A multitude of orthologous and heterologous model systems have been established to study PKD (108). It was hoped that the exploration of links between renal epithelial cell proliferation, differentiation, and transepithelial transport may highlight a common pathway to cystogenesis. However, even extra-renal manifestations set aside, the rather different nature of these disorders and heterologous models has to be kept in mind. For example, autosomal recessive PKD is characterized by nonobstructive fusiform dilatation mainly of the renal collecting ducts leading to enlarged spongiform kidneys (107). In nephronophthisis, by contrast, kidneys present with normal size but exhibit a triad of rather small corticomedullary cysts, tubular basement membrane disruption, and tubulointerstitial nephropathy (106). Bardet–Biedl syndrome, finally, displays mainly dysplastic abnormalities that include communicating calyceal cysts, calyceal clubbing, and blunting together with tubulo-interstitial nephropathy (105). Despite these differences, the analysis of non-ADPKD cyst genes has yielded likely mechanisms to underlie at least part of the ADPKD pathogenesis. Moreover, several indirect lines of evidence indicate that cystic disorders may be more related than previously suspected.

10. The Primary Cilium Represents a Nexus for Cellular Signaling Pathways

A convergence of findings from algae, nematodes, fruit flies, and mice advocate that defects in function or structure of primary cilia are a possible common mechanism in the development of PKD (106, 109–111). In the face of a growing bias toward *translational research*, it is an exciting paradigm that unbiased studies of basic biological concepts can lead to profound insights into a number of human disease syndromes. The cilia hypothesis is based on two observations. First, genetic ablation of primary cilia in model organisms causes cyst formation; second, many proteins implicated in cystic disorders localize to the primary cilium (112–115).

Cilia are highly conserved throughout evolution. Structural analysis established that the cilium is composed of a microtubule-based core structure called the axoneme, which is surrounded by a ciliary membrane (116). Cilia are conventionally classified as primary (9 + 0) or motile (9 + 2), depending on whether the axoneme includes, beside nine peripherally located microtubule doublets, an additional central pair of microtubules (116, 117). Flagella are found on spermatozoa and single-celled eukaryotes and primarily function in cell locomotion (118). The structure of a flagellum is identical to a motile cilium; therefore, the two names are often used interchangeably (118). In vertebrates, virtually all cell types can produce primary cilia (9 + 0), also termed sensory cilia, which are usually immotile but can detect physical and chemical extracellular stimuli (119–124).

Key experimental evidence to connect primary cilia and human pathology arose from the *oak ridge polycystic kidney (orpk)* mouse (109). The *orpk* phenotype is caused by a hypomorphic allele of the Tg737 gene and is characterized by cystic kidneys, liver lesions, and defects in left–right asymmetry (112, 125). The discovery that Tg737 is the mouse homologue of IFT88, a protein required for the assembly of flagella in the green algae, *Chlamydomonas reinhardtii*, was instrumental in linking a ciliary defect to PKD (113).

The primary cilium, a once termed vestigial organelle, now claims central stage. Several studies highlight the localization of Polycystin-1 and TRPP2 in primary cilia (126, 127). However, while there is a clear connection between PKD and dysfunction of ciliary proteins, the precise nature of this relationship is not understood. In contrast to the *orpk* mouse that has stunted cilia, PKD patients form primary cilia that are normal in appearance (114, 115). None of the human PKD genes appears to be required for the assembly of primary cilia. Instead, the human disease gene products seem to be essential for ciliary signaling (70, 110).

To date, multiple cellular mechanisms have been described to act in concert with primary cilia including Ca^{2+}, Hedgehog, JAK-STAT, mTOR, Notch, and Wnt signaling (115, 124). Primary cilia are more and more perceived as nexus for multiple signaling pathways. Apparently, the dynamic interplay of a large number of components determines the output of many biological processes occurring in parallel. Polycystin-1 and TRPP2 have been linked to almost all ciliary signaling cascades (70, 109, 115, 129, 130). However, not all proteins implicated in cystic disorders localize to the primary cilium, most notably the two endoplasmic reticulum (ER) proteins PRKCSH and SEC63 (131–133). The task to decipher the complexity in cellular signal processing is an enormous one.

11. The Contentious Link between Cilia and Wnt Signaling

Abnormalities in cell proliferation and oriented cell division have been proposed as fundamental to the pathogenesis of PKD (134–136). According to that model, dividing cells align along the longitudinal axis of the growing nephron, whereas homozygous loss of PKD1 or PKD2 causes the axis of cell division to be randomized (137). Hence, affected cells may divide along an axis not parallel to the tubule lumen, causing tubule expansion rather than elongation (137). This concept seems plausible but inconsistencies remain.

Canonical and noncanonical Wnt signaling regulate proliferation and oriented cell division, respectively (138, 139). Wnt ligands are a family of secreted lipid-modified glycoproteins that act in paracrine signal transduction (140). The term Wnt is a hybrid of the two founding gene names, mouse *Integration 1* and fruit fly *Wingless* (141–143). Wnt signaling is highly conserved across evolution and indispensable to metazoan development, regeneration, and human malignancies (138, 139, 144). In kidney development, Wnts are essential for both induction of the metanephric mesenchyme and ureteric bud branching (145).

The canonical Wnt pathway converges on transcriptional regulation of its target genes by β-catenin (139). Without stimulation, cytoplasmic β-catenin is targeted for ubiquitin-mediated degradation by a multiprotein complex that includes Apc, Axin, and GSK3β (139).

The constitutive activation of β-catenin, by transgenic overexpression or targeted inhibition of β-catenin degradation by *Apc* knockout, causes PKD in mice (146, 147). An increased expression of Ki-67, a nuclear protein present only in cycling cells, was reported in these animals, which is consistent with dysregulated growth (146, 147). Furthermore, primary cilia have been described to restrain canonical Wnt signaling (148–153). In the absence of KIF3A, a core component of the Kinesin-2 motor protein complex, cells cannot generate cilia and display an up-regulated canonical Wnt pathway (148, 149). In mice, constitutive inactivation of *Kif3a* is characterized by defects in visceral organ arrangement and embryonic lethality (148). The kidney-specific deletion of *Kif3a* results in ESRD due to PKD by postnatal day 21 together with an increased canonical Wnt reporter activity (148, 149). These studies suggest that activation of canonical Wnt signaling may increase epithelial cell proliferation and thereby contribute to PKD. However, this model, though appealing, might require some modification. On the one hand, it is important to note that recent studies have challenged the role for cilia in canonical Wnt signaling. These studies provide evidence that the phenotypes of mouse and zebrafish mutants that lack cilia do not overlap with the phenotypes of Wnt pathway mutants (150, 151). Neither in mice nor in zebrafish did loss of primary cilia via mutation of the intraflagellar transport protein IFT88 cause alterations in the expression of Wnt target genes including *Apc* and *Axin* (150, 151). Similar results were established in mice for IFT172, DYNC2H1 and, surprisingly, KIF3A (151). In addition, *Ift88*$^{-/-}$, *Ift172*$^{-/-}$ *and Dync2h1*$^{-/-}$ mouse embryonic fibroblasts respond to Wnt3a independent of cilia (151). On the other hand, mitotic rates in *Pkd1*$^{-/-}$ kidneys do not seem to be significantly elevated, arguing against sustained canonical Wnt activity as the primary defect in these mice (84). Future studies will have to clarify the link between canonical Wnt signaling, cilia and ADPKD. It will be informative to know whether increased proliferation is causal or secondary to cyst initiation. Sensitive reporter constructs could be used to assay canonical Wnt activity in PKD1 or PKD2 knockout mice (152).

The β-catenin-independent noncanonical branch of Wnt signaling proceeds through a multitude of factors including Frizzled 3/6, Dishevelled 1/2, and Strabismus (144). Noncanonical Wnt signaling is fundamental to planar cell polarity, i.e., the polarity within the plane of an epithelium, and thereby a prerequisite for oriented cell division (160). In mice, inactivation of the planar cell polarity related protein Fat4 results in both loss of oriented cell division and PKD (161). Moreover, random oriented cell division has been established in advance of cyst formation in *Hnf1*β$^{-/-}$ and *Kif3a*$^{-/-}$ mice kidney tubules (142, 162). Surprisingly, the targeted kidney-specific inactivation of *Pkd1* or *Pkd2* does not present evidence of impaired planar cell polarity in advance of cystic formation (157). Oriented cell division is lost only after tubules begin to dilate (157). This is in line with the observation that *Pkd1* deficiency is not associated with an obvious planar cell polarity phenotype in inner ear hair cells (158). Besides, a model of autosomal recessive PKD, *Pkhd1*$^{del4/del4}$, shows random mitotic spindle orientation in elongating tubules but does not develop PKD (159). Apparently, loss of oriented cell division is neither required for cystogenesis nor sufficient to produce cysts in ADPKD (157).

The interconnection of canonical and noncanonical Wnt components may provide an explanation for sustained cyst expansion other than planar cell polarity. It has been hypothesized that breakdown of noncanonical Wnt signaling may induce an excessive canonical Wnt activity (135). The divergence of Wnt pathways relies on the Dishevelled protein (160). In the canonical Wnt signaling cascade Dishevelled shuttles between different cellular compartments including cytoplasm, nucleus, and plasma membrane (161, 162). Whereas in noncanonical Wnt signaling Dishevelled stays plasma membrane associated (162). The noncanonical Wnt pathway protein Inversin was reported to control Dishevelled localization (135). In humans, homozygous loss of Inversin causes an autosomal recessive form of PKD, Nephronophthisis type 2 (163). In vitro experiments suggested that Inversin inhibits the canonical Wnt signaling cascade by down-regulating cytoplasmic Dishevelled (135). Therefore, Inversin was postulated to act as switch between canonical and noncanonical Wnt pathways (135). However, a recent study has challenged that concept. While *Inversin*$^{-/-}$ mice present with random oriented cell division, canonical Wnt reporter constructs do not show an increased activity in these animals (164). The fact remains that many points in the connection of noncanonical Wnt signaling and PKD are imprecise or vague. *Fat4*$^{-/-}$, *Hnf1*β$^{-/-}$, and *Inversin*$^{-/-}$ mice develop PKD but the underlying mechanisms remain elusive (136, 155, 158). To evaluate the impact of noncanonical Wnt signaling on ADPKD it will be crucial to analyze loss of function models of core noncanonical Wnt components. Will these develop PKD, and if so, what does induce cyst formation? The presented data provide ample testable hypotheses for future discovery.

12. PRKCSH and SEC63 Associated with Polycystic Disease But Not with the Cilia-Basal-Body Complex

Kidney cysts are the eponymous manifestation of ADPKD and with good reason most research has focused on the kidney to elucidate the pathophysiology of cystic disorders. Nonetheless, additional extra-renal cysts are very common and deserve closer attention. The recent analysis of hepatic cyst formation has provided valuable evidence of a tissue specific susceptibility to cyst induction together with an ER-dependent mechanism of Polycystin-1 and TRPP2 protein maturation.

Radiological imaging studies show that liver cysts are evident in 94% of ADPKD patients who are older than 35 years (14). These cysts are rare in children but subsequently manifest in an age-dependent manner during late adulthood (165–167). Underlying cyst growth is an increased proliferation and dilatation of biliary ductules and peribiliary glands (168–170). Polycystic liver disease (PLD) is generally asymptomatic even though symptoms may occur due to the mass effect of very large cysts or a large number of cysts that impinge on surrounding organs or from cyst complications such as hemorrhage, rupture, or infection (166, 167). Interestingly, it has become apparent that PLD also exists as a distinct autosomal dominant disorder without kidney cysts (ADPLD) (#174050 in Online Mendelian Inheritance of Man, http://www.ncbi.nlm.nih.gov/omim/) (171, 172). The prevalence of isolated ADPLD is estimated to be 1 per 158,000 (167). Mutations in two genes cause ADPLD, protein kinase C substrate 80K-H (PRKCSH), and SEC63 (173–176). Still, a significant number of ADPLD patients do not carry alterations in either gene, indicating the involvement of another locus (167, 177). Clinically and histologically, liver cysts in ADPLD are indistinguishable from those found in ADPKD (166, 167). Their similarity has inspired studies into the relation of ADPKD and ADPLD.

The autosomal dominant nature of ADPLD together with its focal cyst phenotype suggests that the disorder might occur analogous to ADPKD by a "two-hit" mechanism. Indeed, micro-dissection from affected patients indicates somatic "second hit" mutations in liver cyst epithelia (178). Furthermore, the "two-hit" hypothesis is supported by the observation that the constitutive deletion of either *Prkcsh* or *Sec63* in mice results in early embryonic lethality (55). *Prkcsh*$^{-/-}$ embryos die in utero by embryonic day 11.5; *Sec63*$^{-/-}$ mutants before embryonic day 7.5 (55). Whereas the liver-selective homozygous loss of function in *Prkcsh* or *Sec63* recapitulates the human ADPLD phenotype (55). Remarkably, the targeted inactivation of *Prkcsh* or *Sec63* in mice kidneys presents with pronounced cystogenesis. This PKD phenotype is even aggravated in the *Prkcsh*$^{-/-}$; *Sec63*$^{-/-}$ compound situation (55). Experiments in zebrafish embryos have yielded similar data: the morpholino induced depletion of PRKCSH leads to pronephric cysts, abnormal body curvature, and defects in visceral left–right asymmetry (132).

The availability of orthologous mouse models of ADPKD and ADPLD together with tissue-selective *Cre recombinase* mouse lines made it feasible not only to characterize the function of single genes, but also to combine gene deficiencies and/or gene overexpression to establish molecular interdependencies. For example, the *Ksp-Cre* transgenic line is based on the kidney-specific member of the Cadherin family of cell adhesion molecules (179). *Ksp-Cre* prompts expression of *Cre recombinase* in the thick ascending loops of Henle, the distal convoluted tubule and collecting ducts, and thus allows for a kidney-selective inactivation of alleles flanked by *LoxP* sites (179).

A common biogenetic pathway of *Prkcsh*, *Sec62*, *Pkd1*, and *Pkd2* was established by three sets of genetic experiments. First, the *Ksp-Cre* controlled deletion of *Prkcsh*$^{flox/flox}$ or Sec63$^{flox/flox}$ on a *Pkd1*$^{+/-}$ or *Pkd2*$^{+/-}$ sensitized background shows an increased severity of PKD (55). Second, the transgenic overexpression of *Pkd1* abolishes cysts in both *Prkcsh*$^{flox/flox}$; *Ksp-Cre* and *Sec63*$^{flox/flox}$; *Ksp-Cre* mice (55). Moreover, *Pkd1* overexpression also abrogates ADPLD caused by liver-selective elimination of *Prkcsh* or *Sec63* (55). Only after 6 months in *Prkcsh*$^{flox/flox}$; *Ksp-Cre* and 45 days in *Sec63*$^{flox/flox}$; *Ksp-Cre* did first renal cysts develop, suggesting a rather delayed expansion in tubule diameter (55). Finally, the *Pkd1*-dependent rescue of PKD requires functional TRPP2, because augmented *Pkd1* expression can ameliorate *Prkcsh*$^{flox/flox}$, *Ksp-Cre, Pkd2*$^{+/-}$ but not *Pkd2*$^{-/-}$ genotypes (55). Taken together, it can be concluded that homozygous loss of *Prkcsh* or *Sec63* mediates cystic disease by effectively reducing the activity of functional Polycystin-1–TRPP2 complexes (55, 62, 180).

Biochemical analyses place PRKCSH and SEC63 in the secretory pathway. The two proteins most likely act in cellular protein maturation. SEC63 is an evolutionary highly conserved, ubiquitously expressed integral ER membrane protein of 760 amino acids; in humans it is encoded on chromosome 6q21–6q23 (133, 176, 181, 182). The protein features a coiled-coil domain and an ER luminal DnaJ domain (1330, 183). It constitutes together with SEC61 and SEC62 the core translocon complex, which comprises the ER membrane translocation machinery (182, 184). In eukaryotes, the translocon mediates co-translational passage of nascent peptides into the ER (185). Studies in *Saccharomyces cerevisiae* have shown that the ER luminal DnaJ domain of Sec63p recruits the Hsp70 homolog BiP (binding immunoglobulin protein) to translocation sites and thereby facilitates the ATP-dependent movement of polypeptides into the ER lumen (186, 187). In addition, the translocon has also been observed in the converse *modus operandi*: promoting the ER dislocation of substrates to the cytoplasm during ubiquitin-mediated ER-associated protein degradation (188). Once in the ER, the process of posttranslational protein maturation includes signal peptide cleavage, chaperone binding, protein folding, disulfide bond formation, transmembrane domain integration, and glycosylation (189). PRKCSH is involved in N-linked glycosylation of ER proteins (190). The human PRKCSH gene on chromosome 19p13.2 encodes for a 528 amino acids long protein (PRKCSH also known as 80K-H). In mouse tissue, *Prkcsh* mRNA is ubiquitously expressed (191). PRKCSH is a multifunctional soluble protein, which contains two Ca^{2+}-binding EF hands and a carboxy-terminal mannose 6-phosphate receptor homologous domain in its tertiary structure together with a putative ER targeting sequence, *HDEL*, at its carboxy terminus (181, 191, 192). It has been initially purified from rat liver as part of glucosidase II (GII), an ER luminal enzyme with a key role in the glycoprotein-specific folding and quality control machinery (190). Newly synthesized glycoproteins interact in a glycosylation-dependent manner with calnexin and calreticulin (193). These two ER-resident lectin chaperones enhance folding efficiency and prevent premature protein exit from the ER. GII-mediated glycan cleavage abolishes the glycoprotein-lectin–chaperone interaction, thus allowing mature glycoproteins to pursue their transit through the secretory pathway (194). Aberrant proteins, in contrast, are selected for ER-associated degradation. GII is a heterodimer composed of a catalytic α-subunit (GIIα) and a noncatalytic β-subunit (GIIβ) (190, 195). GIIα bears the glycosyl hydrolase active site, whereas the ADPLD-associated GIIβ (PRKCSH) has been reported to be involved in the ER retention, folding, and modulation of GIIα (190, 192, 196–198).

The combination of genetic and biochemical data makes it likely that ADPLD proteins are required for ADPKD protein maturation. Defects in PRKCSH or SEC63 might reduce levels of Polycystin-1 and TRPP2 by impairment of protein processing along the cellular secretory pathway. The following reduction of ciliary Polycystin-1–TRPP2 complexes might in turn trigger cyst initiation. Data from cell culture experiments further strengthen this model. On the one hand, physical interaction of TRPP2 and PRKCSH has been demonstrated; on the other hand, Endo-H resistant post-Golgi-complex Polycystin-1 is reduced in *Prkcsh*$^{-/-}$ cells (55, 132). But why do ADPLD patients develop only hepatic and no renal cysts? Possibly due to a different threshold for cyst growth in liver and kidney. However, parts of the putative protein interaction network remain penumbral. Primarily, PRKCSH and SEC63 are required for essential steps in protein biogenesis and a global phenotype would be expected upon deletion, still, all but effective Polycystin-1 protein quantity seem to be normal (55). For instance, TRPP2 expression is not changed in *Prkcsh*$^{-/-}$ or *Sec63*$^{-/-}$ cells (55). Furthermore, both ADPLD proteins have been implicated in additional cellular processes: SEC63 in ER-associated protein degradation and PRKCSH in modulation of protein activity, prominently of GIIα but also of Inositol 1,4,5-trisphosphate receptors and others (188, 190, 192, 196–199). While sufficient experimental evidence concerning these functions is missing, speculation prevails whether these impact cystogenesis. Finally, it will be necessary to identify the missing ADPLD causing genes and define their relation to Polycystin-1 and TRPP2. What cellular circuits will these implicate? Unambiguously, extra-renal manifestations of ADPKD can shed light on hitherto unknown disease mechanisms and complete our limited understanding of cystic disorders.

13. TRPP2 and PKD1L1 Are Required for Left–Right Patterning

A mirror reflects the self-evident: we have two eyes, two arms, and two legs, one on the right and one on the left. The apparently bilaterally symmetrical body plan of vertebrates conceals profound asymmetries on the inside (200). There is only a single heart, liver, spleen, and stomach. In response to an intricate cascade of signaling molecules in early embryonic development, these unpaired organs are arranged in a highly ordered pattern along the left–right axis (201). The heart points to the left, the liver is located on the right, and the stomach and spleen are on the left. Thus, the usual asymmetric position of internal organs, *situs solitus viscerum*, is established (202).

Approximately 1 in 10,000 people is born with a complete mirror-image reversal of the intricate asymmetry of visceral organs—a condition known as *situs inversus viscerum* (203, 204). Remarkably, this condition is not often of pathological significance and may be detected only incidentally by radio- or ultrasonography. By contrast, randomness of heart-, lung-, or gut sidedness—a condition known as *situs ambiguus viscerum* or *heterotaxy* (*hetero* meaning "other" and *taxy* meaning "arrangement")—is almost always associated with visceral ailment (202).

In mice the breaking of left–right symmetry involves the unidirectional flow of extra-embryonic fluid in the embryonic node at ~8 days postcoitum (201, 205–207). The primitive node is the organizer for gastrulation in vertebrates (2082). It is a transient embryonic structure that is located at the midline, immediately anterior to the primitive streak. So far, the mechanism by which nodal flow is detected and evokes subsequent asymmetric gene cascades remains unknown (209, 210).

PKD2 is known to be essential for left–right asymmetry formation in vertebrates. The constitutive loss of PKD2 genes presents in mice, medaka, and zebrafish with heterotaxy (50, 211–214). In humans, the association of heterotaxy with PKD2 and ADPKD is masked by a heterozygous PKD2 gene locus but has been suggested for three patients (215).

TRPP2 proteins have been localized in mouse and teleosts to nodal cilia. Their absence does not give the impression of impaired embryonic node morphology and node cilia function, but alteration of subsequent signaling cascades (211–214, 216). Homozygous *Pkd2* mutant mice do not manifest left-sided Ca^{2+} signals and fail to activate detectable levels of *Nodal* expression in the lateral plate mesoderm (50). In medaka and zebrafish, nodal expression is not lost upon PKD2 gene inactivation, yet seems to be randomized (211, 212, 214). In spite of this difference, the collective data argue that PKD2 acts downstream of embryonic nodal flow but upstream of asymmetric gene expression. The obvious candidate protein to directly modulate TRPP2 ion channel activity would be Polycystin-1, but, peculiarly, *Pkd1* knock-out mice exhibit normal left–right determination (217). Polycystin-1 is not found in mouse node cilia (217). Apparently, TRPP2 acts in the establishment of left–right asymmetry in the absence of Polycystin-1. Lately, a Polycystin-1-like protein has emerged to operate in concert with TRPP2 in the embryonic node. Inactivation of *Pkd1l1* phenocopies *Pkd2* heterotaxy in both mouse and medaka (213, 214).

The 57 exon PKD1L1 gene is located on human chromosome 7p12–p13 (67). Its 2,849 amino acids long PKD1L1 protein product shares a 38% degree of sequence homology with both Polycystin-1 and TRPP2, even though structural properties and phylogenetic analysis place PKD1L1 within the Polycystin-1 family (2, 67). PKD1L1 and Polycystin-1 share the signature tripartite combination of (1) G protein-coupled receptor proteolytic site (GPS), (2) receptor for egg jelly (REJ) domain, and (3) lipoxygenase homology/polycystin, lipoxygenase, atoxin (LH2 / PLAT) domain (68, 218–220). All the same, PKD1L1 contains only two PKD domains in its amino terminus versus 16 in Polycystin-1 and its REJ domain is smaller (111 in contrast to 525 amino acid residues) (67).

Expression analysis of PKD1L1 places it prominently in the embryonic node, the testis's Leydig cells, the heart, and in the mammary gland (67, 213, 214, 221). Several splice variants have been described but their functional consequences remain elusive (67).

Genetic screens have generated two independent *Pkd1l1* mouse mutants that both present with *situs ambiguus viscerum*, $Pkd1l1^{10-12}$ and $Pkd1l1^{rks}$, respectively (213, 222). Homozygous disruption of *Pkd1l1* exons 10–12 in mouse ($Pkd1l1^{10-12/10-12}$) manifests in a significantly reduced viability (only 35% of the expected number of homozygotes were identified) and approximately one-third of surviving mice exhibit laterality defects (222). The *rks* mouse phenotype is linked to the genetic region between 8.69 and 8.89 Mb on chromosome 11 and is most likely due to a single amino acid replacement of a charged aspartic acid at Pkd1l1 position 411 by an uncharged nonpolar glycine residue in the second Pkd1l1 PKD domain (213). *Pkd1l1*$^{rks/rks}$ embryos fail to survive beyond 15.5 days postcoitum. They exhibit incidence of heterotaxy, e.g., right pulmonary isomerism, with four lung lobes evident on each side (100%), together with randomization of stomach (25%) and gross cardiac situs (25%) (213). Embryonic node morphology, cilia morphology, and cilia motility do not seem to be altered in *Pkd1l1*$^{rks/rks}$ mutants (213). However, the left-sided genes *Nodal*, *Lefty2*, and *Pitx2* are not expressed in those animals, indicating a failure to activate the symmetry breaking Nodal signaling cascade (213).

Pkd1l1$^{10-12}$ and *Pkd1l1*^{rks} mutations affect the establishment of asymmetry, but the identification of homozygous viable *Pkd1l1*$^{10-12}$ mice represents a clear phenotypic difference from *Pkd1l1*^{rks}. Future studies will have to examine whether this is due to the different genetic background of mice analyzed (*Pkd1l1*$^{10-12}$ on B6;129 and *Pkd1l1*^{rks} on C3H) or because of allelic effects on the phenotype.

The requirement for PKD1L1 during left–right patterning is conserved in vertebrates.

In medaka, *pkd1l1* mutants present, similar to *pkd2* deficient fish, with no overt phenotype other than heterotaxy and randomization of normally side-restricted downstream gene expression, e.g., of *southpaw*, *lefty*, and *charon* (214). Remarkably, the normal expression of *pkd2* in node cilia is lost in those mutants (214).

PKD1L1, like TRPP2, acts downstream of embryonic nodal flow but upstream of asymmetric gene expression, suggesting that PKD1L1 may be the TRPP2 binding partner in left–right determination. Indeed, TRPP2 and PKD1L1 localize to nodal cilia and the PKD1L1 carboxy terminus can physically interact in overexpression systems with TRPP2 (213, 214). It is tempting to speculate that a PKD1L1–TRPP2 complex may operate via a general molecular mechanism of transforming nodal flow into asymmetric gene expression in vertebrates. However, additional studies will have to establish the endogenous relation of PKD1L1 and TRPP2 and characterize the functional properties of a putative receptor–ion-channel complex.

14. Conclusions

PKD1 and PKD2 have been cloned as ADPKD disease genes. Since their initial description the accelerated pace of scientific discovery has produced an impressive body of knowledge. TRPP subunits can form functional homomeric and heteromeric signaling modules, expanding the diversity of TRPP channel functions. TRPP2–Polycystin-1 is required for renal three-dimensional tissue organization. TRPP2–PKD1L1 is essential for establishment of left–right asymmetry. The respective signaling cascades, however, remain elusive. One of the most fundamental open questions remains the identity of upstream and downstream players in the polycystin signaling pathway. Will it be possible to alleviate ADPKD in PKD1 patients by targeted TRPP2 activation? Hopefully, future studies will lead to the development of therapies designed to correct the fundamental cell-biological abnormalities of ADPKD.

Acknowledgments

The author would like to thank Michael Köttgen for helpful discussions. He apologizes to those whose work could not be cited due to space limitations. Funding was awarded by the Deutsche Forschungsgemeinschaft (KFO 201) and the Excellence Initiative of the German Federal and State Governments (GSC-4, Spemann Graduate School).

References

1. Clapham DE (2003) TRP channels as cellular sensors. Nature 426:517–524. doi:10.1038/nature02196
2. Venkatachalam K, Montell C (2007) TRP channels. Annu Rev Biochem 76:387–417. doi:10.1146/annurev.biochem.75.103004.142819
3. Ramsey IS, Delling M, Clapham DE (2006) An introduction to TRP channels. Annu Rev Physiol 68:619–647. doi:10.1146/annurev.physiol.68.040204.100431
4. U.S. Renal Data System (2011) Annual data report: atlas of end-stage renal disease in the United States. National Institutes of Health, Bethesda, MD
5. Longo DL, Harrison TR (2012) Harrison's principles of internal medicine, 18th edn. McGraw-Hill, New York
6. Gabow PA, Chapman AB, Johnson AM, Tangel DJ, Duley IT, Kaehny WD, Manco-Johnson M, Schrier RW (1990) Renal structure and hypertension in autosomal dominant polycystic kidney disease. Kidney Int 38:1177–1180
7. Hateboer N, v Dijk MA, Bogdanova N, Coto E, Saggar-Malik AK, San Millan JL, Torra R, Breuning M, Ravine D (1999) Comparison of phenotypes of polycystic kidney disease types 1 and 2. European PKD1-PKD2 Study Group. Lancet 353:103–107
8. King BF, Reed JE, Bergstralh EJ, Sheedy PF, Torres VE (2000) Quantification and longitudinal trends of kidney, renal cyst, and renal parenchyma volumes in autosomal dominant polycystic kidney disease. J Am Soc Nephrol 11:1505–1511
9. Fick-Brosnahan GM, Belz MM, McFann KK, Johnson AM, Schrier RW (2002) Relationship between renal volume growth and renal function in autosomal dominant polycystic kidney disease: a longitudinal study. Am J Kidney Dis 39:1127–1134
10. Bajwa ZH, Gupta S, Warfield CA, Steinman TI (2001) Pain management in polycystic kidney disease. Kidney Int 60:1631–1644. doi:10.1046/j.1523-1755.2001.00985.x
11. Bajwa ZH, Sial KA, Malik AB, Steinman TI (2004) Pain patterns in patients with polycystic kidney disease. Kidney Int 66:1561–1569. doi:10.1111/j.1523-1755.2004.00921.x
12. Grantham JJ, Chapman AB, Torres VE (2006) Volume progression in autosomal dominant polycystic kidney disease: the major factor determining clinical outcomes. Clin J Am Soc Nephrol 1:148–157
13. Kelleher CL, McFann KK, Johnson AM, Schrier RW (2004) Characteristics of hypertension in young adults with autosomal dominant polycystic kidney disease compared with the general U.S. population. Am J Hypertens

17:1029–1034. doi:10.1016/j.amjhyper.2004.06.020

14. Bae KT, Zhu F, Chapman AB, Torres VE, Grantham JJ, Guay-Woodford LM, Baumgarten DA, King BF, Wetzel LH, Kenney PJ, Brummer ME, Bennett WM, Klahr S, Meyers CM, Zhang X, Thompson PA, Miller JP (2006) Magnetic resonance imaging evaluation of hepatic cysts in early autosomal-dominant polycystic kidney disease: the Consortium for Radiologic Imaging Studies of Polycystic Kidney Disease cohort. Clin J Am Soc Nephrol 1:64–69. doi:10.2215/CJN.00080605
15. Danaci M, Akpolat T, Baştemir M, Sarikaya S, Akan H, Selçuk MB, Cengiz K (1998) The prevalence of seminal vesicle cysts in autosomal dominant polycystic kidney disease. Nephrol Dial Transplant 13:2825–2828
16. Wijdicks EF, Torres VE, Schievink WI (2000) Chronic subdural hematoma in autosomal dominant polycystic kidney disease. Am J Kidney Dis 35:40–43
17. Hossack KF, Leddy CL, Johnson AM, Schrier RW, Gabow PA (1988) Echocardiographic findings in autosomal dominant polycystic kidney disease. N Engl J Med 319:907–912
18. Schievink WI, Huston J, Torres VE, Marsh WR (1995) Intracranial cysts in autosomal dominant polycystic kidney disease. J Neurosurg 83:1004–1007. doi:10.3171/jns.1995.83.6.1004
19. Hadimeri H, Lamm C, Nyberg G (1998) Coronary aneurysms in patients with autosomal dominant polycystic kidney disease. J Am Soc Nephrol 9:837–841
20. Lumiaho A, Ikäheimo R, Miettinen R, Niemitukia L, Laitinen T, Rantala A, Lampainen E, Laakso M, Hartikainen J (2001) Mitral valve prolapse and mitral regurgitation are common in patients with polycystic kidney disease type 1. Am J Kidney Dis 38:1208–1216
21. Leung GKK, Fan YW (2005) Chronic subdural haematoma and arachnoid cyst in autosomal dominant polycystic kidney disease (ADPKD). J Clin Neurosci 12:817–819. doi:10.1016/j.jocn.2004.09.025
22. Igarashi P, Somlo S (2002) Genetics and pathogenesis of polycystic kidney disease. J Am Soc Nephrol 13:2384–2398
23. Grantham JJ (2008) Clinical practice. Autosomal dominant polycystic kidney disease. N Engl J Med 359:1477–1485
24. Harris PC, Torres VE (2011) Polycystic Kidney Disease, Autosomal Dominant. In: Pagon RA, Bird TD, Dolan CR, et al., editors. GeneReviews™ [Internet]. Seattle (WA): University of Washington, Seattle; 1993-. Available from: http://www.ncbi.nlm.nih.gov/books/NBK1246/. Sept 2011
25. Mochizuki T, Wu G, Hayashi T, Xenophontos SL, Veldhuisen B, Saris JJ, Reynolds DM, Cai Y, Gabow PA, Pierides A, Kimberling WJ, Breuning MH, Deltas CC, Peters DJ, Somlo S (1996) PKD2, a gene for polycystic kidney disease that encodes an integral membrane protein. Science 272:1339–1342
26. Reeders ST, Breuning MH, Davies KE, Nicholls RD, Jarman AP, Higgs DR, Pearson PL, Weatherall DJ (1985) A highly polymorphic DNA marker linked to adult polycystic kidney disease on chromosome 16. Nature 317:542–544
27. (1994) The polycystic kidney disease 1 gene encodes a 14 kb transcript and lies within a duplicated region on chromosome 16. The European Polycystic Kidney Disease Consortium. Cell 77:881–894
28. (1995) Polycystic kidney disease: the complete structure of the PKD1 gene and its protein. The International Polycystic Kidney Disease Consortium. Cell 81:289–298
29. Peters DJ, Sandkuijl LA (1992) Genetic heterogeneity of polycystic kidney disease in Europe. Contrib Nephrol 97:128–139
30. Dobin A, Kimberling WJ, Pettinger W, Bailey-Wilson JE, Shugart YY, Gabow P (1993) Segregation analysis of autosomal dominant polycystic kidney disease, *Genet.* Epidemiol 10:189–200
31. Rossetti S, Consugar MB, Chapman AB, Torres VE, Guay-Woodford LM, Grantham JJ, Bennett WM, Meyers CM, Walker DL, Bae K, Zhang QJ, Thompson PA, Miller JP, Harris PC (2007) Comprehensive molecular diagnostics in autosomal dominant polycystic kidney disease. J Am Soc Nephrol 18:2143–2160
32. Barua M, Cil O, Paterson AD, Wang K, He N, Dicks E, Parfrey P, Pei Y (2009) Family history of renal disease severity predicts the mutated gene in ADPKD. J Am Soc Nephrol 20:1833–1838
33. Daoust MC, Reynolds DM, Bichet DG, Somlo S (1995) Evidence for a third genetic locus for autosomal dominant polycystic kidney disease. Genomics 25:733–736
34. Bogdanova N, Dworniczak B, Dragova D, Todorov V, Dimitrakov D, Kalinov K, Hallmayer J, Horst J, Kalaydjieva L (1995) Genetic heterogeneity of polycystic kidney disease in Bulgaria. Hum Genet 95:645–650
35. de Almeida S, de Almeida E, Peters D, Pinto JR, Távora I, Lavinha J, Breuning M, Prata MM (1995) Autosomal dominant polycystic kidney disease: evidence for the existence of a

third locus in a Portuguese family. Hum Genet 96:83–88
36. Turco AE, Clementi M, Rossetti S, Tenconi R, Pignatti PF (1996) An Italian family with autosomal dominant polycystic kidney disease unlinked to either the PKD1 or PKD2 gene. Am J Kidney Dis 28:759–761
37. Paterson AD, Pei Y (1998) Is there a third gene for autosomal dominant polycystic kidney disease? Kidney Int 54:1759–1761
38. Bergmann C, Brüchle NO, Frank V, Rehder H, Zerres K (2008) Perinatal deaths in a family with autosomal dominant polycystic kidney disease and a PKD2 mutation. N Engl J Med 359:318–319
39. Menezes LF, Onuchic LF (2006) Molecular and cellular pathogenesis of autosomal recessive polycystic kidney disease. Braz J Med Biol Res 39:1537–1548
40. Sweeney WE, Avner ED (2006) Molecular and cellular pathophysiology of autosomal recessive polycystic kidney disease (ARPKD). Cell Tissue Res 326:671–685
41. Torra R, Badenas C, Darnell A, Nicolau C, Volpini V, Revert L, Estivill X (1996) Linkage, clinical features, and prognosis of autosomal dominant polycystic kidney disease types 1 and 2. J Am Soc Nephrol 7:2142–2151
42. Magistroni R, He N, Wang K, Andrew R, Johnson A, Gabow P, Dicks E, Parfrey P, Torra R, San-Millan JL, Coto E, van Dijk M, Breuning M, Peters D, Bogdanova N, Ligabue G, Albertazzi A, Hateboer N, Demetriou K, Pierides A, Deltas C, St George-Hyslop P, Ravine D, Pei Y (2003) Genotype-renal function correlation in type 2 autosomal dominant polycystic kidney disease. J Am Soc Nephrol 14:1164–1174
43. Gabow PA, Johnson AM, Kaehny WD, Kimberling WJ, Lezotte DC, Duley IT, Jones RH (1992) Factors affecting the progression of renal disease in autosomal-dominant polycystic kidney disease. Kidney Int 41: 1311–1319
44. Rossetti S, Burton S, Strmecki L, Pond GR, San Millán JL, Zerres K, Barratt TM, Ozen S, Torres VE, Bergstralh EJ, Winearls CG, Harris PC (2002) The position of the polycystic kidney disease 1 (PKD1) gene mutation correlates with the severity of renal disease. J Am Soc Nephrol 13:1230–1237
45. Harris PC, Bae KT, Rossetti S, Torres VE, Grantham JJ, Chapman AB, Guay-Woodford LM, King BF, Wetzel LH, Baumgarten DA, Kenney PJ, Consugar M, Klahr S, Bennett WM, Meyers CM, Zhang QJ, Thompson PA, Zhu F, Miller JP (2006) Cyst number but not the rate of cystic growth is associated with the mutated gene in autosomal dominant polycystic kidney disease. J Am Soc Nephrol 17:3013–3019
46. Paterson AD, Wang KR, Lupea D, St George-Hyslop P, Pei Y (2002) Recurrent fetal loss associated with bilineal inheritance of type 1 autosomal dominant polycystic kidney disease. Am J Kidney Dis 40:16–20
47. Boulter C, Mulroy S, Webb S, Fleming S, Brindle K, Sandford R (2001) Cardiovascular, skeletal, and renal defects in mice with a targeted disruption of the Pkd1 gene. Proc Natl Acad Sci USA 98:12174–12179. doi: 10.1073/pnas.211191098
48. Lu W, Shen X, Pavlova A, Lakkis M, Ward CJ, Pritchard L, Harris PC, Genest DR, Perez-Atayde AR, Zhou J (2001) Comparison of Pkd1-targeted mutants reveals that loss of polycystin-1 causes cystogenesis and bone defects. Hum Mol Genet 10:2385–2396
49. Muto S, Aiba A, Saito Y, Nakao K, Nakamura K, Tomita K, Kitamura T, Kurabayashi M, Nagai R, Higashihara E, Harris PC, Katsuki M, Horie S (2002) Pioglitazone improves the phenotype and molecular defects of a targeted Pkd1. Hum Mol Genet 11:1731–1742
50. Pennekamp P, Karcher C, Fischer A, Schweickert A, Skryabin B, Horst J, Blum M, Dworniczak B (2002) The ion channel polycystin-2 is required for left-right axis determination in mice. Curr Biol 12:938–943
51. Wu G, Markowitz GS, Li L, D'Agati VD, Factor SM, Geng L, Tibara S, Tuchman J, Cai Y, Park JH, van Adelsberg J, Hou H, Kucherlapati R, Edelmann W, Somlo S (2000) Cardiac defects and renal failure in mice with targeted mutations in Pkd2. Nat Genet 24:75–78. doi:10.1038/71724
52. Wu G, Tian X, Nishimura S, Markowitz GS, D'Agati V, Park JH, Yao L, Li L, Geng L, Zhao H, Edelmann W, Somlo S (2002) Trans-heterozygous Pkd1 and Pkd2 mutations modify expression of polycystic kidney disease. Hum Mol Genet 11:1845–1854
53. Pei Y, Paterson AD, Wang KR, He N, Hefferton D, Watnick T, Germino GG, Parfrey P, Somlo S, St George-Hyslop P (2001) Bilineal disease and trans-heterozygotes in autosomal dominant polycystic kidney disease. Am J Hum Genet 68: 355–363
54. Persu A, Duyme M, Pirson Y, Lens XM, Messiaen T, Breuning MH, Chauveau D, Levy M, Grünfeld J-P, Devuyst O (2004) Comparison between siblings and twins supports a role for modifier genes in ADPKD. Kidney Int 66:2132–2136. doi:10.1111/j.1523-1755.2004.66003.x

55. Fedeles SV, Tian X, Gallagher A-R, Mitobe M, Nishio S, Lee SH, Cai Y, Geng L, Crews CM, Somlo S (2011) A genetic interaction network of five genes for human polycystic kidney and liver diseases defines polycystin-1 as the central determinant of cyst formation. Nat Genet 43:639–647
56. Takakura A, Contrino L, Zhou X, Bonventre JV, Sun Y, Humphreys BD, Zhou J (2009) Renal injury is a third hit promoting rapid development of adult polycystic. Hum Mol Genet 18:2523–2531. doi:10.1093/hmg/ddp147
57. Bastos AP, Piontek K, Silva AM, Martini D, Menezes LF, Fonseca JM, Fonseca II, Germino GG, Onuchic LF (2009) Pkdl haploinsufficiency increases renal damage and induces microcyst formation. J Am Soc Nephrol 20:2389–2402. doi:10.1681/ASN.2008040435
58. Prasad S, McDaid JP, Tam FW, Haylor JL, Ong AC (2009) Pkd2 dosage influences cellular repair responses following ischemia-reperfusion. Am J Pathol 175:1493–1503. doi:10.2353/ajpath.2009.090227
59. Qian F, Germino FJ, Cai Y, Zhang X, Somlo S, Germino GG (1997) PKD1 interacts with PKD2 through a probable coiled-coil domain. Nat Genet 16:179–183. doi:10.1038/ng0697-179
60. Tsiokas L, Kim E, Arnould T, Sukhatme VP, Walz G (1997) Homo- and heterodimeric interactions between the gene products of PKD1 and PKD2. Proc Natl Acad Sci USA 94:6965–6970
61. Delmas P, Nauli SM, Li X, Coste B, Osorio N, Crest M, Brown DA, Zhou J (2004) Gating of the polycystin ion channel signaling complex in neurons and kidney cells. FASEB J 18:740–742. doi:10.1096/fj.03-0319fje
62. Hanaoka K, Qian F, Boletta A, Bhunia AK, Piontek K, Tsiokas L, Sukhatme VP, Guggino WB, Germino GG (2000) Co-assembly of polycystin-1 and -2 produces unique cation-permeable currents. Nature 408:990–994. doi:10.1038/35050128
63. Giamarchi A, Feng S, Rodat-Despoix L, Xu Y, Bubenshchikova E, Newby LJ, Hao J, Gaudioso C, Crest M, Lupas AN, Honoré E, Williamson MP, Obara T, Ong ACM, Delmas P (2010) A polycystin-2 (TRPP2) dimerization domain essential for the function of heteromeric polycystin complexes. EMBO J 29:1176–1191. doi:10.1038/emboj.2010.18
64. Köttgen M (2007) TRPP2 and autosomal dominant polycystic kidney disease. Biochim Biophys Acta 1772:836–850. doi:10.1016/j.bbadis.2007.01.003
65. Hughes J, Ward CJ, Peral B, Aspinwall R, Clark K, San Millán JL, Gamble V, Harris PC (1995) The polycystic kidney disease 1 (PKD1) gene encodes a novel protein with multiple cell recognition domains. Nat Genet 10:151–160. doi:10.1038/ng0695-151
66. Nims N, Vassmer D, Maser RL (2003) Transmembrane domain analysis of polycystin-1, the product of the polycystic kidney disease-1 (PKD1) gene: evidence for 11 membrane-spanning domains. Biochemistry 42:13035–13048. doi:10.1021/bi035074c
67. Yuasa T, Venugopal B, Weremowicz S, Morton CC, Guo L, Zhou J (2002) The sequence, expression, and chromosomal localization of a novel polycystic kidney disease 1-like gene, PKD1L1, in human. Genomics 79:376–386. doi:10.1006/geno. 2002.6719
68. Li A, Tian X, Sung S-W, Somlo S (2003) Identification of two novel polycystic kidney disease-1-like genes in human and mouse genomes. Genomics 81:596–608
69. Sutton KA, Jungnickel MK, Ward CJ, Harris PC, Florman HM (2006) Functional characterization of PKDREJ, a male germ cell-restricted polycystin. J Cell Physiol 209:493–500. doi:10.1002/jcp. 20755
70. Zhou J (2009) Polycystins and primary cilia: primers for cell cycle progression. Annu Rev Physiol 71:83–113. doi:10.1146/annurev.physiol.70.113006.100621
71. Hofherr A, Köttgen M (2011) TRPP channels and polycystins. Adv Exp Med Biol 704:287–313
72. Cantiello HF (2004) Regulation of calcium signaling by polycystin-2. Am J Physiol Renal Physiol 286:F1012–F1029. doi:10.1152/ajprenal.00181.2003
73. Delmas P, Padilla F, Osorio N, Coste B, Raoux M, Crest M (2004) Polycystins, calcium signaling, and human diseases. Biochem Biophys Res Commun 322:1374–1383. doi:10.1016/j.bbrc.2004.08.044
74. Wu G, Hayashi T, Park JH, Dixit M, Reynolds DM, Li L, Maeda Y, Cai Y, Coca-Prados M, Somlo S (1998) Identification of PKD2L, a human PKD2-related gene: tissue-specific expression and mapping to chromosome 10q25. Genomics 54:564–568. doi:10.1006/geno.1998.5618
75. Nomura H, Turco AE, Pei Y, Kalaydjieva L, Schiavello T, Weremowicz S, Ji W, Morton CC, Meisler M, Reeders ST, Zhou J (1998) Identification of PKDL, a novel polycystic kidney disease 2-like gene whose murine homologue is deleted in mice with kidney and retinal defects. J Biol Chem 273: 25967–25973

76. Veldhuisen B, Spruit L, Dauwerse HG, Breuning MH, Peters DJ (1999) Genes homologous to the autosomal dominant polycystic kidney disease genes (PKD1 and PKD2). Eur J Hum Genet 7:860–872. doi:10.1038/sj.ejhg.5200383
77. Guo L, Schreiber TH, Weremowicz S, Morton CC, Lee C, Zhou J (2000) Identification and characterization of a novel polycystin family member, polycystin-L2, in mouse and human: sequence, expression, alternative splicing, and chromosomal localization. Genomics 64:241–251. doi:10.1006/geno.2000.6131
78. Boron WF, Boulpaep EL (2009) Medical physiology. A cellular and molecular approach, 2nd edn. Elsevier Saunders, Philadelphia, PA
79. Reeders ST (1992) Multilocus polycystic disease. Nat Genet 1:235–237. doi:10.1038/ng0792-235
80. Pei Y (2001) A "two-hit" model of cystogenesis in autosomal dominant polycystic kidney disease? Trends Mol Med 7:151–156
81. Qian F, Watnick TJ, Onuchic LF, Germino GG (1996) The molecular basis of focal cyst formation in human autosomal dominant polycystic kidney disease type I. Cell 87:979–987
82. Watnick TJ, Torres VE, Gandolph MA, Qian F, Onuchic LF, Klinger KW, Landes G, Germino GG (1998) Somatic mutation in individual liver cysts supports a two-hit model of cystogenesis in autosomal dominant polycystic kidney disease. Mol Cell 2:247–251
83. Menezes LF, Germino GG (2009) Polycystic kidney disease, cilia, and planar polarity. Methods Cell Biol 94:273–297. doi:10.1016/S0091-679X(08)94014-0
84. Piontek K, Menezes LF, Garcia-Gonzalez MA, Huso DL, Germino GG (2007) A critical developmental switch defines the kinetics of kidney cyst formation after loss of Pkd1. Nat Med 13:1490–1495. doi:10.1038/nm1675
85. Baert L (1978) Hereditary polycystic kidney disease (adult form): a microdissection study of two cases at an early stage of the disease. Kidney Int 13:519–525
86. Carone FA, Makino H, Kanwar YS (1988) Basement membrane antigens in renal polycystic disease. Am J Pathol 130:466–471
87. Karihaloo A, Koraishy F, Huen SC, Lee Y, Merrick D, Caplan MJ, Somlo S, Cantley LG (2011) Macrophages promote cyst growth in polycystic kidney disease. J Am Soc Nephrol 22:1809–1814
88. Woo D (1995) Apoptosis and loss of renal tissue in polycystic kidney diseases. N Engl J Med 333:18–25
89. Lanoix J, D'Agati V, Szabolcs M, Trudel M (1996) Dysregulation of cellular proliferation and apoptosis mediates human autosomal dominant polycystic kidney disease (ADPKD). Oncogene 13:1153–1160
90. Wodarczyk C, Distefano G, Rowe I, Gaetani M, Bricoli B, Muorah M, Spitaleri A, Mannella V, Ricchiuto P, Pema M, Castelli M, Casanova AE, Mollica L, Banzi M, Boca M, Antignac C, Saunier S, Musco G, Boletta A (2010) Nephrocystin-1 forms a complex with polycystin-1 via a polyproline motif/SH3 domain interaction and regulates the apoptotic response in mammals. PLoS One 5:e12719. doi:10.1371/journal.pone.0012719
91. Grantham JJ, Geiser JL, Evan AP (1987) Cyst formation and growth in autosomal dominant polycystic kidney disease. Kidney Int 31: 1145–1152
92. Evan AP, Gardner KD, Bernstein J (1979) Polypoid and papillary epithelial hyperplasia: a potential cause of ductal obstruction in adult polycystic disease. Kidney Int 16:743–750
93. Shibazaki S, Yu Z, Nishio S, Tian X, Thomson RB, Mitobe M, Louvi A, Velazquez H, Ishibe S, Cantley LG, Igarashi P, Somlo S (2008) Cyst formation and activation of the extracellular regulated kinase pathway after kidney specific inactivation of Pkd1. Hum Mol Genet 17:1505–1516
94. Grantham JJ (1990) Polycystic kidney disease: neoplasia in disguise. Am J Kidney Dis 15:110–116
95. Ye M, Grantham JJ (1993) The secretion of fluid by renal cysts from patients with autosomal dominant polycystic kidney disease. N Engl J Med 329:310–313
96. Perrone RD (1985) In vitro function of cyst epithelium from human polycystic kidney. J Clin Invest 76:1688–1691
97. Sutters M, Germino GG (2003) Autosomal dominant polycystic kidney disease: molecular genetics and pathophysiology. J Lab Clin Med 141:91–101. doi:10.1067/mlc.2003.13
98. Shillingford JM, Murcia NS, Larson CH, Low SH, Hedgepeth R, Brown N, Flask CA, Novick AC, Goldfarb DA, Kramer-Zucker A, Walz G, Piontek KB, Germino GG, Weimbs T (2006) The mTOR pathway is regulated by polycystin-1, and its inhibition reverses renal cystogenesis in polycystic kidney disease. Proc Natl Acad Sci USA 103:5466–5471. doi:10.1073/pnas.0509694103
99. Wu M, Wahl PR, Le Hir M, Wackerle-Men Y, Wuthrich RP, Serra AL (2007) Everolimus retards cyst growth and preserves kidney function in a rodent model for polycystic kidney disease. Kidney Blood Press Res 30:253–259

100. Serra AL, Poster D, Kistler AD, Krauer F, Raina S, Young J, Rentsch KM, Spanaus KS, Senn O, Kristanto P, Scheffel H, Weishaupt D, Wüthrich RP (2010) Sirolimus and kidney growth in autosomal dominant polycystic kidney disease. N Engl J Med 363:820–829
101. Walz G, Budde K, Mannaa M, Nürnberger J, Wanner C, Sommerer C, Kunzendorf U, Banas B, Hörl WH, Obermüller N, Arns W, Pavenstädt H, Gaedeke J, Büchert M, May C, Gschaidmeier H, Kramer S, Eckardt K-U (2010) Everolimus in patients with autosomal dominant polycystic kidney disease. N Engl J Med 363:830–840
102. Watnick T, Germino GG (2010) mTOR inhibitors in polycystic kidney disease. N Engl J Med 363:879–881
103. Torres VE, Meijer E, Bae KT, Chapman AB, Devuyst O, Gansevoort RT, Grantham JJ, Higashihara E, Perrone RD, Krasa HB, Ouyang JJ, Czerwiec FS (2011) Rationale and design of the TEMPO (Tolvaptan Efficacy and Safety in Management of Autosomal Dominant Polycystic Kidney Disease and its Outcomes) 3-4 Study, Am J Kidney Dis 57: 692–699
104. Higashihara E, Torres VE, Chapman AB, Grantham JJ, Bae K, Watnick TJ, Horie S, Nutahara K, Ouyang J, Krasa HB, Czerwiec FS (2011) Tolvaptan in autosomal dominant polycystic kidney disease: three years' experience, Clin J Am Soc Nephrol 6:2499–2507
105. Beales PL, Elcioglu N, Woolf AS, Parker D, Flinter FA (1999) New criteria for improved diagnosis of Bardet-Biedl syndrome: results of a population survey. J Med Genet 36: 437–446
106. Hildebrandt F, Attanasio M, Otto E (2009) Nephronophthisis: disease mechanisms of a ciliopathy. J Am Soc Nephrol 20:23–35. doi:10.1681/ASN.2008050456
107. Turkbey B, Ocak I, Daryanani K, Font-Montgomery E, Lukose L, Bryant J, Tuchman M, Mohan P, Heller T, Gahl WA, Choyke PL, Gunay-Aygun M (2009) Autosomal recessive polycystic kidney disease and congenital hepatic fibrosis (ARPKD/CHF). Pediatr Radiol 39:100–111
108. Wilson PD (2008) Mouse models of polycystic kidney disease. Curr Top Dev Biol 84:311–350. doi:10.1016/S0070-2153(08)00606-6
109. Harris PC, Torres VE (2009) Polycystic kidney disease. Annu Rev Med 60:321–337. doi:10.1146/annurev.med.60. 101707.125712
110. Gerdes JM, Davis EE, Katsanis N (2009) The vertebrate primary cilium in development, homeostasis, and disease. Cell 137:32–45. doi:10.1016/j.cell.2009.03.023
111. Satir P, Pedersen LB, Christensen ST (2010) The primary cilium at a glance. J Cell Sci 123:499–503. doi:10.1242/jcs.050377
112. Yoder BK, Richards WG, Sweeney WE, Wilkinson JE, Avener ED, Woychik RP (1995) Insertional mutagenesis and molecular analysis of a new gene associated with polycystic kidney disease. Proc Assoc Am Physicians 107: 314–323
113. Pazour GJ, Dickert BL, Vucica Y, Seeley ES, Rosenbaum JL, Witman GB, Cole DG (2000) Chlamydomonas IFT88 and its mouse homologue, polycystic kidney disease gene tg737, are required for assembly of cilia and flagella. J Cell Biol 151:709–718
114. Liu W, Murcia NS, Duan Y, Weinbaum S, Yoder BK, Schwiebert E, Satlin LM (2005) Mechanoregulation of intracellular Ca2+ concentration is attenuated in collecting duct of monocilium-impaired orpk mice. Am J Physiol Renal Physiol 289:F978–F988. doi:10.1152/ajprenal.00260.2004
115. Hildebrandt F, Benzing T, Katsanis N (2011) Ciliopathies. N Engl J Med 364:1533–1543
116. Praetorius HA, Spring KR (2005) A physiological view of the primary cilium. Annu Rev Physiol 67:515–529. doi:10.1146/annurev.physiol.67.040403.101353
117. Rosenbaum JL, Witman GB (2002) Intraflagellar transport. Nat Rev Mol Cell Biol 3:813–825
118. Ishikawa H, Marshall WF (2011) Ciliogenesis: building the cell's antenna. Nat Rev Mol Cell Biol 12:222–234
119. Wheatley DN (1995) Primary cilia in normal and pathological tissues. Pathobiology 63:222–238
120. Praetorius HA, Spring KR (2001) Bending the MDCK cell primary cilium increases intracellular calcium. J Membr Biol 184:71–79
121. Praetorius HA, Spring KR (2003) Removal of the MDCK cell primary cilium abolishes flow sensing. J Membr Biol 191:69–76. doi:10.1007/s00232-002-1042-4
122. Inglis PN, Ou G, Leroux MR, Scholey JM (2007) The sensory cilia of *Caenorhabditis elegans*. WormBook : the online review of C. elegans biology 1–22. doi:10.1895/wormbook.1.126.2
123. Jenkins PM, McEwen DP, Martens JR (2009) Olfactory cilia: linking sensory cilia function and human disease. Chem Senses 34:451–464. doi:10.1093/chemse/bjp020
124. Goetz SC, Anderson KV (2010) The primary cilium: a signalling centre during vertebrate development. Nat Rev Genet 11:331–344. doi:10.1038/nrg2774
125. Murcia NS, Richards WG, Yoder BK, Mucenski ML, Dunlap JR, Woychik RP (2000) The Oak

Ridge Polycystic Kidney (orpk) disease gene is required for left-right axis determination. Development 127:2347–2355

126. Pazour GJ, San Agustin JT, Follit JA, Rosenbaum JL, Witman GB (2002) Polycystin-2 localizes to kidney cilia and the ciliary level is elevated in orpk mice with polycystic kidney disease. Curr Biol 12:R378–R380
127. Yoder BK, Hou X, Guay-Woodford LM (2002) The polycystic kidney disease proteins, polycystin-1, polycystin-2, polaris, and cystin, are co-localized in renal cilia. J Am Soc Nephrol 13:2508–2516
128. Köttgen M, Hofherr A, Li W, Chu K, Cook S, Montell C, Watnick T (2011) Drosophila sperm swim backwards in the female reproductive tract and are activated via TRPP2 ion channels. PLoS One 6:e20031
129. Torres VE, Harris PC (2009) Autosomal dominant polycystic kidney disease: the last 3 years. Kidney Int 76:149–168. doi:10.1038/ki.2009.128
130. Torres VE, Harris PC, Pirson Y (2007) Autosomal dominant polycystic kidney disease. Lancet 369:1287–1301. doi:10.1016/S0140-6736(07)60601-1
131. Sang L, Miller JJ, Corbit KC, Giles RH, Brauer MJ, Otto EA, Baye LM, Wen X, Scales SJ, Kwong M, Huntzicker EG, Sfakianos MK, Sandoval W, Bazan JF, Kulkarni P, Garcia-Gonzalo FR, Seol AD, O'Toole JF, Held S, Reutter HM, Lane WS, Rafiq MA, Noor A, Ansar M, Devi ARR, Sheffield VC, Slusarski DC, Vincent JB, Doherty DA, Hildebrandt F, Reiter JF, Jackson PK (2011) Mapping the NPHP-JBTS-MKS protein network reveals ciliopathy disease genes and pathways. Cell 145:513–528
132. Gao H, Wang Y, Wegierski T, Skouloudaki K, Pütz M, Fu X, Engel C, Boehlke C, Peng H, Kuehn EW, Kim E, Kramer-Zucker A, Walz G (2010) PRKCSH/80K-H, the protein mutated in polycystic liver disease, protects polycystin-2/TRPP2 against HERP mediated degradation. Hum Mol Genet 19:16–24. doi:10.1093/hmg/ddp463
133. Skowronek MH, Rotter M, Haas IG (1999) Molecular characterization of a novel mammalian DnaJ-like Sec63p homolog. Biol Chem 380:1133–1138
134. Gabow PA (1993) Autosomal dominant polycystic kidney disease. N Engl J Med 329: 332–342
135. Simons M, Gloy J, Ganner A, Bullerkotte A, Bashkurov M, Krönig C, Schermer B, Benzing T, Cabello OA, Jenny A, Mlodzik M, Polok B, Driever W, Obara T, Walz G (2005) Inversin, the gene product mutated in nephronophthisis type II, functions as a molecular switch between Wnt signaling pathways. Nat Genet 37:537–543
136. Fischer E, Legue E, Doyen A, Nato F, Nicolas J-F, Torres V, Yaniv M, Pontoglio M (2006) Defective planar cell polarity in polycystic kidney disease. Nat Genet 38:21–23
137. Germino GG (2005) Linking cilia to Wnts. Nat Genet 37:455–457
138. Simons M, Mlodzik M (2008) Planar cell polarity signaling: from fly development to human disease. Annu Rev Genet 42: 517–540
139. Mosimann C, Hausmann G, Basler K (2009) Beta-catenin hits chromatin: regulation of Wnt target gene activation. Nat Rev Mol Cell Biol 10:276–286
140. Hausmann G, Bänziger C, Basler K (2007) Helping Wingless take flight: how WNT proteins are secreted. Nat Rev Mol Cell Biol 8:331–336
141. Nusse R, Varmus HE (1982) Many tumors induced by the mouse mammary tumor virus contain a provirus integrated in the same region of the host genome. Cell 31:99–109
142. Cabrera CV, Alonso MC, Johnston P, Phillips RG, Lawrence PA (1987) Phenocopies induced with antisense RNA identify the wingless gene. Cell 50:659–663
143. Rijsewijk F, Schuermann M, Wagenaar E, Parren P, Weigel D, Nusse R (1987) The Drosophila homolog of the mouse mammary oncogene int-1 is identical to the segment polarity gene wingless. Cell 50:649–657
144. Klaus A, Birchmeier W (2008) Wnt signalling and its impact on development and cancer. Nat Rev Cancer 8:387–398
145. Perantoni AO (2003) Renal development: perspectives on a Wnt-dependent process. Semin Cell Dev Biol 14:201–208
146. Saadi-Kheddouci S, Berrebi D, Romagnolo B, Cluzeaud F, Peuchmaur M, Kahn A, Vandewalle A, Perret C (2001) Early development of polycystic kidney disease in transgenic mice expressing an activated mutant of the beta-catenin gene. Oncogene 20:5972–5981
147. Qian C-N, Knol J, Igarashi P, Lin F, Zylstra U, Teh BT, Williams BO (2005) Cystic renal neoplasia following conditional inactivation of apc in mouse renal tubular epithelium. J Biol Chem 280:3938–3945
148. Lin F, Hiesberger T, Cordes K, Sinclair AM, Goldstein LSB, Somlo S, Igarashi P (2003) Kidney-specific inactivation of the KIF3A subunit of kinesin-II inhibits renal ciliogenesis and produces polycystic kidney disease. Proc Natl Acad Sci USA 100:5286–5291
149. Corbit KC, Shyer AE, Dowdle WE, Gaulden J, Singla V, Chen M-H, Chuang P-T, Reiter JF (2008) Kif3a constrains beta-catenin-dependent

Wnt signalling through dual ciliary and non-ciliary mechanisms. Nat Cell Biol 10:70–76
150. Huang P, Schier AF (2009) Dampened Hedgehog signaling but normal Wnt signaling in zebrafish without cilia, Development 136:3089–3098
151. Ocbina PJR, Tuson M, Anderson KV (2009) Primary cilia are not required for normal canonical Wnt signaling in the mouse embryo, PLoS ONE 4:e6839
152. Maretto S, Cordenonsi M, Dupont S, Braghetta P, Broccoli V, Hassan AB, Volpin D, Bressan GM, Piccolo S (2003) Mapping Wnt/beta-catenin signaling during mouse development and in colorectal tumors, Proc Natl Acad Sci USA 100:3299–3304
153. Gerdes JM, Liu Y, Zaghloul NA, Leitch CC, Lawson SS, Kato M, Beachy PA, Beales PL, DeMartino GN, Fisher S, Badano JL, Katsanis N (2007) Disruption of the basal body compromises proteasomal function and perturbs intracellular Wnt response. Nat Genet 39: 1350–1360
154. Gong Y, Mo C, Fraser SE (2004) Planar cell polarity signalling controls cell division orientation during zebrafish gastrulation. Nature 430:689–693
155. Saburi S, Hester I, Fischer E, Pontoglio M, Eremina V, Gessler M, Quaggin SE, Harrison R, Mount R, McNeill H (2008) Loss of Fat4 disrupts PCP signaling and oriented cell division and leads to cystic kidney disease. Nat Genet 40:1010–1015
156. Patel V, Li L, Cobo-Stark P, Shao X, Somlo S, Lin F, Igarashi P (2008) Acute kidney injury and aberrant planar cell polarity induce cyst formation in mice lacking renal cilia. Hum Mol Genet 17:1578–1590
157. Nishio S, Tian X, Gallagher AR, Yu Z, Patel V, Igarashi P, Somlo S (2010) Loss of oriented cell division does not initiate cyst formation. J Am Soc Nephrol 21:295–302. doi:10.1681/ASN.2009060603
158. Steigelman KA, Lelli A, Wu X, Gao J, Lin S, Piontek K, Wodarczyk C, Boletta A, Kim H, Qian F, Germino G, Géléoc GSG, Holt JR, Zuo J (2011) Polycystin-1 is required for stereocilia structure but not for mechanotransduction in inner ear hair cells. J Neurosci 31:12241–12250
159. Gallagher AR, Esquivel EL, Briere TS, Tian X, Mitobe M, Menezes LF, Markowitz GS, Jain D, Onuchic LF, Somlo S (2008) Biliary and pancreatic dysgenesis in mice harboring a mutation in Pkhd1. Am J Pathol 172:417–429. doi:10.2353/ajpath.2008.070381
160. Gray RS, Roszko I, Solnica-Krezel L (2011) Planar cell polarity: coordinating morphogenetic cell behaviors with embryonic polarity. Dev Cell 21:120–133
161. Itoh K, Brott BK, Bae G-U, Ratcliffe MJ, Sokol SY (2005) Nuclear localization is required for Dishevelled function in Wnt/beta-catenin signaling. J Biol 4:3
162. Gao C, Chen Y-G (2010) Dishevelled: the hub of Wnt signaling. Cell Signal 22:717–727
163. Otto EA, Schermer B, Obara T, O'Toole JF, Hiller KS, Mueller AM, Ruf RG, Hoefele J, Beekmann F, Landau D, Foreman JW, Goodship JA, Strachan T, Kispert A, Wolf MT, Gagnadoux MF, Nivet H, Antignac C, Walz G, Drummond IA, Benzing T, Hildebrandt F (2003) Mutations in INVS encoding inversin cause nephronophthisis type 2, linking renal cystic disease to the function of primary cilia and left-right axis determination. Nat Genet 34:413–420
164. Sugiyama N, Tsukiyama T, Yamaguchi TP, Yokoyama T (2011) The canonical Wnt signaling pathway is not involved in renal cyst development in the kidneys of inv mutant mice. Kidney Int 79(9):957–965
165. Rapola J, Kääriäinen H (1988) Polycystic kidney disease. Morphological diagnosis of recessive and dominant polycystic kidney disease in infancy and childhood. APMIS 96:68–76
166. Qian Q, Li A, King BF, Kamath PS, Lager DJ, Huston J, Shub C, Davila S, Somlo S, Torres VE (2003) Clinical profile of autosomal dominant polycystic liver disease. Hepatology 37:164–171
167. van Keimpema L, de Koning DB, van Hoek B, van Den Berg AP, van Oijen MGH, de Man RA, Nevens F, Drenth JPH (2011) Patients with isolated polycystic liver disease referred to liver centres: clinical characterization of 137 cases. Liver Int 31:92–98
168. Everson GT, Emmett M, Brown WR, Redmond P, Thickman D (1990) Functional similarities of hepatic cystic and biliary epithelium: studies of fluid constituents and in vivo secretion in response to secretin. Hepatology 11:557–565
169. Perrone RD, Grubman SA, Rogers LC, Lee DW, Moy E, Murray SL, Torres VE, Jefferson DM (1995) Continuous epithelial cell lines from ADPKD liver cysts exhibit characteristics of intrahepatic biliary epithelium. Am J Physiol 269:G335–G345
170. Perrone RD, Grubman SA, Murray SL, Lee DW, Alper SL, Jefferson DM (1997) Autosomal dominant polycystic kidney disease decreases anion exchanger activity. Am J Physiol 272:C1748–C1756

171. Qian Q (2010) Isolated polycystic liver disease. Adv Chronic Kidney Dis 17:181–189
172. Janssen MJ, Waanders E, Woudenberg J, Lefeber DJ, Drenth JPH (2010) Congenital disorders of glycosylation in hepatology: the example of polycystic liver disease. J Hepatol 52:432–440
173. Reynolds DM, Falk CT, Li A, King BF, Kamath PS, Huston J, Shub C, Iglesias DM, Martin RS, Pirson Y, Torres VE, Somlo S (2000) Identification of a locus for autosomal dominant polycystic liver disease, on chromosome 19p13.2-13.1. Am J Hum Genet 67:1598–1604
174. Li A, Davila S, Furu L, Qian Q, Tian X, Kamath PS, King BF, Torres VE, Somlo S (2003) Mutations in PRKCSH cause isolated autosomal dominant polycystic liver disease. Am J Hum Genet 72:691–703
175. Drenth JPH, te Morsche RHM, Smink R, Bonifacino JS, Jansen JBMJ (2003) Germline mutations in PRKCSH are associated with autosomal dominant polycystic liver disease. Nat Genet 33:345–347
176. Davila S, Furu L, Gharavi AG, Tian X, Onoe T, Qian Q, Li A, Cai Y, Kamath PS, King BF, Azurmendi PJ, Tahvanainen P, Kääriäinen H, Höckerstedt K, Devuyst O, Pirson Y, Martin RS, Lifton RP, Tahvanainen E, Torres VE, Somlo S (2004) Mutations in SEC63 cause autosomal dominant polycystic liver disease. Nat Genet 36:575–577
177. Waanders E, Venselaar H, te Morsche RHM, de Koning DB, Kamath PS, Torres VE, Somlo S, Drenth JPH (2010) Secondary and tertiary structure modeling reveals effects of novel mutations in polycystic liver disease genes PRKCSH and SEC63. Clin Genet 78:47–56
178. Janssen MJ, Waanders E, Te Morsche RHM, Xing R, Dijkman HBPM, Woudenberg J, Drenth JPH (2011) Secondary, somatic mutations might promote cyst formation in patients with autosomal dominant polycystic liver disease. Gastroenterology 141:2056.e2–2063.e2
179. Shao X, Somlo S, Igarashi P (2002) Epithelial-specific Cre/lox recombination in the developing kidney and genitourinary tract. J Am Soc Nephrol 13:1837–1846
180. Yu Y, Ulbrich MH, Li M-H, Buraei Z, Chen X-Z, Ong ACM, Tong L, Isacoff EY, Yang J (2009) Structural and molecular basis of the assembly of the TRPP2/PKD1 complex. Proc Natl Acad Sci USA 106:11558–11563. doi:10.1073/pnas.0903684106
181. Woollatt E, Pine KA, Shine J, Sutherland GR, Iismaa TP (1999) Human Sec63 endoplasmic reticulum membrane protein, map position 6q21. Chromosome Res 7:77
182. Meyer HA, Grau H, Kraft R, Kostka S, Prehn S, Kalies KU, Hartmann E (2000) Mammalian Sec61 is associated with Sec62 and Sec63. J Biol Chem 275:14550–14557
183. Gasteiger E, Gattiker A, Hoogland C, Ivanyi I, Appel RD, Bairoch A (2003) ExPASy: the proteomics server for in-depth protein knowledge and analysis. Nucleic Acids Res 31:3784–3788
184. Deshaies RJ, Sanders SL, Feldheim DA, Schekman R (1991) Assembly of yeast Sec proteins involved in translocation into the endoplasmic reticulum into a membrane-bound multisubunit complex. Nature 349:806–808
185. Panzner S, Dreier L, Hartmann E, Kostka S, Rapoport TA (1995) Posttranslational protein transport in yeast reconstituted with a purified complex of Sec proteins and Kar2p. Cell 81:561–570
186. Matlack KE, Misselwitz B, Plath K, Rapoport TA (1999) BiP acts as a molecular ratchet during posttranslational transport of prepro-alpha factor across the ER membrane. Cell 97:553–564
187. Vembar SS, Jonikas MC, Hendershot LM, Weissman JS, Brodsky JL (2010) J domain co-chaperone specificity defines the role of BiP during protein translocation. J Biol Chem 285:22484–22494
188. Oyadomari S, Yun C, Fisher EA, Kreglinger N, Kreibich G, Oyadomari M, Harding HP, Goodman AG, Harant H, Garrison JL, Taunton J, Katze MG, Ron D (2006) Cotranslocational degradation protects the stressed endoplasmic reticulum from protein overload. Cell 126:727–739
189. Braakman I, Bulleid NJ (2011) Protein folding and modification in the mammalian endoplasmic reticulum. Annu Rev Biochem 80:71–99
190. Trombetta ES, Simons JF, Helenius A (1996) Endoplasmic reticulum glucosidase II is composed of a catalytic subunit, conserved from yeast to mammals, and a tightly bound noncatalytic HDEL-containing subunit. J Biol Chem 271:27509–27516
191. Gkika D, Mahieu F, Nilius B, Hoenderop JGJ, Bindels RJM (2004) 80K-H as a new Ca2+ sensor regulating the activity of the epithelial Ca2+ channel transient receptor potential cation channel V5 (TRPV5). J Biol Chem 279:26351–26357
192. Stigliano ID, Alculumbre SG, Labriola CA, Parodi AJ, D'Alessio C (2011) Glucosidase II and N-glycan mannose content regulate the half-lives of monoglucosylated species in vivo. Mol Biol Cell 22:1810–1823

193. Ruddock LW, Molinari M (2006) N-glycan processing in ER quality control. J Cell Sci 119:4373–4380
194. Trombetta ES, Parodi AJ (2003) Quality control and protein folding in the secretory pathway. Annu Rev Cell Dev Biol 19:649–676
195. Trombetta ES, Fleming KG, Helenius A (2001) Quaternary and domain structure of glycoprotein processing glucosidase II. Biochemistry 40:10717–10722
196. Pelletier MF, Marcil A, Sevigny G, Jakob CA, Tessier DC, Chevet E, Menard R, Bergeron JJ, Thomas DY (2000) The heterodimeric structure of glucosidase II is required for its activity, solubility, and localization in vivo. Glycobiology 10:815–827
197. Treml K, Meimaroglou D, Hentges A, Bause E (2000) The alpha- and beta-subunits are required for expression of catalytic activity in the hetero-dimeric glucosidase II complex from human liver. Glycobiology 10:493–502
198. Stigliano ID, Caramelo JJ, Labriola CA, Parodi AJ, D'Alessio C (2009) Glucosidase II beta subunit modulates N-glycan trimming in fission yeasts and mammals. Mol Biol Cell 20:3974–3984
199. Kawaai K, Hisatsune C, Kuroda Y, Mizutani A, Tashiro T, Mikoshiba K (2009) 80K-H interacts with inositol 1,4,5-trisphosphate (IP3) receptors and regulates IP3-induced calcium release activity. J Biol Chem 284:372–380
200. Harvey RP (1998) Links in the left/right axial pathway. Cell 94:273–276
201. Shiratori H, Hamada H (2006) The left-right axis in the mouse: from origin to morphology. Development 133:2095–2104
202. Srivastava D (1997) Left, right … which way to turn? Nat Genet 17:252–254
203. Mercola M, Levin M (2001) Left-right asymmetry determination in vertebrates. Annu Rev Cell Dev Biol 17:779–805
204. Satir P, Christensen ST (2007) Overview of structure and function of mammalian cilia. Annu Rev Physiol 69:377–400
205. Nonaka S, Tanaka Y, Okada Y, Takeda S, Harada A, Kanai Y, Kido M, Hirokawa N (1998) Randomization of left-right asymmetry due to loss of nodal cilia generating leftward flow of extraembryonic fluid in mice lacking KIF3B motor protein. Cell 95:829–837
206. Nonaka S, Shiratori H, Saijoh Y, Hamada H (2002) Determination of left-right patterning of the mouse embryo by artificial nodal flow. Nature 418:96–99
207. Hirokawa N, Tanaka Y, Okada Y, Takeda S (2006) Nodal flow and the generation of left-right asymmetry. Cell 125:33–45
208. Arnold SJ, Robertson EJ (2009) Making a commitment: cell lineage allocation and axis patterning in the early mouse embryo. Nat Rev Mol Cell Biol 10:91–103
209. Hamada H (2008) Breakthroughs and future challenges in left-right patterning. Dev Growth Differ 50(Suppl 1):S71–S78
210. Marshall WF, Nonaka S (2006) Cilia: tuning in to the cell's antenna. Curr Biol 16:R604–R614
211. Bisgrove BW, Snarr BS, Emrazian A, Yost HJ (2005) Polaris and Polycystin-2 in dorsal forerunner cells and Kupffer's vesicle are required for specification of the zebrafish left-right axis. Dev Biol 287:274–288. doi:10.1016/j.ydbio.2005.08.047
212. Schottenfeld J, Sullivan-Brown J, Burdine RD (2007) Zebrafish curly up encodes a Pkd2 ortholog that restricts left-side-specific. Development 134:1605–1615. doi:10.1242/dev.02827
213. Field S, Riley K-L, Grimes DT, Hilton H, Simon M, Powles-Glover N, Siggers P, Bogani D, Greenfield A, Norris DP (2011) Pkd1l1 establishes left-right asymmetry and physically interacts with Pkd2. Development 138:1131–1142
214. Kamura K, Kobayashi D, Uehara Y, Koshida S, Iijima N, Kudo A, Yokoyama T, Takeda H (2011) Pkd1l1 complexes with Pkd2 on motile cilia and functions to establish the left-right axis. Development 138:1121–1129
215. Bataille S, Demoulin N, Devuyst O, Audrézet M-P, Dahan K, Godin M, Fontès M, Pirson Y, Burtey S (2011) Association of PKD2 (polycystin 2) mutations with left-right laterality defects. Am J Kidney Dis 58:456–460
216. McGrath J, Somlo S, Makova S, Tian X, Brueckner M (2003) Two populations of node monocilia initiate left-right asymmetry in the mouse. Cell 114:61–73
217. Karcher C, Fischer A, Schweickert A, Bitzer E, Horie S, Witzgall R, Blum M (2005) Lack of a laterality phenotype in Pkd1 knock-out embryos correlates with absence of polycystin-1 in nodal cilia. Differentiation 73:425–432. doi:10.1111/j.1432-0436.2005.00048.x
218. Qian F, Noben-Trauth K (2005) Cellular and molecular function of mucolipins (TRPML) and polycystin 2 (TRPP2). Pflugers Arch 451:277–285. doi:10.1007/s00424-005-1469-4
219. Moy GW, Mendoza LM, Schulz JR, Swanson WJ, Glabe CG, Vacquier VD (1996) The sea urchin sperm receptor for egg jelly is a modular protein with extensive homology to the human polycystic kidney disease protein, PKD1. J Cell Biol 133:809–817

220. Qian F, Boletta A, Bhunia AK, Xu H, Liu L, Ahrabi AK, Watnick TJ, Zhou F, Germino GG (2002) Cleavage of polycystin-1 requires the receptor for egg jelly domain and is disrupted by human autosomal-dominant polycystic kidney disease 1-associated mutations. Proc Natl Acad Sci USA 99:16981–16986. doi:10.1073/pnas.252484899

221. Yuasa T, Takakura A, Denker BM, Venugopal B, Zhou J (2004) Polycystin-1L2 is a novel G-protein-binding protein. Genomics 84:126–138. doi:10.1016/j.ygeno.2004.02.008

222. Vogel P, Read R, Hansen GM, Freay LC, Zambrowicz BP, Sands AT (2010) Situs inversus in Dpcd/Poll-/-, Nme7-/-, and Pkd1l1-/- mice. Vet Pathol 47:120–131. doi:10.1177/0300985809353553

Chapter 11

TRPV1 as a Polymodal Sensor: Potential to Discover TRPV1 Antagonists Selective for Specific Activating Modalities

Olivier Radresa, Stephen Zicha, William Brown, and Jennifer M.A. Laird

Abstract

The transient receptor potential V1, or TRPV1, channel has a complex polymodal activation profile that integrates information from membrane potential changes, heat, and protons in addition to channel gating following ligand binding. TRPV1 is expressed along the peripheral pain pathway from the distal endings of nociceptor neurons to dorsal root ganglia; but has limited expression in the central nervous system and is detected in arteriolar smooth muscle cells and various epithelia. Several TRPV1 antagonist candidate drugs have been tested in clinical trials over the past 5 years with varying results. The reported side-effect profile of these oral drug formulations includes hyperthermia and impaired detection of painful heat. These adverse events have tentatively been linked to characteristic in vitro profiles of the candidate drugs, specifically the ability of these molecules to differentially block one or several opening modalities. Thus the selection of lead candidates that are selective for specific modalities could be key to the development of a successful TRPV1 antagonist drug. This minireview updates the details of TRPV1 activation modes, existing in vitro assays available for screening of novel molecules and the clinical and preclinical profiles of next-generation TRPV1 candidate drugs developed by pharmaceutical companies.

Key words: TRPV1, Capsaicin, Vanilloid, Resiniferatoxin, Hyperthermia, Drug discovery, Drug development

1. Introduction

Transient receptor potential vanilloid 1 (TRPV1) is a member of the transient receptor potential (TRP) family. TRPV1 receptors are cation channels gated by a wide range of stimuli, many of which are damaging or noxious. The endogenous "ligands" for TRPV1 include noxious heat (42–53 degrees), protons (pH 5–6), and lipid derivatives, including lipoxygenase metabolites of arachidonic acid,

Arpad Szallasi and Tamás Bíró (eds.), *TRP Channels in Drug Discovery: Volume I*, Methods in Pharmacology and Toxicology, DOI 10.1007/978-1-62703-077-9_11, © Springer Science+Business Media, LLC 2012

endogenous cannabinoids-like anandamide, and endogenous vanilloids for example *N*-Arachidonoyl-dopamine (NADA) (1–3). Which of these activating agents is responsible for TRPV1 gating depends on the tissue and the physiological or pathophysiological circumstances, but the various stimuli may also potentiate the effects of each other on TRPV1.

Several natural products derived from plants also gate TRPV1. Capsaicin, the pungent substance found in spicy chili peppers (genus *Capsicum)*, is a specialized metabolite produced in the fruits as a deterrent to seed predators (4). The production of capsaicin was positively selected for during evolution because its stimulatory effect is aversive in mammals, but is without effect in birds. Nonetheless, capsaicin-producing *Capsicum* species have been cultivated by humans for more than 6,000 years. Other plants cultivated for culinary or medicinal use also produce TRPV1 agonists, for example, piperine, the pungent ingredient of black pepper corns, and resiniferatoxin (RTX), a very potent TRPV1 agonist secreted in the sap of the succulent plant *Euphorbia resinifera*, which is used as a purgative.

Capsaicin was used as a tool in pain research for many years before the cloning of the TRPV1 receptor (5). At moderate concentrations, capsaicin acutely and intensely excites nociceptive afferent fibers (nociceptors) without causing tissue damage (6), producing nociceptive behavior in animals (7–10) and pain reports in human subjects (11–15). The identification of TRPV1 as “the capsaicin receptor” by Caterina and colleagues (23) generated considerable interest in this area, because TRPV1 was the first transduction protein for painful stimuli to be identified.

In addition to the wide range of potential adequate stimuli that gate TRPV1 channels, the TRPV1 receptor has another very important characteristic, which is that a large range of factors can sensitize the channel to activating stimuli. This sensitization is manifested both as a decrease in the threshold for activation and as an increase in the channel open time in the presence of normally suprathreshold stimuli. Sensitizing factors include phosphorylation of TRPV1 by protein kinase C (PKC) or protein kinase A and release of phosphatidyl-inositol(4,5) biphosphate (PtdIns(4,5) P2)-mediated inhibition by activation of phospholipase C (PLC) (16, 17). Thus, the TRPV1 receptor is activated by mediators present during inflammation and tissue damage-like lipoxygenase products and low pH, but can also be sensitized to heat, acidic stimuli, anandamide, or vanilloids by a range of inflammatory mediators which act on their own membrane receptors on the same cell and activate PKC and/or PLC, for example, ATP via purinergic receptors, bradykinin via B2 receptors, or nerve growth factor (NGF) via TrkA receptors (18–21). The molecular sensitization of TRPV1 likely underlies the sensitization of nociceptor afferent terminals to endogenous chemical and heat stimuli, under inflammatory or ischemic conditions. For example, application of ATP can reduce the heat threshold of native TRPV1 channels from 43 to 35°C, below normal core body temperature, such that the channel would be activated continuously at body temperature. This property of TRPV1 is what makes it such a compelling analgesia target. More than 30 drug programs targeting TRPV1 antagonism are currently being pursued across the industry. Most major pharma companies set up proprietary or collaborative research projects to target pain or inflammation.

Given the wide range of modes of TRPV1 activation, does selective blockade of one or more of these have different physiological effects? Can the different activation modalities be separated molecularly or pharmacologically? Emerging data suggest that different modalities of activation indeed do have distinct molecular mechanisms and can be blocked differentially by small molecules. High and medium throughput assays have been developed that allow the identification and optimization of compounds that interfere with one or more of the activation modalities. Furthermore, differential pharmacological blockade of the different activation modes appears to produce distinct downstream effects. Here we review these novel concepts and the results of drug discovery programs designed to deliver novel TRPV1 antagonists.

2. Structural and Functional Determinants of TRPV1 Opening Modalities

Of the 28 mammalian members of the TRP family, TRPV1 has been the most investigated in terms of pharmacology and the molecular determinants responsible for its modulation. In human the TRPV1 gene is located on chromosome 17, position 13.2, and the most typically studied form is an 839aa protein that is sensitive to capsaicin, low pH, and temperatures >42°C. The rat ortholog and its pharmacology will be discussed below (*NCBI Gene bank* (22, 23)).

TRPV1, like all TRP channels, are putative six-transmembrane (6TM) polypeptide subunits that assemble as tetramers to form cation-permeable pores (Fig. 1). While the topology is similar to other tetrameric 6TM ion channels such as Shaker-like voltage-gated K^+ channels (24), TRPV1 channels differ significantly in their gating mechanisms and pharmacology. Recently, a 3D structure of TRPV1 has been elucidated by electron cryomicroscopy and confirmed the tetrameric structure and spatial conformations of the intracellular N- and C-termini and transmembrane domains (24). The pore-loop region is formed by a short hydrophobic stretch between S5 and S6, making the channel permeable to $Ca^{2+} > Mg^+ > Na^+ = K^+ = Cs^+$ ions (23, 25, 26). Of particular importance for TRPV1 pore function are Glu636 and Asp646 (25) because

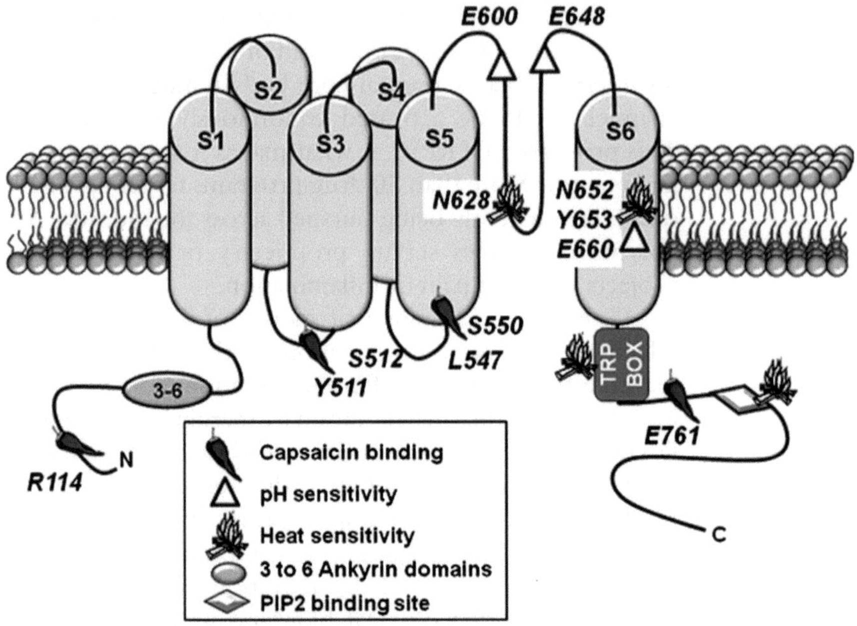

Fig. 1. Topology of the key residues involved in heat, pH and capsaicin sensitivity of TRPV1. The canonical "TRP box," PIP2- and Ankyrin- binding domains, that are further modulating receptors activation or dimerization, are also indicated (details in "Structural and functional determinants of TRPV1 opening modalities").

they sit at the entrance of the pore structure. Mutation of the residues leads to altered permeability of divalent cations (25). The pore region also contains a sequence of amino acids that closely resemble the selectivity filter found on Shaker-like potassium channels (TXGMG vs. TXGYG) just prior to the Asp646 residue. Asp646 and other residues deeper in the pore region such as Glu648 also play a role in the pharmacological responses to RTX and proton-gating, respectively (27, 28).

As recently shown, the intracellular hydrophilic N terminus contains up to six conserved ankyrin repeat domains also acting as calmodulin and ATP binding sites. Of importance also, the ankyrin repeat domains contribute to regulation of the channel function following Ca^{2+} influx after capsaicin activation and to its recovery after desensitization (29–31). The TRP box, which shares homology with other members of the TRP family, is found almost immediately after S6 on the C terminus. Recently, 3 of the 6aa in this region (I696, W697, and R701) were found to play a role in the functional coupling of heat, capsaicin, and low pH sensing and pore opening, as well as channel multimerization (32). The C terminus also contains other sites important for the modulation and activity of the channel. A phosphatydylinsositol-4,5-bisphosphate (PIP2) binding site plays a role in sensitivity to capsaicin and heat, although its detailed function remains somewhat controversial

(17, 29, 33). This site also contains a PKC consensus phosphorylation sequence at S800 that helps sensitize the channel by limiting the interaction between the negatively modulating PIP2 molecule and its C-terminal binding site (34). Even though PKC by itself cannot activate the channel, PKC binding is also important to relieve the Ca^{2+}-dependent desensitization that occurs after channel activation (35). Another PKC and Calcium/Calmodulin-dependent Kinase II (CamKII) consensus sequence found at Thr704 can also alter the channel responses to capsaicin (36).

The activation of TRPV1 by different stimuli can be differentiated by pharmacological tools, supporting the notion that capsaicin, protons, and heat likely activate the channel via different molecular mechanisms (37). The sites believed to be responsible for capsaicin-activation include Arg114 (rat), Tyr511, Ser512, Leu547 (or Met547 in rat), Tyr550 (rabbit), and Glu761 (38–41), all located on the intracellular side of the channel near the S3/S4 region. This is not surprising given the lipophillic nature of the agonist. RTX is believed to interact with Met547 in the TM4 region which is believed to constitute part of the sensor domain for the channel (42). As mentioned previously, two glutamate residues found in the pore region of the rat ortholog, Glu600, and Glu648, are responsible for proton gating (27). Phe660 located in the S6 transmembrane domain of the human ortholog is also responsible for the voltage-dependent proton gating and shifts the voltage dependence of activation to more physiological potentials (43). This is a similar mechanism to the one proposed for heat activation of TRPV1 (44). The C terminus is believed to be involved in modulating the responses to heat (32, 45–47) with elements residing in the TRP box and PIP2 binding domain likely conferring gating activity. Some single amino acid residues have also been identified that alter heat-mediated gating of rat TRPV1, including Asn628, Asn652, and Tyr653, by altering the temperature–voltage coupling mechanism (48).

Several functional splice variants have been characterized and result in altered wild-type TRPV1 channel function. These include deTRPV1b that lacks exon 7 and results in a 60aa deletion in the N terminus, rendering the channel insensitive to heat (49, 50); "VR.5 sv" which has a deletion of 60aa in the ankyrin repeat region and most of the N terminus deleted and also prevents the activation of channels by capsaicin, low pH, and heat (51, 52); an N-terminal splice variant missing exons 2–6 found in the supraoptic nucleus that helps regulate the intrinsic osmotic sensitivity of the cells and is insensitive to capsaicin (53); TRPV1VAR found in renal tissues and has a different N-terminal splice compared to others resulting in a truncated 253aa protein and potentiates responses to RTX (54); and lastly a variant found in taste receptors that regulate responses to NaCl (55).

3. TRPV1 In Vitro Profiling Assays

The polymodal characteristics of TRPV1 activation have enabled the development of a variety of in vitro screening methods. These screening methods and other mechanistic assays in principle allow the design of potential drug candidates with the preferred combination of specific antagonist profiles. The following section exemplifies some of the most important assays that have been used to increase our understanding of TRPV1 modes of action.

3.1. Binding Assays

Although generally not considered high-throughput compliant due to difficulties in handing radioactive ligands in large HTS experimental settings, binding studies have been critical in the characterization of the species-specific pharmacology of TRPV1. In particular for explaining the specific toxicity observed as early as 1980 upon capsaicin dosing in vivo in some species (56). Strangely, in chicken and rabbit, capsaicin toxicity appeared significantly weaker or simply absent compared to rats (41, 57). When RTX binding ((^{3}H)RTX) data became available, this helped identify critical residues in S4 and in the TM3/4 loop that accounted for the species-specific differences in capsaicin sensitivity (58).

On the other hand, binding assays have also fuelled studies of TRPV1 allosteric modulation. For instance, the importance of calmodulin binding to a stretch of 35-aa, and its modulation by Ca^{2+}, via the intracellular C terminus helped explaining how Ca^{2+} could induce desensitization following capsaicin exposure (31). Following this, a dominant-negative study emerged as a new potential therapeutic approach aimed at targeting TRPV1 indirectly, via modulation of the newly identified allosteric site (59). Although interesting, it is not yet clear whether this latest approach could lead to improved clinical profiles of TRPV1 chemical leads.

3.2. Fluorescence Assays

By virtue of its non-selective calcium conductance and ligand-gated properties, TRPV1 makes a natural candidate for high-throughput fluorescence screening. Namely, the industry compliant FLIPR format was used early on for screening assay development and also the comparative study of the pharmacological translation between rat and human of TRPV1 reference agonists. RTX, Capsaicin, Olvanil and Phorbol-12-phenylacetate-13-acetate-20-homovanillate (PPAHV; a vanilloid homolog) were characterized using calcium fluorescence and their potency was shown not to vary significantly between rat and human isoforms. Notably, agonists like capsaicin and olvanil, but not RTX, displayed the expected enhanced activity at lower pH (60). The selection for; and characterization of; the polymodal antagonist properties of Glaxo-Smithkline clinical candidate, SB-705498, was performed using Fluorophore assays based on either capsaicin or pH challenges (61).

3.3. Electrophysiology Profiling Assays

With the advent of the 384-well format, automated patch clamping brought compound screening for ion channels to a new level. Prior to these advances during the past decade, traditional electrophysiology proved instrumental in deciphering the mode of action of the vanilloid receptor and its coupling mechanism. In 1999, soon after the capsaicin-receptor was first cloned, inside-out patch clamping and oocyte recordings pointed at an intracellular binding site for capsaicin molecule to the inner pore of the channel (62). In turn, titration by electrophysiology was also used in the identification of the key extracellular aspartate and glutamate residues as determinants of acidification-induced opening (E600 and E648, see above "structural determinants") (27).

Using voltage ramp and -steps to study channel kinetics, Gunthorpe et al. identified that outward rectification and inward currents exhibited time-dependent gating properties reminiscent of voltage-gated channels, with which the membrane domain seems to share a similar topology (63).

Further on the profiling of TRPV1 activity, currents elicited by large and persistent capsaicin concentration, recorded in the absence of extracellular Na^{+}, revealed the formation of an unusually wide pore within the channel. This large pore accounted for the transport of positively charged molecules such as tetraethyl ammonium (TEA) or *N*-methylglucamine (NMDG), both of high molecular weight (Mw 150 and 195 kD, respectively) (64). So-called pore dilation following persistent activation, allowed the channel to also transport larger cationic probes like 6-Methoxy-*N*-Ethylquinolinium Iodide (MEQ; Mw 188 kD) and its bulkier derivative *N*-(Ethoxycarbonylmethyl)-6-Methoxyquinolinium Bromide (MQAE; Mw 246 kD). Subsequent studies finally demonstrated the transport of a quaternary amine derivative of lidocaine (QX-314; Mw 263 kD) and the still larger fluorescent dyes YO-PRO (Mw 375 kD) and FM1-43 (Mw 451) (65–67).

Finally, considering automated electrophysiology, TRPV1 assays have been reported on all existing platforms from 16 to 384-well formats. Uniquely though, Nanion's Patchliner® was modified by the manufacturer to provide users with a flexible temperature add-on for the study of thermo-regulated ion channels. The integrated pipette heater is used for application of external solution at a controlled temperature directly to the recording chamber (68). Cooling Pelletier elements though have not been commercially available to date. The ease in generating robust, high-quality recombinant cells with high current expression certainly helped for the development of TRPV1 screening assays in many laboratories.

3.4. Other Label-Free Assays

Recently TRPV1 activity in recombinant CHO cells was successfully recorded using the new generation of label-free technologies (69, 70). Dose–responses of capsaicin agonists were obtained in CellKey, BIND, and Epic. These technologies record changes in impedence or visible light diffraction upon modification of cells properties at the bottom of the wells. All responses were followed for 30 min (CellKey, BIND) or 15 min (Epic) after capsaicin addition and as such represented "integrated" cell reactions to the specific TRPV1 current induced by capsaicin addition.

3.5. Targeting the Temperature Sensitivity of TRPV1 in Profiling Assays

In 2009, Evotec (a German Biotech company) reported on a novel temperature assay to assess the potency of TRPV1 antagonists like iodo-resiniferatoxin (I-RTX), capsazepine, AMG-9810, and *N*-(4-tert-butylphenyl)-4-(3-chloropyridin-2-yl)piperazine-1-carboxamide (BCTC) (71). Blockade of heat activation was recorded using a method modified from a QT-PCR assay (TaqMan; Applied Biosystem). A real-time PCR machine was used to provide for rapid heat jump to +50°C, followed by calcium entry. The combined fluorescence readout, using the Ca^{2+} probe Fluo-4, allowed for determination of dose-dependent block. Here though, the agents described here blocked the other opening modalities of the channel as well. At AstraZeneca, a similar assay was developed and extended to other thermo-TRPs, interestingly it proved to be easily translatable to native rat dorsal root ganglia neurons in either 384 or 96 well formats (72).

Very recently, Pfizer (in coll. with Ideas Studios) reported the design of a completely new type of FLIPR-based heat assay. This interesting tool was successfully tested in a study of the species-specificity of TRPV1 compounds using chimeras of rat that included small orthologs of the human pore region, and vice versa (73). The custom-built apparatus allowed for temperature control within each individual well of the 384-well FLIPR plate and for applications of heat ramps of 1°C/s (radiant heat). The 384-pins device included a hollow aluminum chamber, kept at a constant 13°C, acting as a heat/cold sink for the surplus energy. This ensured that a stable temperature control is maintained in each well. As a cross validation, the authors confirmed the pharmacology of some reference antagonists, namely PF-4065463, JYL-1421, BCTC, and MK-2295, both in the Patchliner® automated electrophysiology heat assay (see below) and the traditional patch clamp set with an in-line heat flow system for the recording of this specific TRPV1 activity.

4. Current Views on Clinical Status and Properties of TRPV1 Chemical Leads

There have now been several TRPV1 antagonists that have entered clinical studies, these are shown in Table 1. The primary indication for most of these compounds has been analgesia, targeting inflammatory, neuropathic, or visceral pain as this seems to be where the preclinical data are strongest. There has been no clear clinical analgesic efficacy reported to date, but this may be due to the limitations of these compounds imposed by the contribution of other undesired TRPV1-mediated effects, in particular increased core body temperature (hyperthermia) and/or a reduced heat pain detection leading to a risk of accidental heat injury. These dose-limiting effects on body temperature and decreased heat pain sensitivity were somewhat to be expected since the effects have been shown to be TRPV1 mediated in pre-clinical species (23, 74–77).

However, what has become clear is that all TRPV1 antagonists are not created equally with respect to their effects on hyperthermia and heat pain threshold. In Amgen's phase 1 study a long lasting hyperthermic effect was noted with AMG517 that raised body temperatures as high as 40°C lasting up to 4 days (75). Hyperthermia has also been reported with ABT-102 (78) and AZD1386 (79), but these effects appeared to be much less severe. This finding prompted Amgen to further investigate these effects. They discovered a correlation between blocking the different modes of channel activation and subsequent effects in animals. They examined three primary activators of TRPV1, capsaicin, low pH, and heat. They found that compounds that only blocked activation by the exogenous ligand capsaicin retained their analgesic effects in rodent pain models, but exhibited no effect on body temperature. Amgen then proceeded to build a correlation between the blockade of the pH response with hyperthermia and blockade of capsaicin activation with analgesic effects. These efforts afforded AMG8562 which blocks only capsaicin-induced activation at the rat TRPV1. This last compound did not affect body temperature in rats, although it still showed some hyperthermia in dogs (37).

Abbot has recently presented their clinical data with ABT-102, this compound showed good efficacy in a human experimental pain model and only a modest increase in core body temperature in a phase 1 study. They also reported that the dose-limiting temperature elevation is attenuated after multiple dosing. Abbot have since embarked on a pre-clinical program similar to that of Amgen's where they are seeking compounds that retain the ability to block activation by capsaicin without affecting the activation by low pH. They are currently in the process of optimizing on a new chemotype with these desired properties (80).

The second undesired target-mediated tolerability effect, which is the impairment of basal noxious heat perception, is currently less understood. The role of TRPV1 as a noxious heat sensor in humans has been supported by recent clinical studies. GlaxoSmithKline (GSK) reported that an oral dose of 400 mg of SB-705498 reduced heat perception on non-sensitized skin in healthy volunteers (81). Similar findings have recently been reported for AZD1386 and ABT-102 (78, 79). The most severe effects have been reported by Merck where subjects dosed with MK-2295 could not perceive potentially harmful temperatures. It could be imagined that this effect on heat pain response could be

Table 1
TRPV1 antagonists in clinical development (or close follow up compound): candidate structure, safety and efficacy outcomes are indicated where available, together with indication and originating companies

Compound	Structure	Development stage and indication	Outcome	References
AMG 517 (Amgen)	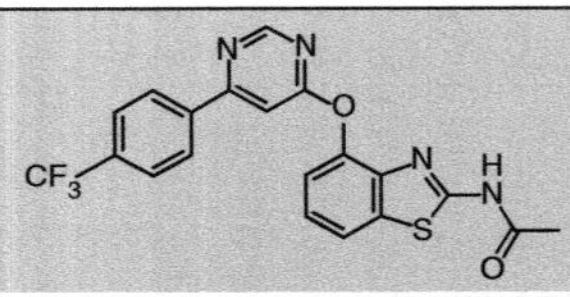	• Phase I molar extraction terminated	• Body temperatures of 39–40.2°C persisted for 1–4 days in 33 % of the subjects	(75)
AMG 8562 (Amgen)	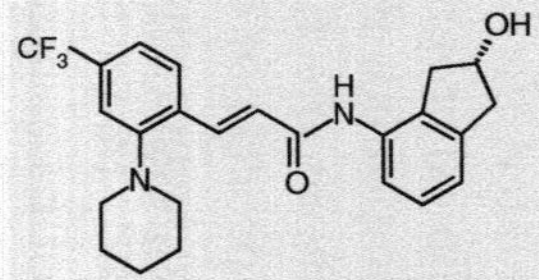	• Preclinical study	• No effect on body temperature observed	(37)
ABT 102 (Abbott)	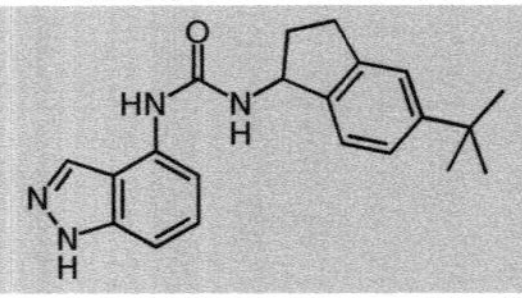	• Phase I in osteoarthritis (OA) completed	• No safety findings • Body temperatures stayed below 39°C in all participants • Increased heat pain threshold	(78, 84)
AZD 1386 (AstraZeneca)	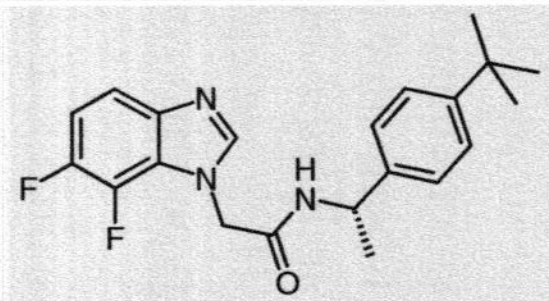	• Phase II OA pain terminated • Phase II gastro-esophageal reflux disease (GERD) completed	• Increased oesophageal and skin heat pain thresholds, and showed safe adverse-event profile in GERD study	(85, 86)
DWP 05195 (Daewoong)	Not disclosed	• Phase I completed	• No information on safety/ efficacy	(87)
GRC 6211 (Glenmark, Eli Lilly)	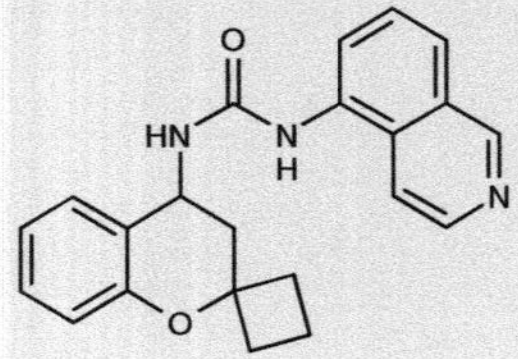	• Phase I completed • Phase II OA pain suspended	• No information on safety/ efficacy	(88), WO07042906

JNJ 39439335 (Jonhson & Jonhson)	Not disclosed	• Phase I in OA patients completed • Phase II in OA ongoing	• No information on safety/efficacy	NCT00933582, NCT01343303
JTS 653 (Japan Tobacco)	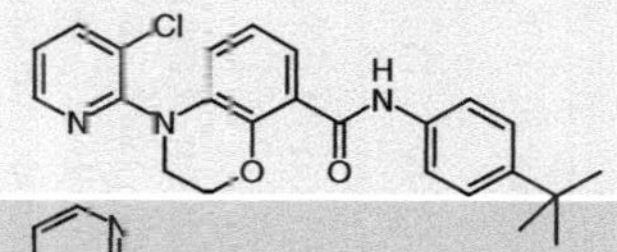	• Phase II post-herpetic neuralgia (PHN) completed • Phase II overactive bladder (OAB)	• No information on safety/efficacy	WO060006741, JapicCTI-101177
MK 2295 (Merck)	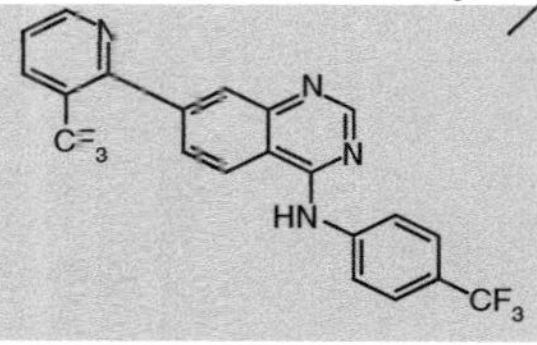	• Phase IIa, post-op dental pain completed	• Potential liability from defective heat perception	(82)
MR 1817 (Mochida Pharmaceutical/Wyeth)	Not disclosed	• Phase I single ascending dose in healthy volunteers completed	• No information on safety/efficacy	NCT00960180
PHE 377 (PharmEste)	Not disclosed	• Phase Ia completed • Phase Ib open	• No information on safety/efficacy	http://www.PharmEste.com
SB 705498 (Glaxo-SmithKline)	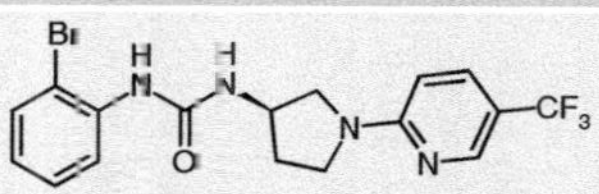	• Phase II topical non-kallergic rhinitis and pruritis	• Increased heat-pain threshold • Reduced flare	(81, 89)
XEN D0501 (Xention/Provesica)	Not disclosed	• Phase Ia completed in healthy volunteers • Phase II for OAB ongoing	• Transient, dose dependent, increase in body temperature <1°C at reported doses (healthy volunteers)	(90), TrialTroveID-118016, XEN-D0501-CL-03

References to clinical trial identification, literature data and patent files are indicated for further detailed information correlated to blockade of channel activation by heat, but there have been no published data to support this assumption. The extreme effect of MK-2295 on noxious heat sensation may be due to its additional blockade of TRPV3, the warm sensing channel (82).

Recently an academic group in Spain has reported discovery compounds that appear to be open-channel blockers. They speculate that these uncompetitive antagonists may serve as activity-dependant blockers, preferentially blocking overactive channels. This approach may result in a better side-effect profile than competitive antagonism, but this remains to be proven and this class of compounds is still a long way from being tested in the clinic (83).

5. Conclusions

Antagonism of TRPV1 remains a desirable approach with many potentially useful clinical therapeutic indications. It remains to be seen whether or not one can successfully separate the desired target-mediated effects from the undesired ones. However, the emerging understanding that first, the different activation modes of the TRPV1 receptor could be responsible for the differential physiological effects observed in the clinic, and second that it is now feasible to address these activation modes separately by selective small molecules, would suggest that a second generation of TRPV1 antagonists has the potential to deliver on the promises of this long-standing analgesia target.

References

1. Sagar DR et al (2004) TRPV1 and CB(1) receptor-mediated effects of the endovanilloid/endocannabinoid N-arachidonoyl-dopamine on primary afferent fibre and spinal cord neuronal responses in the rat. Eur J Neurosci 20:175–184
2. Sexton A et al (2007) 12-Lipoxygenase-derived eicosanoids protect against myocardial ischemia/reperfusion injury via activation of neuronal TRPV1. FASEB J 21:2695–2703
3. Ross RA (2004) Anandamide and vanilloid TRPV1 receptors. Br J Pharmacol 140: 790–801
4. Tewksbury JJ et al (2001) Seed dispersal. Directed deterrence by capsaicin in chilies. Nature 412:403–404
5. Szolcsányi J (2004) Forty years in capsaicin research for sensory pharmacology and physiology. Neuropeptides 38:377–384
6. Dray A (1992) Mechanism of action of capsaicin-like molecules on sensory neurons. Life Sci 51:1759–1765
7. Laird JM et al (2001) A new model of visceral pain and referred hyperalgesia in the mouse. Pain 92:335–342
8. Lin Q, Wu J, Willis WD (1999) Dorsal root reflexes and cutaneous neurogenic inflammation following intradermal injection of capsaicin in rats. J Neurophysiol 82:2602–2611
9. Mitchell K et al (2010) Ablation of rat TRPV1-expressing Adelta/C-fibers with resiniferatoxin: analysis of withdrawal behaviors, recovery of function and molecular correlates. Mol Pain 6:94
10. Caterina MJ et al (2000) Impaired nociception and pain sensation in mice lacking the capsaicin receptor. Science 288:306–313
11. Helme RD, McKernan S (1985) Neurogenic flare responses following topical application of capsaicin in humans. Ann Neurol 18:505–509
12. Simone DA et al (1989) Dose-dependent pain and mechanical hyperalgesia in humans after intradermal injection of capsaicin. Pain 38:99–107

13. Petersen KL, Rowbotham MC (1999) A new human experimental pain model: the heat/capsaicin sensitization model. Neuroreport 10:1511–1516
14. Modir JG, Wallace MS (2010) Human experimental pain models 3: heat/capsaicin sensitization and intradermal capsaicin models. Methods Mol Biol 617:169–174
15. Aykanat V et al (2012) Intradermal capsaicin as a neuropathic pain model in patients with unilateral sciatica. Br J Clin Pharmacol 73(1):37–45. doi:10.1111/1365-2125.2011.04059.x, "Accepted Article"
16. Klein RM et al (2008) Determinants of molecular specificity in phosphoinositide regulation. Phosphatidylinositol (4,5)-bisphosphate (PI(4,5)P2) is the endogenous lipid regulating TRPV1. J Biol Chem 283:26208–26216
17. Lukacs V et al (2007) Dual regulation of TRPV1 by phosphoinositides. J Neurosci 27:7070–7080
18. Zhang X, Huang J, McNaughton PA (2005) NGF rapidly increases membrane expression of TRPV1 heat-gated ion channels. EMBO J 24:4211–4223
19. Chuang HH et al (2001) Bradykinin and nerve growth factor release the capsaicin receptor from PtdIns(4,5)P2-mediated inhibition. Nature 411:957–962
20. Lin Q et al (2007) Roles of TRPV1 and neuropeptidergic receptors in dorsal root reflex-mediated neurogenic inflammation induced by intradermal injection of capsaicin. Mol Pain 3:30–45
21. Ferreira J, da Silva GL, Calixto JB (2004) Contribution of vanilloid receptors to the overt nociception induced by B2 kinin receptor activation in mice. Br J Pharmacol 141: 787–794
22. NCBI GeneBank: Human gene ID 7442. http://www.ncbi.nlm.nih.gov/gene?term=7442. Accessed Feb 2012
23. Caterina MJ et al (1997) The capsaicin receptor: a heat-activated ion channel in the pain pathway. Nature 389:816–824
24. Moiseenkova-Bell VY et al (2008) Structure of TRPV1 channel revealed by electron cryomicroscopy. Proc Natl Acad Sci USA 105: 7451–7455
25. Garcia-Martinez C et al (2000) Identification of an aspartic residue in the P loop of the vanilloid receptor that modulates pore properties. J Biol Chem 275:32552–32558
26. Ferrer-Montiel A et al (2004) Molecular architecture of the vanilloid receptor Insights for drug design. Eur J Biochem 271:1820–1826
27. Jordt S-E, Tominaga M, Julius D (2000) Acid potentiation of the capsaicin receptor determined by a key extracellular site. Proc Natl Acad Sci USA 97:8134–8139
28. Welch JM, Simon SA, Reinhart PH (2000) The activation of rat vanilloid receptor 1 bycaps aicin involves the pore domain and differs from the activation by either acid or heat. Proc Natl Acad Sci USA 25:13889–13894
29. Lishko PV et al (2007) The ankyrin repeats of TRPV1 bind multiple ligands and modulate channel sensitivity. Neuron 54:905–918
30. Schindl R et al (2008) The first ankyrin like repeat is the minimum indispensable key structure for functional assembly of homo- and heteromeric TRPC4/TRPC5 channels. Cell Calcium 43:260–269
31. Rosenbaum T et al (2004) Ca^{2+}/calmodulin modulates TRPV1 activation by capsaicin. J Gen Physiol 123:53–62
32. Valente P et al (2008) Identification of molecular determinants of channel gating in the transient receptor potential box of vanilloid receptor. FASEB J 22:3298–3309
33. Prescott E, Julius D (2003) A modular PIP2 site as a determinant of capsaicin receptor sensitivity. Science 300:1284–1288
34. Bhave G et al (2003) Protein kinase C phosphorylation sensitizes but does not activate the capsaicin receptor potential vanilloid 1 (TRPV1). Proc Natl Acad Sci USA 100: 12480–12485
35. Mandadi S et al (2004) Activation of protein kinase C reverses capsaicin-induced calcium-dependent desensitization of TRPV1 ion channels. Cell Calcium 35:471–478
36. Jun J et al (2004) Phosphorylation of vanilloid receptor 1 by Ca^{2+}/calmodulin-dependent kinase II regulates its vanilloid binding. J Biol Chem 979:7048–7054
37. Lehto SG et al (2008) Antihyperalgesic effects of (R, E)-N-(2-hydroxy-2,3-dihydro-1H-inden-4-yl)-3-(2-(piperidin-1-yl)-4-(trifluoromethyl)phenyl)-acrylamide (AMG8562), a novel transient receptor potential vanilloid type 1 modulator that does not cause hyperthermia in rats. J Pharmacol Exp Ther 326:218–229
38. Pingle SC, Matta JA, Ahern GP (2007) Capsaicin receptor: TRPV1 a promiscuous TRP channel. Handb Exp Pharmacol 179:155–171
39. Vyklicky L et al (2003) Vanilloid receptor TRPV1 is not activated by vanilloids applied intracellularly. Neuroreport 14:1061–1065
40. Sutton KG et al (2005) Functional characterisation of the S512Y mutant vanilloid human TRPV1 receptor. Br J Pharmacol 146:702–711
41. Gavva NR et al (2004) Molecular determinants of vanilloid sensitivity in TRPV1. J Biol Chem 279:20283–20295

42. Chou MZ et al (2004) Resiniferatoxin binds to the capsaicin receptor (TRPV1) near the extracellular side of the S4 transmembrane domain. Biochemistry 43:2501–2511
43. Aneiros E et al (2011) The biophysical and molecular basis of TRPV1 proton gating. EMBO J 30:994–1002
44. Voets T et al (2004) The principle of temperature-dependent gating in cold- and heat-sensitive TRP channels. Nature 430:748–754
45. Garcia-Sanz N et al (2007) A role of the transient receptor potential domain of vanilloid receptor I in channel gating. J Neurosci 27:11641–11650
46. Brauchi S et al (2006) A hot-sensing cold receptor: C-terminal domain determines thermosensation in transient receptor potential channels. J Neurosci 26:4835–4840
47. Vlachova V et al (2003) Functional role of C-terminal cytoplasmic tail of rat vanilloid receptor 1. J Neurosci 23:1340–1350
48. Grandl J et al (2010) Temperature-induced opening of TRPV1 ion channel is stabilized by the pore domain. Nat Neurosci 13:708–714
49. Lu G et al (2005) TRPV1b, a functional human vanilloid receptor splice variant. Mol Pharmacol 67:1119–1127
50. Vos MH et al (2006) TRPV1b overexpression negatively regulates TRPV1 responsiveness to capsaicin, heat and low pH in HEK293 cells. J Neurochem 99:1088–1102
51. Schumacher MA et al (2000) Molecular cloning of an N-terminal splice variant of the capsaicin receptor. Loss of N-terminal domain suggests functional divergence among capsaicin receptor subtypes. J Biol Chem 275: 2756–2762
52. Eilers H et al (2007) The rat vanilloid receptor splice variant VR.5′sv Blocks TRPV1 activation. Neuroreport 18:969–973
53. Sharif Naeini R et al (2006) An N-terminal variant of Trpv1 channel is required for osmosensory transduction. Nat Neurosci 9:93–98
54. Tian W et al (2006) Regulation of TRPV1 by a novel renally expressed rat TRPV1 splice variant. Am J Physiol Renal Physiol 290:F117–F126
55. Lyall V et al (2004) The mammalian amiloride-insensitive non-specific salt taste receptor is a vanilloid receptor-1 variant. J Physiol 558:147–159
56. Glinsukon T et al (1980) Acute toxicity of capsaicin in several animal species. Toxicon 18:215–220
57. Jordt SE, Julius D (2002) Molecular basis for species-specific sensitivity to "hot" chili peppers. Cell 108:421–430
58. Szallasi A, Blumberg PM (1993) [3H]resiniferatoxin binding by the vanilloid receptor: species-related differences, effects of temperature and sulfhydryl reagents. Naunyn Schmiedebergs Arch Pharmacol 347:84–91
59. Numazaki M et al (2003) Structural determinant of TRPV1 desensitization interacts with calmodulin. Proc Natl Acad Sci USA 100:8002–8006
60. Jerman JC et al (2000) Characterization using FLIPR of rat vanilloid receptor (rVR1) pharmacology. Br J Pharmacol 130:916–922
61. Rami HK et al (2006) Discovery of SB-705498: a potent, selective and orally bioavailable TRPV1 antagonist suitable for clinical development. Bioorg Med Chem Lett 16: 3287–3291
62. Jung J et al (1999) Capsaicin binds to the intracellular domain of the capsaicin activated ion channel. J Neurosci 19:529–538
63. Gunthorpe MJ et al (2000) Voltage- and time-dependent properties of the recombinant rat vanilloid receptor (rVR1). J Physiol 525:747–759
64. Hellwig N et al (2004) TRPV1 acts as proton channel to induce acidification in nociceptive neurons. J Biol Chem 279:34553–34561
65. Meyers JR et al (2003) Lighting up the senses: FM1-43 loading of sensory cells through non-selective Ion channels. J Neurosci 23:4054–4065
66. Binshtok AM, Bean BP, Woolf CJ (2007) Inhibition of nociceptors by TRPV1-mediated entry of impermeant sodium channel blockers. Nature 449:607–610
67. Chung M-K, Guler AD, Caterina MJ (2008) TRPV1 shows dynamic ionic selectivity during agonist stimulation. Nat Neurosci 11:555–564
68. Heat activation of TRPV1 on Nanion's Patchliner®. Nanion's application note. http://www.nanion.de. Accessed Sept 2011
69. Peters MF et al (2010) Comparing label-free biosensors for pharmacological screening with cell-based functional assays. Assay Drug Dev Technol 8:219–227
70. Liu J et al (2009) A label-free cell-based functional assay for identification and characterization of TRPV1 channel modulators [abstract]. 15th Annual meeting of the Society for Biomolecular Screening
71. Reubish D et al (2009) Functional assessment of temperature-gated ion-channel activity using a real-time PCR machine. Biotechniques 47(3):iii–ix
72. Aneiros E, Dabrowski M (2009) Novel temperature activation cell-based assay on

thermo-TRP ion channels. J Biomol Screen 14:662–667
73. Papakosta M et al (2011) Chimeric approach reveals that differences in the TRPV1 pore domain determine species-specific sensitivity to block of heat activation. J Biol Chem 286(45):39663–39672. doi:10.1074/jbc.M111.273581
74. Szallasi A, Blumberg PM (1999) Vanilloid (capsaicin) receptors and mechanisms. Pharmacol Rev 51:159–212
75. Gavva NR et al (2008) Pharmacological blockade of the vanilloid receptor TRPV1 elicits marked hyperthermia in humans. Pain 136:202–210
76. Iida T et al (2005) Attenuated fever response in mice lacking TRPV1. Neurosci Lett 378:28–33
77. Toth DM et al (2010) Nociception, neurogenic inflammation and thermoregulation in TRPV1 knockdown transgenic mice. Cell Mol Life Sci 68:2589–25601
78. Rowbotham MC et al (2011) Oral and cutaneous thermosensory profile of selective TRPV1 inhibition by ABT-102 in a randomized healthy volunteer trial. Pain 152:1192–1200
79. Krarup AL et al (2011) Randomised clinical trial: the efficacy of transient receptor potential vanilloid 1 antagonist AZD1386 in human esophageal pain. Aliment Pharmacol Ther 33:113–1122
80. Abbott presentation at Society for Neuroscience Meeting in San Diego (2011)
81. Chizh BA et al (2007) The effects of the TRPV1 antagonist SB-705498 on TRPV1 receptor mediated activity and inflammatory hyperalgesia in humans. Pain 132:132–141
82. Crutchlow M (2009) Pharmacologic inhibition of TRPV1 impairs sensation of potentially injurious heat in healthy subjects. Am Soc Clin Pharmacol Ther (ASCPT)
83. Vidal-Mosquera M et al (2011) Triazine-based vanilloid 1 receptor open channel blockers:design, synthesis, evaluation, and SAR analysis. J Med Chem 54(21):7441–7452. doi:doi.org/10.1021/jm200981s
84. Surowy CS et al (2008) (R)-(5-tert-butyl-2,3-dihydro-1H-inden-1-yl)-3-(1H-indazol-4-yl)-urea (ABT-102) blocks polymodal activation of transient receptor potential vanilloid 1 receptors in vitro and heat-evoked firing of spinal dorsal horn neurons in vivo. J Pharmacol Exp Ther 326:879–888
85. Xia R et al (2011) TRPV1: a therapy target that attracts the pharmaceutical interests. Adv Exp Med Biol 704:637–665
86. Krarup AL et al (2011) Randomised clinical trial: the efficacy of a transient receptor potential vanilloid 1 antagonist AZD1386 in human oesophageal pain. Aliment Pharmacol Ther 33:1113–1122
87. Daewoong Press release (2009) "Daewoong Pharmaceutical (CEO: Jong Wook Lee) has received approval from the KFDA (Korean Food and Drug Administration) to enter into clinical trials for a next generation neural pain killer: DWP05195". http://www.daewoong.co.kr/www_pharm/english_new/aboutus/whatsnews_view.asp?idx=60652. Accessed Feb 2012
88. FierceBiotech Press cut (2008) Posted October 24, 2008. Further Clinical Trials in Osteoarthritis Pain Suspended for GRC 6211. http://www.fiercebiotech.com/press-releases/further-clinical-trials-osteoarthritis-pain-suspended-grc-6211. Accessed Feb 2012
89. Gunthorpe MJ et al (2007) Characterisation of SB-705498, a potent and selective TRPV1 antagonist which inhibits the capsaicin-, acid- and heat-mediated activation of the vanilloid receptor. J Pharmacol Exp Ther 321:1183–1192
90. Round P, Priestley A, Robinson J (2011) An investigation of the safety and pharmacokinetics of the novel TRPV1 antagonist XEN-D0501 in healthy human subjects. Br J Clin Pharmacol 72:921–931

Chapter 12

Nociceptive and Nonnociceptive Roles of TRPV3 and Its "Druggability"

Sungjae Yoo and Sun Wook Hwang

Abstract

Following the discovery of transient receptor potential vanilloid 1 (TRPV1), the founding member of the subfamily, knowledge accumulation on its biological roles in human pathology is being accelerated by its genetic, physiological, and pharmacological approaches. Owing to these scientific efforts, a series of its TRPV cousins have been found, and currently their characterizations are getting gradually active. TRPV3 was discovered as the last member a decade ago. To date, studies concerning TRPV3 revealed the distinct expression profiles in keratinocytes, extended modality to thermal and chemical insults, unique mode of sensitization, and intriguing phenotypes beyond sensory deficits when ablated. This knowledge has stimulated production of heterogeneous therapeutic hypotheses. However, in this 10th year of the TRPV3 cloning, substantial advances in TRPV3 pharmacological research is concentrated on nociception. This review not only focuses on the understanding of the pain pathology involving TRPV3 activation, but also highlights emerging information that is helping to define the roles of TRPV3 in skin defects. Multiple linkages between TRPV3 and other cellular and molecular signaling pathways are also discussed. Although the functions of TRPV3 in human physiology and disease still remain to be fully resolved at this point, insights gained from the present overview of outcomes from different fields may help unearth novel therapeutic implications and a better understanding of TRPV3 as a pharmacological target.

Key words: TRPV3, Pain, Keratinocyte, Sensory neuron, Knockout, Ligand

1. Introduction

TRPV3 made its debut as the last member of the vanilloid subfamily of TRP channels. Its discovery was noteworthy because the genome location of the gene is as close as ~10 kb to that of TRPV1, the founding member of TRPVs and a big drug target and three different research group reported TRPV3 simultaneously (1–3). Despite ~40% identity to the TRPV1 gene and such a nearby location, the major tissues expressing TRPV1 and TRPV3 differ greatly: TRPV1 is expressed in sensory neurons and TRPV3 in keratinocytes. The possible implications regarding transcriptional

Arpad Szallasi and Tamás Bíró (eds.), *TRP Channels in Drug Discovery: Volume I*, Methods in Pharmacology and Toxicology, DOI 10.1007/978-1-62703-077-9_12, © Springer Science+Business Media, LLC 2012

regulation could have been recognized but nothing tangible has been revealed. TRPV3 thermal threshold that is close to the body surface temperature would prompt researchers to be more deliberate on its homeostatic function and the data currently begin to be released. To date, expectedly, the major body of information on TRPV3 has been obtained from the sensory and the dermatological fields. Many diverse natural and endogenous ligands for TRPV3 have been reported. The collective pharmacological data infers its sensory roles in the detection of external or internal environmental cues. Less specific properties of these ligands to other TRP channels and likewise promiscuous selectivity of TRPV3 itself sometimes make it difficult to sharply translate a central role of TRPV3. Recent transgenic and pharmacological studies have greatly advanced the definition of the roles of TRPV3. These two approaches are contributing to a construction of TRPV3 proof of concept. Here we summarize the current information on the nociceptive and nonnociceptive roles of TRPV3 proposed by these efforts. This update will help which physiology and pathophysiology are modifiable by future pharmacological TRPV3 targeting.

2. Nociceptive Roles of TRPV3

Peripheral environmental signals are relayed by somatosensory neurons (dorsal root ganglion (DRG) and trigeminal neurons). These neurons innervate the skin structure with their termini as a probe in order to detect environmental changes near the body's tegument. Since those neurons are the initial place where nociceptive signals occur, molecules to ignite excitation or to propagate it present in the sensory neuronal termini are posited as important peripheral druggable targets (4, 5). Examples are TRPV1, tetrodotoxin-resistant voltage-gated Na^+ channels, and a variety of inflammatory mediator receptors including the bradykinin B2 receptors and prostaglandin receptors. However, cloning of TRPV3 and its scrutiny of expresser tissues enabled the modification of this hypothesis. In rodent species, TRPV3 expression in DRG neurons is not obvious and minor detection has been reported in human sensory neurons (1, 3, 6, 7). On the other hand, an abundance of TRPV3 in the skin keratinocytes was being confirmed. Together with functional investigations such as intracellular Ca^{2+} imaging or electrophysiology using TRPV3 activators and cultured keratinocytes, studies have shown ~80% TRPV3-positive population at maximum (6–9). Sensing modality of TRPV3 is another aspect that could put into question the DRG-exclusive hypothesis. First, TRPV3 is a thermo-TRP sensing heat over 33 °C. Although the threshold seems to be innocuous, TRPV3 shows bigger responses to higher temperatures and even sensitizes along iterating stimuli

(1, 3, 6). Therefore, TRPV3 is a detector of potentially noxious heat. Besides, TRPV3 is a chemosensor. A large population of botanical terpenoid substances can activate TRPV3 (see below). The number of these TRPV3 activating natural terpenes is far more than the number of natural ligands for other sensory TRP channels. This unique polymodal feature may explain that TRPV3 is also a sensor for natural environment. Collectively, there appears to be another front guard located more peripherally than DRG termini to initiate alert signals, namely, keratinocytes and their TRPV3.

In fact, it has been hard to conclude that TRPV3 is among nociceptive sensors before behavioral approaches involving transgenics including knockouts. For TRPV1, DRG neurons were already well characterized for the correlation between their morphology and functions even prior to the discovery of TRPV1. For example, unmyelinated C-fibers or thinly myelinated Aδ fibers with small-to-medium diameter cell bodies are the nociceptive subpopulation (10). In addition, its well-known ligand capsaicin has been a standard noxious stimulus in in vivo and in vitro pain research. Even the specific antagonist capsazepine was developed without knowing whether the putative capsaicin receptor was a membrane receptor or an ion channel (11). In contrast, it is unknown whether keratinocytes can be classified according to their sensory roles and it has remained unclear which intercellular signal dominates communication from keratinocytes to adjacent sensory neurons. Special synaptic or nonsynaptic structures between keratinocytes and sensory neurons are still elusive. Specific TRPV3 activators began to be found 3 years after its cloning (6). Under this situation, genetic approaches have mainly provided firm evidence for thermal nociceptive roles of TRPV3.

2.1. Knockouts in 2005 and 2011

The first TRPV3-ablated mice were generated in 2005 (6). The study showed not only thermotaxis and thermal preference behaviors (30–38 °C preferred) but also the thermal withdrawal behaviors involve TRPV3. In hot plate assays and tail immersion assays, the escape behaviors of TRPV3-null mice in response to extremely hot temperatures (55 °C and 52 °C, respectively) were significantly blunted compared with wild-type mice. These results suggest that the noxious heat insults were detected by TRPV3. For mice species, TRPV3 is not expressed in DRGs but in skin keratinocytes. Thus, these altered heat nociceptions very likely implicate the thermosensory roles of the TRPV3 in keratinocytes, and subsequent relay of the sensory signals toward the DRG neurons. More refined behavioral analysis by the same group confirmed the heat nociceptive role of TRPV3 (9). Interestingly, species-specificity and sex-specificity of TRPV3 roles in thermosensitive behaviors were suggested in this recent study. The initial knockouts were of the hybrid background of C57BL/6J and R1 ES cell derived 129X1/SvJ and 129S1/Sv-+^{p}+Try-c Kitl^{Sl-J}/+ (6). In the 2011 study, the same group backcrossed TRPV3-null mice to pure C57BL/6J and 129S1/SvImJ backgrounds. While the C57BL/6J TRPV3-null mice were not impaired in thermosensation, 129S1/SvImJ mice clearly showed thermosensory deficits again in nociceptive hot plate and nonnociceptive thermal preference assays. Concerning thermal preference, females were more prominent than males. Difference in the nociceptive hotplate phenotypes was not noticeable. With similar pure backgrounded TRPV3-knockout mice (of C57BL/6J and 129S6), another sophisticated behavioral results were reported (12). In this study C57BL/6J knockouts were unimpaired in both noxious and innocuous heat sensation and 129S6 knockouts displayed moderately blunted warm preference as shown earlier. Behavioral indices of noxious heat assays, however, were not different between 129S6 knockouts and its wild types. The discrepancy in heat nociception results from the two groups remains to be solved. Furthermore, considering that painkilling might be the pharmacologically most feasible strategy through TRPV3 modulation, the possible species or sex difference in TRPV3 roles in variable nociceptive phenotypes should be carefully considered for development of preclinical and clinical tools.

2.2. TRPV3-Overexpressing Transgenics

More evidence on TRPV3 contribution to thermal nociception was obtained from a study using TRPV3-overexpressing transgenic mice (13). Although their sex-independent assays with the C57BL/6J strain might have detected less dramatic phenotypes according to the recent study (9), the overexpressing transgenic mice gave hints for pain-related TRPV3 function. In thermal preference behavioral assays, TRPV3 overexpression did not influence on locations occupied by mice along the continuous floor temperature gradient or in their rates for preferred selection. On the other hand, although it is varying from cohort to cohort, acute escape behaviors in the nociceptive tail immersion and radiant heat assays of this study were blunted in the TRPV3-overexpressing transgenic mice. Moreover, when TRPV1 was pharmacologically incapacitated, these changes in the escape behaviors of the transgenic mice were more obvious. Since TRPV3 shares the temperature sensing band in noxious ranges with TRPV1, there might be a functional redundancy. Therefore, it is tempting to speculate that TRPV1 likely masks TRPV3 function. In fact, such functional overlap may be predictable because multiple alert systems in response to noxiously high temperatures enough for tissue death might have been evolutionarily advantageous. Indeed, even in TRPV1-deficient mice or another heat receptor TRPV4-deficient mice, changes in thermal phenotypes were more modest than expected (14–17). TRPV3 is not conclusively regarded as a major nociceptive heat sensor like TRPV1 without further and more refined analyses in the future such as studies with double transgenic animals.

However, it seems obvious that TRPV3 is among the target molecules for pain modulation.

2.3. Lessons from Endogenous Substances

Compared to acute thermal pain, information on whether pathological pain involves TRPV3 remains scant. Moreover, inflammatory conditions have been tested in terms of TRPV3-mediated nociceptive phenotypes but yielded equivocal results. In thermal hypersensitivity assays following local bradykinin injection, no difference was detected in wild-type and TRPV3-null mice (6). Also in TRPV3-overexpressing transgenic mice, complete Freund's adjuvant (CFA)-induced inflammation did not cause any difference in heightened nociception compared to wild-type mice (13). In our hands, however, pharmacological blockade of TRPV3 using newly found endogenous inhibitory substances, 17(S)-resolvin D1, 17(R)-resolvin D1, and isopentenyl pyrophosphate (IPP), were all effective at reversing lowered thermobehavioral thresholds caused by CFA inflammation (18–20). Furthermore, in a carrageenan-induced acute inflammation model, intraplantarly administration with a potent and specific TRPV3 agonist, farnesyl pyrophosphate (FPP), elicited acute irritative behaviors although noninflamed animals lacked a response to the injection with the same agent (19, 21). The FPP-evoked nociceptive behaviors were attenuated by TRPV3 knockdown with specific small hairpin RNAs. Different experimental conditions including genetic backgrounds of mice or deviated inflammatory milieu may have different potential to sensitize TRPV3. Gq G-protein-coupled receptor (GPCR) mediated and receptor-tyrosine kinase mediated TRPV3 modulation may occur under inflammation (7, 8) and further studies pinpointing upstream pathological signaling for TRPV3 may help clarify this debate. Interestingly, recently developed synthetic TRPV3 antagonists have shown efficacies for inflammatory pain (see below).

The potent TRPV3 agonist FPP is an endogenous substance produced by FPP synthase in the mevalonate pathway (see below). Nitrogen-containing bisphosphonates such as alendronate and zolendronate inhibit FPP synthase, which accounts for their therapeutic mechanism for osteoporosis (22). Intriguingly, bisphosphonate treatment can mitigate bone cancer pain and neuropathic pain (23, 24). FPP or the mevalonate pathway has not been mechanistically investigated regarding pain pathology. The roles of FPP and whether pharmacological manipulation of the body FPP level can result in beneficial outcomes in the TRPV3-related context remain to be defined. It would be worthwhile to investigate whether certain painful states involve abnormally balanced FPP metabolism or whether this is closely connected to TRPV3 overactivation.

2.4. Pathological Pain

In humans, TRPV3 is present in DRG. Human TRPV3 was shown to be co-expressed with TRPV1 in small-to-medium diameter sensory neurons implicating that human TRPV3 may play a nociceptive role as TRPV1 does or as a subunit for TRPV1 heteromultimers (2). However, documentation on molecular approaches as to neuropathic pain involvement of TRPV3 is still limited. Imperial College and GlaxoSmithKline colleagues have confirmed human DRG expression again and have also shown increased immunoreactivities of TRPV3 together with TRPV1 in injured peripheral fibers and DRGs (25). Interestingly, motor nerves and spinal ventral horn likewise exhibited a temporal increase in TRPV3 expression. Novartis researchers have shown elevation of TRPV3 mRNA in injured DRG of the spinal nerve ligation model (Chung model) (26). Inflammatory mediators and neurotrophins secreted from injured nerves or surrounding tissues might be a cause. Otherwise, blockade of distal transport of the TRPV3 mRNA could occur. The subsequent increase in human sensory TRPV3 expression may possibly account for initiation or persistence of pathological pain. However, functional studies and simulations with multiple neuropathic pain models are still needed. In contrast, human skin epidermis from patients with diabetic neuropathy displayed a decreased expression of TRPV3 (25). This result might be more relevant to skin proliferation/differentiation-associated TRPV3 fate mediating barrier dysfunction than to pain pathway because thinner epidermal layers were also detected. And again, pharmacological evidence on TRPV3-related neuropathic pain from studies on novel synthetic antagonists is now increasing (see below).

2.5. Keratinocyte Messengers

Since TRPV3 in skin keratinocytes is considered to play a nociceptive role, messengers secreted from keratinocytes upon TRPV3 activation for excitation of DRG termini have been sought. A strong candidate, nitric oxide (NO) was proposed in a recent study (9). Unexpectedly, NO production of keratinocytes upon TRPV3 activation does not depend on the canonical biosynthetic pathway via NO synthase. Instead, the authors suggested that intracellular nitrite as the NO precursor is required for TRPV3-mediated NO production. It remains unknown which enzymatic or nonenzymatic process is involved in the NO production (Fig. 1). The process seems to require acidic condition (protons) by TRPV3 activation, but the mechanism for the intracellular acidification is unclear because TRPV3 does not robustly permeate proton. Further, the in vivo extrapolation was successful by comparing phenotypes of the 129S1/SvImJ-backgrounded TRPV3-decient mice and nitrite/nitrate-deprived wild-type mice. Virtually no difference between the two cohorts occurred either in thermal nociception or thermal preference assays and moreover, TRPV3-deficient mice, when deprived of nitrite/nitrate, did not show further thermosensory deficit. This indicates that the nitrite/nitrate-dependent NO signaling is on the same afferent pathway for TRPV3-mediated modality and that TRPV3 ablation had saturated the phenotype

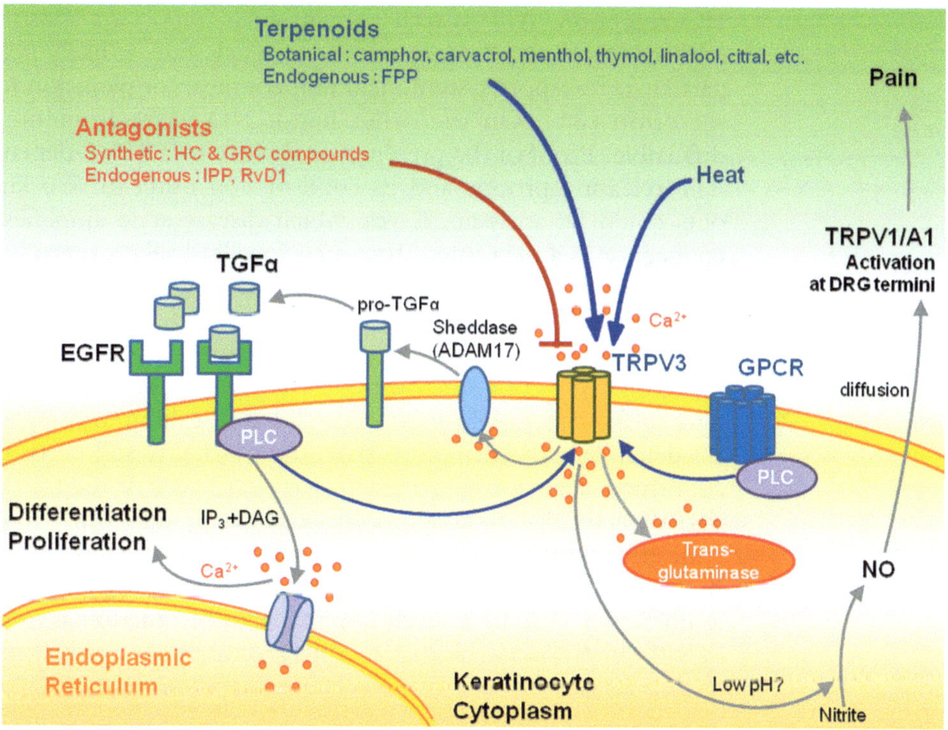

Fig. 1. TRPV3 activation in keratinocytes. Heat and terpenoid agonists activate TRPV3. Upstreams of TRPV3 activation are represented by *blue arrows* and downstream signalings by *gray arrows*. TRPV3 activation promotes NO production and NO subsequently stimulates DRG termini leading to nociception. Other mediators such as ATP or PGE2 may also be involved. TRPV3-sheddase-TGFα-EGFR-PLC activation form a positive feedback circuit and resultant intracellular Ca^{2+} elevations likely contribute to differentiation, proliferation, and barrier function of keratinocytes. Antagonists for TRPV3 may interfere with these processes.

deficit. Interestingly, NO is known to activate TRPV3 itself, TRPV1 and TRPA1 (27, 28). TRPV1 and TRPA1 are both specifically expressed in nociceptive DRG neurons and thus TRPV3-dependent painful modality is possibly mediated by TRPV1/TRPA1-positive sensory afferents. It remains unknown which TRP subpopulation is predominant or whether TRPV1/TRPA1-coexpressing-subpopulation is solely responsible.

Several other possible mediators released from keratinocytes have been also suggested. ATP seems to be a candidate. ATP secretion upon intracellular Ca^{2+} elevation of keratinocyte was shown (29). The P2Y ATP receptor on vicinal DRG neurons appears to be activated by the released ATP according to their keratinocyte-DRG neuron co-culture system. Another group confirmed this ATP secretion and further showed its TRPV3-dependent releases upon thermal stimulation and P2X ion channel involvement in the subsequent DRG excitation (30). Their in vitro communication pathway works in co-culture even with DRG neurons from TRPV1-null mice, which may partly diverge from the aforementioned NO-mediated pathway in which TRPV1 is a major candidate for

NO reception. ATP is well known as an important constituent released at cellular damage but in an intact state, it likely needs a particular transport system including connexin or pannexin for its secretion (31). On the other hand, NO is freely membrane diffusible. Proof of the presence or absence of TRPV3-dependent ATP-releasing process and its confirmation using in vivo knockouts could be an issue. If yes, which disease state amplifies this connection is a next interesting question. Basically, a reproducible co-culture system of keratinocyte and DRGs needs to be established first (32).

Prostaglandin E2 was proposed as another possible candidate based on results of a study with TRPV3-overexpressing transgenic mice (13). Since wild-type mice did not exhibit prostaglandin E2-dependent thermal phenotype, the possibility still needs to be confirmed. Transforming growth factor-α (TGFα) is released upon TRPV3 activation in keratinocytes and has an autocrine and paracrine effect leading to keratinocyte proliferation and migration (8). Interleukin-1α was also suggested by the same group, but its role is unknown (7). In a study on the gain-of-function transgenic ($TRPV3^{Gly573Ser}$) DS mice (33), a heightened nerve growth factor (NGF) release from epidermal sheets was observed at 33 °C, the temperature threshold for TRPV3 activation, compared to normal DS mice. The contributions of all these candidates to TRPV3-mediated nociception through DRG excitation should be more closely assessed, with considering specific diseases, locations and timing of the release in the future. Taken together, the data support the view that TRPV3 serves nociceptive functions in thermal and chemical pain, which indicates that TRPV3 can be a peripheral analgesic target. Growing knowledge on related specific diseases may extend its therapeutic implications.

3. Pathological Roles of TRPV3 in Skin

3.1. Loss of Function

Information on nonnociceptive roles of TRPV3 has been generated mainly from knockout and transgenic studies. A recent knockout study reported morphologically and anatomically prominent phenotypes (8). Their TRPV3-decient mice of either global or keratinocyte-specific ablation exhibited wavy hair and curled whiskers. Follicular keratinocytes without TRPV3 seem to become defective in terminal differentiation, leading to abnormal hair follicle formation. The resulting misaligned and curved follicles with deranged angles likely caused wavy hair and curly whiskers. The molecular mechanism whereby TRPV3 loss of function brings about the malfunction of follicular kerantinocyte appears to be aberrant epidermal growth factor receptor (EGFR) signaling. EGFR signaling is regulated by positive feedback via Ca^{2+} influx in keratinocytes. Increased intracellular Ca^{2+} activates ADAM17, a sheddase enzyme required for TGFα release. Released TGFα activates EGFR, resulting in further Ca^{2+} influx. Thus, TGFα is released more and then EGFR is activated again (Fig. 1). During this process the increased Ca^{2+} and branched enzymatic cascades promote keratinocyte terminal differentiation and also proliferation. Cheng et al. demonstrated that TRPV3 ablation produces a defect in this positive feedback cycle (8). That is, TRPV3 is shown to be a part of this signaling complex as the Ca^{2+} influx channel. In fact, the defective EGFR signaling-induced trouble occurs in keratinocytes in other skin regions. Because the developmental fate of keratinocytes stay less differentiated stages and transglutaminase (a Ca^{2+}-dependent enzyme that crosslinks proteins) activity is lessened, skin epidermis has thickened keratin K1/K10 layers and defective barrier function. Collectively, these results imply that TRPV3 might be a target for restoring hair morphology and barrier function. Ordinarily, EGFR-tyrosine kinase-induced phosphorylation and constitutive body surface temperature (32–33 °C) seems to be mild endogenous stimulatory factors for TRPV3 activation. Additional robust activation, e.g., by administrating ligands, might have a therapeutic implication at some skin defects.

Different aspects of skin healthiness affected by TRPV3 function were emphasized in other knockout studies. The TRPV3 knockouts exhibited slower wound healing than wild types, which was observed, although the keratinocyte proliferation was not affected by the loss of TRPV3 (9). Surprisingly, in this process, the downstream NO also contributes to keratinocyte migration and thereby TRPV3-dependent wound closure. Some discrepancy still remains because in corneal epithelial cells, Ca^{2+} influx via TRPV3 activation by a nonspecific agonist carvacrol contributes to both proliferation and migration thereby enhancing wound healing (34). In contrast to the above TRPV3 knockout study (8), neither critical deficits in histology or abnormal epidermal barrier formation was detected in these TRPV3-knockout lines (6, 9, 35). Some of these TRPV3-deficient mice have wavy hair. The largely conflicting results in the morphological assays might be due to methodological variations or genetically different backgrounds. Taken together, despite some controversy, TRPV3 activation is likely to improve skin barrier function and wound recovery.

3.2. Gain of Function

Two rodent TRPV3 mutant lines were revealed by Shionogi researchers (36). The autosomal dominant DS-Nh (Non-hair) mice and WBN/Kob-Ht (hypotrichosis) rats have been studied as animal examples resembling alopecia and atopic dermatitis because those are both spontaneously hairless and with allergic and pruritic dermatitis when raised in the presence of superantigen-producing *Staphylococcus aureus* and not under SPF (specific pathogen-free) conditions (37, 38). Asakawa et al. demonstrated that one substitution point mutation in the same regions of intracellular S4–S5 linker of TRPV3 channel protein (G573S in DS-Nh; G573C in WBN/Kob-Ht) was responsible for the pathologic skin phenotypes of both strains (36). Like normal TRPV3, the significant expression of the mutant proteins was observed in keratinocytes. Electrophysiological approaches for the mutant channel revealed that this is the gain-of-function mutation (39). Either G573S or G573C mutation renders TRPV3 constitutively active. Results from the electrophysiology of heterologously coexpressed mutants with wild-type TRPV3 implicate that the mechanism for the spontaneous activation is likely due to lowered temperature thresholds, with which TRPV3 should be activated by the ambient or skin temperature. Ca^{2+} overload via this constitutive activation might perturb normal functions and fates of the skin keratinocytes causing the dermatitic and alopecic symptoms.

The Shionogi researchers also newly constructed $TRPV3^{Gly573Ser}$ transgenic DS mice (with a putative promoter sequence in the 5′ region of TRPV3) and C57BL-Nh mice (by crossing C57BL with DS-Nh mice) (33). The transgenic DS mice exhibited typical signs of dermatitis: erythema, edema, dry skin, skin erosions, excoriations, pachyderma, hyperkeratosis, mast cell and CD4+ T cell infiltration, increased peripheral nerve density, and scratching behaviors, which closely resembles human atopic dermatitis. Serum levels of total IgE, chemokines, interleukins (IL-13, IL-17), interferon-γ, and monocyte chemoattractant protein-1 were also elevated in the transgenic mice. Furthermore, NGF production from epidermal sheets was elevated upon temperature stimulation (33 °C). Exposure to *Staphylococcus aureus* is essential for the development of dermatitis for DS-Nh mice (37). By contrast, $TRPV3^{Gly573Ser}$ transgenic DS mice exhibit dermatitis under SPF condition, which seems to be caused by the overexpressed mutant TRPV3 in the transgenic line. Although it needs to be experimentally proven, the higher expression and activation of TRPV3 may exacerbate dermatitis symptoms. Interestingly, the C57BL-Nh mice obtained by crossing C57BL with DS-Nh mice did not develop dermatitis, although they still exhibited scratching behaviors, thickened epidermis, higher number of mast cells, and increased peripheral nerve density in skin. This is possibly because each strain has a different quality or quantity of cellular immunity, for example, a Th1-type response which is relatively resistant to allergen stimulation. Such variation leads to the notion that hyperactive TRPV3 play a crucial role for dermatitis development but is not always solely adequate. Collectively, TRPV3 is likely to be a newly found and essential therapeutic target for dermatitis. Despite a relatively unclear mechanism, alopecia or hirsutism is another candidate disease for which TRPV3 modulation may work. For example, fever or dermatitis may cause pathologic alopecia and the heightened temperature and dermatitic microenvironment are the very activators and sensitizers for TRPV3. Indeed, the activation of TRPV3 in organ

Table 1
Summary of important neuronal and nonneuronal findings described in this review

Neuronal	Nonneuronal
Expression in DRG and trigeminal neurons and other brain regions (2, 3, 25)	Warm temperature activation of TRPV3 in keratinocytes (1, 3, 6, 30, 58)
Expression in vagal nerves (44, 45)	Heat nociception mediated by keratinocyte TRPV3 (6, 9)
Increased expression in DRG in neuropathic injury models (25, 26)	Decreased expression in human skin epidermis with diabetic neuropathy (25)
	ATP secretion upon TRPV3 activation (30)
	PGE2 secretion upon TRPV3 activation (13)
	TGFα/IL-1α secretion upon TRPV3 activation (7, 8)
	NO production upon TRPV3 activation (9)
	NGF secretion from epidermal sheet in TRPV3 gain of function (33)
	Wavy hair in TRPV3 knockouts due to follicular keratinocyte malfunction and aberrance EGFR signaling (8)
	Defective skin barrier functions in TRPV3 knockouts (8)
	Slow wound healing in TRPV3 knockouts (9)
	Spontaneous hairlessness, allergic/pruritic dermatitis in gain-of-function mutants (36–38)
	Polymorphism correlated with high colorectal cancer risk (51)
	Cold activation of Xenopus TRPV3 (42)

References were in parentheses

cultures of human hair follicles was shown to attenuate hair growth (40). Androgenic alopecia is a common form of male hair loss. In fact, robust upregulation of TRPV3 transcription in response to androgen receptor activation was observed (41). Altogether, aberrant TRPV3 function might practically mediate a variety of skin pathologies. Thus, correcting TRPV3 is a possible strategy. Otherwise, compromising symptoms in diseases involving abnormal Ca^{2+} mobilization, even if TRPV3 is not a cause, by adjusting TRPV3 activity to a certain level possibly possess therapeutic implications. See Table 1 for a summary of important neuronal and nonneuronal findings.

4. Other Assignments

Recently, extensive comparative analysis of TRPV genes throughout multiple vertebrates was performed (42). Surprisingly, one of the key findings of this study is that authors first cloned the full length of western clawed frog (*Xenopus tropicalis*) TRPV3, which is activated by cold temperatures (<16 °C). Despite different sensing temperature bands, roles for frog TRPV3 appear to be similar to those for mammalian ones in that both TRPV3s are able to sense noxious temperatures of uninhabitable ranges and that skin is the major expresser tissue. Because sequences of coding DNA and its translated amino acids at N- and C-termini are evolutionarily diversified from mammalian ones but not those of the transmembrane region, the temperature sensing domain must be located at the terminal regions. This diversification also negatively affects ligand sensitivity, cooperativity, and sensitization kinetics to known mammalian TRPV3 activators: for example, frog TRPV3 is unresponsive to botanical ligands such as camphor and eucalyptol. Thus, the implications from such comparative approaches with processing evolutionary information might be helpful in quest of the raison d'etre for TRPV3 and also drug-binding pockets of TRPV3.

TRPV3 expression in other tissues and its functions currently require further evidence. When initially cloned, TRPV3 mRNA was moderately detected in the central nervous system (1–3). A Boswellia diterpene incensole acetate was shown to activate TRPVs with a highest potency for TRPV3 (43). Incensole acetate caused anxiolytic-like and antidepressive-like behaviors, which was ablated in the female TRPV3-knockout mice. It would be interesting to further investigate the psychoactive roles of TRPV3 and the related brain location. Peripherally, TRPV3 is detected in vagal nerves (44, 45). TRPV3 is detected in mouse olfactory and tongue epithelia where it may play a chemoreceptive role (7, 46). The same group also demonstrated the expression in the organ of Corti, spiral ganglion cells, laryngeal epithelial cells (47, 48).

Skin warming is well known to cause local vasodilation. Epidermal TRPV3 may regulate thermal homeostasis in the skin with changes in local blood flow by communicating with neighboring vascular structure in a keratinocyte to DRG-like manner. Intriguingly, cerebral artery endothelial cells express functional TRPV3 by themselves. TRPV3 activation by carvacrol or eugenol elicits endothelium-dependent vasodilation. Ca^{2+} increases in vascular endothelium through TRPV3 might influence on vascular tone (49). This is not affected by NO synthase or cyclooxygenase blockade and the NO synthase-independent NO production upon TRPV3 activation might be involved (9). Initial studies of the TRPV3 cloning have shown TRPV3 mRNA in the gastrointestinal tract (2, 3). Additional studies confirmed TRPV3 expression in the murine stomach, small intestine, and distal colon, but lacking functional implications (44, 50). Of note, a single-nucleotide polymorphism study on 1,225 patients with cancer and 2,032 controls demonstrated that TRPV3 is associated with higher risk for colorectal cancer (51). The critical polymorphisms were synonymous or intronic, which implicates other unknown functions of TRPV3 gene.

5. TRPV3 Pharmacology

5.1. Natural Compounds

Few research articles have addressed on synthetic TRPV3 ligands (see below). On the other hand, the literature for botanical or endogenous substances is currently growing. The collective data indicate that TRPV3 appears to be largely sensitive to terpenoids

(Fig. 1). Many terpenoids have been known to irritate skin, sometimes leading to unpleasant sensations, or inversely, can soothe and relieve irritations, any of which might involve TRPV3 action. Camphor, thymol, menthol, citral, and carvacrol are among the terpenoids activating TRPV3 (6, 7, 52, 53). Scrutiny of the monoterpenoid efficacy has been once performed and borneol, thujone, and linalool were also shown to activate TRPV3 (54). As mentioned earlier, the diterpenoid incensole acetate is a TRPV3 agonist (43). A few nonterpene amino acid derivatives were also found to activate TRPV3 such as vanillin and eugenol (7). Polyunsaturated fatty acids like arachidonic acid are able to sensitize TRPV3 activities (55). All the natural activators have low potencies (with tens of micromolar to millimolar EC50s). Some of these substances induce mild irritation but contradictorily have been traditional pain relievers. Although their nonspecific actions on other analgesic targets may occur, it is tempting to speculate that, with a low potency and efficacy, those might act as partial antagonists that compete with endogenous activators for TRPV3 binding, otherwise that gate control theory may work by stimulating quiescent afferent pathways which neighbor disease sites. Different from TRPV3 sensitization upon repeated stimulation, prolonged exposure upon monoterpene agonists was reported to desensitize TRPV3 (56), which might be another explanation. Further studies are required to clarify these notions.

5.2. Endogenous Ligands

Promiscuous TRPV3 interaction with many of terpenes raises the question of whether human versions of these substances might also act on TRPV3. A terpenoid synthesis pathway is conserved in our body, namely the mevalonate pathway, where cholesterol, dolichol, heme A, retinoids etc. are produced. We focused on intermediates rather than end products and FPP was shown to be a potent TRPV3-specific agonist (EC50 = 131 nM) (21). Although the metabolism operates intracellularly, it seems to bind to a certain extracellular site, indicating that there might be an unknown cell leakage mechanism such as via transporters or damaged regions. The left shift of voltage dependence of TRPV3 is the gating mechanism by FPP as predicted in previous TRPV3 results and other sensory TRPVs (57–59). FPP is able to evoke pain responses under carrageenan inflammation, which supports the view that TRPV3 is a pain transducer. Interestingly, an upstream precursor of FPP, isopentenyl pyrophosphate (IPP) is a potent inhibitor for TRPV3 (IC50 = 239 nM) (19). IPP attenuated TRPV3-mediated chemical and thermal pain phenotypes in this study. Regarding the biosynthesis pathway of IPP and FPP, there might be an endogenous feedback circuit for TRPV3 activation related to pain states, but studies measuring actual tissue concentrations and determining if dynamic production at pathological stages occurs are still few.

Table 2
Summary of TRPV3 ligands described in this review

Natural	Agonists: Terpenoids (camphor, thymol, menthol, citral, carvacrol, dihydrocarveol, borneol, linalool, incensole acetate, thujone, etc.) (6, 7, 43, 52–54) Amino acid derivatives (vanillin, ethyl-vanillin eugenol) (7)	Sensitizers: Poly-unsaturated fatty acids (arachidonic acid, linoelaidic acid, linoleic acid, etc.) (55)
Endogenous	Agonists: FPP (21), NO (27)	Antagonists: IPP (19), 17(S)- and 17(R)-resolvin D1 (18, 20)
Synthetic	Agonist: 2-APB (53, 55, 58, 64)	Antagonists: Glenmark compounds (GRC 15133, GRC 15300, GRC 17173) (61, 62), Hydra compounds (HC-001403, #64, #82) (61, 63), 2,2-Diphenyltetrahydrofuran (58)

References were in parentheses

In addition to IPP, two other endogenous TRPV3 antagonists were reported by our group: 17(S)- and 17(R)-resolvin D1 (RvD1). Resolvin D1, like other types of resolvins, is generated by the ω-3 fatty acid metabolism pathway in our body and has potent pro-resolving and anti-inflammatory activities (60). 17(S)-RvD1 showed less selectivity, inhibiting TRPA1, TRPV3, and TRPV4 but has a striking efficacy for all those TRP-mediated pain phenotypes (18). 17(R)-RvD1 specifically inhibits TRPV3, which is confirmed by in vivo pain tests (20). Stimulating their endogenous production or exogenous administration with RvD1 might be a possible strategy for painkilling. Furthermore, their known pro-resolving effects via GPCRs expressed in immune cells including microglia could have synergy for pain therapy. See Table 2 for a summary of TRPV3 ligands.

5.3. Synthetic Ligands

The development of small molecule TRPV3-specific antagonists is increasingly of industrial interest. Despite the small number of candidates to date, those seem to broaden TRPV3 concept to prove: with the synthetic antagonists, although they are specific, extended analgesic outcomes are generated over the phenotypes or modalities obtained from the genetic and electrophysiological approaches earlier. For example, a former Glenmark compound GRC 15300 exhibits dose-dependent reversal of CFA-induced mechanical hyperalgesia in rats (ED50 = 1.66 mg/kg), a phenotype which was not confirmed in the physiological studies (see (61)). Surprisingly, the effect of GRC 15300 was more potent than that of diclofenac at 2 and 4 h after oral administration. In the chronic constriction injury (CCI), neuropathic pain model which is also not thoroughly studied in the TRPV3 physiology, GRC 15300 reversed mechanical hypersensitivity. GRC 15300 also attenuated thermal hyperalgesia in a postoperative incision model (ED50 = 0.92 mg/kg) and osteoarthritic mechanical hypersensitivity. This compound was licensed out to Sanofi-Aventis and recently became the first TRPV3 drug of which the phase I clinical trial was completed. Other compounds from the same company, GRC 15133 and GRC 17173, also show diclofenac- and pregabalin-comparable efficacies in CFA and CCI models (62).

A compound from Hydra Biosciences, HC-001403, was effective at CFA-inflammatory hyperalgesia, formalin-induced pain, and pain from thermal injury or intrathecal substance P without efficacy in Chung model of neuropathic pain (63). Hydra patents on their compound #64 and #82 were recently disclosed and those were shown to alleviate thermal hyperalgesia in burn injury and a carrageenan-inflammation model in spite of their relatively low in vitro potencies (200–1,000 nM) (61). Chemical structures of many of these potent TRPV3-specific antagonists that are in preclinical or clinical phase are yet to be disclosed. According to disclosed chemical pools of synthetic TRPV3 antagonists in patents, they appear to have some common features (61). A heterocyclic ring with a carbonyl moiety for hydrogen bonds is positioned in the center and it spreads one or two hydrophobic branches adjacently. Designing a certain pharmacophore reflecting this feature is possibly anticipated.

Binding pockets for the synthetic antagonists are yet to be known. However, structural information for TRPV3 begins to be increased. Two cytoplasmic residues (H426 and R696) of human TRPV3 are shown to be important for a synthetic TRPV activator 2-aminoethoxydiphenyl borate (2-APB) (64). Interestingly, swapping the two relevant residues in TRPV4, which is a 2-APB-insensitive TRP, by TRPV3 sequences, renders TRPV3-like 2-APB sensitivity. 2,2-Diphenyltetrahydrofuran, a 2-APB analog blocks TRPV3, which implicates that the chemical is a competitive antagonist sharing the binding site with 2-APB (58). The unbiased mutagenesis study by the Scripps team demonstrated that amino acids clustered in 6th transmembrane helix and the adjacent extracellular loop are responsible for heat sensation (59). Structural approaches might give an insight for ligand binding. Although the TRPV3 protein structure is not known, mapping its N-terminal ankyrin-repeat domain by simulating that of TRPV1 was once attempted (65). An ATP/calmodulin binding site of TRPV3 on the ankyrin repeats was predicted and the presence of intracellular ATP was shown to interfere with TRPV3 sensitization by 2-APB. Mutagenesis confirmed this ATP effect. Future extensive challenges on binding pocket sequence candidates in this way would help constructing more straightforward rationale for selecting potent ligands.

6. Conclusion

Pathological phenotypes in nociception and skin morphology from genetic and physiological studies implicate that TRPV3 is an important therapeutic target. Of note, pharmacology using endogenous or synthetic specific ligands is now contributing to extending the boundary of TRPV3 proof of concept that has not been shown even by the genetic and physiological approaches. Inflammatory and neuropathic pain and related diseases appear to be overriding issues to challenge by a TRPV3 modulation strategy. TRPV3 involvement in dermatitic dysfunction and hair defects demonstrates its therapeutic implication. However, either overactive or hypoactive TRPV3, when modulated, should be carefully considered because those may promote adverse effects at TRPV3 expressers out of the main target. For example, selecting administration doses and routes could be dependent on this information. Many aspects of the molecular mechanism of TRPV3 activation in disease remain to be clarified. The list of upstream stimulators for TRPV3 activation including native GPCR ligands should be more updated particularly in terms of nociception. Which specific diseases configure the skin microenvironment boosting the endogenous ligand secretion and subsequent TRPV3 activation/sensitization need to be answered Nociceptive signals from activated keratinocytes appear to flow to the sensory nerve in a specific fashion through specialized synaptic contacts, reminiscing the labeled line theory in the nervous system (66). Thorough examination is still needed about DRG subtypes, keratinocyte fate dependence, and transmitter utilization pathways upon TRPV3 activation. Researchers are intensively challenging TRP channel structuralization. Given that numbers of natural ligands modulate multiple TRPs (7, 18, 19, 52, 53), it is possible that part of the structures might be conserved among the TRPs in terms of ligand interaction. Thus, growing TRP structural data will contribute to our knowledge expansion on TRPV3 ligand recognition and accelerating its drug development.

Acknowledgments

This work was supported by a grant of the Korea Health technology R&D Project (A111373), Ministry of Health & Welfare and by Kil Chung Hee Fellowship Fund, Republic of Korea.

References

1. Peier AM, Reeve AJ, Andersson DA, Moqrich A, Earley TJ, Hergarden AC, Story GM, Colley S, Hogenesch JB, McIntyre P, Bevan S, Patapoutian A (2002) A heat-sensitive TRP channel expressed in keratinocytes. Science 296(5575):2046–2049
2. Smith GD, Gunthorpe MJ, Kelsell RE, Hayes PD, Reilly P, Facer P, Wright JE, Jerman JC, Walhin JP, Ooi L, Egerton J, Charles KJ, Smart D, Randall AD, Anand P, Davis JB (2002) TRPV3 is a temperature-sensitive vanilloid receptor-like protein. Nature 418(6894): 186–190
3. Xu H, Ramsey IS, Kotecha SA, Moran MM, Chong JA, Lawson D, Ge P, Lilly J, Silos-Santiago I, Xie Y, DiStefano PS, Curtis R, Clapham DE (2002) TRPV3 is a calcium-permeable temperature-sensitive cation channel. Nature 418(6894):181–186
4. Hwang SW, Oh U (2007) Current concepts of nociception: nociceptive molecular sensors in sensory neurons. Curr Opin Anaesthesiol 20(5):427–434
5. Patapoutian A, Tate S, Woolf CJ (2009) Transient receptor potential channels: targeting pain at the source. Nat Rev Drug Discov 8(1):55–68
6. Moqrich A, Hwang SW, Earley TJ, Petrus MJ, Murray AN, Spencer KS, Andahazy M, Story GM, Patapoutian A (2005) Impaired thermosensation in mice lacking TRPV3, a heat and camphor sensor in the skin. Science 307(5714):1468–1472
7. Xu H, Delling M, Jun JC, Clapham DE (2006) Oregano, thyme and clove-derived flavors and skin sensitizers activate specific TRP channels. Nat Neurosci 9(5):628–635
8. Cheng X, Jin J, Hu L, Shen D, Dong XP, Samie MA, Knoff J, Eisinger B, Liu ML, Huang SM, Caterina MJ, Dempsey P, Michael LE, Dlugosz AA, Andrews NC, Clapham DE, Xu H (2010) TRP channel regulates EGFR signaling in hair morphogenesis and skin barrier formation. Cell 141(2):331–343
9. Miyamoto T, Petrus MJ, Dubin AE, Patapoutian A (2011) TRPV3 regulates nitric oxide synthase-independent nitric oxide synthesis in the skin. Nat Commun 2:369
10. Dubin AE, Patapoutian A (2010) Nociceptors: the sensors of the pain pathway. J Clin Invest 120(11):3760–3772
11. Dickenson AH, Dray A (1991) Selective antagonism of capsaicin by capsazepine: evidence for a spinal receptor site in capsaicin-induced antinociception. Br J Pharmacol 104(4): 1045–1049
12. Huang SM, Li X, Yu Y, Wang J, Caterina MJ (2011) TRPV3 and TRPV4 ion channels are not major contributors to mouse heat sensation. Mol Pain 7:37
13. Huang SM, Lee H, Chung MK, Park U, Yu YY, Bradshaw HB, Coulombe PA, Walker JM, Caterina MJ (2008) Overexpressed transient receptor potential vanilloid 3 ion channels in skin keratinocytes modulate pain sensitivity via prostaglandin E2. J Neurosci 28(51):13727–13737, Erratum in: J Neurosci 2009;29(3):29
14. Caterina MJ, Leffler A, Malmberg AB, Martin WJ, Trafton J, Petersen-Zeitz KR, Koltzenburg M, Basbaum AI, Julius D (2000) Impaired nociception and pain sensation in mice lacking the capsaicin receptor. Science 288(5464):306–313
15. Davis JB, Gray J, Gunthorpe MJ, Hatcher JP, Davey PT, Overend P, Harries MH, Latcham J, Clapham C, Atkinson K, Hughes SA, Rance K, Grau E, Harper AJ, Pugh PL, Rogers DC, Bingham S, Randall A, Sheardown SA (2000) Vanilloid receptor 1 is essential for inflammatory thermal hyperalgesia. Nature 405(6783): 183–187
16. Liedtke W, Friedman JM (2003) Abnormal osmotic regulation in trpv4-/- mice. Proc Natl Acad Sci USA 100(23):13698–13703
17. Suzuki M, Mizuno A, Kodaira K, Imai M (2003) Impaired pressure sensation in mice lacking TRPV4. J Biol Chem 278(25): 22664–22668
18. Bang S, Yoo S, Yang TJ, Cho H, Kim YG, Hwang SW (2010) Resolvin D1 attenuates activation of sensory transient receptor potential channels leading to multiple anti-nociception. Br J Pharmacol 161(3):707–720
19. Bang S, Yoo S, Yang TJ, Cho H, Hwang SW (2011) Isopentenyl pyrophosphate is a novel antinociceptive substance that inhibits TRPV3 and TRPA1 ion channels. Pain 152(5): 1156–1164
20. Bang S, Yoo S, Yang T, Cho H, Hwang S (2012) 17(R)-resolvin D1 specifically inhibits TRPV3 leading to peripheral antinociception. Br J Pharmacol 165(3):683–92. doi:10.1111/j.1476-5381.2011.01568.x
21. Bang S, Yoo S, Yang TJ, Cho H, Hwang SW (2010) Farnesyl pyrophosphate is a novel pain producing molecule via specific activation of TRPV3. J Biol Chem 285(25):19362–19371
22. Russell RG, Xia Z, Dunford JE, Oppermann U, Kwaasi A, Hulley PA, Kavanagh KL, Triffitt JT, Lundy MW, Phipps RJ, Barnett BL, Coxon FP, Rogers MJ, Watts NB, Ebetino FH (2007) Bisphosphonates: an update on mechanisms of

action and how these relate to clinical efficacy. Ann N Y Acad Sci 1117:209–257

23. Sharma A, Williams K, Raja SN (2006) Advances in treatment of complex regional pain syndrome: recent insights on a perplexing disease. Curr Opin Anaesthesiol 19(5):566–572
24. von Moos R, Strasser F, Gillessen S, Zaugg K (2008) Metastatic bone pain: treatment options with an emphasis on bisphosphonates. Support Care Cancer 16(10):1105–1115
25. Facer P, Casula MA, Smith GD, Benham CD, Chessell IP, Bountra C, Sinisi M, Birch R, Anand P (2007) Differential expression of the capsaicin receptor TRPV1 and related novel receptors TRPV3, TRPV4 and TRPM8 in normal human tissues and changes in traumatic and diabetic neuropathy. BMC Neurol 7:11
26. Bevan S, Ganju P, McIntyre P, Patapoutian A, Peier A, Song C (2009) Transient receptor potential channel TRPV3 and its use. US/7396910 B2
27. Yoshida T, Inoue R, Morii T, Takahashi N, Yamamoto S, Hara Y, Tominaga M, Shimizu S, Sato Y, Mori Y (2006) Nitric oxide activates TRP channels by cysteine S-nitrosylation. Nat Chem Biol 2(11):596–607
28. Miyamoto T, Dubin AE, Petrus MJ, Patapoutian A (2009) TRPV1 and TRPA1 mediate peripheral nitric oxide-induced nociception in mice. PLoS One 4(10):e7596
29. Koizumi S, Fujishita K, Inoue K, Shigemoto-Mogami Y, Tsuda M, Inoue K (2004) Ca^{2+} waves in keratinocytes are transmitted to sensory neurons: the involvement of extracellular ATP and P2Y2 receptor activation. Biochem J 380(Pt 2):329–338
30. Mandadi S, Sokabe T, Shibasaki K, Katanosaka K, Mizuno A, Moqrich A, Patapoutian A, Fukumi-Tominaga T, Mizumura K, Tominaga M (2009) TRPV3 in keratinocytes transmits temperature information to sensory neurons via ATP. Pflugers Arch 458(6):1093–1102
31. Thompson RJ, Macvicar BA (2008) Connexin and pannexin hemichannels of neurons and astrocytes. Channels (Austin) 2(2):81–86
32. Tsutsumi M, Goto M, Denda S, Denda M (2011) Morphological and functional differences in coculture system of keratinocytes and dorsal-root-ganglion-derived cells depending on time of seeding. Exp Dermatol 20(6):464–467
33. Yoshioka T, Imura K, Asakawa M, Suzuki M, Oshima I, Hirasawa T, Sakata T, Horikawa T, Arimura A (2009) Impact of the Gly573Ser substitution in TRPV3 on the development of allergic and pruritic dermatitis in mice. J Invest Dermatol 129(3):714–722
34. Yamada T, Ueda T, Ugawa S, Ishida Y, Imayasu M, Koyama S, Shimada S (2010) Functional expression of transient receptor potential vanilloid 3 (TRPV3) in corneal epithelial cells: involvement in thermosensation and wound healing. Exp Eye Res 90(1):121–129
35. Sokabe T, Fukumi-Tominaga T, Yonemura S, Mizuno A, Tominaga M (2010) The TRPV4 channel contributes to intercellular junction formation in keratinocytes. J Biol Chem 285(24):18749–18758
36. Asakawa M, Yoshioka T, Matsutani T, Hikita I, Suzuki M, Oshima I, Tsukahara K, Arimura A, Horikawa T, Hirasawa T, Sakata T (2006) Association of a mutation in TRPV3 with defective hair growth in rodents. J Invest Dermatol 126(12):2664–2672
37. Yoshioka T, Hikita I, Matsutani T, Yoshida R, Asakawa M, Toyosaki-Maeda T, Hirasawa T, Suzuki R, Arimura A, Horikawa T (2003) DS-Nh as an experimental model of atopic dermatitis induced by Staphylococcus aureus producing staphylococcal enterotoxin C. Immunology 108(4):562–569
38. Asakawa M, Yoshioka T, Hikita I, Matsutani T, Hirasawa T, Arimura A, Sakata T, Horikawa T (2005) WBN/Kob-Ht rats spontaneously develop dermatitis under conventional conditions: another possible model for atopic dermatitis. Exp Anim 54(5):461–465
39. Xiao R, Tian J, Tang J, Zhu MX (2008) The TRPV3 mutation associated with the hairless phenotype in rodents is constitutively active. Cell Calcium 43(4):334–343
40. Borbíró I, Lisztes E, Tóth BI, Czifra G, Oláh A, Szöllosi AG, Szentandrássy N, Nánási PP, Péter Z, Paus R, Kovács L, Bíró T (2011) Activation of transient receptor potential vanilloid-3 inhibits human hair growth. J Invest Dermatol 131(8):1605–1614
41. Jariwala U, Prescott J, Jia L, Barski A, Pregizer S, Cogan JP, Arasheben A, Tilley WD, Scher HI, Gerald WL, Buchanan G, Coetzee GA, Frenkel B (2007) Identification of novel androgen receptor target genes in prostate cancer. Mol Cancer 6:39
42. Saito S, Fukuta N, Shingai R, Tominaga M (2011) Evolution of vertebrate transient receptor potential vanilloid 3 channels: opposite temperature sensitivity between mammals and western clawed frogs. PLoS Genet 7(4):e1002041
43. Moussaieff A, Rimmerman N, Bregman T, Straiker A, Felder CC, Shoham S, Kashman Y, Huang SM, Lee H, Shohami E, Mackie K, Caterina MJ, Walker JM, Fride E, Mechoulam R (2008) Incensole acetate, an incense

component, elicits psychoactivity by activating TRPV3 channels in the brain. FASEB J 22(8): 3024–3034
44. Zhang L, Jones S, Brody K, Costa M, Brookes SJ (2004) Thermosensitive transient receptor potential channels in vagal afferent neurons of the mouse. Am J Physiol Gastrointest Liver Physiol 286(6):G983–G991
45. Zhao H, Simasko SM (2010) Role of transient receptor potential channels in cholecystokinin-induced activation of cultured vagal afferent neurons. Endocrinology 151(11):5237–5246
46. Ahmed MK, Takumida M, Ishibashi T, Hamamoto T, Hirakawa K (2009) Expression of transient receptor potential vanilloid (TRPV) families 1, 2, 3 and 4 in the mouse olfactory epithelium. Rhinology 47(3):242–247
47. Hamamoto T, Takumida M, Hirakawa K, Takeno S, Tatsukawa T (2008) Localization of transient receptor potential channel vanilloid subfamilies in the mouse larynx. Acta Otolaryngol 128(6):685–693
48. Ishibashi T, Takumida M, Akagi N, Hirakawa K, Anniko M (2009) Changes in transient receptor potential vanilloid (TRPV) 1, 2, 3 and 4 expression in mouse inner ear following gentamicin challenge. Acta Otolaryngol 129(2): 116–126
49. Earley S, Gonzales AL, Garcia ZI (2010) A dietary agonist of transient receptor potential cation channel V3 elicits endothelium-dependent vasodilation. Mol Pharmacol 77(4): 612–620
50. Ueda T, Yamada T, Ugawa S, Ishida Y, Shimada S (2009) TRPV3, a thermosensitive channel is expressed in mouse distal colon epithelium. Biochem Biophys Res Commun 383(1): 130–134
51. Hoeft B, Linseisen J, Beckmann L, Müller-Decker K, Canzian F, Hüsing A, Kaaks R, Vogel U, Jakobsen MU, Overvad K, Hansen RD, Knüppel S, Boeing H, Trichopoulou A, Koumantaki Y, Trichopoulos D, Berrino F, Palli D, Panico S, Tumino R, Bueno-de-Mesquita HB, van Duijnhoven FJ, van Gils CH, Peeters PH, Dumeaux V, Lund E, Huerta Castaño JM, Muñoz X, Rodriguez L, Barricarte A, Manjer J, Jirström K, Van Guelpen B, Hallmans G, Spencer EA, Crowe FL, Khaw KT, Wareham N, Morois S, Boutron-Ruault MC, Clavel-Chapelon F, Chajes V, Jenab M, Boffetta P, Vineis P, Mouw T, Norat T, Riboli E, Nieters A (2010) Polymorphisms in fatty-acid-metabolism-related genes are associated with colorectal cancer risk. Carcinogenesis 31(3):466–472
52. Macpherson LJ, Hwang SW, Miyamoto T, Dubin AE, Patapoutian A, Story GM (2006) More than cool: promiscuous relationships of menthol and other sensory compounds. Mol Cell Neurosci 32(4):335–343
53. Stotz SC, Vriens J, Martyn D, Clardy J, Clapham DE (2008) Citral sensing by Transient (corrected) receptor potential channels in dorsal root ganglion neurons. PLoS One 3(5): e2082
54. Vogt-Eisele AK, Weber K, Sherkheli MA, Vielhaber G, Panten J, Gisselmann G, Hatt H (2007) Monoterpenoid agonists of TRPV3. Br J Pharmacol 151(4):530–540
55. Hu HZ, Xiao R, Wang C, Gao N, Colton CK, Wood JD, Zhu MX (2006) Potentiation of TRPV3 channel function by unsaturated fatty acids. J Cell Physiol 208(1):201–212
56. Sherkheli MA, Benecke H, Doerner JF, Kletke O, Vogt-Eisele AK, Gisselmann G, Hatt H (2009) Monoterpenoids induce agonist-specific desensitization of transient receptor potential vanilloid-3 (TRPV3) ion channels. J Pharm Pharm Sci 12(1):116–128
57. Voets T, Droogmans G, Wissenbach U, Janssens A, Flockerzi V, Nilius B (2004) The principle of temperature-dependent gating in cold- and heat-sensitive TRP channels. Nature 430(7001):748–754
58. Chung MK, Güler AD, Caterina MJ (2005) Biphasic currents evoked by chemical or thermal activation of the heat-gated ion channel, TRPV3. J Biol Chem 280(16):15928–15941
59. Grandl J, Hu H, Bandell M, Bursulaya B, Schmidt M, Petrus M, Patapoutian A (2008) Pore region of TRPV3 ion channel is specifically required for heat activation. Nat Neurosci 11(9):1007–1013
60. Ji RR, Xu ZZ, Strichartz G, Serhan CN (2011) Emerging roles of resolvins in the resolution of inflammation and pain. Trends Neurosci 34(11):599–609
61. Reilly RM, Kym PR (2011) Analgesic potential of TRPV3 antagonists. Curr Top Med Chem 11(17):2210–2215
62. Khairatkar Joshi N, Maharaj N, Thomas A (2010) The TRPV3 receptor as a pain target: a therapeutic promise or just some more new biology? Open Drug Discov J 2:88–95
63. Broad LM, Mogg AJ, Beattie RE, Ogden AM, Blanco MJ, Bleakman D (2009) TRP channels as emerging targets for pain therapeutics. Expert Opin Ther Targets 13(1):69–81
64. Hu H, Grandl J, Bandell M, Petrus M, Patapoutian A (2009) Two amino acid residues

determine 2-APB sensitivity of the ion channels TRPV3 and TRPV4. Proc Natl Acad Sci USA 106(5):1626–1631
65. Phelps CB, Wang RR, Choo SS, Gaudet R (2010) Differential regulation of TRPV1, TRPV3, and TRPV4 sensitivity through a conserved binding site on the ankyrin repeat domain. J Biol Chem 285(1):731–740
66. Zylka MJ, Rice FL, Anderson DJ (2005) Topographically distinct epidermal nociceptive circuits revealed by axonal tracers targeted to Mrgprd. Neuron 45(1):17–25

Chapter 13

TRPV4 and Drug Discovery

Fabien Vincent and Matthew A.J. Duncton

Abstract

Transient receptor potential vanilloid 4 (TRPV4) was first identified in 2000 as an osmolarity sensor. Further investigations rapidly revealed this ion channel to be a polymodal receptor with additional activating or modulating stimuli including warm temperatures, endogenous lipids, and phosphorylation. The broad tissue and cell type distribution of TRPV4, coupled with its varied activation profile, lead to a wide variety of physiological roles. These include sheer stress detection in blood vessels, osteoclast differentiation control in bone, along with temperature monitoring in skin keratinocytes and osmolarity sensing in kidneys. Recent work has also implicated TRPV4 mutations in multiple genetic disorders such as brachyolmia and Charcot–Marie–Tooth disease 2C. Characterization of its roles in disease states naturally led to a rising interest in the modulation of TRPV4 for therapeutic purposes. Therapeutic areas of interest are diverse and include several with significant unmet medical needs such as inflammatory and neuropathic pain, bladder dysfunctions, as well as mechanical lung injury. Herein we review the roles of TRPV4 in pathologies and summarize the progress made in identifying small molecule modulators of its activity for target validation and therapeutic purposes.

Key words: TRPV4, Agonist, Antagonist, Disease, Drug, Indication, Inhibitor, Activator, Pain, Bladder, Lung, Injury, Inflammation, Neuropathy, Chemistry

1. Introduction

The TRP ion channel superfamily is composed of the TRPA, TRPC, TRPM, TRPN, TRPP, and TRPV families, with a total of 28 members in mammals (1, 2). TRPV4 belongs to the TRPV (vanilloid) channel family, itself subdivided into two subfamilies with nonselective cation permeable channels ($P_{Ca}/P_{Na} \leq 10$) highly sensitive to temperature changes (TRPV1-4) and Ca^{2+} selective channels ($P_{Ca}/P_{Na} > 100$) insensitive to temperature variations (TRPV5-6) (3–5). TRPV1, the canonical member of this family, was first identified as the receptor of the active ingredient of chili peppers, capsaicin (6). Additional studies uncovered other activating modalities such as noxiously high temperatures (>42°C), acidity

Arpad Szallasi and Tamás Bíró (eds.), *TRP Channels in Drug Discovery: Volume I*, Methods in Pharmacology and Toxicology, DOI 10.1007/978-1-62703-077-9_13, © Springer Science+Business Media, LLC 2012

(<pH 5), and lipids (e.g., anandamide) (7, 8). As activation by a given stimulus will modulate the sensitivity of the ion channel to other inputs, TRPV1 was shown to function as a polymodal sensor, integrating a variety of activating signals to produce a resulting cellular output (8).

TRPV4 is closely related to the TRPV1 receptor, sharing 40.9% of its sequence. The human gene is located on chromosome 12q23-q24.1, with five proposed splice variants produced from 12 exons (9, 10). The 871 amino acids of full length hTRPV4 define six transmembrane domains, with the pore region being located between TM5 and TM6, as well as six ankyrin repeats near the N terminus, a PRD domain close to the first ankyrin domain, and putative calmodulin domains near the C terminus (4, 5, 9). As with TRPV1, TRPV4 is believed to form homotetramers based on cryo-electron microscopy studies using detergent-solubilized rat TRPV4 (4, 11). Helpfully for drug discovery work, rat and mouse sequences exhibit 95% identity with their human counterpart (4, 5, 9). The seminal studies which led to the discovery of TRPV4 revealed its modulation by osmolarity (12–14). Additional stimuli were soon identified, including warm temperatures (>27°C) (15), phosphorylation by Src, PKC, and PKA (16–18), and endogenous lipids (4, 19, 20).

TRPV4 displays a wide expression pattern with its presence being documented in kidney, lung, brain, dorsal root ganglia (DRG), bladder, skin, vascular endothelium, liver, testis, fat, inner ear, pancreas, cornea, and heart (4, 13, 14, 21–29). Correspondingly, TRPV4 is found in a variety of cell types, both excitable and non-excitable, including peripheral, hippocampal, and subfornical organ neurons, renal epithelial and urothelial cells, airway smooth muscle cells, chondrocytes, osteoblasts and osteoclasts, smooth muscle cells of the aorta, insulin-producing β cells, astrocytes, and others (3, 4, 12, 23, 30–40).

This widespread expression pattern, coupled with multiple activating modalities, leads to a diversity of biological roles for TRPV4. Detailing these is beyond the scope of this chapter—the reader is referred to several recent reviews (4, 9, 41)—while some areas of interest for drug discovery will be expanded upon. Briefly, functional studies investigating TRPV4 functions were conducted using a range of tools including small molecule agonists and antagonists, antisense oligodeoxynucleotides, KO mice, and specific genetic mutations linking TRPV4 to rare genetic disorders (12, 15, 26, 30, 33, 42–47). Using these, TRPV4 was demonstrated to play a role in osmolarity sensing and regulation in the CNS (12–14, 30, 48), thermosensation and possibly thermoregulation (15, 21, 49), bladder function (43, 47), bone formation and remodeling (38, 40, 45, 46, 50), and mechanosensation in the vascular endothelium (37, 51).

A number of small molecules have been described as being either agonists or antagonists of TRPV4. As a detailed discussion of every such compound is not possible here, the reader is directed towards an earlier review providing a more detailed focus on this topic (52). This mini review will nonetheless contain a succinct account of the major TRPV4 modulators, together with some applications of these compounds in vivo.

2. Areas of Interest for Drug Discovery

2.1. Inflammatory and Neuropathic Pain

TRPV4 is expressed in peripheral nociceptive neurons (DRG and trigeminal (TG) neurons), with data suggesting its transport to peripheral nerve endings (32, 33). As TRPV4 was originally identified as an osmosensor, early work using TRPV4 KO investigated whether this channel was also involved in mechanosensation. These studies provided the first evidence of TRPV4 involvement in nociception with KO mice displaying an increase in mechanical nociceptive threshold while baseline mechanical sensitivity remained unaffected (12, 44).

Using complementary approaches to lower TRPV4 activity, TRPV4 KO mice, and TRPV4 knockdown in rats using antisense oligodeoxynucleotides (AS ODN), Alessandri-Haber, Levine, and coworkers have produced a convincing body of work documenting the involvement of TRPV4 in pain under pathological conditions (33, 34, 53–58). Prostaglandin E2, a known inflammatory mediator, was originally observed to sensitize this ion channel to both hypotonic and hypertonic media, conditions relevant to diabetes and asthma, for example (33, 56, 59). This finding was later extended through the use of several additional inflammatory mediators, leading to the discovery that increased cAMP levels as well as PKA and PKCε activation were necessary for TRPV4-mediated mechanical hyperalgesia (54). Further studies investigated its connection with protease-activated receptor 2 (PAR2), a GPCR co-expressed with TRPV4 on a subset of primary afferent neurons. The activation of PAR2 by proteases during inflammation sensitizes TRPV4 through the stimulation of PLCβ, PKA, PKC, and PKD (60). Several groups have now confirmed the link between PAR2 activation and TRPV4-mediated mechanical hyperalgesia and release of nociceptive peptides substance P and CGRP (60–62). Besides its modulation by PAR2, TRPV4 was documented to be highly present on colonic afferent neurons in the same studies and its role in visceral pain was demonstrated using both TRPV4 KO and AS ODN approaches (62, 63). Interestingly, PAR4, a relative of PAR2, was recently documented to antagonize PAR2 and TRPV4-induced hyperalgesia in response to colorectal distension (64).

Peripheral nerve damage from diverse etiologies can lead to allodynia or hyperalgesia in patients. Alessandri-Haber, Levine, and coworkers demonstrated that TRPV4 mediates mechanical hyperalgesia in multiple rodent models of painful peripheral neuropathy (34, 55). The wide range of reagents used for nerve injury in these models (paclitaxel, vincristine, streptozocin, ddC, and alcohol) suggests a potentially broad role for TRPV4 in neuropathic pain (55). Mechanistic investigations revealed a requirement for both Src tyrosine kinase and α2β1 integrin in TRPV4-mediated mechanical hyperalgesia (55). TRPV4 expression levels were observed to increase in a rat model of chronic constriction of the DRG (CCD). TRPV4-specific AS ODN reversed this increase and partially blocked the observed mechanical allodynia while leaving baseline nociception unaffected (32). Additionally, the PAR2–TRPV4 relationship was recently found to extend beyond inflammatory conditions and to include paclitaxel-induced mechanical and thermal hyperalgesia (65).

Overall, data appear to suggest that TRPV4 may selectively mediate (mechanical) nociception under pathological conditions, without altering baseline mechanosensation (54). TRPV4 antagonists may therefore prove therapeutically useful in the treatment of inflammatory and neuropathic pain, potentially including visceral pain related to Crohn's disease and inflammatory bowel syndrome.

2.2. Bladder Dysfunction

TRPV4 is expressed in the bladder urothelium (both basal and intermediate cells), in bladder smooth muscle cells, as well as in the urothelium cells lining the renal pelvis, ureters, and urethra (26, 27, 43, 66, 67). Importantly, both mRNA and protein were documented to be present in the urothelium, with increases in intracellular calcium being observed in presence of well-known TRPV4 activating modalities such as 4αPDD, GSK1016790A, and hypotonicity (26, 27, 43). This pharmacological response was, however, reduced (hypotonicity) or absent (4αPDD and GSK1016790A) in TRPV4 KO cells stimulated in the same fashion (27, 43).

Mice lacking TRPV4 displayed a perturbed urine-voiding pattern, with longer intervoiding intervals, larger voided volumes, fewer productive voiding contractions, and a disturbed spatial distribution of urine (27). These data are consistent with TRPV4 ablation leading to higher bladder filling and delayed micturition (68). Conversely, use of TRPV4 agonists 4αPDD in rats (but not mice) and GSK1016790A led to increased intravesical pressure and bladder overactivity, respectively (26, 43). Notably, this effect of GSK1016790A was not observed in KO mice. More recently, the decrease in bladder capacity and voided volume prompted by GSK1016790A infusion were negated by the application of selective TRPV4 antagonist RN-1734 (69).

Other TRP channels are expressed in the bladder and are known to contribute to the regulation of its function (68, 70–72). TRPV1 expression levels in TRPV4 WT and KO mice were found to be similar (27). Conversely, no significant change in TRPV4 expression was observed in TRPV1 KO animals suggesting these ion channels do not compensate for each other (27, 68).

Following the activation of the urothelium mechanosensory system, ATP is released to activate P_2X_3 and $P_2X_{2/3}$ purinergic channels on sensory afferents. This pressure-initiated ATP release is decreased in TRPV4 null mice, a result consistent with the postulated mechanosensory function of TRPV4. Furthermore, the action of TRPV4 agonists 4αPDD and GSK1016790A can be counteracted by P_2X_3 and $P_2X_{2/3}$ antagonists TNT-ATP, A317491, and PPADS, further validating the upstream involvement of TRPV4 in this signaling pathway (26, 69).

Further validation of the potential of TRPV4 antagonism for the treatment of bladder disorders involving reduced bladder function and increased micturition frequency was published recently (47). These researchers observed that TRPV4 deletion significantly reduced cystitis-induced bladder dysfunction. Furthermore, similar results were obtained using a novel TRPV4 antagonist, HC-067047 (47). The lack of effect of HC-067047 in TRPV4 KO animals strongly suggests its efficacy was due to TRPV4 antagonism.

Expressed in the bladder urothelium and smooth muscle cells, TRPV4 is a major component of the mechanosensor monitoring bladder distension and its activity controls, in part, voiding of the bladder (26, 27, 43). Evidence is accumulating that TRPV4 antagonism is correlating with decreased micturition frequency and increased voided volumes. Furthermore, due to the expression of TRPV4 in the urethra and ureters, it has been suggested that it may be involved in the urethra-to-bladder reflex facilitating complete voiding (27). Overall, these data paint a promising profile for the treatment of overactive bladder syndrome, cystitis, and prostate hyperplasia through antagonism of TRPV4.

2.3. Acute Lung Injury

Acute lung injury and acute respiratory distress syndrome affect 200,000 patients every year in the United States, with limited treatment options leading to a high mortality rate (73, 74). These ailments are characterized by disruption of the alveolar septal barrier, hypoxemia, and patchy alveolar flooding (75). Mechanical ventilation is responsible for a significant part of these injuries (76). Previous studies have revealed the increase in lung vasculature permeability to be initiated by intracellular calcium influx, which itself can be prompted by high airway pressure (mechanical stress) (77).

Hamanaka et al. hypothesized that TRPV4 may be the mechanosensor at the origin of this injury process (78). The permeability increase linked to high airway pressure was indeed observed to be abolished in TRPV4 KO mice. This result could be replicated in TRPV4 WT animals by using ruthenium red, a broad ion channel inhibitor (52), and inhibitors of arachidonic acid production (methanandamide) and its metabolism to EETs (miconazole) (20, 78) (Note: EETs are endogenous agonists of TRPV4- see section 3.1.4). Similarly, after confirming the expression of TRPV4 in situ, Alvarez and coworkers monitored lung endothelium permeability in response to TRPV4 agonists 4αPDD and 5,6-EET in TRPV4 WT and null mice (79). Notably, the increase in permeability observed in response to the application of the agonists in TRPV4 WT mice was abolished in TRPV4 KO animals. Additionally, application of ruthenium red to WT lungs stimulated by channel agonists produced similar results, further confirming the involvement of TRPV4 (79). As macrophage depletion can attenuate ventilator-induced lung injury, subsequent work focused on the role of alveolar macrophages, believed to be key early actors in the injury process (80, 81). Instillation of TRPV4$^{+/+}$ macrophages in TRPV4$^{-/-}$ mice restored the permeability increase observed in response to high airway pressure, confirming the involvement of both macrophages and TRPV4.

In summary, while early reports linked TRPV4 to high airway pressure-mediated lung vasculature permeability, recent studies provided additional mechanistic information on the pathways involved (80, 82). Given the role of this ion channel as an early enabler of injury, TRPV4 antagonism holds promise for the treatment of acute lung injury, especially that caused by mechanical ventilation.

3. TRPV4 Agonists and Antagonists

3.1. TRPV4 Agonists

3.1.1. Phorbol Esters as TRPV4 Agonists

The phorbol ester, 4α phorbol 12,13-didecanoate (4 αPDD) **1**, was the first TRPV4 agonist to be described in the literature (83). Unlike many phorbol esters, 4αPDD does not serve as an activator of protein kinase C (PKC) at concentrations up to 25 μM (83, 84).

Analogues related to 4αPDD have also been evaluated as TRPV4 agonists and have revealed important facets of the binding of phorbol esters to TRPV4 (Fig. 1) (85). For example, 4αPDD has an EC_{50} of 0.37 μM (±0.08 μM) at TRPV4. Shortening the ester chain length from 10 carbons found in 4αPDD, to 6 carbons, giving dihexanoate ester **2**, resulted in an increase of EC_{50} to

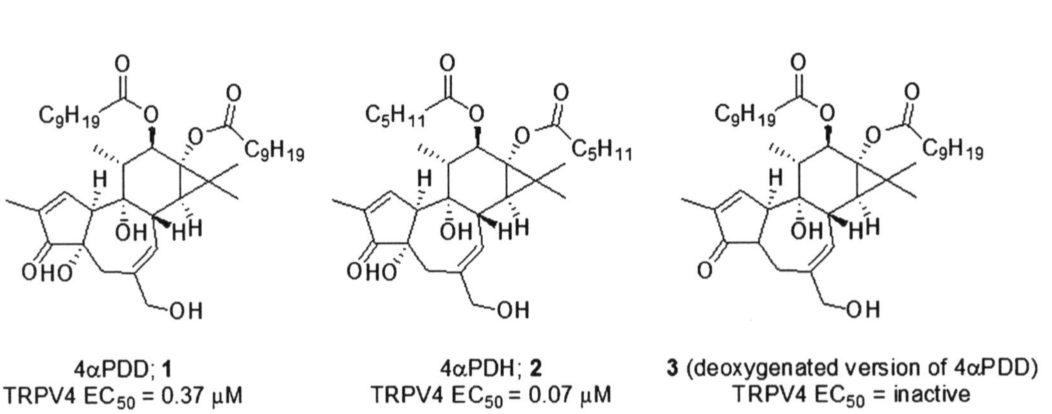

Fig. 1. Phorbol esters as agonists of TRPV4.

0.07 μM (±0.01 μM). Interestingly, other lengths of ester side chain resulted in a dramatic decrease in TRPV4 activity. During these structure–activity studies, it was also found that a hydroxyl group at the 5,7-ring junction of the phorbol skeleton was very important for TRPV4 activity, since its removal to give compound **3** abolished agonism of TRPV4.

Phorbol esters are reported to interact with the TRPV4 ion channel directly (5, 83–85), and in contrast to the binding of phorbols to PKC, do not use a "typical" cysteine-rich phorbol binding site (85).

4αPDD has been a very important ligand for investigation of the role of TRPV4, being used to study effects upon pain (53, 86), bladder function (26), and control of blood pressure (87, 88) among others.

3.1.2. GSK1016790A

GSK1016790A **4** (from GlaxoSmithKline) was recently disclosed as a very potent TRPV4 agonist (EC_{50} hTRPV4 = 5 nM) (43). Significantly, GSK1016790A evoked a much greater current density at TRPV4 when compared with 4αPDD, indicating greater relative efficacy. The starting point for the discovery of GSK1016790A was the Cathepsin K inhibitor **5**, which was found to be a submicromolar agonist of TRPV4 (Fig. 2). Subsequent optimization of compound **5** gave rise to GSK1016790A, and a whole series of structurally related agonists (42, 43, 89, 90).

GSK101670A is reported to be selective against TRPV1, but may interact with other receptors, as exposure of non-TRPV4 expressing HEK293 cells to GSK1016790A results in calcium uptake at low concentrations (50–100 nM) (42, 43). GSK101670A has been utilized for a number of in vivo investigations. For example, it induced bladder hyperactivity in mice (43). However, work

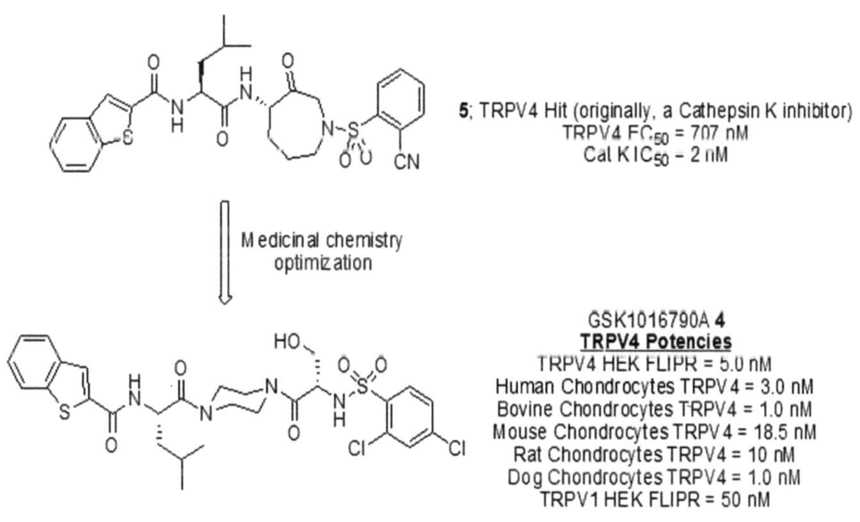

Fig. 2. GSK101670A as a TRPV4 agonist.

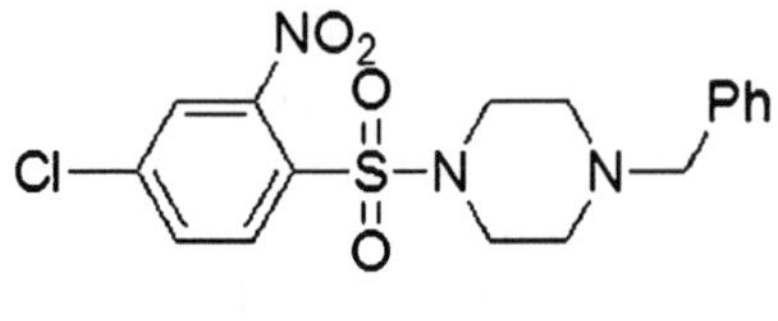

RN-1747 6
hTRPV4 EC_{50} = 0.77 μM

Fig. 3. RN-1747 as a TRPV4 agonist.

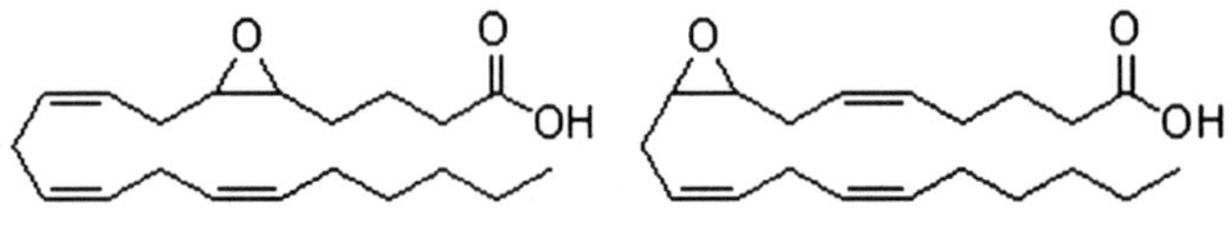

5,6-EET 7 **8,9-EET 8**

Fig. 4. Lipid metabolites of the arachidonic acid serving as TRPV4 agonists.

with GSK1016790A has been somewhat limited, as a result of lethal TRPV4-mediated circulatory collapse in mice, rats, and dogs, upon exposure to an intravenous injection of this TRPV4 agonist (42). Naturally, this result indicates that the use of systemic TRPV4 agonists in human clinical trials may be limited.

3.1.3. RN-1747

RN-1747 **6** (Fig. 3) was identified as a submicromolar TRPV4 agonist during a screen of commercial arylsulfonamides as TRPV4 ligands (EC_{50} = 0.77 μM) (91). RN-1747 is relatively selective for TRPV4, although it serves as a TRPM8 antagonist at higher concentrations (IC_{50} = 4 μM).

3.1.4. Endogenous Agonists of TRPV4

Lipid metabolites of the arachidonic acid pathway have been shown to serve as endogenous agonists of TRPV4 (5, 19, 20, 83). For example, 5,6-epoxyeicosatrienoic acid **7** (5,6-EET) and 8,9-epoxyeicosatrienoic acid **8** (8,9-EET) were identified by Nilius and coworkers as TRPV4 agonists, with an EC_{50} of 0.15 μM versus hTRPV4 (Fig. 4). Whether they interact directly with the ion channel is still uncertain, however (4).

3.2. TRPV4 Antagonists

3.2.1. Ruthenium Red

Ruthenium red **9** (Fig. 5), a metal-based dye acting as a pore blocker, was one of the first potent TRPV4 antagonists to be identified (15, 83). However, ruthenium red is an unselective ligand as activity has been reported with more than 20 other ion channels and biological targets (52). Nonetheless, ruthenium red has been used extensively as a probe of TRPV4 activity in the in vivo setting (86, 88, 92).

3.2.2. HC-067047

In 2010, Hydra Biosciences disclosed the structure of HC-067047 **10** as a potent TRPV4 antagonist (Fig. 6) (47). This compound is

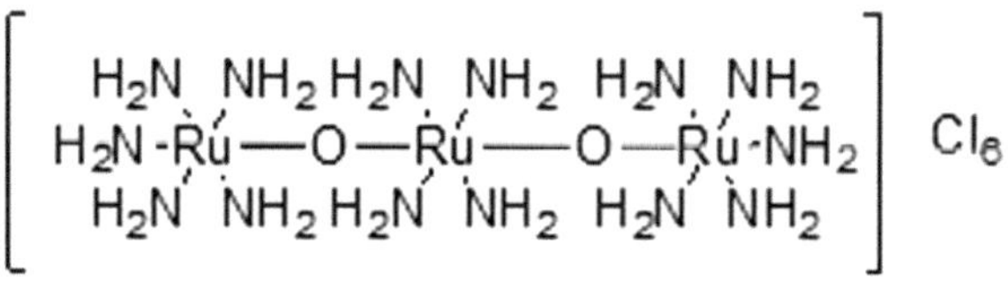

Ruthenium Red **9**

Fig. 5. Ruthenium red, a TRPV4 antagonist.

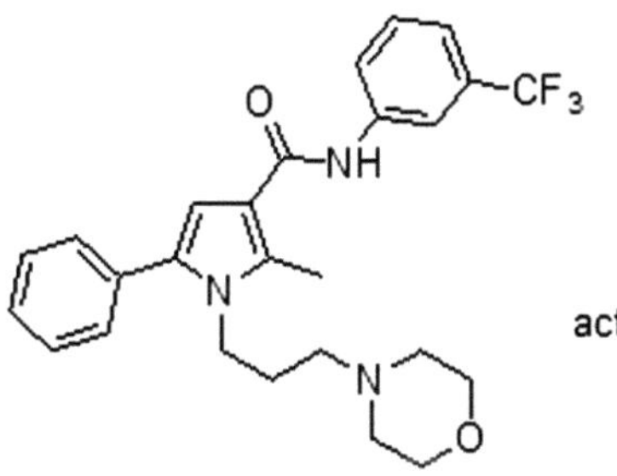

HC-067047 **10**
hTRPV4 IC_{50} = 48 nM
rTRPV4 IC_{50} = 133 nM
mTRPV4 IC_{50} = 17 nM
active at 10 mg/kg i.p. in model of cystitis

Fig. 6. HC-067047, a TRPV4 antagonist active in rodent models of cystitis.

a potent inhibitor of the human, mouse, and rat ortholog of TRPV4 (IC_{50} of 48 nM, 17 nM, and 133 nM, respectively). HC-067047 also inhibits TRPM8 and the hERG potassium channel, with an approximate tenfold window of selectivity at TRPV4 over these other ion channels. The activity at the hERG potassium channel may limit the use of HC-067047 to the preclinical setting, as inhibition of hERG has been associated with QTc prolongation, a potentially severe cardiovascular side effect (93). In vivo, HC-067047 was shown to improve bladder function in rodent models of cystitis, induced by cyclophosphamide (47). Other in vivo measurements indicated that HC-067047 did not affect core body temperature, thermal selection behavior, water intake, heart rate, locomotion, or motor coordination. Thus, HC-067047 may serve as an important tool compound to probe the functions of TRPV4 in vivo.

3.2.3. GSK-205 and Other Antagonists from GlaxoSmithkline

Extending their work beyond TRPV4 agonists, GlaxoSmithkline (GSK) have also reported on the design and synthesis of TRPV4 antagonists. For example, GSK have reported on the use of an aminothiazole ligand **11**, named GSK-205, as a submicromolar selective inhibitor of TRPV4 in porcine chondrocytes (IC_{50} = 0.6 μM) (94). Additionally, GSK have also detailed a number of different ligands as TRPV4 antagonists. Representative structures of some compounds described in recent patent applications are shown in Fig. 7 (95–98).

3.2.4. RN-1734 and RN-9893

RN-1734 **13** (Fig. 8) has been identified in the same focused arylsulfonamide screen as TRPV4 agonist RN-1747 described earlier (91).

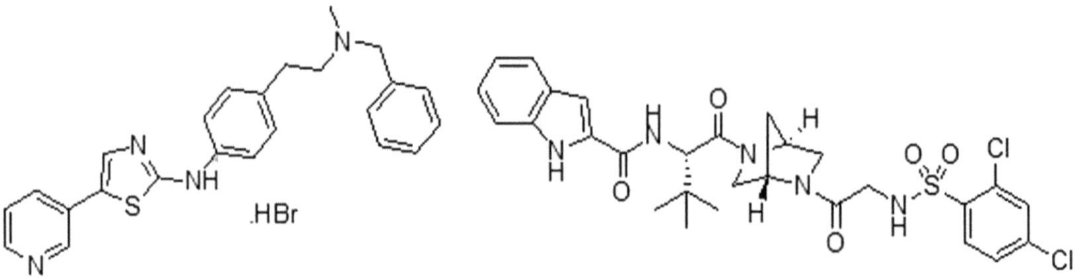

GSK-205 **11**
IC_{50} porcine TRPV4 = 0.6 μM

Representative TRPV4 antagonist **12** (from GSK)

Fig. 7. GSK-205 and other representative TRPV4 antagonists from GlaxoSmithKline.

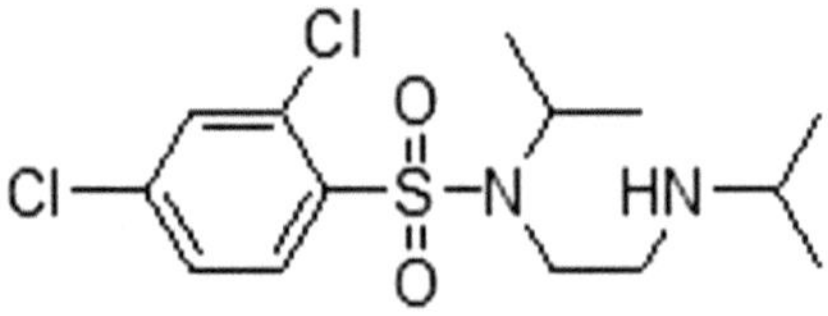

RN-1734 **13**
IC_{50} hTRPV4 = 2.3 μM

Fig. 8. RN-1734 as a TRPV4 antagonist.

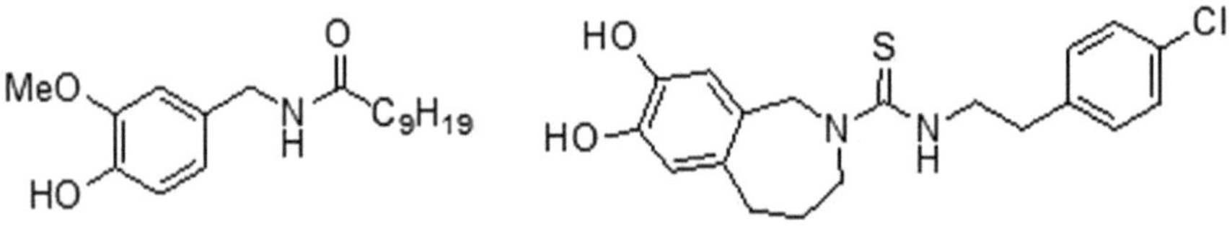

Vanilloid derivative **14**

Capsazepine **15**

Fig. 9. Vanilloid derivative and the well-known TRPV1 antagonist, capsazepine as TRPV4 antagonists.

RN-1734 inhibits TRPV4 in the micromolar range when 4αPDD, or hypotonicity, is used as the activating stimuli (IC_{50} hTRPV4 ca. 2.3 μM). A somewhat more potent TRPV4 antagonist, RN-9893, has also been reported, but the structure has not been disclosed at the time of writing this chapter (99).

3.2.5. Vanilloids and TRPV1 Ligands as TRPV4 Antagonists

It comes as no surprise that certain vanilloids and literature TRPV1 antagonists may have activity at TRPV4. For example, antagonist activity at TRPV4 has been described for vanilloid **14** and the quintessential TRPV1 antagonist, capsazepine **15** (91, 100). However, in the case of capsazepine, the IC_{50} for inhibition of TRPV4 (15 μM) is significantly higher than its IC_{50} against TRPV1 (5–50-fold, depending on the assay) (91, 101) (see Fig. 9).

4. Conclusion

With its wide expression pattern and multiple activating modalities, TRPV4 plays a variety of physiological roles, including several related to disease states. Potent, bioavailable, and relatively selective TRPV4 modulators such as agonist GSK1016790A **4** and antagonist HC-067047 **10** are now available to help validate TRPV4 modulation against specific diseases. Of special interest are the potential therapeutic benefits of TRPV4 antagonism for the treatment of inflammatory and neuropathic pain, bladder and urinary disorders, as well as ventilator-induced lung injury, all areas with significant unmet medical needs. While TRPV4-directed compounds are often active against other TRP channels, selectivity appears to be an achievable goal. On the other hand, the multiplicity of physiological roles carried out by this ion channel suggests that careful evaluation of potential on-target toxicity will be necessary in any drug discovery project targeting TRPV4.

References

1. Damann N et al (2008) TRPs in our senses. Curr Biol 18:R880–R889
2. Venkatachalam K, Montell C (2007) TRP channels. Annu Rev Biochem 76:387–417
3. Vennekens R et al (2008) Vanilloid transient receptor potential cation channels: an overview. Curr Pharm Des 14:18–31
4. Everaerts W et al (2009) The vallinoid transient receptor potential channel Trpv4: from structure to disease. Prog Biophys Mol Biol 103(1):2–17. doi:10.1016/j.pbiomolbio.2009.1010.1002
5. Nilius B et al (2004) TRPV4 calcium entry channel: a paradigm for gating diversity. Am J Physiol Cell Physiol 286:C195–C205
6. Caterina MJ et al (1997) The capsaicin receptor: a heat-activated ion channel in the pain pathway. Nature 389:816–824
7. Cortright DN, Szallasi A (2004) Biochemical pharmacology of the vanilloid receptor TRPV1. An update. Eur J Biochem 271:1814–1819
8. Szallasi A et al (2007) The vanilloid receptor TRPV1: 10 years from channel cloning to antagonist proof of concept. Nat Rev Drug Discov 6:357–372
9. Plant TD, Strotmann R (2007) Trpv4. Handb Exp Pharmacol (179):189-205
10. Arniges M et al (2006) Human TRPV4 channel splice variants revealed a key role of ankyrin domains in multimerization and trafficking. J Biol Chem 281:1580–1586
11. Shigematsu H et al (2010) A 3.5-nm structure of rat TRPV4 cation channel revealed by zernike phase-contrast cryo-EM. J Biol Chem 285(15):11210–11218
12. Liedtke W, Friedman JM (2003) Abnormal osmotic regulation in trpv4-/- mice. Proc Natl Acad Sci USA 100:13698–13703
13. Strotmann R et al (2000) OTRPC4, a nonselective cation channel that confers sensitivity to extracellular osmolarity. Nat Cell Biol 2:695–702
14. Liedtke W et al (2000) Vanilloid receptor-related osmotically activated channel (VR-OAC), a candidate vertebrate osmoreceptor. Cell 103:525–535
15. Guler AD et al (2002) Heat-evoked activation of the ion channel, TRPV4. J Neurosci 22:6408–6414
16. Xu H et al (2003) Regulation of a transient receptor potential (TRP) channel by tyrosine phosphorylation. SRC family kinase-dependent tyrosine phosphorylation of TRPV4 on TYR-253 mediates its response to hypotonic stress. J Biol Chem 278:11520–11527
17. Fan HC et al (2009) Activation of the TRPV4 ion channel is enhanced by phosphorylation. J Biol Chem 284:27884–27891
18. Peng H et al (2010) Identification of a Protein Kinase C-dependent phosphorylation site involved in sensitization of TRPV4 channel. Biochem Biophys Res Commun 391: 1721–1725

19. Vriens J et al (2005) Modulation of the Ca2 permeable cation channel TRPV4 by cytochrome P450 epoxygenases in vascular endothelium. Circ Res 97:908–915
20. Watanabe H et al (2003) Anandamide and arachidonic acid use epoxyeicosatrienoic acids to activate TRPV4 channels. Nature 424:434–438
21. Chung MK et al (2003) Warm temperatures activate TRPV4 in mouse 308 keratinocytes. J Biol Chem 278:32037–32046
22. Fernandez-Fernandez JM et al (2002) Maxi K+channel mediates regulatory volume decrease response in a human bronchial epithelial cell line. Am J Physiol Cell Physiol 283:C1705–C1714
23. Jia Y et al (2004) Functional TRPV4 channels are expressed in human airway smooth muscle cells. Am J Physiol Lung Cell Mol Physiol 287:L272–L278
24. Wissenbach U et al (2000) Trp12, a novel Trp related protein from kidney. FEBS Lett 485:127–134
25. Delany NS et al (2001) Identification and characterization of a novel human vanilloid receptor-like protein, VRL-2. Physiol Genomics 4:165–174
26. Birder L et al (2007) Activation of urothelial transient receptor potential vanilloid 4 by 4alpha-phorbol 12,13-didecanoate contributes to altered bladder reflexes in the rat. J Pharmacol Exp Ther 323:227–235
27. Gevaert T et al (2007) Deletion of the transient receptor potential cation channel TRPV4 impairs murine bladder voiding. J Clin Invest 117:3453–3462
28. Shen J et al (2006) Functional expression of transient receptor potential vanilloid 4 in the mouse cochlea. Neuroreport 17: 135–139
29. Mergler S et al (2011) Characterization of transient receptor potential vanilloid channel 4 (TRPV4) in human corneal endothelial cells. Exp Eye Res 93:710–719
30. Mizuno A et al (2003) Impaired osmotic sensation in mice lacking TRPV4. Am J Physiol Cell Physiol 285:C96–C101
31. Shibasaki K et al (2007) Effects of body temperature on neural activity in the hippocampus: regulation of resting membrane potentials by transient receptor potential vanilloid 4. J Neurosci 27:1566–1575
32. Zhang Y et al (2008) A transient receptor potential vanilloid 4 contributes to mechanical allodynia following chronic compression of dorsal root ganglion in rats. Neurosci Lett 432:222–227
33. Alessandri-Haber N et al (2003) Hypotonicity induces TRPV4-mediated nociception in rat. Neuron 39:497–511
34. Alessandri-Haber N et al (2004) Transient receptor potential vanilloid 4 is essential in chemotherapy-induced neuropathic pain in the rat. J Neurosci 24:4444–4452
35. Wu L et al (2007) Dual role of the TRPV4 channel as a sensor of flow and osmolality in renal epithelial cells. Am J Physiol Renal Physiol 293:F1699–F1713
36. Watanabe H et al (2002) Heat-evoked activation of TRPV4 channels in a HEK293 cell expression system and in native mouse aorta endothelial cells. J Biol Chem 277:47044–47051
37. Kohler R et al (2006) Evidence for a functional role of endothelial transient receptor potential V4 in shear stress-induced vasodilatation. Arterioscler Thromb Vasc Biol 26:1495–1502
38. Kumar S et al (2009) WO2006029209. Chem Abstr 144:286221
39. Masuyama R et al (2008) TRPV4-mediated calcium influx regulates terminal differentiation of osteoclasts. Cell Metab 8:257–265
40. Mizoguchi F et al (2008) Transient receptor potential vanilloid 4 deficiency suppresses unloading-induced bone loss. J Cell Physiol 216:47–53
41. Nilius B, Owsianik G (2010) Channelopathies converge on TRPV4. Nat Genet 42:98–100
42. Willette RN et al (2008) Systemic activation of the transient receptor potential vanilloid subtype 4 channel causes endothelial failure and circulatory collapse: Part 2. J Pharmacol Exp Ther 326:443–452
43. Thorneloe KS et al (2008) N-((1S)-1-{[4-((2S)-2-{[(2,4-dichlorophenyl)sulfonyl] amino}-3-hydroxypropa noyl)-1-piperazinyl] carbonyl}-3-methylbutyl)-1-benzothiophene-2-carboxamid e (GSK1016790A), a novel and potent transient receptor potential vanilloid 4 channel agonist induces urinary bladder contraction and hyperactivity: Part I. J Pharmacol Exp Ther 326:432–442
44. Suzuki M et al (2003) Impaired pressure sensation in mice lacking TRPV4. J Biol Chem 278:22664–22668
45. Rock MJ et al (2008) Gain-of-function mutations in TRPV4 cause autosomal dominant brachyolmia. Nat Genet 40:999–1003
46. Krakow D et al (2009) Mutations in the gene encoding the calcium-permeable ion channel TRPV4 produce spondylometaphyseal dysplasia, Kozlowski type and metatropic dysplasia. Am J Hum Genet 84:307–315

47. Everaerts W et al (2010) Inhibition of the cation channel TRPV4 improves bladder function in mice and rats with cyclophosphamide-induced cystitis. Proc Natl Acad Sci USA 107:19084–19089
48. Liedtke W (2005) TRPV4 as osmosensor: a transgenic approach. Pflugers Arch 451:176–180
49. Lowry CA et al (2009) That warm fuzzy feeling: brain serotonergic neurons and the regulation of emotion. J Psychopharmacol 23: 392–400
50. Muramatsu S et al (2007) Functional gene screening system identified TRPV4 as a regulator of chondrogenic differentiation. J Biol Chem 282:32158–32167
51. Gao X et al (2003) Temperature-modulated diversity of TRPV4 channel gating: activation by physical stresses and phorbol ester derivatives through protein kinase C-dependent and -independent pathways. J Biol Chem 278: 27129–27137
52. Vincent F, Duncton MA (2011) TRPV4 agonists and antagonists. Curr Top Med Chem 11:2216–2226
53. Alessandri-Haber N et al (2009) TRPC1 and TRPC6 channels cooperate with TRPV4 to mediate mechanical hyperalgesia and nociceptor sensitization. J Neurosci 29:6217–6228
54. Alessandri-Haber N et al (2006) A transient receptor potential vanilloid 4-dependent mechanism of hyperalgesia is engaged by concerted action of inflammatory mediators. J Neurosci 26:3864–3874
55. Alessandri-Haber N et al (2008) Interaction of transient receptor potential vanilloid 4, integrin, and SRC tyrosine kinase in mechanical hyperalgesia. J Neurosci 28:1046–1057
56. Alessandri-Haber N et al (2005) TRPV4 mediates pain-related behavior induced by mild hypertonic stimuli in the presence of inflammatory mediator. Pain 118:70–79
57. Chen X et al (2007) Marked attenuation of inflammatory mediator-induced C-fiber sensitization for mechanical and hypotonic stimuli in TRPV4-/- mice. Mol Pain 3:31
58. Levine JD, Alessandri-Haber N (2007) TRP channels: targets for the relief of pain. Biochim Biophys Acta 1772:989–1003
59. Vakili C et al (1970) Chemical and osmolar changes of interstitial fluid in acute inflammatory states. Surg Forum 21:227–228
60. Grant AD et al (2007) Protease-activated receptor 2 sensitizes the transient receptor potential vanilloid 4 ion channel to cause mechanical hyperalgesia in mice. J Physiol 578:715–733
61. Cenac N et al (2008) Transient receptor potential vanilloid-4 has a major role in visceral hypersensitivity symptoms. Gastroenterology 135:937–946, 946.e1-2
62. Sipe WE et al (2008) Transient receptor potential vanilloid 4 mediates protease activated receptor 2-induced sensitization of colonic afferent nerves and visceral hyperalgesia. Am J Physiol Gastrointest Liver Physiol 294:G1288–G1298
63. Brierley SM et al (2008) Selective role for TRPV4 ion channels in visceral sensory pathways. Gastroenterology 134:2059–2069
64. Auge C et al (2009) Protease-activated receptor-4 (PAR 4): a role as inhibitor of visceral pain and hypersensitivity. Neurogastroenterol Motil 21:1189–e1107
65. Chen Y et al (2011) Proteinase-activated receptor 2 sensitizes transient receptor potential vanilloid 1, transient receptor potential vanilloid 4, and transient receptor potential ankyrin 1 in paclitaxel-induced neuropathic pain. Neuroscience 193:440–451
66. Janssen DA et al (2011) The mechanoreceptor TRPV4 is localized in adherence junctions of the human bladder urothelium: a morphological study. J Urol 186:1121–1127
67. Mochizuki T et al (2009) The TRPV4 cation channel mediates stretch-evoked Ca2+ influx and ATP release in primary urothelial cell cultures. J Biol Chem 284:21257–21264
68. Skryma R et al (2011) From urgency to frequency: facts and controversies of TRPs in the lower urinary tract. Nat Rev Urol 8:617–630
69. Aizawa N et al (2012) Effects of TRPV4 cation channel activation on the primary bladder afferent activities of the rat. Neurourol Urodyn 31:148–155
70. Boudes M et al (2011) TRPV4, new therapeutic target for urinary problems. Med Sci (Paris) 27:232–234
71. Angelico P, Testa R (2010) TRPV4 as a target for bladder overactivity. F1000 Biol Rep 2
72. Everaerts W et al (2008) On the origin of bladder sensing: Tr(i)ps in urology. Neurourol Urodyn 27:264–273
73. Ware LB, Matthay MA (2000) The acute respiratory distress syndrome. N Engl J Med 342.1334–1349
74. Ware LB, Matthay MA (2005) Clinical practice. Acute pulmonary edema. N Eng J Med 353. 2788–2796
75. Bernard GR et al (1994) The American-European Consensus Conference on ARDS. Definitions, mechanisms, relevant outcomes, and clinical trial coordination. Am J Respir Crit Care Med 149:818–824

76. Dos Santos CC, Slutsky AS (2000) Invited review: mechanisms of ventilator-induced lung injury: a perspective. J Appl Physiol 89:1645–1655
77. Tiruppathi C et al (2002) Role of Ca^{2+} signaling in the regulation of endothelial permeability. Vasc Pharmacol 39:173–185
78. Hamanaka K et al (2007) TRPV4 initiates the acute calcium-dependent permeability increase during ventilator-induced lung injury in isolated mouse lungs. Am J Physiol Lung Cell Mol Physiol 293:L923–L932
79. Alvarez DF et al (2006) Transient receptor potential vanilloid 4-mediated disruption of the alveolar septal barrier: a novel mechanism of acute lung injury. Circ Res 99:988–995
80. Hamanaka K et al (2010) TRPV4 channels augment macrophage activation and ventilator-induced lung injury. Am J Physiol Lung Cell Mol Physiol 299:L353–L362
81. Eyal FG et al (2007) Reduction in alveolar macrophages attenuates acute ventilator induced lung injury in rats. Intensive Care Med 33:1212–1218
82. Jian MY et al (2008) High vascular pressure-induced lung injury requires P450 epoxygenase-dependent activation of TRPV4. Am J Respir Cell Mol Biol 38:386–392
83. Watanabe H et al (2002) Activation of TRPV4 channels (hVRL-2/mTRP12) by phorbol derivatives. J Biol Chem 277:13569–13577
84. Vriens J et al (2007) Determinants of 4 alpha-phorbol sensitivity in transmembrane domains 3 and 4 of the cation channel TRPV4. J Biol Chem 282:12796–12803
85. Klausen TK et al (2009) Modulation of the transient receptor potential vanilloid channel TRPV4 by 4alpha-phorbol esters: a structure-activity study. J Med Chem 52:2933–2939
86. Ding XL et al (2010) Involvement of TRPV4-NO-cGMP-PKG pathways in the development of thermal hyperalgesia following chronic compression of the dorsal root ganglion in rats. Behav Brain Res 208:194–201
87. Gao F, Wang DH (2010) Hypotension induced by activation of the transient receptor potential vanilloid 4 channels: role of Ca^{2+}-activated K+ channels and sensory nerves. J Hypertens 28:102–110
88. Gao F et al (2009) Salt intake augments hypotensive effects of transient receptor potential vanilloid 4: functional significance and implication. Hypertension 53:228–235
89. Jeong J et al (2009) In: 238th ACS national meeting, Washington, DC, pp MEDI-392
90. Casillas LT et al (2009) In: 238th ACS national meeting, Washington, DC, pp MEDI-451
91. Vincent F et al (2009) Identification and characterization of novel TRPV4 modulators. Biochem Biophys Res Commun 389:490–494
92. Tsushima H, Mori M (2006) Antidipsogenic effects of a TRPV4 agonist, 4alpha-phorbol 12,13-didecanoate, injected into the cerebroventricle. Am J Physiol Regul Integr Comp Physiol 290:R1736–R1741
93. Pollard CE et al (2008) Strategies to reduce the risk of drug-induced QT interval prolongation: a pharmaceutical company perspective. Br J Pharmacol 154:1538–1543
94. Phan MN et al (2009) Functional characterization of TRPV4 as an osmotically sensitive ion channel in porcine articular chondrocytes. Arthritis Rheum 60:3028–3037
95. Cheung M et al (2009) WO2009146182. Chem Abstr 152:27389
96. Cheung M et al (2009) WO2009146177. Chem Abstr 152:27388
97. Cheung M, Eidam HS (2010) WO2010011912. Chem Abstr 152:215581
98. Cheung M et al (2009) WO2009111680. Chem Abstr 151:359104
99. Wei Z-L et al (2007) Discovery of a proof-of-concept TRPV4 antagonist. In: Keystone symposia on the transient receptor potential ion channel superfamily, Breckenridge, CO
100. Tsukamoto S, Sawamura S (2009) JP2009084290. Chem Abstr 150:438732
101. Reubish D et al (2009) Functional assessment of temperature-gated ion-channel activity using a real-time PCR machine. Biotechniques 47:3–9

Part III

TRPs and Airways

Chapter 14

TRP Expression and Function in the Lung Airways

Alexander Dietrich and Thomas Gudermann

Abstract

The airways of the lungs are composed of the epithelium, smooth muscle cells, and sensory nerve cells as well as resident immune cells. Transient receptor potential (TRP) channels are expressed in all of these tissues and cells and there is increasing evidence that they play a key role in respiratory diseases such as asthma and chronic obstructive pulmonary disease (COPD). This chapter focuses on what is currently known about TRP expression and function in lung airways by studying specific channel blockers as well as TRP-deficient mouse models. Targeting selected TRP functions may have important therapeutic benefits for respiratory diseases in the future.

Key words: TRP channels, Lung, Asthma, COPD, Cough, Smooth muscle cells, Epithelial cells, Sensory neurons, Immune cells

1. Introduction

The lungs are functionally divided into the lung airways and the pulmonary vasculature. While lung airways transport gases to and from alveoli, pulmonary blood vessels ensure oxygen supply to the circulating blood. The airways are composed of bronchi which by themselves branch many times into smaller airways, ending in the narrowest ones, the bronchioles. Large airways are held open by semiflexible, fibrous connective tissue called cartilage, while smaller airways are supported by the lung tissue that surrounds and is attached to them. Both contain circular airway smooth muscle, which can dilate or constrict, thus changing the airway size. At the end of each bronchiole are thousands of small air sacs called alveoli, where the gas exchange with the circulating blood takes place. Lung airways are the gateway not only to gas exchange, but also to infectious agents like bacteria and viruses as well as allergens invading the body. Therefore, immune cells in the lung provide a very effective immune defense and are responsible for allergic diseases like

Arpad Szallasi and Tamás Bíró (eds.), *TRP Channels in Drug Discovery: Volume I*, Methods in Pharmacology and Toxicology, DOI 10.1007/978-1-62703-077-9_14,

asthma. Chronic obstructive pulmonary diseases (COPD) including chronic bronchitis and emphysema are also very common, whose suggested prevalence is increasing worldwide. In addition, pulmonary fibrosis after inhalation of environmental and occupational pollutants like cigarette smoke results in a gradual exchange of lung parenchyma with fibrotic tissues. The main target of all of these diseases is the lung airways, and pathophysiological mechanisms like bronchoconstriction, inflammation, and airway remodeling may involve ion channels of the mammalian superfamily of transient receptor potential (TRP) channels expressed in different tissues of the lung airways. They have been described as polymodal cellular sensors and as candidates for receptor- and probably store-operated calcium influx in different cell types.

A rise in the cytosolic free calcium concentration ($[Ca^{2+}]_i$) is a powerful stimulus to cell activation and plays an important role in the regulation of cell types found in the lung airways like epithelial cells, airway smooth muscle cells, and sensory nerve endings. Moreover, eosinophils, neutrophils, and T lymphocytes migrate to the lungs in response to injury and infection after an initial rise in intracellular Ca^{2+} levels. In summary, changes in $[Ca^{2+}]_i$ play a crucial role in functional responses like mucus production, generation of *r*eactive *o*xygen *s*pecies (ROS), cell proliferation, as well as airway smooth muscle contraction.

Members of the transient receptor potential (TRP) superfamily can be divided into six families consisting of 28 known members: The TRPA (for Ankyrin), TRPC (for Canonical or Classical), TRPM (for Melastain), TRPP (for Polycystin), TRPML (for Mucolipin), and the TRPV (for Vanilloid) family. They form homo- or heterotetramers in between their TRP families with varying ion selectivity from unselective to Ca^{2+}-selective (see (1) for a recent review). Members of the TRPML and TRPP families are not found in lung airways and will not be described in this chapter. In between the other families only members are listed that have been identified in the cells of the lung airways.

2. The TRPA Channel

The family of transient receptor potential A is identified by a high number of ankyrin repeats and comprises only one mammalian member, TRPA1. These aminoterminal repeats, which are also found in TRPC and TRPV channels, may be relevant to a proposed mechanosensory role of the channel (2). TRPA1 is a calcium permeable channel which was originally cloned from human lung fibroblasts (3). The channel is also expressed in vagal sensory neurons innervating the airways (4, 5), and is stimulated by noxious cold (<17°C) (6, 7) and isothiocyanates, the compounds of mustard oil and horseradish (7, 8). Most interestingly, the TRPA1 channel is also activated by toxic substances like acrolein and other unsaturated aldehydes which are inhaled with cigarette smoke and may be sensed by chemosensory (C fiber) neurons in the lung. Chemosensory respiratory reflexes triggered in the upper airways may limit the exposure of the lower airways to the toxic substances. On the other hand, this activation may induce inflammatory processes like asthma. Moreover, oxidative stress induced by cigarette smoke is an important risk factor in the development of COPD. It was recently demonstrated that hypochlorite- and hydrogen peroxide-activated Ca^{2+} influxes and membrane currents in a subpopulation of chemosensory cells which are sensitive to reactive oxygen species released by these substances. These membrane currents were absent in TRPA1-deficient mice, indicating the essential role of this channel in inducing the calcium influx (9). Along these lines, TRPA1$^{-/-}$ mice displayed profound deficiencies in hypochlorite- and hydrogen peroxide-induced respiratory depression as well as decreased pain behavior (9). For these reasons, it is not surprising that potent selective TRPA1 antagonists like AP-18 and HC-030031 as potential therapeutics were already tested in animal models. AP-18 was able to inhibit acrolein-induced depolarization of isolated human vagus nerves (10) while HC-030031 attenuated the tussive responses induced by inhalation of TRPA1 agonists by >50% (10). Although a complete inhibition was not possible due to insufficient dosing of HC-030031, which for solubility reasons could not be increased, TRPA1 blockers may have a huge therapeutic benefit in the treatment of allergic asthma and COPD in the near future.

3. TRPC Channels

The classical or canonical family of TRPC channels shares the most homologies to the initially discovered *trp* channels of *Drosophila melanogaster*. They include seven members (TRPC1-7), but only six are functional in humans, because TRPC2 is a pseudogene in the human genome. In the case of the TRPC3/6/7 subfamily of TRPC channels, their activation is due to the diacylglycerol (DAG) formed from the phospholipase C (PLC)-mediated cleavage of plasma membrane phosphatidyl-inositol-4,5-bisphosphate (PIP_2).

Although TRPC1 was the first mammalian TRP channel to be cloned, its physiological functions are still elusive. It is not even clear if TRPC1 is an ion channel or merely an adaptor protein regulating ion channel activity of other TRPC members. Homomeric TRPC1 complexes do not translocate to the plasma membrane (11) but TRPC1 as a member of heteromeric TRPC1/5 or TRPC1/4 complexes have a high impact on TRPC4/5 currents (12). It has been reported that TRPC1 is important for the proliferation of bronchial smooth muscle cells, which may induce thickening of the airways facilitating airway hyperreactivity (summarized in (13)). Along these lines, TRPC1 is up-regulated in proliferating airway smooth muscle cells while TRPC1 down-regulation inhibits cell proliferation (14, 15). However, it is not clear if proliferation is influenced exclusively by TRPC1 channels or by TRPC1/4 channel complexes, because heteromeric TRPC1/4 channel complexes seem to be more important for the lung extra-alveolar endothelial barrier than for the capillary endothelial barrier (summarized in (16)).

TRPC3 expression is up-regulated by tumor necrosis factor (TNF)-α in human airway smooth muscle cells (17). TNF-α is a pro-inflammatory cytokine which has been implicated with asthma and COPD because it increases airway hyperresponsiveness (AHR) by increasing airway smooth muscle Ca^{2+}. Therefore, TNF-α blockage might represent a promising pharmacological target for the treatment of asthma and COPD (summarized in (18)).

Many studies were already published on the role of TRPC6 channels in the lung airways. Indeed TRPC6 function in the lung might be most important because it is the most prominently expressed TRPC channel in lungs (19). Therefore, we set out to analyze allergic responses in a TRP6-deficient mouse model expecting a reduced response in comparison to wild-type (WT) mice. However, we found an increased methacholine-induced AHR in TRPC6-deficient mice compared to WT mice. This surprising finding is most probably due to compensatory up-regulation of TRPC3 in airway smooth muscle cells (20). By experimental inflammation induced by intraperitoneal ovalbumin (OVA) sensitization followed by OVA aerosol challenges we were able to detect a decreased level of T-helper type 2 (Th2) cytokines (IL-5 and IL-13) as well as reduced IgE levels in the blood of TRPC6$^{-/-}$ mice compared to WT mice. Mucus production in goblet cells from challenged mice however was not altered in TRPC6-deficient mice (20). These data point to an important role of TRPC6 in the immune responsive rather than in airway smooth muscle tissues itself. Because neutrophil numbers are increased in the sputum of COPD and asthma patients, TRPC6 function in these cells might be also important for the progress of the diseases. Migration of TRPC6$^{-/-}$ neutrophils in response to macrophage inflammatory protein-2 (MIP2 also known as CXCL2) was reduced compared to WT neutrophils (21). In the same report, an involvement of TRPC6 in cytoskeletal rearrangements during neutrophil migration was demonstrated suggesting an important role of TRPC6 in migrating lung neutrophils. TRPC6 was also identified in other immune cells of the lung like alveolar macrophages (22). Most interestingly, TRPC6 mRNA expression was significantly increased in macrophages obtained from COPD patients compared to healthy controls, while TRPC3 and TRPC7 levels remained unchanged (22). In summary, TRPC channels represent good pharmacological targets but specific TRPC channel blockers still need to be identified.

4. TRPM Channels

The mammalian TRPM family consists of seven channels with melastatin as the founding member, originally described as a tumor suppressor protein. They contain an important temperature sensor (TRPM8) and exhibit variable permeability ranging from Ca^{2+} impermeable (TRPM4/5) to highly Ca^{2+} and Mg^{2+} permeable (TRPM6/7). TRPM2 and TRPM6/7 are also exceptional in having associated enzymatic activities associated with their C-terminal domains.

A unique function of TRPM2 is its intracellular C-terminal ADP-ribose (ADPR) hydrolase activity and its activation by intracellular ADPR (summarized in (23)). TRPM2 is also described as an H_2O_2-activated channel expressed in lung macrophages (22). Generation of reactive oxygen species (ROS) and especially H_2O_2 is important in host defense eliminating invading bacteria. However, high levels of oxidative stress from extracellular sources like cigarette smoke or endogenous inflammatory processes are also important in the pathogenesis of asthma and COPD. Moreover, the *N*-formylmethionyl-leucyl-phenylalanine (fMLP) peptide, which are released from invading bacteria, induce only a decreased Ca^{2+} and migration response in neutrophils from TRPM2-deficient mice compared to WT mice while the CXCL2-induced responses were unaltered (24).

A TRPM4-deficient mouse model uncovered its function in mast cells. Bone marrow-derived mast cells from TRPM4$^{-/-}$ showed increased Ca^{2+} entry and therefore enhanced release of histamine, leukotrienes, and TNF-α compared to WT mast cells. Mast cell migration was also impaired (25, 26).

In summary, both TRPM2 and TRPM4 channels are expressed in immune cells (see Fig. 1) whose exact involvement in allergic and nonallergic lung inflammation still needs to be determined.

TRPM8 is a menthol and thermally regulated channel activated by breathing cold air and inducing respiratory autonomic responses like airway constriction, cough, and mucosal secretion (27, 28). While the TRPM8 channel is expressed in only very few bronchopulmonary C fibers, its truncated but functional variant has been identified in human bronchial epithelial cells (29). Stimulation of these cells with both cold (18°C) and menthol caused an enhanced expression of pro-inflammatory cytokines (e.g., IL-1α and β, IL-4, IL-6, IL-8, and IL-13) as well as granulocyte-macrophage colony-stimulating factor (GM-CSF) and TNF-α (30). It will be exciting to test TRPM8 inhibitors in cold-induced asthma and to evaluate TRPM8 function in gene-deficient mouse models.

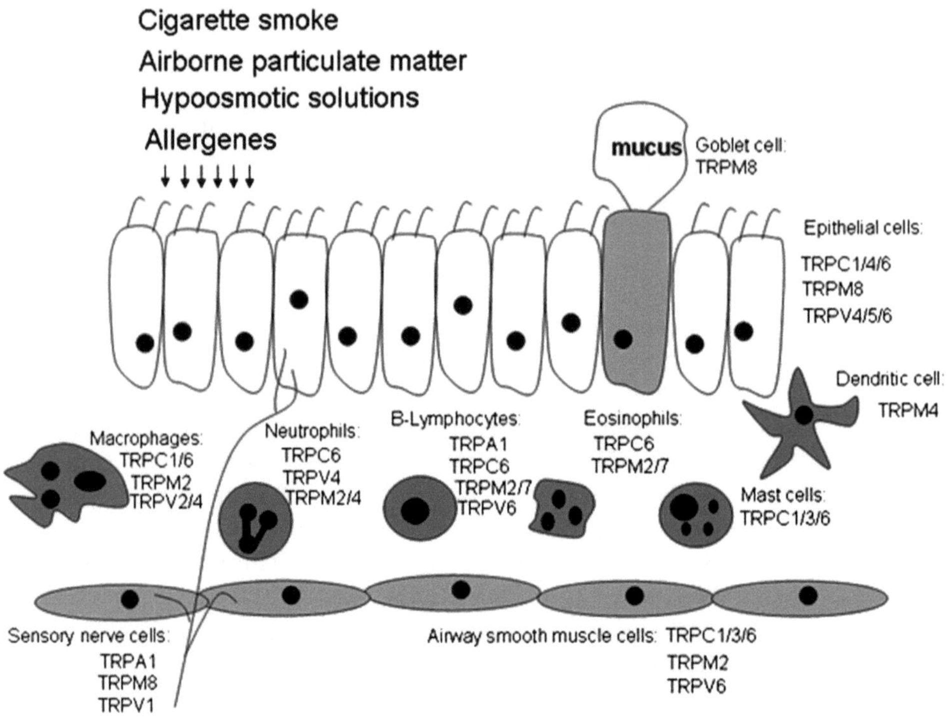

Fig. 1. Expression profiles of TRP channels in airway smooth muscle cells (adapted from (34]). See the text for more details.

5. TRPV Channels

Six members were identified for the TRPV family, whose founding channel TRPV1 is activated by vanilloids including capsaicin, the ingredient of red chili pepper. Only TRPV5 and 6 show Ca^{2+} selectivity, while TRPV1-4 are described as nonselective thermo- and chemosensitive channels.

TRPV1 is expressed in sensory neurons, pulmonary smooth muscle cells, bronchial and tracheal epithelial cells, human larynx epithelial cells, as well as in dendritic cells of the lung (31–33). Next to capsaicin, TRPV1 is also activated by protons and heat as well as agonists like NGF, bradykinin, and prostaglandins which remove inhibitory PIP_2 by cleavage of receptor-activated PLCs (summarized in (34)). In contrast to other TRP channels, a strong and specific inhibitor capsazepine has been identified and several other antagonists are also tested in clinical phases, because many studies have linked TRPV1 to respiratory diseases (34). TRPV1 was up-regulated in capsaicin-sensitive patients with chronic persistent cough (35, 36), asthma, and COPD (37). Blockage of TRPV1 by its inhibitors prevents damage of airway epithelium by airborne particulate matter (38, 39) and reduces cough induced by citric acid or antigen challenge (40, 41) as well as OVA-induced cough response in guinea pigs (41). In human bronchial epithelial

(BEAS-2B) and alveolar (A549) cells, TRPV1 activation induced cell death by endoplasmic reticulum stress which was, however, not inhibited by TRPV1 antagonists (42, 43). Most interestingly, the analysis of TRPV1$^{-/-}$ mice also identified a counterregulatory mechanism by activated TRPV1. Airway inflammation induced by intranasal administration of lipopolysaccharides (LPS) from *Escherichia coli* was significantly increased in TRPV1-deficient mice compared to WT mice. Most interestingly, LPS was able to induce somatostatin secretion in WT but not in TRPV1$^{-/-}$ mice. Thus, TRPV1-activated somatostatin production may protect mice from bronchial hyperreactivity, peribroncheal edema, neutrophil/macrophage infiltration, as well as goblet cell hyperplasia during airway inflammation (44). Therefore, TRPV1 may be most important in the development and protection from airway diseases and TRPV1 agonists as well as antagonists are currently under development by several companies (see (34) for a recent review).

The TRPV4 channel is activated by hypotonic solutions as well as mechanical stress and may play a role in osmotic and mechanical sensing in the lung airways (45, 46). Channel activity of TRPV4 is also stimulated by 4α-phorbol-12,13-didecnoate (4αPDD) and inhibited by low temperatures (<30°C) (47). In the lung, TRPV4 is expressed in ciliated tracheal and airway epithelial cells as well as in alveolar septal walls (48, 49). Moreover, expression on mononuclear cells (50) and alveolar macrophages (45, 51) has been reported. The cilia beat frequency can be increased by activation of TRPV4 and inhibited by a TRPV4-specific antibody in vitro pointing to an important role of TRPV4 in mucocilia transport (52). The permeability of the epithelial layers of the gas exchanging septal regions was increased by 4αPDD and this response was largely decreased in TRPV4$^{-/-}$ mice (49, 53). This mechanism can induce alveolar flooding and dramatically decrease gas exchange. Hypotonic solutions can initiate smooth muscle contraction in isolated human and guinea pigs airways (54), a process which might be important for bronchial hyperresponsiveness in asthma patients. TRPV4 is also an important candidate gene for COPD, because in a genetic approach including 953 COPD patients and 956 healthy persons, 7 out of 20 single nucleotide polymorphisms (SNPs) in the TRPV4 gene were associated with COPD (55). So far there are no specific TRPV4 inhibitors identified, although TRPV4 is an important pharmacological target for the treatment of asthma and COPD.

6. Conclusion

TRP channels are expressed in all of the different cell types in lung airways. However, we are just beginning to understand their functions in lung airway physiology by analyzing gene-deficient mouse models. Specific activators or inhibitors for TRPM8 as well as TRPV1 and TRPV3 already launched by certain pharmaceutical companies may also be beneficial in the pharmacotherapy of airway diseases like asthma and COPD. TRPC and TRPA1 channels are also important pharmacological targets but specific blockers still need to be identified.

References

1. Nilius B, Owsianik G (2011) The transient receptor potential family of ion channels. Genome Biol 12:218
2. Nagata K, Duggan A, Kumar G, Garcia-Anoveros J (2005) Nociceptor and hair cell transducer properties of TRPA1, a channel for pain and hearing. J Neurosci 25:4052–4061
3. Jaquemar D, Schenker T, Trueb B (1999) An ankyrin-like protein with transmembrane domains is specifically lost after oncogenic transformation of human fibroblasts. J Biol Chem 274:7325–7333
4. Nassenstein C, Kwong K, Taylor-Clark T, Kollarik M, Macglashan DM, Braun A, Undem BJ (2008) Expression and function of the ion channel TRPA1 in vagal afferent nerves innervating mouse lungs. J Physiol 586:1595–1604
5. Brooks SM (2008) Irritant-induced chronic cough: irritant-induced TRPpathy. Lung 186(Suppl 1):S88–S93
6. Story GM, Peier AM, Reeve AJ, Eid SR, Mosbacher J, Hricik TR, Earley TJ, Hergarden AC, Andersson DA, Hwang SW, McIntyre P, Jegla T, Bevan S, Patapoutian A (2003) ANKTM1, a TRP-like channel expressed in nociceptive neurons, is activated by cold temperatures. Cell 112:819–829
7. Bandell M, Story GM, Hwang SW, Viswanath V, Eid SR, Petrus MJ, Earley TJ, Patapoutian A (2004) Noxious cold ion channel TRPA1 is activated by pungent compounds and bradykinin. Neuron 41:849–857
8. Jordt SE, Bautista DM, Chuang HH, McKemy DD, Zygmunt PM, Hogestatt ED, Meng ID, Julius D (2004) Mustard oils and cannabinoids excite sensory nerve fibres through the TRP channel ANKTM1. Nature 427:260–265
9. Bessac BF, Sivula M, von Hehn CA, Escalera J, Cohn L, Jordt SE (2008) TRPA1 is a major oxidant sensor in murine airway sensory neurons. J Clin Invest 118:1899–1910
10. Birrell MA, Belvisi MG, Grace M, Sadofsky L, Faruqi S, Hele DJ, Maher SA, Freund-Michel V, Morice AH (2009) TRPA1 agonists evoke coughing in guinea pig and human volunteers. Am J Respir Crit Care Med 180:1042–1047
11. Hofmann T, Schaefer M, Schultz G, Gudermann T (2002) Subunit composition of mammalian transient receptor potential channels in living cells. Proc Natl Acad Sci USA 99:7461–7466
12. Strubing C, Krapivinsky G, Krapivinsky L, Clapham DE (2001) TRPC1 and TRPC5 form a novel cation channel in mammalian brain. Neuron 29:645–655
13. Li S, Westwick J, Poll C (2003) Transient receptor potential (TRP) channels as potential drug targets in respiratory disease. Cell Calcium 33:551–558
14. Sweeney M, Yu Y, Platoshyn O, Zhang S, McDaniel SS, Yuan JX (2002) Inhibition of endogenous TRP1 decreases capacitative Ca^{2+} entry and attenuates pulmonary artery smooth muscle cell proliferation. Am J Physiol Lung Cell Mol Physiol 283:L144–L155
15. Sweeney M, McDaniel SS, Platoshyn O, Zhang S, Yu Y, Lapp BR, Zhao Y, Thistlethwaite PA, Yuan JX (2002) Role of capacitative Ca^{2+} entry in bronchial contraction and remodeling. J Appl Physiol 92:1594–1602
16. Cioffi DL, Lowe K, Alvarez DF, Barry C, Stevens T (2009) TRPing on the lung endothelium: calcium channels that regulate barrier function. Antioxid Redox Signal 11:765–776
17. White TA, Xue A, Chini EN, Thompson M, Sieck GC, Wylam ME (2006) Role of transient receptor potential C3 in TNF-alpha-enhanced calcium influx in human airway myocytes. Am J Respir Cell Mol Biol 35:243–251
18. Amrani Y (2007) TNF-alpha and calcium signaling in airway smooth muscle cells: a never-ending story with promising therapeutic relevance. Am J Respir Cell Mol Biol 36: 387–388
19. Hofmann T, Schaefer M, Schultz G, Gudermann T (2000) Transient receptor potential channels as molecular substrates of receptor-mediated cation entry. J Mol Med (Berl) 78:14–25
20. Sel S, Rost BR, Yildirim AO, Sel B, Kalwa H, Fehrenbach H, Renz H, Gudermann T, Dietrich A (2008) Loss of classical transient receptor potential 6 channel reduces allergic airway response. Clin Exp Allergy 38:1548–1558

21. Damann N, Owsianik G, Li S, Poll C, Nilius B (2009) The calcium-conducting ion channel transient receptor potential canonical 6 is involved in macrophage inflammatory protein-2-induced migration of mouse neutrophils. Acta Physiol (Oxf) 195:3–11
22. Finney-Hayward TK, Popa MO, Bahra P, Li S, Poll CT, Gosling M, Nicholson AG, Russell RE, Kon OM, Jarai G, Westwick J, Barnes PJ, Donnelly LE (2010) Expression of transient receptor potential c6 channels in human lung macrophages. Am J Respir Cell Mol Biol 43:296–304
23. Dietrich A, Gudermann T (2008) Another TRP to endothelial dysfunction: TRPM2 and endothelial permeability. Circ Res 102:275–277
24. Yamamoto S, Shimizu S, Kiyonaka S, Takahashi N, Wajima T, Hara Y, Negoro T, Hiroi T, Kiuchi Y, Okada T, Kaneko S, Lange I, Fleig A, Penner R, Nishi M, Takeshima H, Mori Y (2008) TRPM2-mediated Ca2+influx induces chemokine production in monocytes that aggravates inflammatory neutrophil infiltration. Nat Med 14:738–747
25. Shimizu T, Owsianik G, Freichel M, Flockerzi V, Nilius B, Vennekens R (2009) TRPM4 regulates migration of mast cells in mice. Cell Calcium 45:226–232
26. Barbet G, Demion M, Moura IC, Serafini N, Leger T, Vrtovsnik F, Monteiro RC, Guinamard R, Kinet JP, Launay P (2008) The calcium-activated nonselective cation channel TRPM4 is essential for the migration but not the maturation of dendritic cells. Nat Immunol 9:1148–1156
27. Yoshihara S, Geppetti P, Hara M, Linden A, Ricciardolo FL, Chan B, Nadel JA (1996) Cold air-induced bronchoconstriction is mediated by tachykinin and kinin release in guinea pigs. Eur J Pharmacol 296:291–296
28. Carlsen KH, Carlsen KC (2002) Exercise-induced asthma. Paediatr Respir Rev 3:154–160
29. Sabnis AS, Shadid M, Yost GS, Reilly CA (2008) Human lung epithelial cells express a functional cold-sensing TRPM8 variant. Am J Respir Cell Mol Biol 39:466–474
30. Sabnis AS, Reilly CA, Veranth JM, Yost GS (2008) Increased transcription of cytokine genes in human lung epithelial cells through activation of a TRPM8 variant by cold temperatures. Am J Physiol Lung Cell Mol Physiol 295:L194–L200
31. Yamamoto Y, Sato Y, Taniguchi K (2007) Distribution of TRPV1- and TRPV2-immunoreactive afferent nerve endings in rat trachea. J Anat 211:775–783
32. Tominaga M, Caterina MJ, Malmberg AB, Rosen TA, Gilbert H, Skinner K, Raumann BE, Basbaum AI, Julius D (1998) The cloned capsaicin receptor integrates multiple pain-producing stimuli. Neuron 21:531–543
33. Watanabe N, Horie S, Michael GJ, Keir S, Spina D, Page CP, Priestley JV (2006) Immunohistochemical co-localization of transient receptor potential vanilloid (TRPV)1 and sensory neuropeptides in the guinea-pig respiratory system. Neuroscience 141:1533–1543
34. Banner KH, Igney F, Poll C (2011) TRP channels: emerging targets for respiratory disease. Pharmacol Ther 130:371–384
35. Groneberg DA, Niimi A, Dinh QT, Cosio B, Hew M, Fischer A, Chung KF (2004) Increased expression of transient receptor potential vanilloid-1 in airway nerves of chronic cough. Am J Respir Crit Care Med 170:1276–1280
36. Mitchell JE, Campbell AP, New NE, Sadofsky LR, Kastelik JA, Mulrennan SA, Compton SJ, Morice AH (2005) Expression and characterization of the intracellular vanilloid receptor (TRPV1) in bronchi from patients with chronic cough. Exp Lung Res 31:295–306
37. Doherty MJ, Mister R, Pearson MG, Calverley PM (2000) Capsaicin induced cough in cryptogenic fibrosing alveolitis. Thorax 55:1028–1032
38. Agopyan N, Bhatti T, Yu S, Simon SA (2003) Vanilloid receptor activation by 2- and 10-microm particles induces responses leading to apoptosis in human airway epithelial cells. Toxicol Appl Pharmacol 192:21–35
39. Agopyan N, Head J, Yu S, Simon SA (2004) TRPV1 receptors mediate particulate matter-induced apoptosis. Am J Physiol Lung Cell Mol Physiol 286:L563–L572
40. Materazzi S, Nassini R, Gatti R, Trevisani M, Geppetti P (2009) Cough sensors. II. Transient receptor potential membrane receptors on cough sensors. Handb Exp Pharmacol 49–61
41. McLeod RL, Fernandez X, Correll CC, Phelps TP, Jia Y, Wang X, Hey JA (2006) TRPV1 antagonists attenuate antigen-provoked cough in ovalbumin sensitized guinea pigs. Cough 2:10
42. Reilly CA, Taylor JL, Lanza DL, Carr BA, Crouch DJ, Yost GS (2003) Capsaicinoids cause inflammation and epithelial cell death through activation of vanilloid receptors. Toxicol Sci 73:170–181
43. Thomas KC, Sabnis AS, Johansen ME, Lanza DL, Moos PJ, Yost GS, Reilly CA (2007) Transient receptor potential vanilloid 1 agonists cause endoplasmic reticulum stress and cell death in human lung cells. J Pharmacol Exp Ther 321:830–838

44. Helyes Z, Elekes K, Nemeth J, Pozsgai G, Sandor K, Kereskai L, Borzsei R, Pinter E, Szabo A, Szolcsanyi J (2007) Role of transient receptor potential vanilloid 1 receptors in endotoxin-induced airway inflammation in the mouse. Am J Physiol Lung Cell Mol Physiol 292:L1173–L1181
45. Liedtke W, Choe Y, Marti-Renom MA, Bell AM, Denis CS, Sali A, Hudspeth AJ, Friedman JM, Heller S (2000) Vanilloid receptor-related osmotically activated channel (VR-OAC), a candidate vertebrate osmoreceptor. Cell 103:525–535
46. Strotmann R, Harteneck C, Nunnenmacher K, Schultz G, Plant TD (2000) OTRPC4, a nonselective cation channel that confers sensitivity to extracellular osmolarity. Nat Cell Biol 2: 695–702
47. Watanabe H, Vriens J, Suh SH, Benham CD, Droogmans G, Nilius B (2002) Heat-evoked activation of TRPV4 channels in a HEK293 cell expression system and in native mouse aorta endothelial cells. J Biol Chem 277: 47044–47051
48. Willette RN, Bao W, Nerurkar S, Yue TL, Doe CP, Stankus G, Turner GH, Ju H, Thomas H, Fishman CE, Sulpizio A, Behm DJ, Hoffman S, Lin Z, Lozinskaya I, Casillas LN, Lin M, Trout RE, Votta BJ, Thorneloe K, Lashinger ES, Figueroa DJ, Marquis R, Xu X (2008) Systemic activation of the transient receptor potential vanilloid subtype 4 channel causes endothelial failure and circulatory collapse: Part 2. J Pharmacol Exp Ther 326:443–452
49. Alvarez DF, King JA, Weber D, Addison E, Liedtke W, Townsley MI (2006) Transient receptor potential vanilloid 4-mediated disruption of the alveolar septal barrier: a novel mechanism of acute lung injury. Circ Res 99: 988–995
50. Delany NS, Hurle M, Facer P, Alnadaf T, Plumpton C, Kinghorn I, See CG, Costigan M, Anand P, Woolf CJ, Crowther D, Sanseau P, Tate SN (2001) Identification and characterization of a novel human vanilloid receptor-like protein, VRL-2. Physiol Genomics 4:165–174
51. Hamanaka K, Jian MY, Townsley MI, King JA, Liedtke W, Weber DS, Eyal FG, Clapp MM, Parker JC (2010) TRPV4 channels augment macrophage activation and ventilator-induced lung injury. Am J Physiol Lung Cell Mol Physiol 299:L353–L362
52. Andrade YN, Fernandes J, Vazquez E, Fernandez-Fernandez JM, Arniges M, Sanchez TM, Villalon M, Valverde MA (2005) TRPV4 channel is involved in the coupling of fluid viscosity changes to epithelial ciliary activity. J Cell Biol 168:869–874
53. Hamanaka K, Jian MY, Weber DS, Alvarez DF, Townsley MI, Al-Mehdi AB, King JA, Liedtke W, Parker JC (2007) TRPV4 initiates the acute calcium-dependent permeability increase during ventilator-induced lung injury in isolated mouse lungs. Am J Physiol Lung Cell Mol Physiol 293:L923–L932
54. Jia Y, Wang X, Varty L, Rizzo CA, Yang R, Correll CC, Phelps PT, Egan RW, Hey JA (2004) Functional TRPV4 channels are expressed in human airway smooth muscle cells. Am J Physiol Lung Cell Mol Physiol 287:L272–L278
55. Zhu G, Gulsvik A, Bakke P, Ghatta S, Anderson W, Lomas DA, Silverman EK, Pillai SG (2009) Association of TRPV4 gene polymorphisms with chronic obstructive pulmonary disease. Hum Mol Genet 18:2053–2062

Chapter 15

Hypoxia-Dependent TRP Channel Function in Pulmonary Arterial Smooth Muscle Cells

Beate Fuchs, Hermann Kalwa, Norbert Weissmann, Thomas Gudermann, and Alexander Dietrich

Abstract

Hypoxic pulmonary vasoconstriction (HPV) is an essential physiological mechanism of the lung which matches perfusion to ventilation to optimize gas exchange. Pulmonary arterial smooth muscle cells (PASMC) are the effector and possibly also the sensor cells of HPV. Contraction of these cells under hypoxia is induced by an increase in intracellular Ca^{2+} concentration ($[Ca^{2+}]_i$). To investigate details of the hypoxia-induced increase in $[Ca^{2+}]_i$, we use a live cell imaging procedure with fura-2 in isolated wild-type (WT) and gene-deficient mouse PASMC. We also describe here the manganese-quenching method to determine the role of nonselective cation influx through transient receptor potential (TRP) channels in response to hypoxia.

Key words: Hypoxic pulmonary vasoconstriction, Hypoxia, Lung, Pulmonary artery smooth muscle cells, Calcium imaging, Manganese-quenching, Fura-2, TRPC6

1. Introduction

Oxygen is indispensable for generation of ATP during the energy fixation process in mitochondria. To prevent life-threatening arterial hypoxemia under pathophysiological conditions of severe regional alveolar hypoxia and to optimize pulmonary gas exchange under physiological conditions within the lung, a mechanism known as hypoxic pulmonary vasoconstriction (HPV) is essential. HPV redirects blood flow from poorly or nonventilated to well ventilated areas of the lung to assure optimal uptake of oxygen during ventilation (1–4).

Although this principle is well known, the underlying oxygen sensor and signal transduction pathway(s) has/have not been fully elucidated yet. Pulmonary arterial smooth muscle cells (PASMC) have been suggested as sensor and effector cell at least for acute

Arpad Szallasi and Tamás Bíró (eds.), *TRP Channels in Drug Discovery: Volume I*, Methods in Pharmacology and Toxicology, DOI 10.1007/978-1-62703-077-9_15, © Springer Science+Business Media, LLC 2012

HPV, lasting seconds to some minutes since isolated PASMC respond to hypoxia by an increase of intracellular Ca^{2+} concentration ($[Ca^{2+}]_i$) and subsequent contraction (5–11). This increase in $[Ca^{2+}]_i$ can be caused by Ca^{2+} release from intracellular stores or Ca^{2+} influx from extracellular space. Besides voltage-operated Ca^{2+} channels, receptor-operated cation channels, which are activated by agonists, or store-operated cation channels, which are activated by depletion of intracellular Ca^{2+} stores, are identified to mediate the entry of cations like Ca^{2+} and Na^+ (12). Both, receptor-operated cation channels and store-operated cation channels, can be formed by transient receptor potential (TRP) proteins which are divided in six subfamilies (13, 14). In 1995, the first TRP subfamily was identified in mammals, called the *c*lassical or *c*anonical TRP (TRPC) subfamily consisting of seven members (14). One member of the TRPC family, TRPC6, is highly expressed in the lung and in many smooth muscle tissues and plays an important role for regulation of smooth muscle contractility (14).

Although the analysis of intact lungs has the advantage of closely reflecting the in vivo situation, investigations of isolated PASMC allow a more detailed insight into cell-specific processes. Thus, a combination of both has the potential to significantly improve our knowledge about HPV.

HPV occurs predominantly in the precapillary region of the pulmonary vascular tree (9, 15). Therefore, the regions from which PASMC are isolated (e.g., precapillary versus main pulmonary arterial) to investigate HPV are very important. Indeed, PASMC from the distal pulmonary arterial region show a stronger contraction in response to hypoxia or pharmacological vasoconstrictors than from the proximal region (16, 17). Even more contradictory, hypoxia can cause dilatation instead of vasoconstriction of proximal or larger pulmonary arteries (18, 19).

For imaging of changes in free $[Ca^{2+}]_i$, different fluorescent indicators are available. One widely used selective Ca^{2+} indicator is fura-2 acetoxymethylester (fura-2 AM). Compared to nonratiometric Ca^{2+} dyes, e.g., fluo-3 or fluo-4, fura-2 as a ratiometric Ca^{2+} dye is independent of loading efficiency or other factors influencing the fluorescent signal efficiency (e.g., temperature).

Fura-2 AM is membrane permeable due to its lipophilic ester groups. After crossing the cell membrane, cytosolic esterases cleave the ester groups generating a hydrophilic and membrane-impermeable fura-2. In the cytosol, fura-2 is a selective Ca^{2+} chelator and its excitation spectrum depends on changes in $[Ca^{2+}]_i$. Increasing $[Ca^{2+}]_i$ leads to an increase of fluorescence intensity after excitation at 340 nm and a decrease of fluorescence intensity after excitation at 380 nm. Therefore, the ratio of fluorescence intensity after excitation at 340 nm to fluorescence intensity after excitation at 380 nm is proportional to $[Ca^{2+}]_i$. Ratiometric indicators like fura-2 have the advantage to depend solely on changes in the indicator-specific ion, independently of loading efficiencies.

Moreover, fura-2 also exhibits its isosbestic point at an excitation wavelength of 360 nm where the fluorescence is independent of $[Ca^{2+}]_i$. In contrast to Ca^{2+}, Mn^{2+} has a much higher affinity to fura-2 and leads to a quenching effect after binding and thereby to a reduction of fluorescence intensity. For this reason, a method called manganese-quenching allows the analysis of Mn^{2+} ion influx from the extracellular into the intracellular space through nonselective cation channels after extracellular application of Mn^{2+}. A decrease of fura-2 fluorescence intensity at the isosbestic point is therefore directly correlated to channel activity.

Against this background, we here describe isolation technique of primary PASMC from the precapillary region of the lung and the investigation of hypoxia-induced modulations in $[Ca^{2+}]_i$ and nonselective channel activity using primary PASMC from wild-type (WT) and gene-deficient mouse models.

2. Materials

2.1. Isolation of Pulmonary Arterial Smooth Muscle Cells

For isolation and culturing of PASMC from mice, the following materials and solutions are needed.

2.1.1. Isolation and Culture Procedure of PASMC

Materials

1. Low-melting-point agarose, iron-oxide, collagenase type IV, and collagen solution (type 1) (Sigma-Aldrich, Steinheim, Germany)
2. Medium 199 (Invitrogen, Karlsruhe, Germany)
3. DPBS and penicillin/streptomycin (PAN Biotech, Aidenbach, Germany)
4. Fetal calf serum (FCS) (PAA Laboratories, Coelbe, Germany)
5. Smooth muscle cell growth medium 2 (Promocell, Heidelberg, Germany)
6. Aqua dest. and 0.9% NaCl solution (B.Braun, Melsungen, Germany)
7. Tracheal cannula: intubation cannula (Hugo Sachs, March-Hugstetten, Germany) connected to a tube connector (1.6 × 2.3 mm, Novodirect, Kehl/Rhein, Germany) and then connected to a combifix adapter (B.Braun) by tubes (Novodirect)
8. Pulmonary arterial cannula from IL-1 setup (Hugo Sachs) connected to a tube connector (1.6 × 2.3 mm, Novodirect) and then connected to a combifix adapter (B.Braun) by tubes (Novodirect)

9. Twines (Coats, Kenzingen, Germany)
10. Crocodile clamp (Becker GmbH Netproshop, Röthenbach, Germany) in a tool holder with magnetic base (Kanetec, Tokyo, Japan) on a tripod base-plate (Roth, Karlsruhe, Germany) for precise positioning of the pulmonary arterial cannula
11. Three 2 ml syringes (B.Braun)
12. Magnetic device (Dynal® MPC™-1) (Invitrogen)
13. 15-gauge needle (Dispomed Witt, Gelnhausen, Germany) and 18-gauge needle (HMD Healthcare LTD., Horsham, UK)
14. Round glass coverslips (24 mm diameter, Menzel, Braunschweig, Germany) (see Note 1)
15. Magnetic plate, e.g., for pin board (BR Technik Kontor, Satrup, Germany) (see Note 2)

Solutions

1. 1% low-melting-point agarose + 1% penicillin/streptomycin in 3 ml Medium 199 and 0.5% low-melting-point agarose + 0.5% iron oxide + 1% penicillin/streptomycin in 3 ml Medium 199 must be prepared freshly at the day of isolation
2. 7 ml of 80 U/ml collagenase type IV in Medium 199 must be prepared freshly at the day of isolation
3. 50 ml of 10% FCS + 1% penicillin/streptomycin in Medium 199
4. Collagen solution (type 1) diluted 1:10 in aqua dest. to receive 0.01% collagen solution
5. Culture medium: smooth muscle cell growth medium 2 + 15% FCS + 1% penicillin/streptomycin

2.1.2. Characterization of PASMC

For characterization of the isolated cells from mouse lungs by immunostaining, the following materials and solutions are needed.

Materials

1. 8-well Permanox® slide chambers (Nalgene Nunc, Wiesbaden, Germany)
2. Anti-α-smooth muscle actin, DAPI, collagen solution (type 1), acetone, and methanol (Sigma-Aldrich)
3. Anti-myosin heavy chain, anti-von Willebrand factor, FITC-conjugated secondary antibody, and Dako fluorescent mounting medium (Dako Diagnostic, Hamburg, Germany)
4. Bovine serum albumin (BSA) (PAA Laboratories)
5. DPBS (PAN Biotech)
6. Coverglass (24 × 50 mm) (Menzel)

Solutions

1. Washing buffer: 0.1% BSA in DPBS
2. Blocking buffer: 3% BSA in DPBS

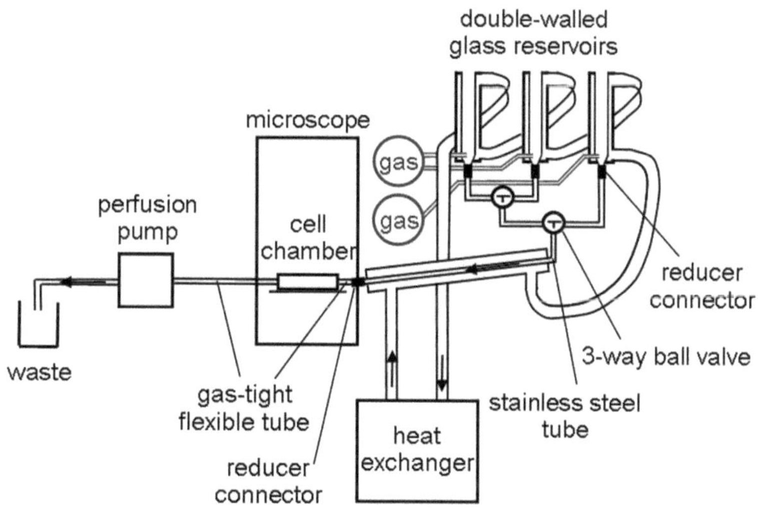

Fig. 1. Scheme of the experimental setup for live cell imaging of pulmonary arterial smooth muscle cells (PASMC) under hypoxic conditions using a perfusion system. PASMC grown on a glass coverslip are placed in a cell chamber on a microscope connected to a perfusion system. The perfusion solution is forwarded to the cell chamber from double-walled glass reservoirs through stainless steel tubes and finally through a short gas-tight flexible tube to the cell chamber. Double-walled glass reservoirs are filled with perfusion solution. For hypoxic condition the perfusion solution is gassed with 100% N_2, and for normoxic condition with 21% O_2, in N_2. Air bubbles in the perfusion system have to be avoided. Stainless steel three-way valves are used to allow immediate switching between normoxic and hypoxic perfusion solutions. Double-walled glass reservoirs as well as stainless steel tubes between three-way valves and gas-tight flexible tube are heated to 32°C by a heat exchanger in a counter flow principle. Perfusion rate is set to 0.5 ml/min by use of a perfusion pump downstream of the perfusion chamber. Schema is modified from [23].

2.2. Live Cell Imaging of PASMC with Fura-2

2.2.1. Hypoxic Perfusion

For a custom-made perfusion system for hypoxic perfusion of the cells, the following materials are needed. A scheme of the perfusion system is shown in Fig. 1.

Materials

1. Three double-walled glass reservoirs with 38 mm inner diameter, 56 mm outer diameter, and 150 ml volume. Dimension of outlet is 4 mm inner diameter and 8 mm outer diameter. Gas inlet should be positioned as deeply as possible into the reservoir to create a small hole for bubbling.
2. Two stainless steel three-way ball valves (1/8 inch tube fitting), three stainless steel reducer connectors (from 1/4 inch to 1/8 inch), one stainless steel reducer connector (from 1/8 inch to 1/16 inch), one stainless steel tube (1/8 inch outer diameter, 0.055 inch inner diameter) (B.E.S.T. Ventil+Fitting, Frankfurt, Germany)
3. Heat exchanger (Julabo, Seelbach, Germany) (see Note 4)
4. Gas-tight flexible tube (0.89 mm inner diameter) (Novodirect)
5. Gas-tight flexible tubes (ca. 6 mm and ca. 18 mm inner diameter) (Masterflex, Gelsenkirchen, Germany)

6. Custom-made temperature-controlled closed chamber for round glass coverslips (diameter 24 mm)
7. Perfusion pump for a perfusion rate of 0.5 ml (Novodirect)
8. N_2 for bubbling the perfusion solution to achieve hypoxic conditions, and gas mixture of 21% O_2 and rest N_2 for bubbling normoxic perfusion solution (Airliquide, Siegen, Germany)

2.2.2. Preparation of PASMC

To prepare PASMC grown on glass coverslips (see Note 1) for live cell imaging with fura-2 (see Note 3), the following materials and solutions are needed.

Materials

1. Fura-2 AM (Invitrogen) diluted in DMSO (1 mM stock) (Merck, Darmstadt, Germany); aliquots are kept in the dark at -20°C
2. Bovine serum albumin (BSA) (PAA Laboratories)
3. KCl, $CaCl_2$, $MgCl_2$, and glucose (Sigma-Aldrich)
4. NaCl and Hepes (Roth, Karlsruhe, Germany)

Solutions

1. Hepes-Ringer solution: 5.6 mM KCl, 136.4 mM NaCl, 1 mM $MgCl_2$, 2.2 mM $CaCl_2$, 5 mM glucose, 10 mM Hepes, adjusted to pH 7.4
2. 5 μM fura-2 in Hepes-Ringer solution with 0.1% BSA must prepared freshly at the day of the experiments and kept in the dark

2.2.3. Measurement of Intracellular Ca^{2+} Concentration and Manganese-Quenching with Fura-2

After loading the cells with fura-2 (see Note 3), the following materials are needed for measurement of $[Ca^{2+}]_i$ and manganese-quenching experiments.

Materials

1. Microscopic setup consisting of a microscope (IX70 WI, Olympus, Hamburg, Germany) connected to a monochromator (polychrome II, TILL Photonics, Gräfelfing, Germany); fura-2 filter set (TILL Photonics); 20x oil objective (0.80 numerical aperture) (Olympus, Hamburg, Germany); camera (IMAGO CCD camera, TILL Photonics); personal computer with the software TILLvisION v3.3 (TILL Photonics)
2. Glass coverslips (28 mm diameter) (Schilder, Giessen, Germany)
3. Endothelin-1 (Merck)
4. Ethylene glycol-bis(2-aminoethylether)-tetraacetic acid, nicardipine, and $MnCl_2$ (see Note 5) (Sigma-Aldrich)

3. Methods

3.1. Isolation of PASMC

3.1.1. Isolation and Culture Procedure of PASMC

The following protocol describes the procedure for PASMC isolation and culture from mice by a technique based on magnetic separation of intrapulmonary arteries. The protocol is a modification of the method from Waypa et al. (20), originally described by Marshall et al. (21].

1. 0.5% agarose-iron oxide solution and 1% agarose solution (see Sect. 2.1.1) should be heated to 70–80°C to dissolve the low-melting-point agarose. Then, the solutions should be kept at 40°C in syringes.
2. Collagenase solution (see Sect. 2.1.1) is kept in a water bath at 37°C.
3. 20 ml DPBS is cooled on ice.
4. Sacrificed heparinized mice are fixed in a dorsal position and the cervical, thoracic, and abdominal coat is removed.
5. Dissect the trachea free of surrounding tissue, put a twine around the trachea, open the trachea by cross incision, and introduce the tracheal cannula with adapter (see Sect. 2.1.1) which is then fixed by the twine.
6. The abdominal wall must cut open parallel to the costal arch; dissect the diaphragm edgeless to prevent destruction of the lung. Then, after a median cut along the sternum open the thorax completely and bend and fix the ribs.
7. Put a twine around the pulmonary artery and aorta ascendens.
8. Connect the pulmonary arterial cannula with the adapter (see Sect. 2.1.1) to a 2 ml syringe filled completely with DPBS. Pay attention that both are free from air-bubbles. Fix the cannula with adapter in the crocodile clamp in a tool holder with magnetic base on a tripod base-plate.
9. After removing thymus and pericardium, position the pulmonary arterial cannula into the main pulmonary artery via an incision in the right ventricle of the heart and knot the twine around the pulmonary artery and aorta ascendens.
10. After incision of the left ventricle, flush the pulmonary vasculature slowly, first with 3 ml DPBS, then after changing the syringe (avoid air bubbles) with 2–3 ml of the 40°C mixture consisting of 0.5% low-melting-point agarose, 0.5% iron oxide, and 1% penicillin/streptomycin in Medium 199. The iron particles do not pass the capillaries and therefore, accumulate in the precapillary arteries. The slow perfusion of DPBS as well as of the mixture is important and critical to keep the vessels intact and to maintain the mixture in the pulmonary arteries.
11. Remove the cannula from the pulmonary artery and knot the twine immediately.

12. Fill the airways via the tracheal cannula carefully with the 40°C warm mixture consisting of 1% low-melting-point agarose and 1% penicillin/streptomycin in Medium 199. Cut the trachea above the cannula, remove the heart-lung en bloc from the chest cavity and put it in ice-cold DPBS (from step 3) to gel the agarose. Thereby, iron particles are fixed in the precapillary pulmonary arteries.
13. After 15 min dissect the lung lobes under sterile conditions, put them in 4 ml DPBS in a 50 ml sterile tube, mince thoroughly until lung tissue pieces have a size of around 1 mm^3 (needs 5–10 min).
14. Fill the tube to 30 ml with DPBS and put it in the magnetic device to separate lung tissue pieces with iron particles. After removing DPBS, repeat this washing step two times.
15. Re-suspend lung tissue pieces in preheated collagenase solution (from step 2) and incubate in an atmosphere of 5% CO_2 at 37°C for 1 h.
16. Coat six glass coverslips, each in 35 mm dish, with 0.01% collagen solution for at least 1 h at room temperature.
17. Prewarm culture medium and 10% FCS + 1% penicillin/streptomycin in Medium 199 (see Sect. 2.1.1) in a water bath (37°C).
18. To remove extravascular tissue, force lung tissue pieces in collagenase-solution first through a 15-gauge needle and subsequently through an 18-gauge needle after a 1-h incubation time (from step 15). Transfer the extract into a plastic tube filled with 10 ml preheated Medium 199 containing 10% fetal calf serum and 1% penicillin/streptomycin. Expose the solution in the plastic tube to the magnetic device and wash three times with the same solution.
19. Remove collagen solution from glass coverslips and wash once with 0.9% NaCl solution.
20. Add 3 ml preheated culture medium to isolated lung tissue pieces.
21. Place 500 μl of the lung tissue pieces in culture medium in the middle of a coverslips in a cell culture dish on a magnetic plate and carefully add 1 ml culture medium.
22. Carefully transfer dishes into the incubator (37°C and 5% CO_2). PASMC will grow out of isolated lung tissue pieces and are ready to use for live cell imaging after 3–5 days (see Note 1). Do not move the dishes during this time (see Note 2).

3.1.2. Characterization of PASMC

Isolated cells can be characterized by indirect immunofluorescence staining. The following protocol describes the respective staining procedure.

For characterization of PASMC, cells are stained with primary smooth muscle-specific antibodies (anti-α-smooth muscle actin and anti-smooth muscle myosin heavy chain). To exclude endothelial cells, cells are stained with a primary endothelium-specific antibody (anti-von Willebrand factor). A FITC-conjugated secondary antibody is used and nuclear staining is done with DAPI. To identify a staining by the secondary antibody itself, primary antibodies are omitted in control experiments.

All steps are done at room temperature.

1. Coat an 8-well Permanox® slide chamber with 0.01% collagen solution (type 1) for 1 h at room temperature, seed freshly isolated lung tissue pieces (as described in Sect. 3.1.1) in these wells and incubate them for 5 days.
2. Wash the cells with warm DPBS and fix as well as permeabilize them with an ice-cold mixture of acetone and methanol (proportion 1:1) for 5 min at room temperature.
3. Wash the cells with washing buffer (0.1% BSA) four times and block unspecific binding sites with blocking buffer (3% BSA) for 30 min.
4. Wash the cells three times, dilute primary antibodies in washing buffer and apply for 1 h.
5. Wash the cells four times, dilute secondary antibody in washing buffer and apply for 1 h in the dark.
6. Wash the cells four times, dilute DAPI in washing buffer and apply for 5 min in the dark.
7. Wash the cells three times and after removing the chambers, cover the slide with Dako fluorescent mounting medium and then with a coverglass.
8. Take fluorescent images using a fluorescent microscope.

Figure 2 shows a recently published example of stained PASMC (22). Images were taken with a fluorescent microscope (Leica DMLA with camera Leica DC300FX) and the software Leica QWin V3 (Leica Microsystems, Wetzlar, Germany). Excitation of FITC-conjugated secondary antibody was performed by the use of a 450-490 nm filter. The maximum of the emitted light was at 530 nm. The filter for excitation of DAPI had a band width of 340-380 nm, and the maximum of the emission spectrum was at 461 nm.

3.2. Live Cell Imaging of PASMC with fura-2

3.2.1. Hypoxic Perfusion

For the analysis of hypoxia-induced ion influx by live cell imaging, hypoxic solution is perfused through a closed chamber with PASMC.

The following protocol describes the setup of a perfusion system for application of hypoxic solutions to PASMC (22). A scheme of the perfusion system is given in Fig. 1.

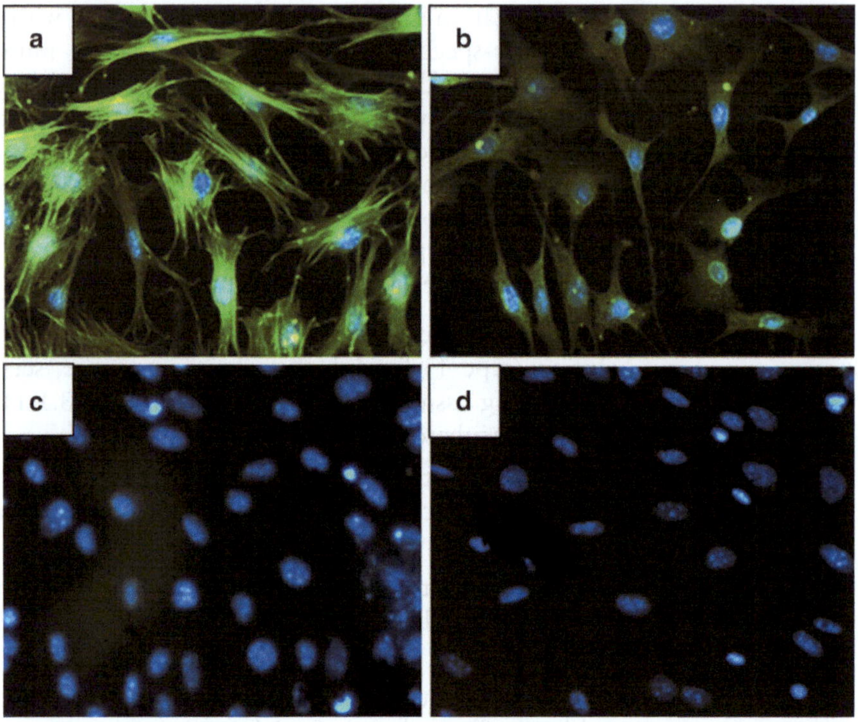

Fig. 2. Characterization of pulmonary arterial smooth muscle cells (PASMC) cultured for five days. PASMC are stained with primary antibodies (**a**) against α-smooth muscle actin and (**b**) against smooth muscle myosin heavy chain. To exclude the presence of endothelial cells, primary antibody against von Willebrand factor is used (**c**). All stainings are performed with FITC-conjugated secondary antibody (*green*). Nuclear staining is done with DAPI (*blue*). To exclude staining by the secondary antibody, primary antibodies are omitted in control experiments (**d**). Images are taken at a magnification of 400-fold. The example is taken from [22–24].

1. Connect each of three double-walled glass reservoirs to a reducer connector (from 1/4 inch to 1/8 inch) by a gas-tight flexible tube (ca. 6 mm inner diameter).
2. Connect two reducer connectors which are side by side to one three-way ball valve by stainless steel tubes. Connect this three-way ball valve and the third reducer connector to the second three-way ball valve by stainless steel tubes. Keep the stainless steel tubes as short as possible.
3. Fix one side of a stainless steel tube in the outlet of the second three-way ball valve, put the other side through two waterproof T-pieces which are connected by a tube (approx. 18 mm inner diameter), and finally, connect it to a reducer connector (from 1/8 to 1/16 inch).
4. Connect this reducer connector to the closed chamber on the microscope by a gas-tight flexible tube (0.89 mm inner diameter). Keep the gas-tight flexible tube as short as possible.
5. Connect the T-piece which is closed to the microscope with the heat exchanger and the other T-piece to the double-walled glass reservoirs one after the other and then back to the heat

exchanger. Subsequently, water for temperature control is pumped in counter flow principle.

6. Connect cell chamber downstream to the perfusion pump by gas-tight flexible tubes (0.89 mm inner diameter).

The following protocol describes the installation of the perfusion system for the production of hypoxic solutions.

1. Fill Hepes-Ringer solution in double-walled glass reservoirs.
2. To generate a hypoxic perfusion solution bubble Hepes-Ringer solution with 100% N_2 for at least 1 h. For normoxic perfusion bubble Hepes-Ringer solution with 21% O_2, rest N_2.
3. The temperature of the Hepes-Ringer solution in the perfusion system should be kept at 32°C (see Note 4).

3.2.2. Preparation of PASMC for Fura-2 Measurements

For live cell imaging, we use PASMC cultured on collagen-coated glass coverslips for 3–5 days after isolation (see Note 1).

The following protocol describes the loading procedure of PASMC with fura-2 as well as the preparation of PASMC for live cell imaging.

1. Wash PASMC twice with Hepes-Ringer solution.
2. Apply 1 ml of 5 μM fura-2 in Hepes-Ringer solution with 0.1% BSA for 1 h at 37°C.
3. Wash PASMC twice with Hepes-Ringer solution.
4. Place the glass coverslip with PASMC into the heated cell chamber (32°C) and close the chamber by a second glass coverslip (28 mm diameter). Connect the chamber with the perfusion system by gas-tight flexible tubes (0.89 mm inner diameter) and fill the cell chamber with Hepes-Ringer solution free of air bubbles. Start perfusion pump with a perfusion rate of 0.5 ml/min.
5. Using a 20× oil objective, select fura-2-loaded PASMC in the display window of the TILLvisION v3.3 software.

The analysis of $[(Ca^{2+})_i]$ and the quantification of channel activity by manganese-quenching is described separately (see below).

3.2.3. Quantification of the Intracellular Ca^{2+} Concentration by Fura-2

For the analysis of $[Ca^{2+}]_i$ with fura-2, we prepare PASMC as described in Sect. 3.2.2. To obtain hypoxic conditions for PASMC, we perfuse PASMC in a closed cell chamber with a hypoxic perfusion solution as described in Sect. 3.2.1 (22–24).

Analysis of the Intracellular Ca^{2+} Concentration Under Hypoxic Conditions

For the quantification of $[Ca^{2+}]_i$ under hypoxic conditions, the perfusion system is prepared as described in Sect. 3.2.1.

Hepes-Ringer solution is filled in one double-walled glass reservoir and Hepes-Ringer solution containing 4 nM ET-1 is filled in the other two double-walled glass reservoirs (for normoxic and hypoxic conditions).

1. Select and mark fura-2-loaded PASMC (see Note 3) as region of interest after excitation with 340 nm and 380 nm in one of both images by the TILLvisION v3.3 software.
2. Excite fura-2 dye in the PASMC by 340 nm and 380 nm every 6 s and detect the emitted fluorescence signals with a camera and by the TILLvisION v3.3 software.
3. Switch perfusion to normoxic Hepes-Ringer solution containing 4 nM ET-1 as a priming stimulus. ET-1 is applied in the cell chamber 4.5 min after starting the experiment.
4. Switch perfusion to hypoxic Hepes-Ringer solution containing 4 nM ET-1. Onset of hypoxia in the cell chamber starts after additional 7.5 min and results in a pO_2 of≈18 mmHg.
5. The experiment is finished after additional 25 min.
6. Calculate ratios of the fluorescence signal intensity after excitation at 340 nm and after excitation at 380 nm (ratio (F340/F380)) for the marked PASMC (region of interest, ROI). First, create a ROI of the cell-free background in the first recorded image of one wavelength and copy it into the first recorded image of the second wavelength.
7. This ROI is used for background subtraction during calculation of the ratios (F340/F380) of all marked PASMC for each time-point. If the live cell imaging setup is calibrated to $[Ca^{2+}]_i$, ratios can be directly calculated as $[Ca^{2+}]_i$ in TILLvisION v3.3 software.

Figure 3a (23) is an example for $[Ca^{2+}]_i$ measurements in PASMC under normoxic and hypoxic conditions. In contrast to normoxia, hypoxia induced an increase of $[Ca^{2+}]_i$.

To address the role of TRPC channels in hypoxia-induced increase of $[Ca^{2+}]_i$, PASMC from TRPC6$^{-/-}$ mice are compared to WT mice (Fig. 3b) (22–24).

Role of Intracellular Ca^{2+} Stores Under Hypoxic Conditions

In general, increases of $[Ca^{2+}]_i$ can be generated by Ca^{2+} release from intracellular stores or by the Ca^{2+} influx from the extracellular space. To distinguish between both sources, experiments can be performed without extracellular Ca^{2+} to analyze only the release from intracellular Ca^{2+} stores.

For the quantification of $[Ca^{2+}]_i$ under hypoxic conditions, the perfusion system is prepared as described in Sect. 3.2.1.

Hepes-Ringer solution with Ca^{2+}, Ca^{2+}-free Hepes-Ringer solution containing 2 mM of Ca^{2+}-chelator EGTA for normoxic conditions and Ca^{2+}-free Hepes-Ringer solution with 2 mM EGTA and 4 nM ET-1 for hypoxic condition are filled into the double-walled glass reservoirs. 4 nM ET-1 is added to normoxic Ca^{2+}-free Hepes-Ringer solution containing 2 mM EGTA during the experiment. The experiment is done as described in Sect. 3.2.3.1 with the solutions mentioned above.

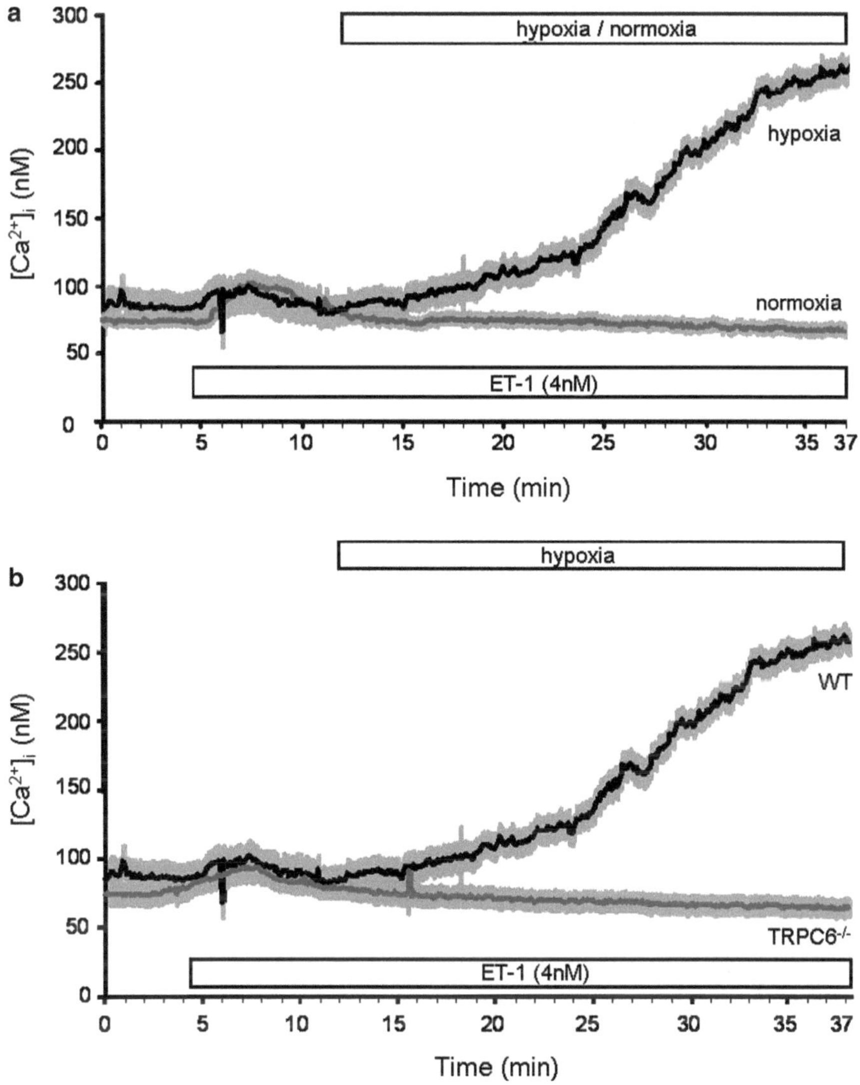

Fig. 3. Hypoxia-induced changes of the intracellular Ca^{2+} concentration ($[Ca^{2+}]_i$) in pulmonary arterial smooth muscle cells (PASMC) after priming with endothelin-1 (ET-1). In these examples primary cultured PASMC (3–5 days after isolation) on coated glass coverslips are loaded with fura-2 and $[Ca^{2+}]_i$ is analyzed by fluorescence live cell imaging. Application of ET-1 to PASMC starts after 4.5 min and is present during the remainder of the experiment. Exposure of PASMC from WT mice to hypoxia starts after additional 7.5 min and was compared to normoxic perfusion of PASMC from WT mice (**a**) or hypoxic perfusion of PASMC from TRPC6$^{-/-}$ mice (**b**). Values are mean (±SEM). Images are taken from [22–24].

As previously published (22–24) and shown in Fig. 4a, hypoxia-induced increase of $[Ca^{2+}]_i$ is not detectable in the absence of extracellular Ca^{2+} and therefore not dependent on Ca^{2+}-release from intracellular stores.

Involvement of Voltage-Operated Ca^{2+} Channels Under Hypoxic Conditions

One concept for the mechanism underlying HPV proposes a hypoxia-induced inhibition of voltage-gated K^+ (K_V) channels leading to membrane depolarization and Ca^{2+} entry through voltage-operated Ca^{2+} channels (25). Therefore, we investigated the role of voltage-operated Ca^{2+} channels for hypoxia-induced increase of $[Ca^{2+}]_i$ by inhibition of these channels with nicardipine.

The experimental procedure is the same as described in Sect. 3.2.3.1. Only 10 μM nicardipine was applied additionally to the normoxic and hypoxic Hepes-Ringer solution containing 4 nM ET-1.

As previously published (22–24), Fig. 4b shows that an increase of $[Ca^{2+}]_i$ was suppressed by nicardipine indicating an important role of voltage-operated Ca^{2+} channels for the acute hypoxic

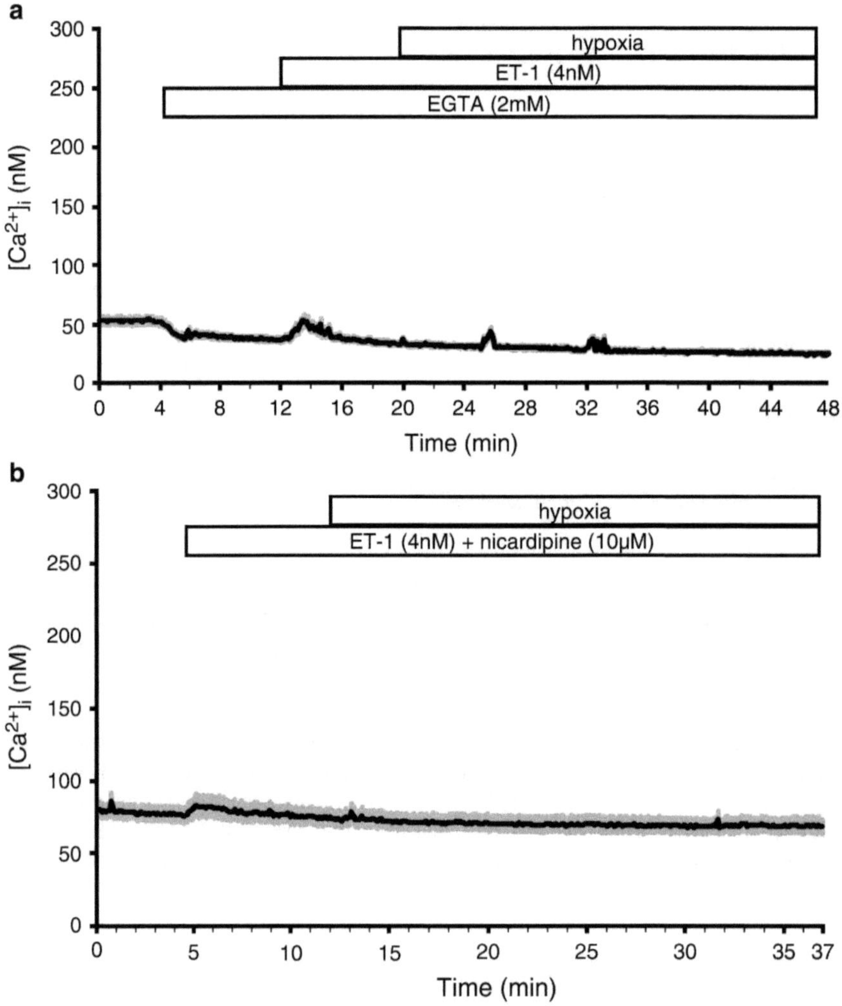

Fig. 4. Hypoxia- and agonist-induced changes of the intracellular Ca^{2+} concentration ($[Ca^{2+}]_i$) during conditions of Ca^{2+}-free extracellular medium and effect of nicardipine on hypoxia-induced changes of $[Ca^{2+}]_i$ in pulmonary arterial smooth muscle cells (PASMC). Primarily cultured PASMC (3–5 days after isolation) from WT mice on coated glass coverslips are loaded with fura-2 and $[Ca^{2+}]_i$ is analyzed by fluorescence live cell imaging. (**a**) Ca^{2+}-free perfusion (with EGTA) of PASMC starts after 4 min and application of ET-1 to PASMC starts after additional 8 min. Both are present during the remaining time of the experiment. Hypoxia is applied after additional 8 min. (**b**) Application of nicardipine and ET-1 to PASMC starts after 4.5 min. Exposure of PASMC to hypoxia starts after additional 7.5 min. Values are mean (±SEM). Images are taken from [22–24].

response of PASMC. As described in the literature, TRPC6 channels are predominantly permeable to Na^+, and only a small percentage of the whole-cell current is caused by Ca^{2+} in the presence of extracellular Na^+ (26). However, Na^+-influx through TRPC6 channels initiate activation of voltage-operated Ca^{2+} channels and subsequent influx of Ca^{2+} which is completely inhibited in TRPC6-deficient PASMC. Since TRPC6 channels, but not voltage-operated Ca^{2+} channels, are permeable for Mn^{2+}, hypoxia-induced influx of ions through TRPC6 channels can be analyzed by a method called manganese-quenching which is described in Sect. 3.2.4.

3.2.4. Analysis of Nonselective Cation Channel Activity Under Hypoxic Conditions with Fura-2 by Manganese-Quenching

To analyze nonselective cation channel activity, manganese-quenching experiments can be performed. In this chapter, we describe manganese-quenching with fura-2 for PASMC from WT mice compared to $TRPC6^{-/-}$ mice under hypoxic conditions to investigate the role of TRPC6 channels for hypoxia-induced cellular response of PASMC.

To obtain hypoxic conditions for PASMC, we perfuse PASMC in a closed cell chamber by hypoxic perfusion solution. First prepare the perfusion system as described in Sect. 3.2.1. Hepes-Ringer solution without additional substances is filled in one double-walled glass reservoir and Hepes-Ringer solution containing 4 nM ET-1 is filled in two double-walled glass reservoirs (for normoxic and hypoxic conditions).

After positioning fura-2-loaded PASMC into the perfusion system as described in Sect. 3.2.2, the manganese-quenching experiments under hypoxic conditions were carried out in the following way:

1. Select and mark fura-2-loaded PASMC as region of interest after excitation with 360 nm in the recorded image by the TILLvisION v3.3 software.
2. Excite fura-2 dye in the PASMC by 360 nm every six seconds and detect emitted fluorescence signals with a camera and record by the TILLvisION v3.3 software.
3. Switch perfusion to normoxic Hepes Ringer solution containing 300 μl $MnCl_2$ (see Note 5) and 4 nM of the priming stimulus ET-1, that $MnCl_2$ and ET-1 are applied to the PASMC 4.5 min after starting the experiment.
4. Switch perfusion to hypoxic Hepes-Ringer solution containing 300 μl $MnCl_2$ (see Note 5) and 4 nM ET-1. Onset of hypoxia in the cell chamber starts after additional 7.5 min and results in a pO_2 of $\approx$ 18 mmHg.
5. The experiment is finished after additional 11 min.
6. Analyze the fluorescence signal intensities after excitation at 360 nm (F360) for the marked PASMC after background subtraction and normalize to the initial value (100%) of each PASMC.

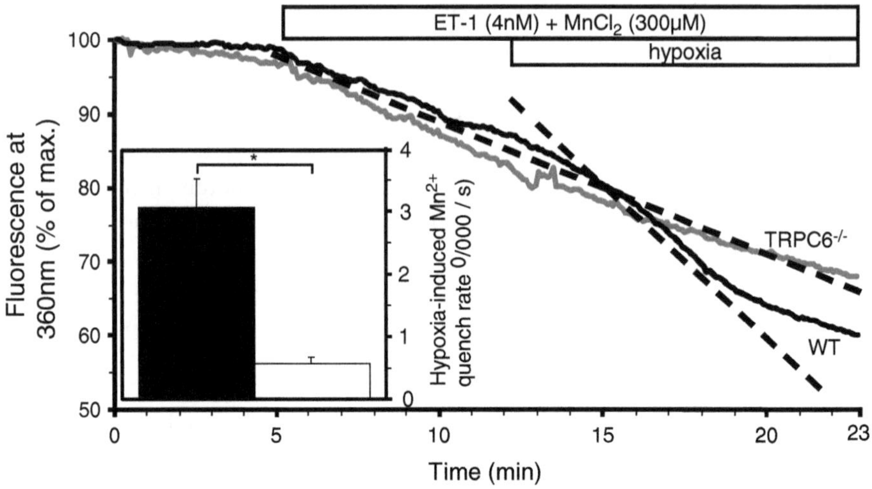

Fig. 5. Hypoxia-induced changes of Mn^{2+} influx in pulmonary arterial smooth muscle cells (PASMC) after priming with endothelin-1 (ET-1). Primary cultured PASMC (3–5 days after isolation) on coated glass coverslips are loaded with fura-2 and manganese-quenching is analyzed by fluorescence live cell imaging. Application of Mn^{2+} and ET-1 to PASMC starts after 4.5 min and is present during the remainder of the experiment. Exposure of PASMC to hypoxia starts after additional 7.5 min. The insert summarizes hypoxia-induced Mn^{2+} quench rates. Differences in the linear range of the Mn^{2+} quench rate after Mn^{2+} addition and after application of hypoxic conditions are calculated for PASMC from WT (*filled bars*) and TRPC6$^{-/-}$ mice (*open bars*). Values are mean (±SEM). *Dashed lines* indicate the slopes of the curves. Images are taken from [22–24].

As recently shown (22–24), fluorescence quenching at 360 nm is significantly increased under hypoxic conditions in PASMC from WT mice while fluorescence at 360 nm was unchanged in PASMC from TRPC6$^{-/-}$ mice under the same conditions (Fig. 5).

4. Notes

1. To minimize morphological and functional changes of cultured PASMC, they should grow directly on coated glass coverslips and should be used freshly (three to five days after isolation).
2. During culturing dishes should not be moved and magnetic plates under the dishes improve the contact of the isolated arteries containing iron particles to the bottom of the cell culture dish.
3. For live cell imaging with fluorescence dyes, efficient loading of cells with fura-2 is essential to receive sufficient high fluorescence signals in comparison to background signals.
4. Pre-warm all solutions to 32°C, because higher temperatures, e.g., 37°C cause stronger bleaching of fura-2.
5. The amount of manganese added in the manganese-quenching experiments needs to be adapted to the cells, because some cell types are more leaky for ions than others.

References

1. Fishman AP (1976) Hypoxia on the pulmonary circulation. How and where it acts. Circ Res 38:221–231
2. Orchard CH, de Sanchez L, Sykes MK (1983) The relationship between hypoxic pulmonary vasoconstriction and arterial oxygen tension in the intact dog. J Physiol 338:61–74
3. Voelkel NF (1986) Mechanisms of hypoxic pulmonary vasoconstriction. Am Rev Respir Dis 133:1186–1195
4. Naeije R, Brimioulle S (2001) Physiology in medicine: importance of hypoxic pulmonary vasoconstriction in maintaining arterial oxygenation during acute respiratory failure. Crit Care 5:67–71
5. Cornfield DN, Stevens T, McMurtry IF, Abman SH, Rodman DM (1993) Acute hypoxia increases cytosolic calcium in fetal pulmonary artery smooth muscle cells. Am J Physiol 265:L53–L56
6. Cornfield DN, Stevens T, McMurtry IF, Abman SH, Rodman DM (1994) Acute hypoxia causes membrane depolarization and calcium influx in fetal pulmonary artery smooth muscle cells. Am J Physiol 266:L469–L475
7. Kato M, Staub NC (1966) Response of small pulmonary arteries to unilobar hypoxia and hypercapnia. Circ Res 19:426–440
8. Salvaterra CG, Goldman WF (1993) Acute hypoxia increases cytosolic calcium in cultured pulmonary arterial myocytes. Am J Physiol 264:L323–L328
9. Staub NC (1985) Site of hypoxic pulmonary vasoconstriction. Chest 88:240S–245S
10. Vadula MS, Kleinman JG, Madden JA (1993) Effect of hypoxia and norepinephrine on cytoplasmic free Ca2+ in pulmonary and cerebral arterial myocytes. Am J Physiol 265: L591–L597
11. Zhang F, Carson RC, Zhang H, Gibson G, Thomas HM III (1997) Pulmonary artery smooth muscle cell [Ca2+]i and contraction: responses to diphenyleneiodonium and hypoxia. Am J Physiol 273:L603–L611
12. Dietrich A, Kalwa H, Fuchs B, Grimminger F, Weissmann N, Gudermann T (2007) In vivo TRPC functions in the cardiopulmonary vasculature. Cell Calcium 42:233–244
13. Ward JP, Robertson TP, Aaronson PI (2005) Capacitative calcium entry: a central role in hypoxic pulmonary vasoconstriction? Am J Physiol Lung Cell Mol Physiol 289:L2–L4
14. Fuchs B, Dietrich A, Gudermann T, Kalwa H, Grimminger F, Weissmann N (2010) The role of classical transient receptor potential channels in the regulation of hypoxic pulmonary vasoconstriction. Adv Exp Med Biol 661:187–200
15. Hillier SC, Graham JA, Hanger CC, Godbey PS, Glenny RW, Wagner WW Jr (1997) Hypoxic vasoconstriction in pulmonary arterioles and venules. J Appl Physiol 82:1084–1090
16. Madden JA, Vadula MS, Kurup VP (1992) Effects of hypoxia and other vasoactive agents on pulmonary and cerebral artery smooth muscle cells. Am J Physiol 263:L384–L393
17. Sham JS, Crenshaw BR Jr, Deng LH, Shimoda LA, Sylvester JT (2000) Effects of hypoxia in porcine pulmonary arterial myocytes: roles of K(V) channel and endothelin-1. Am J Physiol Lung Cell Mol Physiol 279:L262–L272
18. Archer SL, Souil E, nh-Xuan AT, Schremmer B, Mercier JC, El YA, Nguyen-Huu L, Reeve HL, Hampl V (1998) Molecular identification of the role of voltage-gated K+ channels, Kv1.5 and Kv2.1, in hypoxic pulmonary vasoconstriction and control of resting membrane potential in rat pulmonary artery myocytes. J Clin Invest 101:2319–2330
19. Madden JA, Keller PA, Kleinman JG (2000) Changes in smooth muscle cell pH during hypoxic pulmonary vasoconstriction: a possible role for ion transporters. Physiol Res 49: 561–566
20. Waypa GB, Marks JD, Mack MM, Boriboun C, Mungai PT, Schumacker PT (2002) Mitochondrial reactive oxygen species trigger calcium increases during hypoxia in pulmonary arterial myocytes. Circ Res 91:719–726
21. Marshall C, Mamary AJ, Verhoeven AJ, Marshall BE (1996) Pulmonary artery NADPH-oxidase is activated in hypoxic pulmonary vasoconstriction. Am J Respir Cell Mol Biol 15:633–644
22. Weissmann N, Dietrich A, Fuchs B, Kalwa H, Ay M, Dumitrascu R, Olschewski A, Storch U, Schnitzler M, Ghofrani HA, Schermuly RT, Pinkenburg O, Seeger W, Grimminger F, Gudermann T (2006) Classical transient receptor potential channel 6 (TRPC6) is essential for hypoxic pulmonary vasoconstriction and alveolar gas exchange. Proc Natl Acad Sci USA 103:19093–19098
23. Fuchs B (2007) Die essentielle Bedeutung des "Classical Transient Receptor Potential 6"

(TRPC6)-Ionenkanals für die akute vaskuläre Hypoxiereaktion der Lunge - Untersuchung an isolierten pulmonalarteriellen glatten Muskelzellen. VVB Laufersweiler, Giessen, Germany

24. Kalwa H (2009) The physiological function of TRPC6 in endothelial and smooth muscle cells of the lung. Philipps-University Marburg, Marburg, Germany

25. Aaronson PI, Robertson TP, Knock GA, Becker S, Lewis TH, Snetkov V, Ward JP (2006) Hypoxic pulmonary vasoconstriction: mechanisms and controversies. J Physiol 570:53–58

26. Estacion M, Sinkins WG, Jones SW, Applegate MA, Schilling WP (2006) TRPC6 forms non-selective cation channels with limited Ca^{2+} permeability. J 572(Pt 2):359–377

Chapter 16

Endotoxin-Induced Airway Inflammation and Asthma Models

Zsuzsanna Helyes and Zsófia Hajna

Abstract

Lung inflammation models in experimental animals are particularly important to study the mechanisms and complex neuroimmune interactions involved in the pathophysiological processes, to identify key mediators and target molecules, as well as to test novel drug candidates. Endotoxin (lipopolysaccharide) administration locally into the airways (intranasally or intratracheally) is often used in a variety of laboratory animals for translational research to explore nonallergic inflammatory pathways, as well as to provide information on important mediators and their potential drug targets, although these conditions are not considered to be disease models. Allergic airway inflammation and asthma can be mimicked in rodents and larger animals by sensitization and then elicitation with ovalbumin, house dust mite, cockroach, plant, or helminth antigens. Mouse, rat, and guinea pig models have the major advantage of being easily available and appropriate for genetic modulations, but larger animals (cats, dogs, pigs, sheep, horse, or even primates) are often structurally and functionally closer to human conditions. A broad range of experimental protocols and assessments are used worldwide by different research groups. Differences in technical details greatly influence the results and the conclusions. Although the basic pathophysiology is similar after certain inflammatory stimuli, the effects depend on the animal species, strains, gender and age, the type and dose of the inflammatory or allergic agent, as well as the route of administration and the duration of exposure and investigation. In the present chapter we summarize the currently used research protocols and experimental paradigms of nonallergic and allergic lung inflammation focusing on the major advantages and disadvantages.

Key words: Acute lung injury, Allergic airway inflammation, Asthma, Cockroach antigen, Grass pollen, House dust mite, Lipopolysaccharide, Ovalbumin, Pneumonitis

1. Introduction

Inflammatory mechanisms of the lung involve complex neuroimmune interactions, which are important in bronchial hyperresponsiveness. The airways are densely innervated by capsaicin-sensitive peptidergic afferents which are activated by a variety of immune cell-derived mediators leading to the release of sensory neuropeptides. The importance of Transient Receptor Potential (TRP) ion channels localized on these nerves, particularly TRP ankyrin 1 (TRPA1),

Arpad Szallasi and Tamás Bíró (eds.), *TRP Channels in Drug Discovery: Volume I*, Methods in Pharmacology and Toxicology,
DOI 10.1007/978-1-62703-077-9_16, © Springer Science+Business Media, LLC 2012

vanilloid 1 (TRPV1), and melastatin 2 (TRPM2), has been described and emphasized in inflammatory respiratory diseases (1–6). Reliable and well-reproducible animal models are particularly important to investigate the pathophysiological processes in airway inflammation, to find the mediators which play crucial roles in certain mechanisms, to identify promising drug targets for potential novel pharmacotherapy, as well as to test new drug candidates during preclinical pharmacological studies. Predictive in vivo model systems are essential for translational medicine which offers insights to human disease mechanisms. Gene-deficient and human gene-expressing transgenic mice, as well as immunological and pharmacological modulation of receptors and enzymes helped to identify several potential targets for clinical pharmacological investigation. Both acute mechanism models, such as lipopolysaccharide (LPS)-induced pneumonitis, and real disease models like ovalbumin-induced asthma, are valuable for complex understanding of the molecular basis of the pathophysiology and evaluate drug effects. However, similarly to most clinical conditions, there is no animal model that fully reproduces the characteristics of the human diseases; rather, all available models are used for their relative biological relevance (7–10). Rodent models (mice, rats, guinea pigs) obviously have the major advantage of being easily available and appropriate for genetic modulations (11, 12); however, larger animals (cats, dogs, pigs, sheep, horse, or even primates) are structurally and functionally closer to humans (9, 13).

2. Animal Models of Acute Airway Inflammation to Study Mechanisms

2.1. Endotoxin-Induced Airway Inflammation

2.1.1. Mechanisms of the Inflammatory Process

Endotoxins are constituents of the cell wall of Gram-negative bacteria and are found in the entire environment, such as house dust, tap water, and milk. They cause acute reversible airflow obstruction and airway inflammation. The main component of endotoxin, LPS, is formed by a phosphoglycolipid (lipid A), which is covalently bound to a hydrophilic heteropolysaccharide (14). The receptor for LPS on monocytes and macrophages is the Toll-like receptor 4 (TLR4) forming a complex with the CD14 glycoprotein (15–18). LPS has a specific acute phase protein transporter in the circulation called LPS Binding Protein (LBP); the formed complex then binds to the TLR4–CD14 structure and causes macrophage activation.

They produce and release a broad range of inflammatory cytokines such as tumor necrosis factor–α (TNFα), interleukin-1β (IL-1β), IL-6, IL-8, IL-12, interferon-γ (IFN-γ), and keratinocyte-derived chemokine in the alveolar space (19–25). This induces increased expression of cell adhesion molecules on endothelial cells and after intranasal (i.n.), intratracheal (i.t.), or aerosol inhalation of LPS administration also by an intensive accumulation of neutrophils (21, 23, 26–29). Neutrophils recruited to the alveoli, interstitium, and bronchial/bronchiolar epithelial/ subepithelial regions cause tissue damage via the production and release of oxygen free radicals, proteases, cytokines, and chemokines (30), which attract and stimulate mononuclear cells and lymphocytes. All these inflammatory and immune cells release several other inflammatory mediators, such as leukotriens, prostaglandins, bradykinin, etc., which can directly activate sensory nerve endings in the airways or can induce epithelial damage resulting in the exposure of sensory nerves (31, 32). The released neuropeptides in turn influence the inflammatory process by acting at receptors localized on these peripheral nerve terminals themselves, vascular endothelial, bronchial epithelial, and inflammatory cells.

Other LPS-induced effects include augmentation of basal airway resistance due to airway remodeling, increased mucus production and vascular permeability, as well as increased bronchial responsiveness to inhaled muscarinic receptor agonists like methacholine (33, 34) or carbamoyl-choline (35–40) and even to histamine or serotonin in guinea pigs (41, 42). Bronchial hyperresponsiveness is also observed after systemic injection of LPS, but no neutrophil recruitment to the airways is observed. These effects of LPS are reduced by glucocorticoids, such as dexamethasone (33), but cytokine production is also reduced by p38 mitogen-activated protein kinase inhibition (23, 24).

Participation of capsaicin-sensitive peptidergic sensory fibers, the significance of neuroimmune interactions, and the relationship between inflammation and bronchopulmonary functional disturbances have also been established in LPS-induced nonallergic airway inflammation (25, 35, 36, 38, 39, 43). Neurogenic inflammatory mediators, such as Substance P (SP), neurokinin A (NKA), and calcitonin gene-related peptide (CGRP), are released from capsaicin-sensitive afferents in response to their stimulation/sensitization by a variety of inflammatory mediators (protons, leukotrienes, prostaglandins, bradykinin, inflammatory cytokines, etc.). Inhaled LPS strongly promotes the innervation of sensory C-fibers and the expression of tachykinin NK1 receptors in the airways, which is likely to result in enhancement of neurogenic inflammation (25). In intranasal LPS-evoked murine airway inflammation, functional roles of the released tachykinins and CGRP were established with respective receptor antagonists, but only concerning granulocyte accumulation and the production of the inflammatory cytokine IL 1β. Therefore, these proinflammatory sensory neuropeptides play only a minor role in the overall severity of endotoxin-induced airway inflammation (37, 38). Furthermore, data in rats revealed that pretreatment with high doses of capsaicin at the neonatal age to abolish neurogenic inflammatory factors did not diminish, but enhanced LPS-evoked plasma protein extravasation and neutrophil accumulation in the bronchoalveolar lavage (43). In agreement with this, in the i.n. LPS-evoked pulmonary inflammation model of the mouse, we found that resiniferatoxin (RTX) pretreatment inducing functional impairment of all capsaicin-sensitive sensory nerve endings (similar to the action of capsaicin) markedly increased the severity of airway inflammation providing multiple evidence for a protective role of these fibers (38). Our group described that this protective function is mediated by the activation of the TRPV1 channel, the receptor of capsaicin and RTX, localized on these sensory afferents. We also showed that somatostatin is released from the capsaicin-sensitive afferents innervating the lung upon TRPV1 receptor stimulation, which mediates inhibitory actions via the sst_4 somatostatin receptor (35, 36). In agreement with our conclusion, Tsuji and colleagues have recently described that the TRPV1 receptor agonist SA13353 inhibited leukocyte infiltration and inflammatory cytokine production in both the LPS-induced acute lung injury and the ovalbumin-induced allergic airway inflammation models (44). Earlier data obtained in rats have indicated similar protective role of capsaicin-sensitive fibers in other airway inflammation models (43, 45–47). Besides the TRPV1 receptor, another TRP channel, TRPA1, is also densely expressed on the pulmonary sensory fibers, and they are activated by a variety of endogenous inflammatory stimuli, such as peroxides, prostaglandins, isoprostans, peroxynitrites, etc. (48). Therefore, the role of TRPA1 in triggering sensory nerve activation in airway inflammation can strongly be suggested (5), but it has not yet been investigated in any LPS models. These complex mechanisms and the neuroimmune interactions involved in LPS-induced pneumonitis are summarized in Fig. 1.

LPS-evoked animal models are not classical disease models; yet, they are widely used for translational research to explore pulmonary inflammatory processes and to provide information on important mediators and their potential drug targets (10). However, different groups use different experimental protocols and assessments, which is likely to influence the results. Although the basic pathophysiology is the same, the effect can largely depend on the animal species, strains, gender and age, the type and dose of LPS, as well as the route of administration and the duration of the exposure and the investigation. Therefore, we aim to summarize the presently used LPS protocols here focusing on advantages, as well as potential difficulties.

2.1.2. Strains/Species/Gender/Age

Besides the complex, but relatively well-established pathophysiological mechanisms of LPS-induced airway inflammation, the other main advantage of this model is that it works in a variety of animal species. There are several papers using mice (10, 34–40, 49), rats (23, 50–52), as well as guinea pigs (42) and rabbits (53). The susceptibility to LPS challenge varies due to differences in the production

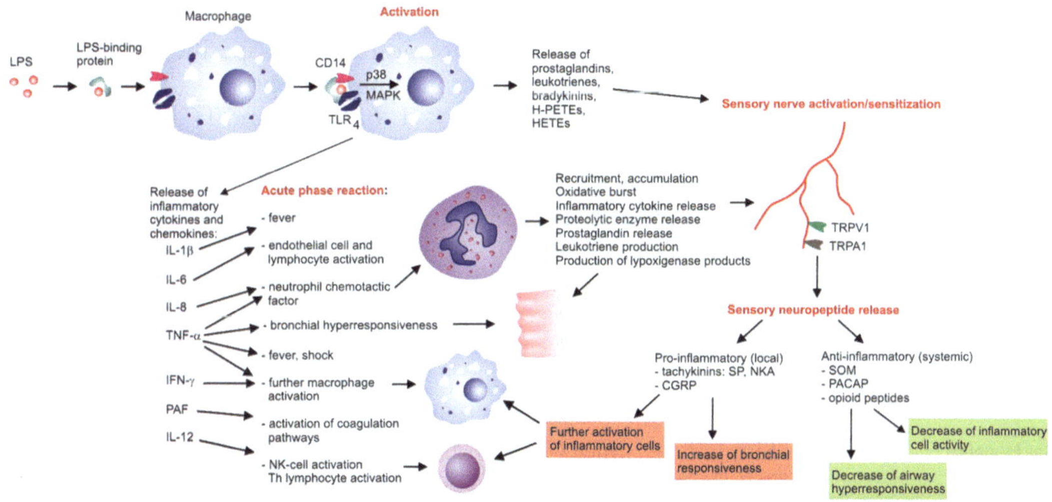

Fig. 1. The complex mechanisms and neuroimmune interactions involved in endotoxin (lipopolysaccharide, LPS)-induced nonallergic acute airway inflammation. LPS binds to TLR4 on macrophages forming a complex with the CD14 glycoprotein and causes their activation through the p38 MAPK pathway. They produce and release a broad range of inflammatory cytokines, chemokines inducing an intensive accumulation of neutrophils. Neutrophils recruited to the alveoli, interstitium, and bronchial/bronchiolar epithelial/subepithelial regions cause tissue damage via the production and release of free radicals, proteases, cytokines, and chemokines, which attract and stimulate mononuclear cells and lymphocytes. All these inflammatory and immune cells release several further inflammatory mediators, such as leukotrienes, prostaglandins, bradykinin, etc., which directly activate or sensitize the sensory nerve endings and induce epithelial damage resulting in the lack of protection of these fibers in the subepithelial region. The released sensory neuropeptides in turn modulate the inflammatory process by acting at receptors localized on these peripheral nerve terminals themselves, vascular endothelial, bronchial epithelial, and inflammatory cells. LPS also increases basal lung resistance due to airway remodeling; enhances mucus production and vascular permeability, as well as produces bronchial responsiveness. *TLR4* Toll-like receptor 4, *MAPK* mitogen-activated protein kinase, *IL-1*β interleukin 1-beta, *TNF-*α tumor necrosis factor-alpha, *IL-6, IL-8, IL-12* interleukin-6, 8 and 12, *IFN-*γ interferon-gamma, *PAF* platelet activating peptide, *NK-cell* natural killer cell, *Th lymphocyte* T helper lymphocyte, *SP* substance P, *NKA* neurokinin A, *CGRP* calcitonin gene-related peptide, *SOM* somatostatin, *PACAP* pituitary adenylate-cyclase activating polypeptide, *TRPV1* Transient Receptor Potential Vanilloid 1, *TRPA1* Transient Receptor Potential Ankyrin 1, *HETE* hydroxyeicosatetraenoic acid, *H-PETE* hydroperoxy-eicosatetraenoic acid.

of inflammatory mediators according to cases of distinct genetic backgrounds, inflammation duration, and strains. Gender-based response is another important factor; male mice are more likely to develop severe LPS-evoked lung inflammation whereas female mice have a survival advantage (10).

Mouse Models

Several authors use C57BL/6 mice in their experiments (34–36, 54–56), but the CD1 strain also gives similar results (39, 40). This is particularly important, since genetically modified, gene deleted, and transgenic mice are very useful in physiological, pathophysiological, and pharmacological research and they are mainly on these two backgrounds (35, 36, 40). Others work with Balb/C mice in the same model with no major difference in their LPS sensitivity (49, 57, 58). However, sensitivity of other mouse strains can largely differ: C3H/HeBFeJ mice are particularly sensitive, but the C3H/HeJ ones are resistant (30, 59, 60). Differences in airway responsiveness, inflammation, and fibroproliferative responses between these strains suggest that the expression of some specific key molecules playing a crucial role in this inflammatory process (TLR4 which is the gene product of *Lps* locus, IL-6, IL-1β, and TGF-β1) is genetically determined (16, 60, 61).

There does not seem to be a remarkable gender difference in most of the LPS-evoked actions; both males (17, 30, 34, 36, 62) and females (35, 38–40, 49, 55) have been investigated. Likewise, the age is also not a very crucial factor as young adult mice within the weight range of 20–30 g are appropriate. Responses of 6–8 week old (17, 30, 34, 62), 8–10 week old (37, 63), and 10–12 week old (55) did not markedly differ.

Rat Models

Similarly to mice, LPS induces a well-defined neutrophilic lung inflammation in different strains, such as Wistar rats (19, 25, 64–66), Sprague–Dawley rats (50, 67, 68), Fischer 344 rats (52), and Lewis rats (20, 21). There are no remarkable sensitivity differences between these strains. Most groups employ young male subjects weighing 150–200 g. However, a recent paper describes data on preweaning, 2-week-old males for the investigation of sensory C-fiber proliferation, tachykinin receptor expression, and neurogenic inflammation in the airways (25).

Guinea Pig Models

The guinea pig bronchial smooth is highly reactive to allergens, histamine, and lipid mediators; therefore, they are predominantly used for allergic models. However, young male Hartley guinea pigs (300–400 g) exposed to nebulized LPS proved to be appropriate model animals for studying airway inflammation, hyperresponsiveness, and fibrosis (42, 69). Both a single aerosolization in an acute, 24 h experiment (42) and repeated inhalations in a chronic setup (for 1 h 15 times at 48 h intervals; (69)) can be used. Others also work with i.v. (70) and i.n. (71) LPS applications in guinea pigs to evoke acute endotoxinemia and respiratory distress (33).

Rabbit Models

Intratracheal instillation of LPS followed by an i.v. injection 24 h later in anesthetized New Zealand white rabbits is also used as a good model of lung inflammation (53). However, in some models, LPS can be given via inhalation for 10 min (1,600 μg/100 ml saline diluted 1 to 50: 0.1 ml into 4.9 ml saline) or i.v. infusion for the same duration (0.4 μg/kg, 0.5 ml/min) (72). I.v. infusion can also be applied for a longer, 30 min period, in 500–5,000 μg/kg doses to examine acute airway hyperresponsiveness and inflammatory alterations (73, 74).

Larger Animal Models

Although *E. coli* and *Salmonella enteritidis* LPS-induced acute respiratory distress is less widely studied in the sheep (75, 76), pig (77–79), and dog (80, 81), there are some papers describing the development of stable lung injury 6–24 h after i.v. endotoxin infusion in these species, as a remarkable component of multiorgan failure (13). The injected doses are 1–100 μg/kg in sheep and pigs, but more than 1 mg/kg is needed in dogs. Despite rodents, LPS is not administered locally in these species.

2.1.3. Type and Dose of LPS

Different research groups publishing data on LPS-induced lung inflammation use different types of LPS which largely influences the results. They basically work through the same mechanism, and the conclusions drawn from these different experiments can obviously be joined together. However, there are remarkable differences in the kinetics, potencies, and efficacies between the different LPS types. Therefore, direct comparisons are only possible if the same serotype is used the same way in the same dose or concentration.

Szarka and colleagues compared the lung injury-inducing actions of i.n. *E. coli serotype O55:B5* and *P. aeruginosa F-D type 1* LPS in mice. They studied concentration–response correlations within the 33 μg–1,667 μg/ml range and investigated different time intervals after a single challenge. The results clearly showed that the 167 mg/ml dose evokes the most prominent inflammatory reaction at the 24–48 h time point. The extent of the inflammatory reaction was different due to the differences of their biological activities based on the distinct chemical structures. *P. aeruginosa* endotoxin caused remarkably stronger inflammation (49). Differences in the signaling of distinct LPS types and serotypes are supported by the data that isolated macrophages derived from the LPS-resistant C3H/HeJ did not synthesize any detectable level of TNF-α in response to *Salmonella minnesota Re-595* LPS, but did produce low levels of TNF-α in response to a high concentration of *E. coli O55:B5* LPS (16). *Salmonella typhosa* LPS was intratracheally instilled in rats by Ulich and colleagues to study changes in cytokine expression in the lung (20, 21).

Several groups use the *E. coli serotype O55:B5* LPS dissolved in sterile saline or PBS for i.n. application in rats (200 μg/kg in 0.5 ml/kg; (50)), as well as intraperitoneal injection (i.p., 0.6–1 mg/animal; (57)) or i.n. instillation (0.25–3 mg/ml in 20–50 μl) in anesthetized mice, aerosol exposure in unanesthetized animals (0.3–3 mg/ml for 10 min; (34, 54, 56)), or even in i.v. infusion in rabbits (0.5–5 mg/kg over 30 min, (73, 74)). The *E. coli serotype O111:B4* LPS is also commonly used in both rats (i.t. 100 μg/animal; (68)), mice (aerosolized for 2.5–4 h per day chronically to 17.5 μg/ml dissolved in sterile PBS for 1 day-8 weeks; (17, 30, 60, 62, 63)), and even rabbits (i.v. or aerosolized; (72)). Others publish data with *E. coli serotype O26:B6* (2.5 mg/kg i.p.; (82)), *serotype O128:B12* nebulized in mice to a nose-only exposure chamber (for 15 min, 100 μg/ml in distilled water; (55)), and also with i.n. *serotype O83* (60 μl, 167 μg/ml in sterile PBS; (35–40)). In one experiment, the LPS dose was increased tenfold to provoke epithelial cell damage (55).

2.1.4. Ways of Administration

As mentioned earlier, in animal models LPS can be administered both systemically (i.p., i.v.) or locally (i.t., i.n. or inhalation). Its i.p. and i.v. injection results in a systemic inflammatory response including pulmonary damage, the most severe condition of which is endotoxin shock with multiorgan failure (57, 65, 70, 83–86). In contrast, local LPS administration predominantly affects the airways, although symptoms of systemic inflammatory changes, such as fever, sedation, decreased appetite, and weight loss, are observed. The i.t. administration is more precisely localized, does not affect the upper airways, but it is technically more difficult, cannulation takes a bit longer time, and anesthesia is needed for the procedure. I.n. application is technically easy and quick, and although it cannot be in conscious animals, a light ether or isoflurane/sevoflurane anesthesia for a few minutes is sufficient. Its main disadvantages, however, are that (1) some of the solution can get into the gastrosintestinal tract; (2) its distribution might not be homogenous; and (3) similarly to aerosol inhalation, it might affect the pharyngeal and laryngeal mucosa, and its influence on bronchial responsiveness and the inflammatory reaction in the lower airways cannot be excluded. Nebulization and inhalation is performed under unrestrained conditions in conscious animals; it can be applied chronically, it is closer to the human conditions, and the distribution of the LPS particles in the lung is more homogenous (10, 13).

Systemic LPS Exposure

Prenatal i.p. application one and two days before birth on days 20–21 of the pregnancy did not trigger TH17 cell differentiation in the offspring. Furthermore, prenatal LPS exposure reduced ovalbumin-induced (TH2-mediated) airway inflammation, eosinophilia, and airway responsiveness. Thus, in utero exposure to endotoxin promotes a TH1 immune environment, which suppresses the development of allergic airway disease later in life (82).

Formerly the major routes of LPS administration to induce pulmonary damage in adult mice, rats, and guinea pigs were i.p. and i.v. injections (57, 65, 70, 83–85, 87). In these studies large doses (0.6–1 mg) of LPS were used to evoke lung injury, but these sublethal or even lethal doses resulted in a relatively moderate edema formation and small increase of proteins in the bronchoalveolar lavage fluid (57). Furthermore, it was also shown that systemic LPS induces firm neutrophil adhesion onto the pulmonary vascular endothelium, but not to the bronchoalveolar lavage fluid in mice (88). When systemic and local LPS applications were directly compared, massive recruitment of leukocytes in the bronchi, specific and severe pulmonary damage only developed in case of the latter mode of administration (49). Therefore, it can be concluded that specific pulmonary inflammation, its precise mechanisms, and consequent hyperresponsiveness cannot be appropriately investigated after systemic administration.

Intratracheal Instillation

This way of administration induces a well-localized inflammatory reaction selectively in the lower airways. The model has recently become popular in mice. LPS (30–100 μg i.t.) evokes a reliable and well-reproducible lung injury, in which the importance of the MAPK and NFκB signal transduction pathways (89), regulation of different cytokines (90), and the correlation between magnetic resonance imaging (MRI) and histological findings (58) were investigated 24–48 h after the challenge.

This instillation has been longer used in rats: 100–1,000 μg LPS provide an appropriate model to investigate protease–antiprotease imbalance being important in asthma and chronic obstructive pulmonary disease (COPD) (68), to study alterations of inflammatory and anti-inflammatory cytokines (20, 21), to elucidate the mechanisms of proteinase and reactive oxygen species secretion by alveolar neutrophils (50), as well as to study corticosteroid action on TNF-α and macrophage inflammatory protein 2 (MIP-2) production (19, 64).

In the guinea pig i.t. LPS model, evidence has been provided for macrophage-derived, secretory type II phospholipase A2 in a TNF-α-dependent manner (91).

I.t. instillation of low doses (50 μg/kg) in rabbits caused a significant increase in bronchoalveolar lavage polymorphonuclear leukocytes without an increase in mononuclear cells or an enhancement of lung permeability. However, i.v. injection of 10 μg/kg LPS at 24 h after i.t. LPS caused significant increases of all inflammatory cell types, IL-8 and monocyte chemotactic protein-1 concentrations, as well as pulmonary edema formation (53).

Intanasal Administration

Intranasal LPS administration is the most commonly used, confined model to study acute interstitial lung inflammation and injury which is known to cause a primarily neutrophil accumulation (30). Low LPS doses evoke maximal inflammation (inflammatory cell infiltration and cytokine production) 24 h after its instillation (36, 37, 39, 56). There are some systemic inflammatory symptoms, such as fever, decreased motility, and appetite, but despite systemic administration, there are no shock-like symptoms and severe toxicity signs. The increase in microvascular permeability during lung inflammation leading to perivascular/peribronchial edema formation is due to at least two distinct mechanisms: an initial one related to the rapid neutrophil influx and a delayed one occurring even under neutropenic conditions (54).

Besides neutrophil recruitment and enhanced production of TNF-α, there is a remarkably increased response to the bronchoconstrictor effect of aerosolized methacholine and by a late augmentation of vasopermeation (33, 34). These effects are inhibited by increasing cAMP concentration in macrophages (28), but only some actions are reduced by the standard glucocorticosteroid dexamethasone (34). In contrast, all symptoms are markedly enhanced by the COX-inhibitor NSAIDs (28). According to the comparison of LPS administrations performed by Szarka and colleagues, this model greatly helps to study the precise pathophysiological mechanisms of respiratory distress and elucidate the therapeutical potential of cytokine-induced pathways. They have shown that acute lung injury develops within 2–4 h and reaches maximal damage at 24–48 h. After 24 h, LPS caused a 29% increase of wet lung weight and a strong influx of granulocytes into the bronchi, as revealed with the flow cytometry analysis of the bronchoalveolar lavage fluid and the histopathological results. After 72 h, pneumonitis dramatically decreases and the acute lung injury recovers (49). Similar time course and kinetics for the intensity of several inflammatory parameters and hyperreactivity were described by other groups (34, 56).

Inhalation of Aerosolized LPS

In these experiments animals are usually unrestrained and placed in individual compartments of cage whole-body exposure chambers; therefore, it is the most appropriate model for chronic experiments. In this model, Inter-Cellular Adhesion Molecule 1 ICAM-1 was shown to play a pivotal role in the development of airway hyperresponsiveness and airway inflammation, but through distinct mechanisms. Airway remodeling, as a fibroproliferative process, was dependent on IL-1β, TGF-β, and IL-6 (60, 61). Inhaled LPS also caused neutrophil-dependent emphysematous changes in the lung architecture associated with apoptosis, differently from those induced by cigarette smoke (63).

A *single LPS nebulization* for 10–30 min is also commonly used in both rat and mouse *acute* models (0.3–3 mg/ml; (23, 34, 54, 56, 66)). This way LPS inhalation induces a remarkable neutrophilic inflammation predominantly in the bronchi, as assessed by the cellular profile and myeloperoxidase content of the bronchoalveolar lavage fluid as well as by histology 4 h postinhalation. These changes are abrogated by low intensity laser therapy at a similar extent to the synthetic glucocorticosteroid dexamethasone (34, 66). In addition, acute LPS aerosolization can also be followed 30 min later by the i.t. instillation to enhance the inflammatory reaction (68).

Nose-only exposure chambers are also appropriate in mice for shorter periods (15 min), in which they are restrained, but anesthesia is not needed. The aerosol is generated by a compressed air nebulizer using 0.1 mg/ml LPS dissolved in endotoxin-free distilled water. If the LPS dose is increased tenfold, it induces severe epithelial cell damage. In this case the particle size distribution is between 0.1 and 0.3 μm, with a total deposition of 18.3 ± 1.23% for small laboratory animals (0.1 mg/ml; (55)). This protocol evokes a transient inflammatory response with a peak of recovered neutrophils in the bronchoalveolar lavage fluid 12–24 h after the exposure. (92). A detectable increase of lactate dehydrogenase activity, as an indicator of cell damage is observed 22 h after the provocation, but only when mice are exposed to a relatively high dose of LPS (1 mg/ml; (55)).

Longer aerosolization periods can vary; some authors apply 2.5–4 h per day only for 1 day whereas others use it chronically during a period up to 8 weeks (17.5 μg/ml LPS dissolved in sterile PBS; (17, 30, 60–63)). Nebulized LPS can be used for 4 h per day for 4 weeks followed by either a 3-day or a 4-week recovery period (63). Investigations are performed immediately after the 4-week exposure and 4 weeks after the end of exposure (30). The mechanisms responsible for switching the predominantly edematous and neutrophilic acute inflammation to the chronic process with lymphocyte infiltration and lung structure remodeling, as well as potential regenerations and recovery can be appropriately examined in these long-lasting models.

2.2. Other Chemical Stimuli for the Induction of Nonallergic Airway Inflammation

LPS is obviously the most commonly used inflammatory agent for studying nonallergic airway inflammation in rodents and also in larger laboratory animal models. However, there are some other chemical stimuli, which are appropriate for investigating specific mechanisms, such as matrix metalloproteinase activity, oxidative stress of airway epithelial cells, or mucus cell metaplasia, in pulmonary injury particularly in rats. These are briefly summarized here:

2.2.1. Air Born Particulate Matter

It is a complex mixture of chemicals, such as oxides of nitrogen, sulfur, carbon, dioxins furans, metals, chlorinated hydrocarbons, and polycyclic aromatic hydrocarbons. Increased asthma risk and exacerbation, particularly in children and infants are associated with exposure to elevated levels of ultrafine particulate matter and oxidant pollutants (93). Nose-only inhalation exposure of 7-day-old Brown Norway rat pups to environmentally persistent free radical (EPFR)-containing combustion generated ultrafine particles (CGUFP) with a mean diameter of 0.2 μm (200 μg/m^3; 20 min per day for 1 week) for 7 consecutive days induces pulmonary oxidative stress and lung dysfunction 24 h following the final exposure. Cytokines can be additionally assessed at 72 h. This is based on alterations in the expression of various proteins associated with the response to oxidative stress and the regulation of glucocorticoid receptor translocation in T lymphocytes (94).

2.2.2. Carbon-Nanoparticle-Induced Noutrophilic Lung Inflammation

Carbon particles having 14 nm diameter are well-accepted model particles representing the carbonaceous core of combustion-derived nanoparticles (95). These carbon nanoparticles were suspended in PBS by sonication (1 ml aliquots, 120 W, 50–60 Hz) for at least 1 h) followed by further dilution and administered through i.n. instillation (0.5–2.5 mg/kg in 0.4 ml) to anesthetized animals (96). TNF-α playing a critical role in lung fibrosis, as well as IL-4, IL-6, IFN-γ being a major activator of macrophages and IL-10 serving as an autoregulatory factor that reduces IL-8 production in neutrophils are elevated 48 h after this challenge. The activation of mitogen-activated kinases (MAPK ERK1/2) is specific for nanoparticle-induced lung epithelial cell damage (97), and the production of IL-8 is strictly dependent on the activation of the MAPK P38 (98). In contrast, NFk-B activation is of minor importance in this process (98).

2.2.3. Cadmium-Induced Pulmonary Inflammation

Cadmium, one of the numerous components of tobacco smoke and a toxic ambient pollutant, has widely been used in several animal species to mimic inflammatory changes observed in the human lung (99–101). Male Sprague–Dawley rats exposed to nebulized $CdCl_2$ 0.1% solution in saline in a whole-body aerosol-exposure chamber develop an acute pulmonary inflammation characterized by a significant increase in the total number of cells in the bronchoalveolar lavage fluid (102). This heavy metal cadmium is suspected to contribute to tobacco related lung diseases involving neutrophilic inflammation resistant to corticosteroids and associated with an imbalance between the activities of matrix metalloproteinases (MMP) and tissue inhibitors of MMPs (103). Similar pathological changes can be experimentally induced in rats exposed to cadmium inhalation (104). A single cadmium inhalation induces a prominent increase in neutrophil and macrophage counts, an increased protein concentration in bronchoalveolar lavage fluid, and a high lung wet-to-dry weight ratio. A significant increase in airway resistance is also observed. The lung injury is associated with a marked MMP-2 and MMP-9 activation greatly contributing to acute inflammatory processes. In contrast, the absence of widely expressed inflammatory cytokines, such as IL-1β, TNF-α, and GM-CSF, in the bronchoalveolar lavage fluid indicates that despite the LPS model, these mediators do not play a major role in this pathophysiology (105). Although it is not a disease model of COPD, acute exposure of rats to cadmium can be considered relevant to some of the main features of the disease and to investigate the influence of pharmacological agents. In such diseases, the inhibition of MMP activities could thus be beneficial to prevent the lung against inflammation and tissue remodeling due to the enhanced activity of these enzymes (105, 106).

2.2.4. Ozone

Ozone is an important air pollutant of photochemical smog. Repeated ozone exposures alone induce site-specific lesions in the centriacinar regions of all lung lobes in rats. These are characterized by a mild neutrophilic inflammation (aveolitis/bronchiolitis), increased number of alveolar macrophages, and regenerative hyperplasia localized predominantly in the terminal bronchioles and proximal alveolar ducts. Mucous cell metaplasia (MCM), a principal feature in endotoxin-instilled rats, is not present in the airway epithelium lining the large- and small-diameter conducting airways

Table 1 Experimental protocols and paradigms in the endotoxin (lipopolysaccharide: LPS)-induced non-airway inflammation models

Species	Strain	Gender/age	Stimulus	Duration	References
Mouse	C57BL/6	Female	I.n. 60 µl/animal, 167 mg/ml LPS *(E. coli serotype O83)*	24 h	(35)
	C57BL/6	Male	I.n. 60 µl/animal, 167 mg/ml LPS *(E. coli serotype O83)*	24 h	(36)
	C57BL/6	Male and female	I.n. 40-50 µl/animal, 250 µg/ml LPS *(E. coli serotype O55:B5)*	24–96 h	(56)
	C57BL/6	10–12-week-old female	Inhaled 0.1–1 mg/ml LPS for 15 min *(E. coli serotype O128:B12)*	2–22 h	(55)
	C57BL/6	5–8-week-old male	I.n. 20–40 µl/animal, 33–3,300 µg/kg LPS or inhaled 33–3,300 µg/ml LPS *(E. coli serotype O55:B5)*	2–24 h	(34)
	C57BL/6	7-week-old male	I.n. 50 µl/animal, 3.3–330 µg/kg LPS or i.p. 1 mg/kg LPS *(E. coli serotype O55:B5)*	3–168 h	(54)
	C57BL/6	3-week-old male	Inhaled 17.5 µg/ml LPS for 2.5–4 h/day *(E. coli serotype O111:B4)*	4 weeks	(36)
	CD1	Female	I.n. 60 µl/animal, 167 mg/ml LPS *(E. coli serotype O83)*	24 h	(39, 40)
	Balb/c	Male	I.p. 0.6–1 mg/animal LPS in 0.2 ml *(E. coli serotype O55:B5)*	1–96 h	(57)
	Balb/c	Female	I.n. 33–1,667 µg/ml LPS *(E. coli serotype O55:B5 and P. aeruginosa F-D type 1)*	24–72 h	(49)
	Balb/c	Female	I.n. 60 µl/animal, 167 mg/ml LPS *(E. coli serotype O83)*	24 h	(39)
	Balb/c	6-week-old male	I.t. 100 µl/animal 0.3 mg/ml LPS *(E. coli serotype O111:B4)*	24 h	(58)
	C3H/HeBFeJ, C3H/HeJ	Male	Inhaled 17.5 µg/ml LPS for 4 h/day *(E. coli serotype O111:B4)*	1 day–8 weeks	(60)

(continued)

Table 1 (continued)

Species	Strain	Gender/age	Stimulus	Duration	References
Rat	Lewis	Male	I.t. 1–10 μg/animal LPS *(S. typhosa)*	2–48 h	(20)
	Lewis	Male	I.t. 100 μg/animal LPS *(S. typhosa)*	2–96 h	(21)
	Wistar	Male	I.v. 5 mg/kg LPS *(S. abbortus equi)*	1.5–48 h	(65)
	Wistar	Male	Inhaled 0.3 mg/ml LPS *(E. coli O111:B4)*	4 h	(66)
	Wistar	Pregnant female	I.p. 2.5 mg/kg LPS *(E. coli serotype O26:B6)* on fetal days 20–21 (term = 22 days)	Investigations on newborn pups	(82)
	Wistar	2-week-old male	Inhaled 0.1 mg/ml LPS for 30 min *(E. coli serotype O55:B5)*	14 h–28 days	(25)
	Wistar	Male	Inhaled 0.3 mg/ml LPS for 30 min or i.p. 1.5 mg/kg LPS	0.5–48 h	(23)
	Sprague-Dawley	No data	I.t. 220 μg/kg LPS	24–40 h	(67)
	Sprague-Dawley	Male	I.t. 200 μg/kg in 0.5 ml/kg LPS *(E. coli serotype 055:B5)*	4–24 h	(50)
	Sprague-Dawley	No data	I.t. 100 μg/animal LPS *(E. coli serotype O111:B4)*	24 h	(68)
	Fischer 344	12–14-week-old female	I.n. 50 μl/animal, 5 mg/ml LPS *(E. coli serotype O111:B4)*	1–7 days	(52)
	Fischer 344	10–12-week-old male	I.n. 2–20 μg/animal LPS in 150 μl *(P. aeruginosa serotype 10)*	96 h	(108)
	Fischer 344	10–12-week-old male	Inhalation of O_3 1 ppm for 8 h	96 h	(108)
	Fischer 344	8-week-old female	I.t. 0.4 ml/animal, 0.5–2.5 mg/kg CNP (14 nm in diameter)	48 h	(96)
	Sprague-Dawley	Male	Inhaled 0.1% $CdCl_2$	24 h	(102)
	Brown Norway	7 day old	Inhalation of EPFR-containing CGUFP 200 μg/m^3 for 20 min/day	1 week	(94)
Guinea pig	Dunkin-Hartley	Male	I.v. 0.1–1 mg/kg LPS *(E. coli serotype O55:B5)*	1–24 h	(70)
	Dunkin-Hartley	Male	I.t. 330 μg/kg LPS *(E. coli serotype O55:B5)*	8 h	(71)
	Dunkin-Hartley	Male	Inhaled 30 μg/ml LPS for 1 h	1–24 h	(42)
	Dunkin-Hartley	Male	Inhaled 30 μg/kg LPS for 1 h, every 2 days *(E. coli serotype O55:B5)*	30 days	(69)

Rabbit	White New Zealand	Adult	I.v. 10 mg/kg LPS at 24 h after i.t. LPS *(E. coli)*	30 h	(53)
	White New Zealand	Adult	I.v. 500 μg/kg LPS *(E. coli serotype O55:B5)*	8 h	(73)
	White New Zealand	No data	Inhaled 3.2 mg/ml LPS or i.v. 0.4 μg/kg LPS *(E. coli serotype O111:B4)*	5 h	(72)
	Japanese white	Adult male	I.v. 5,000 μg/kg LPS over 30 min *(E. coli serotype O55:B5)*	6 h	(74)
Pig		No data	I.v. 25 μg/kg LPS *(E. coli serotype O111:B4)*	6 h	(77)
		No data	I.v. 100 μg/kg LPS for 1 h *(E. coli serotype O111:B4)*	6 h	(78)
		Male and female	I.v. 30 μg/kg/h LPS for 2 h *(E. coli serotype O111:B4)*	6 h	(79)
Sheep		1-year old	I.v. 15 ng/kg/min LPS for 6 h *(E. coli serotype O26:B6)*	6 h	(76)
Dog		Adult	I.v. 2 mg/kg LPS *(E.coli serotype O55:B5)*	6 h	(80)
		No data	I.v. 4 μg/kg LPS *(S. enteridis)*	2 h	(81)

I.p intraperitoneal, *S.c* subcutaneous, *I.t* intratracheal, *I.n* intranasal, *I.v* intravenous, *LPS* lipopolysaccharide, O_3 ozone, *ppm* particles per million, *CNP* carbon nanoparticles, $CdCl_2$ cadmium-chloride, *EPFR* environmentally persistent free radical, *CGUFP* combustion generated ultrafine particles

(107, 108). Harkema and Wagner also compared the actions of inhaled LPS and ozone, in a two toxicant-induced rat models to study the epithelial and inflammatory factors involved in the transformation of mucous cell-free normal airway epithelium to secretory epithelium containing numerous mucus-secreting cells. Dosimetry studies suggest that rats require four- to fivefold higher doses of ozone than humans to create an equal deposition and pulmonary inflammatory response (109). Therefore, 1 ppm is a reasonable exposure level from which to make comparisons with humans. In the above-mentioned ozone–endotoxin co-exposure model, male F344/N rats were first instilled with endotoxin, and 6 h later they were exposed to air or 1 ppm ozone for 8 h. This dosing-exposure regimen was chosen on the basis of endotoxin-elicited airway neutrophils influx peak at 6 and 12 h. One day later, endotoxin instillation and ozone exposures were repeated. Mucin-specific gene expression and MCM were induced in the nasal transitional epithelium, but not in the bronchiolar epithelium in response to ozone. In contrast, inhalation of LPS induces MCM in the bronchiolar epithelium, but not in the nasal transitional epithelium. Both ozone- and endotoxin-induced MCM are dependent on neutrophilic inflammation. Interestingly, each toxicant enhances MCM induced by the other stimulus in a synergistic way through neutrophil-mediated mechanisms (108, 110). Technical details of the LPS-induced lung inflammation models are summarized in Table 1.

3. Animal Models of Asthma and Allergic Pulmonary Inflammation: Disease Models

3.1. Pathophysiological Mechanisms

Allergic (extrinsic) and nonallergic (intrinsic) asthma are of basically different pathophysiological mechanisms, particularly regarding the triggering stimuli. There are several common points in the underlying processes with special emphasis on the complex neuroimmune interactions, but most animal models mimic the allergic responses. In contrast to LPS-induced animal models described earlier, several good reviews have recently focused on animal models of chronic allergic lung inflammation and asthma (8, 9, 11, 12, 33, 111–113).

Several animal species have been used to investigate the allergic processes of the respiratory tract; indeed, there are guinea pig, mouse, rat, sheep, and dog models (8). The guinea pig model was the first described model of asthma and contributed greatly to the development of corticosteroid and β_2 receptor agonist therapies (111), but it has recently lost from its value due to the lack of genetic modifications and specific immunological reagents (33). Sheep and dogs greatly differ from humans and they cost much more than rodents (8, 112). In a very comprehensive recent review, the preclinical in vivo models that recapitulate many of the features of asthma have been described, the advantages and disadvantages of the standard models have been summarized, and recently developed novel experimental systems have also been highlighted (114).

Despite having a broad knowledge about the role of immunity in allergic asthma (12), the complex pathophysiology of the disease with special emphasis on the neurogenic inflammatory components is not that well understood. The importance of peptidergic sensory nerves and sensory neuropeptides is often under-evaluated and not appropriately taken into consideration. Similarly to the endotoxin-induced models, new studies have revealed an essential role for sensory neuronal TRP ion channels, particularly TRPA1 and TRPV1, but also the "canonical" TRPC, the "melastatin" TRPM, the "polycystin" TRPP, and the "mucolipin" TRPML receptors in airway chemosensation, inflammation and cough, excessive mucus secretion, and hyperresponsiveness related to asthma, COPD, and reactive airway dysfunction syndrome (5, 115). TRPC6 deficiency inhibits specific allergic immune responses, pointing to an important immunological function of this cation channel in Th2 cells, eosinophils, mast cells, and B cells. (116). TRPA1 is targeted by a series of by-products of oxidative and nitrative stress, including acrolein, 4-hydroxy-2-nonenal, and hydrogen peroxide. Proinflammatory neuropeptides are released into the area of innervation from the nociceptive nerve terminals in response to TRPV1/TRPA1 stimulation and cause airway neurogenic inflammation. The up-regulation of TRPV1 expression and function has been reported in asthma and other inflammatory conditions (1, 3). A key role for airway sensory neuronal functions in late asthmatic responses was confirmed by the inhibition observed after the nonselective cation channel blocker ruthenium red, the TRPA1 inhibitor HC-030031, and the muscarinic receptor antagonist tiotropium bromide, but not the TRPV1 inhibitor JNJ-17203212 in the rat. It can be proposed that allergen challenge triggers airway sensory nerves via the activation of TRPA1 channels which initiates a central reflex event leading to a parasympathetic cholinergic constrictor response (117). However, the number of TRPV1-immunoreactive axons in the trachea within the epithelium and around areas of smooth muscle increases under allergic inflammatory conditions in the guinea pig ovalbumin-induced asthma model (118).

Neurogenic inflammation developing as a result of TRPV1/TRPA1 activation is suggested to participate in the pathogenesis of bronchial asthma. Proinflammatory neuropeptides, mediating neurogenic inflammation (e.g., SP and NKA) have been proposed to have potent effects on the tone of airway smooth muscle, airway glandular secretions, bronchial circulation, and inflammatory and immune cells by the activation of the neurokinin-1 (NK-1) and neurokinin-2 (NK-2) receptors (119), as well as on airway remodeling and hyperresponsiveness (120).

Meanwhile, there is also a systemic release of anti-inflammatory peptides such as somatostatin and pituitary adenylate-cyclase activating polypeptide, their inhibitory function is also proposed, although clear evidence has not yet been provided in these chronic models (Helyes et al. unpublished data).

Airway remodeling is an important human pathophysiological process which is not easy to mimic and appropriately investigate in animal models repeatedly and noninvasively (9). Here we summarize the asthma models according to the evoking stimulus and relation to human pathophysiology, as disease models. In asthma models, adjuvants, most commonly aluminum hydroxide and heat-killed Bordetella pertussis, are used to initiate and enhance the immune system to react in the desired fashion (121). Aluminum hydroxide, administered together with antigen exposure, promotes the Th2 phenotype (122). Furthermore, lipooligosaccharide from Bordetella pertussis drives a Th2-based response (123). There are also adjuvants that promote a Th1 response, such as Freund's complete adjuvant (124). The disadvantage of adjuvant use is the influence on the immune response, preventing a direct comparison between humans and animals after exposure to a certain allergen (112).

3.2. Animal Models of Allergic Airway Inflammation

There is a broad range of animal models available; the most popular models are rodents (mice and rats). This is mainly due to the fact that they are easy to handle, have a short gestational period, and they are relatively cheap compared to larger animals. Transgenic technology and the development of species-specific probes, which are particularly available in mice, have opened great perspectives for a variety of mechanistic studies in this species (8, 112).

Most asthma-related experiments are nowadays performed in mice. The development of transgenic mouse models that exhibit various lung pathologies is now a huge research enterprise, and mouse models of lung disease have been the subject of a number of recent reviews that cover their various pathophysiological features in detail (11, 12, 113).

Bronchial hyperresponsiveness is mainly associated with airway inflammation, but it was noted in female Balb/c mice without inflammation (125). The intensity of the inflammatory response greatly differs between strains and even within the same strain depending on the immune status of an animal. In contrast to guinea pigs, in CBA, Swiss, IL-5 transgenic mice (126), and BALB/c mouse strains eosinophil recruitment to the airways is not sufficient to induce bronchial hyperresponsiveness (127). Furthermore, more leukotriene C4 is released from the antigen-challenged lung of Swiss mice than of Balb/c mice (128). The immunization protocol and antigen boosting also significantly modify the responses (129).

3.2.1. Ovalbumin-Induced Asthma Models

Mouse Models

The most commonly used sensitizing agent is *ovalbumin* (OVA); it is typically injected i.p. along with an adjuvant, usually aluminum hydroxide (AlumnInject), that serves to enhance the immunogenicity (122, 130). This initiates a systemic immune response against the antigenic during some days, and then a boosting antigen exposure is performed either directly aerosolized into the lung or intranasally instilled. This elicits a localized pulmonary with eosinophil infiltration, mucosal edema, and bronchial hyperresponsiveness. Although there are plenty of groups using the ovalbumin model, there is a great variation between the sensitization/elicitation protocols, boosting, doses, adjuvants, ways of administration, durations, and time intervals.

A widely used protocol is to sensitize the mice with one or two i.p. or subcutaneous (s.c.) injections of 10-100 μg OVA emulsified in aluminum hydroxide with a 7- or 14-day interval in the latter case (127, 128, 131–137). Then the animals are challenged with 1–5% OVA dissolved in sterile PBS *aerosolized* on days 28, 29, 30 for 5–30 min each day with an ultrasonic nebulizer (39, 138, 139). Others use daily or every other day exposures between days 15 and 21 for 20 min (131, 140, 141). Others sensitize by seven i.p. injections of 10 μg OVA in 0.5 ml pyrogen-free saline without adjuvant on alternate days. Two weeks later, mice are exposed to OVA (2 mg/ml) aerosol challenges for 5 min on 8 consecutive days and studied 24 h after the last challenge (142). The intensity of this allergic inflammation reaches its maximum 1–2 days after the final exposure (12). From day 32 increased eosinophil infiltration, bronchial hyperresponsiveness, remodeling, and Th2 cytokine induction occurs, some groups examined up to 74 days (138, 139).

Others use i.p. immunization every other day for 2 weeks (10 μg OVA in 0.1 ml sterile saline), and mice are challenged three times, each 3 days apart, with 20 μg of OVA in 50 μl of saline delivered *intranasally* 40 days after the beginning of immunization. Control animals are similarly immunized i.p. with OVA and challenged i.n. three times with 50 μl of saline. Mice are used 3 days after the last i.n. challenge (143, 144). Intranasal OVA challenge can also be performed 4 times (twice a day) 1 week after the second s.c. immunization, and pulmonary functions, bronchoalveolar lavage examination, as well as lung histology and blood analysis are done 1, 3, 6, 24, 48, and 72 h afterwards (127, 132). The sensitization and the exposure doses, as well as the administration paradigms of the i.n. challenge vary between groups. Elicitation is often done on days 14, 15, 18, and 19 (25 μg in 35 μl saline), the bronchoalveolar lavage is done 2 days after the last application (145–147). Henderson and colleagues use i.n. OVA application on days 14, 25, 26, and 27 (100 μl, and later 50 μg in 50 μl saline). They found increased eosinophil influx and airway hyperresponsiveness from day 28 (137).

Some groups use *i.n.* instillation of 10 μl OVA on 3 days with 3-day periods apart by a nonsurgical technique after the immunization with an adjuvant-free protocol (i.p. 10 μg OVA on each of seven alternate days). Systemic levels of OVA-specific Ig E were allowed to increase for 40 days after the first sensitizing injection. 24 h postfinal OVA challenge: increased number of eosinophils was found in the bronchoalveolar lavage and the lung tissue (144).

However, some authors conclude that although ovalbumin-sensitized mice are good models for airway hyperresponsiveness, this is not appropriate to study acute physiological responses to allergen inhalation. The lack of the early phase response within the first hour following exposure and the limited late phase response that is more prolonged and may occur several hours later, as well as the absence of a link between the late phase response and airway hyperresponsiveness highlight the limitations of this mouse model as a completely appropriate model of lung dysfunction associated with asthma (148).

Variations in the intensity of the inflammatory response according to the immune status of an animal within the same strain and to the strain itself have also been reported. Most commonly standard pathogen-free 4–8-week-old female mice are used in this model, since they proved to be the most sensitive (135, 136). However, more leukotriene C4 was shown to be released from isolated antigen-challenged lungs of Swiss mice than of BALB/c mice (128). It has also been demonstrated in CBA, Swiss, BALB/c, and mice also in IL-5 transgenic mice (126), that similarly to guinea pigs, eosinophil recruitment to the airways is not sufficient to induce bronchial hyperresponsiveness (33, 127).

Rat Models

The immediate and late asthmatic responses after an allergen challenge; responses after a nonspecific challenge with methacholine, acetylcholine, or serotonin; IgE production; as well as inflammatory cell accumulation are similar in rats and humans (149). Therefore, airway hyperresponsiveness, inflammation, and obstruction characteristic of human asthma can be well mimicked in rat models. According to several authors, the main advantage of the rat models over the mouse ones is the clear differentiation of the early and late phases of the asthmatic reaction (8, 112).

Rats are easily sensitized by OVA (112, 121, 150, 151), but they are less responsive to bronchoconstrictor agents than guinea pigs. Therefore, rat asthma models are mainly used to investigate the mechanisms of the inflammatory processes (8, 152). There are, however, significant differences between respective rat strains. The Brown Norway strain is the most appropriate for studying allergic inflammation, since they naturally present a more pronounced IgE and inflammatory response to allergen challenges following sensitization (112, 117, 153, 154). Wistar rats can also be sensitized and challenged with OVA producing similar, but less pronounced symptoms than Brown Norway ones (112, 121). In contrast, Sprague–Dawley and Fischer and Lewis rats do not develop an allergic reaction or an increase in IgE production under the same conditions (112, 155–157). The OVA sensitization protocols in rats slightly differ from that of usually used in mice: in Wistar rats, it is commonly done by an i.p. or s.c. injection precipitated with aluminum hydroxide (1.0 ml of 1.0 mg/ml OVA mixed with 10% aluminum hydroxide suspension in saline). This injection is repeated by an administration of heat-killed *Bordetella pertussis* intraplantarly or i.p., as an adjuvant. A booster injection of 1.0 ml of 1.0 mg/ml OVA suspended in 10% aluminum hydroxide and saline can be injected after 7 days. The rats are used 14 days after the first injection, IgE titer significantly increases by this time point (33, 112, 151, 158). OVA can also be injected s.c. (1 mg OVA and 200 mg/ml aluminum hydroxide in 1 ml PBS) and i.p. injection of 1 ml heat-killed *Bordetella pertussis* bacteria (6×10^{-9}/ml) can be applied for boosting on days 0 and 7. Two weeks later 2% OVA is nebulized for 20 min/day for 6 days (120, 159). Intratracheal instillation of SiO_2 nanoparticles exacerbates airway hyperresponsiveness and remodeling after OVA immunization, presumably due to a remarkable eosinophilia and cytokine imbalance, particularly the increase of tissue IL-4 (160).

Guinea Pig Models

The main advantage of the guinea pig allergic model is the easy sensitization, i.e., the multifactorial, complex immunological response involving sensory neuropeptides, eicosanoids, cytokines, and chemokines (111). A direct anaphylactic bronchoconstriction upon antigen challenge, as well as a late asthmatic response and airway hyperresponsiveness develop under certain conditions. Moreover, the inflammatory response is comparable with the human situation (161). Dunkin–Hartley guinea pigs are commonly used to investigate allergic pulmonary reactions due to the remarkable reactivity of their bronchial smooth muscle to allergens, histamine, and lipid mediators (33).

They can be sensitized with a single i.p or s.c. injection of 10 μg OVA mixed with 100 mg aluminum hydroxide, and elicitation of the lung inflammation is performed with nebulized OVA (100 mg/ml for 1 h) dissolved in pathogen-free saline 14 days later (42, 111). The most commonly used experimental protocols in the OVA-induced asthma models are summarized in Table 2.

3.2.2. House Dust Mite- and Cockroach Antigen-Induced Allergic Lung Inflammation

The house dust extract most commonly contains several allergens including six indoor and three outdoor allergens: German cockroach (*Blattella germanica*, Bla g1, and Bla g2), house dust mite (*Dermatophagoides pteronyssinus* Der p1, and *Dermatophagoides farinae* Der f1), cat (*Felis domesticus*, Fel d1), dog (*Canis familiaris*, Can f1), meadow fescue (*Festuca pratensis*), short ragweed (*Ambrosia artemisiifolia*), and mold (*Alternaria alternata*), as well as endotoxins (162).

Table 2 Experimental protocols and paradigms in the ovalbumin (OVA)-induced allergic asthma models

Species	Strain	Gender/age	Sensitization	Adjuvant	Exposure/challenge	End of the experiment	References
Mouse	Balb/C, BP2	8–10-week-old male	S.c. 0.4 ml of 250 μg/ml OVA on days 0 and 7	S.c. 0.4 ml of 4 mg/ml $Al(OH)_3$ on days 0 and 7	I.n. 10 μg/50 μl OVA on day 14, or on days 14–23 (once a day), or on days 14–15 (twice a day)	24 h after last OVA-challenge	(127)
	Balb/C, Swiss, CBA	8-week-old male	S.c. 100 μg OVA on days 0 and 7	S.c. 1.6 mg $Al(OH)_3$ on days 0 and 7	I.t. 1 and 100 μg OVA on day 14 (*in vitro isolated lung experiment*)	On day 14 (*in vitro isolated lung experiment*)	(128)
	Balb/C	No data	I.p. 10 μg OVA every 2 days for 14 days	–	I.n. 20 μg OVA on days 40, 43 and 46	On day 49	(143)
	Balb/C, BP2	6–8-week-old	S.c. 100 μg OVA on days 0 and 7	S.c. 1.6 mg $Al(OH)_3$	I.n. 10 μg/50 μl OVA on day 14 (once a day), or on days 14–15 (twice a day)	At 1 h, 3 h, 6 h, 24 h, 48 h, 72 h and on days 6 and 11 after last OVA-challenge	(132)
	Balb/C, C57BL/6	Male	I.p. 10 μg OVA on days 0 and 14	$Al(OH)_3$	Inhaled 5% OVA for 20 min/day on day 21, or on days 21–25	At 6 h and on days 1, 3, 7, 14, 21 after last OVA-challenge	(134)
	Balb/C	4-week-old female	I.p. 10 μg OVA on days 0 and 14	$Al(OH)_3$	Inhaled 5% OVA for 20 min/day on day 21	On day 23	(135)
	Balb/C	8-week-old female	I.p. 10 μg OVA on days 1 and 14	$Al(OH)_3$	Inhaled 5% OVA on days 21, or on days 21 and 22, or on days 21, 22, and 23	At 24 h and 48 h after last OVA-challenge	(136)
	Balb/C	6–8-week-old female	I.p. 100 μg OVA on days 0 and 14	$Al(OH)_3$	I.n. 100 μg OVA on day 14 and 50 μg OVA on days 25, 26 and 27	On day 28	(137)
	Balb/C	6–8-week-old male	I.p. 10 μg OVA on 7 alternate days	–	Inhaled 2% OVA for 5 min/day on 8 consecutive days, two weeks after sensitization	24 h after last OVA-challenge	(142)

Balb/C	4–5-week-old male	I.p. 10 μg OVA on 7 alternate days	–	I.t. 20–80 μg OVA on 3 days, each 3 days apart, starting on or after day 40	24 h after last OVA-challenge	(144)
Balb/C	8–10-week-old	I.p. 0.1 mg/animal OVA on day 0	–	Inhaled 2% OVA for 5 min/day on day 8 and inhaled 1% OVA for 20 min/day on days 19–24	6 h, 24 h, 48 h and 72 h after last OVA-challenge	(131)
Balb/C	8–10-week-old	I.p. 0.01 mg/animal OVA on day 0	I.p. 0.2 ml $Al(OH)_3$ on day 0	I.p. 0.01 mg/animal in 0.2 ml $Al(OH)_3$ on day 10 and inhaled 5% OVA for 20 min/day on days 19–24	6 h, 24 h, 48 h and 72 h after last OVA-challenge	(131)
Balb/C	10–12-week-old female	I.p. 20 μg OVA on days 1 and 14	I.p. 2.25 mg $Al(OH)_3$ on days 1 and 14	Inhaled 1% OVA for 20 min/day on days 28–30	On day 32	(138)
Balb/C	8–12-week-old female	I.p. 20 μg OVA on days 0 and 14	I.p. 2.25 mg $Al(OH)_3$ on days 0 and 14	Inhaled 1% OVA for 20 min on days 28–30 and on day 72	On day 74	(139)
Balb/C	Male and female	I.p. 80 μg OVA at 11 and 13 weeks of age	I.p. 2 mg $Al(OH)_3$ at 11 and 13 weeks of age	Inhaled 1% OVA for 20 min/day every other day at 14 weeks of age, plus 1 day before sacrifition	At 15 weeks of age	(140, 141)
C57BL/6	6–7-week-old female	I.n. 10 μg OVA on days 0-2	–	I.n. 25 μg OVA on days 14, 15, 18 and 19	On day 21	(145)
C57BL/6	No data	I.n. 10 μg OVA on day 0	–	I.n. 25 μg OVA on day 14	On day 16	(204)

(continued)

Table 2 (continued)

Species	Strain	Gender/age	Sensitization	Adjuvant	Exposure/challenge	End of the experiment	References
Rat	Brown Norway, Fischer 344, Lewis	Male	S.c. 1 mg OVA on day 0	S.c. 22.5 mg $Al(OH)_3$ and i.p. 1×10^{10}/0.5 ml *B. pertussis* on day 0	Inhaled 0.5% OVA for 1 h on day 14	3 h, 14 h, 24 h, 48 h, 72 h, 6 days, and 1 day after OVA-challenge	(155)
	Brown Norway	7–8-week-old	S.c. 1 mg OVA on day 0	S.c. 200 mg $Al(OH)_3$ and i.p. $6x10^9$/ml *B. pertussis* on day 0	Inhaled 5% OVA for 5 min/day on days 14, 19 and 24	1 day after last OVA-challenge	(154)
	Brown Norway, Sprague Dawley	72–82-day-old	I.p. 1 ml of 1 mg/ml OVA on day 0	I.p. 1 ml of 100 mg/ml $Al(OH)_3$ on day 0	Inhaled 5% OVA for 5 min on day 21	On day 21	(157)
	Brown Norway, Sprague Dawley	6–8-week-old male	Intradermally 1 mg OVA on day 0	S.c. 22.5 mg $Al(OH)_3$ on day 0	Inhaled 1% OVA for 30 min on days 28–29	18 h after last OVA-challenge	(150)
	Wistar	Male	I.p. 0.5 ml of 400 µg/ml OVA on days 1–3 and 11	I.p. 0.5 ml of 13 mg/ml $Al(OH)_3$ on days 1–3 and 11; i. pl. 10^{10} *B. pertussis* in 50 µl on day 0	Inhaled 1% OVA on days 19–25	24 h after last OVA-challenge	(121)
	Wistar	6–7-week-old male	S.c. 1 ml of 200 µg/ml OVA on days 4, 18 and 25	S.c. 1 ml of 6.5 mg/ml $Al(OH)_3$ on days 4, 18 and 25	Inhaled 1% OVA for 30 min/day on days 31–37	On day 38	(160)
	Wistar	Female	S.c. 1 ml of 1 mg/ml OVA on days 0 and 7	S.c. 1 ml of 200 mg/ml $Al(OH)_3$ and i.p. 6×10^9/ml *B. pertussis* on days 0 and 7	Inhaled 2% OVA for 20 min/day on days 21–26	24 h after last OVA-challenge	(120)

	Wistar	Infant	I.p. 50 μg OVA on days PN1 and PN7	I.p. 20 μg $Al(OH)_3$ on days PN1 and PN7	Inhaled 20 mg/ml OVA for 30 min on PN days 14-20	On day PN21	(82)
	Wistar	3–4-months-old male	S.c. 1 ml of 1 mg/ml OVA on days 0 and 7	S.c. 10% $Al(OH)_3$ on days 0 and 7; i.p. 1×10^9/ml *B. pertussis* on day 0	Inhaled 2% OVA for 10 min on day 14	0.5 h, 8 h and 24 h after OVA-challenge	(151)
	Sprague Dawley	8–10-week-old male	S.c. 1 mg/0.8 ml OVA on day 0	S.c. 3.48 mg/0.2 ml $Al(OH)_3$ and i.p. 1×10^{10}/0.5 ml *B pertussis* on day 0	Inhaled 5% OVA for 5 min on day 14	On day 14	(156)
Guinea pig	Dunkin-Hartley	Male	I.p. 10 μg OVA on day 0	I.p. 100 mg $Al(OH)_3$ on day 0	Inhaled 100 μg/ml OVA for 1 h on day 14	On day 14	(42)

I.p intraperitoneal, *S.c* subcutaneous, *I.t* intratracheal, *I.n* intranasal, *OVA* ovalbumin, *B. pertussis* heat-killed *Bordetella pertussis*, *PNU* protein nitrogen unit, *PN1 PN7, PN21* postnatal day 1, 7, 21

Mouse Models

Some groups use the *whole house dust extract* (HDE), others administer only one component of this mixture, such as the mite or the cockroach antigen, to induce allergic pulmonary reactions in female Balb/c mice. The complex extract emulsified in 50 μl TiterMax Gold, as a booster, is administered i.p. for sensitization on day 0, and i.n. challenges are performed on days 14 and 21. Asthma-like responses were confirmed by measuring airway hyperresponsiveness by whole-body plethysmography and methacholine challenge 24 and 48 h after the challenge (162–165).

Several protocols are published for the immunization of mice with *house dust mite antigen* (HDMA). Female Balb/c mice are immunized i.p. with 2 μg of HDMA in 0.2 ml of aluminum hydroxide (alumn) on day 0 and boosted with HDMA/alumn on day 14 or 21. Usually seven days after the second immunization, animals are challenged with 0.1–0.2% HDMA by the i.n. route under isoflurane anesthesia or by aerosolization of the antigen. The inflammatory response is investigated 48–72 h after the elicitation (166, 167).

C57Bl/6 mice are immunized with 10 μg of *cockroach allergen* in incomplete Freund's adjuvant on day 0, and given an i.n. administration of the same dose in 10 ml of diluent on day 14. This initial i.n. antigen induces little cellular infiltration into the lungs, but a second challenge 6 days later by an i.n. administration of 10 mg of cockroach allergen in 50 ml of sterile PBS results in a significant inflammatory response including mononuclear cell and eosinophil infiltration (168). In female Balb/c mice, the sensitization is performed similarly with the same dose emulsified in 2.25 mg aluminum hydroxide in a total volume of 100 ml on days 0 and 14, followed by aerosol sensitization with 1% antigen for 30 min on days 28, 29, and 30. Mice are again challenged with 5% antigen for 30 min on day 32 and pulmonary functions are measured 24 h later to establish hyperresponsiveness to methacholine (169). Others administer cockroach extract to BALB/c mice i.n. on 5 consecutive days for sensitization and a week later for 4 more days for elicitation (170).

Some groups immunize with the purified dust mite allergen *Dermatophagoides pteronyssinus* Der p1 i.p. and from day 14 to 20, and induce the inflammation in the lung with dust mite extract aerosol exposure daily for 30 min house (171). Specific cockroach antigen Bla g2, specific mite antigen *Dermatophagoides farinae* Der f1, and combined immunization on days 0 and 7 can also be used. Challenges by orotracheal inhalation with these allergens are performed on day 14, and then mice are studied and sacrificed on day 17. Both antigens administered alone induced similar dose-dependent inflammatory cell accumulation, total protein increase, and hyperreactivity. Enhanced airway inflammation and epithelial damage, but not hyperresponsiveness, developed in response to the combination of these allergens (172). Others use the combination of HDE and purified cockroach or mite antigen administration. On days 14 and 21 after a conventional HDE sensitization i.p., mice are given an airway challenge of HDE under anesthesia with methoxyflurane. For the second airway exposure either purified cockroach allergen, Bla g2 or purified dust mite allergen, Der p1 is used and mice are investigated 48 h from the last airway challenge (173).

Rat Models

Rats are also easily sensitized by house dust mite extracts (112). Female Brown Norway rats are intratracheally sensitized with 10 μg HDMA (5 μg each on days 1 and 3) and challenged with 10 μg antigen 14 days later (174). After using a similar protocol others found that strong eosinophilia and neutrophilia was associated with an early release of proinflammatory cytokines (IL-13 and IL-1beta, eotaxin and thymus, and activation-regulated chemokine). This response was not attenuated by removal of HDM-associated protease activity (175). Some authors use weanling or young, 3–5-week-old Brown Norway female rats, sensitized with 10 μg HDM i.t. or i.p. and challenged with 5 μg HDM via the trachea 10 days later or for 10 consecutive days to establish an allergy model in rats to more closely reflect the developing immune system (176, 177).

Some groups compared the sensitivity of different rat strains in this model. Brown Norway, Lewis, and Sprague–Dawley rats are sensitized with HDMA by either s.c. or local i.t routes. Systemic injection induces maximal immunization, while mucosal sensitization is a more relevant exposure route. I.t. instillation is a quick, easily reproducible procedure and shows results comparable with those based on the inhalation route of administration. On days 1 and 3 systemically sensitized rats receive an s.c. injection of 5 μg of HDM in 0.5 ml of aluminum hydroxide adjuvant. Two weeks later these rats are challenged i.t. with 10 μg of HDM in 250 μl of saline. Locally sensitized rats received 5 μg of HDM in 250 μl of saline by i.t. instillation on days 1 and 3. Two weeks later, these rats are i.t. challenged with 10 μg of HDM in 250 μl of saline one, two, or five times, each successive challenge separated by 1 week. Airway responses to HDM are evaluated in all locally sensitized rats on the last day of allergen challenge. Airway responses to HDM are measured on the day of challenge, and rats are assessment of allergic responses, pulmonary inflammation, and lung injury 2 days later, since pulmonary inflammation and injury at this time point. Brown Norway rats develop stronger clinical hallmarks of allergic asthma under various immunization regimes compared to other strains; therefore, they are considered to be the most appropriate strain for studying allergic asthma-like responses in rats. Phenotypic differences in response to HDM were associated with differences in the Th1/Th2 cytokine balance and antioxidant capacity (178).

Guinea Pig Models

Two i.p. injections of 100 μg crude mite extract (CME), but not multiple aerosol inhalations cause sensitization in guinea pigs. The sensitization to mites was confirmed by the measurement of serum antimite antibody titer and the detection of anaphylactic bronchoconstriction after i.v. injection of CME solution. The inhalation challenge of CME aerosol in sensitized animals causes prolonged eosinophilia in the bronchoalveolar lavage fluid which persists for at least 7 days after a single challenge. Neither inhalation at higher concentrations of CME aerosol nor repeated inhalation challenges increases the degree of eosinophilia compared to a single application. There is a greater eosinophilia in the mite model than in response to the same OVA protocol (179).

Cat Models

Spontaneous development of asthma with prominent features similar to humans has been documented to occur relatively frequently in *cats*. In an extensive study using HDMA and Bermuda grass, allergens (selected by screening ten privately owned pet cats with spontaneous asthma) were applied by parenteral sensitization and aerosol challenges to replicate the naturally developing disease in the experimental cats. Sensitization with HDMA or the grass allergen in cats led to allergen-specific IgE production, allergen-specific serum and BALF IgG and IgA production, airway hyperreactivity, airway eosinophilia, an acute TH2 cytokine profiles in peripheral blood mononuclear cells and bronchoalveolar lavage fluid cells, and histological evidence of airway remodeling (180).

Sheep Models

Sheep can be actively sensitized with HDMA. In this case, animals develop allergen-specific IgE responses, inflammation of the airways including eosinophilia, and mucus hypersecretion of the airways (9). Immunization is done s.c. with 50-μg HDMA and elicitation with a single bronchial challenge with the antigen. This induces an allergen-specific IgE response in 50–60% of sheep (allergic sheep), with higher antigen doses increasing specific IgG1, but not IgE. Lung challenge of allergic sheep with HDMA leads to the initial recruitment of neutrophils (at 6 h postchallenge) followed by eosinophils and activated lymphocytes into the lung, similar to the late phase allergic response seen in human asthma. Eosinophils, CD4+ T cells, and CD45R+ B cells are the most prominent leucocytes found in lung tissue 48 h after allergen challenge (181).

3.2.3. Plant (Ragweed and Bermuda Grass) and Helminth Antigen-Induced Allergic Lung Inflammation

Mouse and Rat Models

Mice are sensitized and challenged with endotoxin-free ragweed extract (150 μg/100 μl, combined with aluminum hydroxide adjuvant in a 3:1 ratio), on days 0 and 4. On day 11, they are challenged intranasally with the ragweed extract (100 μg) dissolved in 50 μl of PBS. This results in the development of airway hyperresponsiveness, metaplasia of airway epithelial cells, mucus hypersecretion, infiltration of eosinophils into subepithelium of lung, as well as the release of TH2 cytokines (182, 183).

Wistar, Lewis, and Fischer 344 rats can be actively sensitized with dinitrophenylated Ascaris antigens (2 mg protein) and killed *Bordetella pertussis* vaccine. After 5–8 days, asthmatic response is provoked by inhalation of the antigen (0.5 mg protein; (112)). Lewis rats have been proved to be the most suitable for studying the effects of antiallergic agents (184).

Rabbit Models

The allergic rabbit demonstrates many of the pathophysiological features of human asthma including acute bronchoconstriction, late phase airway obstruction, and airway hyperresponsiveness. The sensitizing agent in New Zealand white rabbits of both sexes is commonly the *Alternaria tenius* antigen (0.5 ml, 40,000 protein nitrogen units/ml) in sterile 0.5 ml $Al(OH)_3$ gel and 1 ml saline (185, 186) mixed to an adjuvant aluminum hydroxide and injected i.p. within 24 h after birth to ensure the predominant production of IgE antibodies. If delayed, sensitization produces both IgE and IgG antibodies and a subsequent diminished pulmonary response to antigen provocation. Weekly booster injections of the antigen/alum mixture out to 4 months are necessary to ensure full sensitization. Extreme care must be taken when sensitizing pups because the rabbit doe may reject the offspring under stressful conditions (111, 187).

Larger Animal Models

The larger species display many of the hallmark features of allergic airways disease in humans which are responsive to several antiasthma drugs. Their main advantages are (1) the structural and functional arrangement of the airways, which show closer similarities with humans; (2) their use allows relatively longer term studies; and (3) their ability to perform the type of procedures routinely used in humans, such as lung function measurements, aerosolized allergen challenges, and bronchoscopic procedures for lavage and endobronchial biopsy sampling (188). These are very valuable to explore the kinetics of the diseases and also the structural and functional alterations associated with airway remodeling. However, they are expensive, their use represents several ethical considerations, and there are very few appropriate immunological tools available in these species (8).

Cats can develop idiopathic asthma (189) and therefore, they are unique for experimental model of human asthma following sensitization and challenge with the intestinal helminth parasite *Ascaris suum* (190) or with *house dust mite extract* or *Bermuda grass allergens* (180). Experimental asthma in cats represents many of the features of asthma in humans including allergen-specific IgE, airway eosinophilia, and airway hyperresponsiveness in response to allergen, as well as the acute clinical signs of cough, tachypnea, or respiratory distress. Airway remodeling also develops following frequent allergen exposure. Similarly to what is observed with HDMA, Bermuda grass extract also evokes allergic lung inflammation in cats (180).

Dogs have also been used as a model to study human lung physiology under normal and inflammatory conditions (7, 191, 192). As a model for human allergic asthma, dogs sensitized to *Ascaris suum* (193) or *ragweed* allergen (194) display prominent pulmonary inflammation and airway responsiveness following allergen challenge (9). Furthermore, the standard antiasthma drugs, such as beta-adrenergic agonists, corticosteroids, and leukotriene inhibitors are effective in dog models. However, there are no data on chronic airway challenges and on airway remodeling in dogs. Similar to other larger animals, the lack of immunological reagents and ethical concerns are the main drawbacks to the use of dogs for asthma research (9). The important features of canine models and how they have increased the understanding of the pathology, pathophysiology, and control of human asthma have been recently reviewed (111, 192).

Pigs actively sensitized with *Ascaris suum* (or ovalbumin) develop airway recruitment of eosinophils and bronchoconstrictive responses following airway allergen challenge (195). However, there are no available data on morphological or histological changes in airway tissues in response to allergens. The main problems with this model are the individual variations in sensitization and the differences in the extent of bronchoconstriction probably due to their rapid growth and high endogenous cortisol levels. Therefore, pigs have considerably little significance in asthma research (7, 9).

The use of *sheep* as a model for human allergic asthma has included animals naturally sensitized to *Ascaris suum* (196). Furthermore, a proportion of allergic sheep display early and late phases of the asthmatic reaction similarly to man (9, 197–199).

Airway responses to aerosolized *Ascaris suum* antigen can be measured in naturally allergic sheep with a history of early and late airway responses and airway hyperresponsiveness after antigen challenge immediately after the antigen challenge and then serially for 8 h afterwards (198, 200–202).

Monkey experiments are very rare, but they can be useful to bridge the gap to human relevance more than rodents (9, 111). Technical details of these allergic airway inflammation models are summarized in Table 3.

3.3. Link Between Endotoxin Exposure and Allergic Mechanisms

Endotoxins are present in the whole environment, and it has been well demonstrated that the severity of asthma is related to their amount in house dust (203). Interactions between inhaled endotoxins and allergens in lungs of asthmatic patients could amplify the local inflammatory response, rendering the clinical manifestations more severe and difficult to control. LPS is an important adjuvant related to asthma severity through aggravation of bronchial inflammation; however, on the other hand, it may modulate the Th1:Th2 balance to the Th1 side. This is based on the fact that the lack of LPS exposure or naturally occurring infections in children may delay maturation of the immune system towards a

Table 3 Experimental protocols and paradigms in other allergic pulmonary inflammation models

Species	Strain	Gender/age	Sensitization	Adjuvant	Exposure/challenge	End of the experiment	References
Mouse	Balb/C	4-week-old female	I.p. HDM *(Der f 1)* on days 0 and 21	$Al(OH)_3$	Inhaled 1 mg/ml HDM *(Der f 1)* for 30 min/day on days 29, 31, 33, 36, 38 and 40	On day 42	(167)
	Balb/C	No data	I.p. 2 μg HDM on days 0 and 14	$Al(OH)_3$	I.n. 100 μg HDM on day 21	On day 24	(166)
	C57BL/6	Female	I.p. 10 μg cockroach allergen on day 0 and i.n. 10 μg cockroach allergen on day 14	IFA	I.t. 10 μg cockroach allergen on day 20 or on days 20 and 22	8 h, 24 h or 48 h after last allergen-challenge	(168)
	C57BL/6	6–8-week-old male	I.p. 10 μg purified HDM *(Der p 1)* on day 0	I.p. 1 mg $Al(OH)_3$ on day 0	Inhaled 3, 30 and 300 μg/ml HDM extract *(Der p 1)* for 30 min/day on days 14–20	24 h after last allergen-challenge	(171)
	Balb/C	Female	I.p. 50 μl HDE containing high concentrations of cockroach antigens *(Bla g 1 and Bla g 2)* on day 0	I.p. 50 μl TiterMax Gold on day 0	I.t. 50 μl HDE containing high concentrations of cockroach antigens (Bla g 1 and Bla g 2) on days 14 and 21	On days 22 or 23	(162–164, 173)
	**A/J	6–8-week-old male	I.p. 0.01, 0.1, 1, 10 μg/animal *r Der f 1* or *r Bla g 2*, or 0.05 μg *r Der f 1* and 0.05 μg *r Bla g 2* on days 0 and 7	I.p. 1 mg of 0.2 ml $Al(OH)_3$ on days 0 and 7	Inhaled 50 μg *r Der f 1* or *r Bla g 2*, or 0.25 μg *r Der f 1* and 0.25 μg *r Bla g 2* on day 14	On day 17	(172)
	Balb/C	4–5-week-old female	10 μg CRA in 100 μl on days 0 and 14; inhaled 1% CRA for 30 min on days 28–30 and 5% CRA for 30 min on day 32	2.25 mg $Al(OH)_3$ in 100 μl on day 0 and 14	Inhaled 5% CRA on day 43	On day 45	(169)

(continued)

Table 3 (continued)

Species	Strain	Gender/age	Sensitization	Adjuvant	Exposure/challenge	End of the experiment	References
Rat	Fischer 344, Lewis, Wistar	Male	2 mg DNP-As on day 0 and 0.5 mg DNP-As on day 5	1×10^{10}/ml B. pertussis on day 0	Inhaled 3 mg DNP-As for 5 min on day 8	On day 8	(184)
	Brown Norway	8–10-week-old female	I.t. 5 μg HDM *(Der f 1)* on days 1 and 3	–	I.t. 10 μg HDM *(Der f 1)* 14 days later	2 or 7 days after antigen-challenge	(174)
	Brown Norway, Lewis, Sprague Dawley	Female	S.c. 5 μg HDM *(Der f 1)* on days 1 and 3	0.5 ml 1.3% $Al(OH)_3$ on days 1 and 3	I.t. 10 μg HDM *(Der f 1)* 14 days later	2 days after allergen-challenge	(178)
	Brown Norway	3-week-old female	I.p. 10 μg purified HDM extract *(Der f 1)* on day 0 or or i.t. 3.3 μg purified HDM extract *(Der f 1)* in 100 μl on days 1–3	I.p. 10^8 *B. pertussis* on day 1	I.t. 5 μg HDM antigen 10 days later	2 or 4 days after antigen-challenge	(176)
	Brown Norway	5-week-old female	I.t. 0.15, 15, 150 μg purified HDM extract in 100 μl on days 1–10	–	–	48 h after last allergen-challenge	(177)
Guinea pig	Camm-Hartley	Male	I.p. 100 μg crude mite extract *(Der f 1)* on days 0 and 7	I.p. 4 mg $Al(OH)_3$ on days 0 and 7	Inhaled 5 ml of 1 mg/ml crude mite extract for 10 min on day 21	On days 22, 24 and 28	(179)
Rabbit	New Zealand white	Male and female	I.p. 0.125 ml of 40,000 PNU/ml *A. tenius* within 24 h of birth, and weekly in the first 4 weeks, and then biweekly until 3 months of age	I.p. 0.125 ml of $Al(OH)_3$ within 24 h of birth, and weekly in the first 4 weeks, and then biweekly until 3 months of age	Inhaled 20,000 PNU/ml *A. tenius* for 20 min at 3 months of age	At 3 months of age	(187)

Cat		6–24-month-old	S.c. 10 μg HDM *(Der f 1)* or BGA on days 0 and 21 and i.n. 0.2 ml HDM (Der f 1) or BGA on day 14	S.c. 10 mg $Al(OH)_3$ on days 0 and 21; i.m. 10^5 *B. pertussis* on day 0	Inhaled 0.07 mg/ml HDM *(Der f 1)* and BGA twice weekly from day 28 (7 times), then weekly (for 1 month), then monthly	At 1 year	(180)
		7–13-month-old male and female	I.m. *A. suum* antigen on days 0 and 14	I.m. TiterMax on days 0 and 14	Inhaled 0.01% *A. suum* for 5 min, 5 times at 3-month intervals, starting 2–4 weeks later	72 h after last allergen-challenge	(190)
Pig		Male and female	S.c. 0.6–1.0 mg *A. suum* allergen 3 times	–	Inhaled 2 ml of 7 mg/ml *A. suum* extract 1 week later	8 h after allergen-challenge	(195)
Sheep	Merino-cross lambs	4–5-month-old female	S.c. 5–500 μg HDM or i.t. 1 mg HDM in 5 ml on days 0, 14 and 28	S.c. $Al(OH)_3$ on days 0, 14 and 28	I.t. 1 mg HDM in 5 ml on day 42	48 h after allergen-challenge	(181)
	Florida Native sheep	Female	Naturally sensitized	–	Inhaled 82,000 PNU/ml *A. suum* extract	24 h after allergen-challenge	(201)
Dog	Mongrel dog	Newborn male and female	I.p. 500 μg ragweed allergen within 24 h of birth, and weekly in the first 8 weeks, and then biweekly until 16 weeks of age	I.p. 30 mg $Al(OH)_3$ within 24 h of birth, and weekly in the first 8 weeks, and then biweekly until 16 weeks of age	I.t. ragweed solution at 4, 6, 8, 10, 12 and 15 months of age	2 months after last allergen-challenge	(194)

I.p intraperitoneal, *S.c* subcutaneous, *I.t* intratracheal, I.n intranasal, *OVA* ovalbumin, *IFA* incomplete Freund's adjuvant, *HDM* house dust mite, *HDE* house dust extract, *Der p 1 Dermatophagoides pteronyssinus 1* (European house dust mite), *Der f 1 Dermatophagoides farinae 1* (American house dust mite), *Bla g 1 and 2 Blatella germanica* (German cockroach) antigens, *CRA* cockroach mix allergen: ***Periplaneta americana*** (American cockroach) and *Blatella germanica* (German cockroach), *B. pertussis* heat-killed *Bordetella pertussis, DNP-As* 2,4-dinitrophenylated *Ascaris* extract, *PNU* protein nitrogen unit, *BGA* Bermuda grass

Th1-skewed response, thus increase the risks of allergy and asthma. Although favoring a Th1 cytokine response that could be beneficial to asthmatics, LPS aggravates bronchopulmonary inflammation by several mechanisms. These include neutrophil and eosinophil recruitment, and release of proinflammatory cytokines and nitric oxide from activated macrophages. LPS exerts its biological actions through its interaction with CD14. The genetic locus of CD14 is close to the genomic region controlling levels of IgE. A polymorphism in the CD14 promoter region seems to favor high serum IgE levels. Genetic influences control circulating levels of sCD14 and by this mechanism modulate Th1:Th2 balance and IgE synthesis. LPS exposure, although hazardous to the asthmatic, seems to exert a role in the maturation of the immune system in children towards a Th1-skewed pattern (15, 33). Based on these important human observations, these interactions have recently been aimed to be investigated in animal models as well, particularly in the developing lung. In a mouse experimental asthma model, the administration of LPS, particularly at low doses, enhanced the levels of OVA-induced eosinophilic airway inflammation. The OVA-induced eosinophilic inflammation in the lung was dramatically increased by the administration of LPS in wild-type mice, whereas such increase was not observed in mast-cell-deficient or TLR4-deficient mice. Thus, mast cells appear to control allergic airway inflammation after their activation and modulation through TLR4-mediated increase in Th2 cytokine production (145, 204). Furthermore, repeated inhalation exposures of neonatal and juvenile mice to endotoxin and cockroach allergen increased the pulmonary inflammatory and systemic immune responses in a synergistic manner and enhanced alveolar remodeling in the developing lung (205).

Acknowledgments

This work was sponsored by Hungarian Grants OTKA K73044, ETT 03-380/2009, Developing Competitiveness of Universities in the South Transdanubian Region (SROP-4.2.1.B-10/2/KONV-2010-0002). The authors are grateful to Dr. Ágnes Kemény for drawing Fig. 1.

References

1. Banner KH, Igney F, Poll C (2011) TRP channels: emerging targets for respiratory disease. Pharmacol Ther 130:371–384
2. Jiang LH, Gamper N, Beech DJ (2011) Properties and therapeutic potential of transient receptor potential channels with putative roles in adversity: focus on TRPC5, TRPM2 and TRPA1. Curr Drug Targets 12:724–736
3. Nassini R et al (2010) Transient receptor potential channels as novel drug targets in respiratory diseases. Curr Opin Investig Drugs 11:535–542

4. Lee LY, Gu Q (2009) Role of TRPV1 in inflammation-induced airway hypersensitivity. Curr Opin Pharmacol 9:243–249
5. Bessac BF, Jordt SE (2008) Breathtaking TRP channels: TRPA1 and TRPV1 in airway chemosensation and reflex control. Physiology (Bethesda) 23:360–370
6. Nilius B, Voets T, Peters J (2005) TRP channels in disease. Sci STKE 2005:re8
7. Kurucz I, Szelenyi I (2006) Current animal models of bronchial asthma. Curr Pharm Des 12:3175–3194
8. Zosky GR, Sly PD (2007) Animal models of asthma. Clin Exp Allergy 37:973–988
9. Allen JE et al (2009) Animal models of airway inflammation and airway smooth muscle remodelling in asthma. Pulm Pharmacol Ther 22:455–465
10. Chen H, Bai C, Wang X (2010) The value of the lipopolysaccharide-induced acute lung injury model in respiratory medicine. Expert Rev Respir Med 4:773–783
11. Kips JC et al (2003) Murine models of asthma. Eur Respir J 22:374–382
12. Bates JH, Rincon M, Irvin CG (2009) Animal models of asthma. Am J Physiol Lung Cell Mol Physiol 297:401–410
13. Wang HM, Bodenstein M, Markstaller K (2008) Overview of the pathology of three widely used animal models of acute lung injury. Eur Surg Res 40:305–316
14. Rietschel ET et al (2000) Bacterial endotoxin: molecular relationships of structure to activity and function. FASEB J 8:217–225
15. Lapa e Silva JR et al (2000) Endotoxins, asthma, and allergic immune responses. Toxicology 152:31–35
16. Hoshino K et al (1999) Cutting edge: Toll-like receptor 4 (TLR4)-deficient mice are hyporesponsive to lipopolysaccharide: evidence for TLR4 as the Lps gene product. J Immunol 162:3749–3752
17. Savov JD et al (2005) Toll-like receptor 4 antagonist (E5564) prevents the chronic airway response to inhaled lipopolysaccharide. Am J Physiol Lung Cell Mol Physiol 289: 329–337
18. Lu YC, Yeh WC, Ohashi PS (2008) LPS/TLR4 signal transduction pathway. Cytokine 42:145–151
19. O'Leary EC, Marder P, Zuckerman SH (1996) Glucocorticoid effects in an endotoxin-induced rat pulmonary inflammation model: differential effects on neutrophil influx, integrin expression, and inflammatory mediators. Am J Respir Cell Mol Biol 15:97–106
20. Ulich TR et al (1991) Intratracheal injection of endotoxin and cytokines. II. Interleukin-6 and transforming growth factor beta inhibit acute inflammation. Am J Pathol 138: 1097–1101
21. Ulich TR et al (1991) The intratracheal administration of endotoxin and cytokines. I. Characterization of LPS-induced IL-1 and TNF mRNA expression and the LPS-, IL-1-, and TNF-induced inflammatory infiltrate. Am J Pathol 138:1485–1496
22. Wesselius LJ et al (1995) Synergism of intratracheally administered tumor necrosis factor with interleukin-1 in the induction of lung edema in rats. J Lab Clin Med 125: 618–625
23. Haddad EB et al (2001) Role of p38 MAP kinase in LPS-induced airway inflammation in the rat. Br J Pharmacol 132:1715–1724
24. Togbe D et al (2007) Toll-like receptor and tumour necrosis factor dependent endotoxin-induced acute lung injury. Int J Exp Path 88:387–391
25. Takeda M et al (2011) Proliferation of sensory C fibers and subsequent neurogenic inflammation in rat airway induced by inhaled lipopolysaccharide. Neurotoxicology 32:954–962
26. Pauwels RA et al (1990) The effect of endotoxin inhalation on airway responsiveness and cellular influx in rats. Am Rev Respir Dis 141:540–545
27. Harmsen AG (1988) Role of alveolar macrophages in lipopolysaccharide-induced neutrophil accumulation. Infect Immun 56: 1858–1863
28. Goncalves de Moraes VL et al (1996) Effect of cyclo-oxygenase inhibitors and modulators of cyclic AMP formation on lipopolysaccharide-induced neutrophil infiltration in mouse lung. Br J Pharmacol 117:1792–1796
29. Gordon T et al (1991) Airway oedema and obstruction in guinea pigs exposed to inhaled endotoxin. Br J Ind Med 48:629–635
30. Savov JD et al (2002) Neutrophils play a critical role in development of LPS-induced airway disease. Am J Physiol Lung Cell Mol Physiol 283:952–962
31. Barnes PJ (2001) Neurogenic inflammation in the airways. Respir Physiol 125:145–154
32. Kraneveld AD, Nijkamp FP (2001) Tachykinins and neuro-immune interactions in asthma. Int Immunopharmacol 1: 1629–1650
33. Vargaftig BB (1997) Modifications of experimental bronchopulmonary hyperresponsiveness. Am J Respir Crit Care Med 156: 97–102

34. Lefort J, Motreff L, Boris Vargaftig B (2001) Airway administration of *Escherichia coli* endotoxin to mice induces glucocorticosteroid-resistant bronchoconstriction and vasopermeation. Am J Respir Cell Mol Biol 24:345–351
35. Helyes Z et al (2007) Role of transient receptor potential vanilloid 1 receptors in endotoxin-induced airway inflammation in the mouse. Am J Physiol Lung Cell Mol Physiol 292:1173–1181
36. Helyes Z et al (2009) Impaired defense mechanism against inflammation, hyperalgesia, and airway hyperreactivity in somatostatin 4 receptor gene-deleted mice. Proc Natl Acad Sci USA 106:13088–13093
37. Helyes Z et al (2010) Involvement of preprotachykinin A gene-encoded peptides and the neurokinin 1 receptor in endotoxin-induced murine airway inflammation. Neuropeptides 44:399–406
38. Elekes K et al (2007) Role of capsaicin-sensitive afferents and sensory neuropeptides in endotoxin-induced airway inflammation and consequent bronchial hyperreactivity in the mouse. Regul Pept 141:44–54
39. Elekes K et al (2008) Inhibitory effects of synthetic somatostatin receptor subtype 4 agonists on acute and chronic airway inflammation and hyperreactivity in the mouse. Eur J Pharmacol 578:313–322
40. Elekes K et al (2011) Pituitary adenylate cyclase-activating polypeptide plays an anti-inflammatory role in endotoxin-induced airway inflammation: in vivo study with gene-deleted mice. Peptides 32:1439–1446
41. Vincent D et al (1993) Intratracheal *E. coli* lipopolysaccharide induces platelet-dependent bronchial hyperreactivity. J Appl Physiol 74:1027–1038
42. Toward TJ, Smith N, Broadley KJ (2004) Effect of phosphodiesterase-5 inhibitor, sildenafil (Viagra), in animal models of airways disease. Am J Respir Crit Care Med 169:227–234
43. Long NC, Frevert CW, Shore SA (1996) Role of C fibers in the inflammatory response to intratracheal lipopolysaccharide. Am J Physiol 271:L425–L431
44. Tsuji F et al (2010) Effects of SA13353, a transient receptor potential vanilloid 1 agonist, on leukocyte infiltration in lipopolysaccharide-induced acute lung injury and ovalbumin-induced allergic airway inflammation. J Pharmacol Sci 112:487–490
45. Bowden JJ et al (1996) Sensory denervation by neonatal capsaicin treatment exacerbates Mycoplasma pulmonis infection in rat airways. Am J Physiol 270:393–403
46. Franco-Penteado CF et al (2004) Mechanisms involved in the enhancement of allergic airways neutrophil influx by permanent C-fiber degeneration in rats. J Pharmacol Exp Ther 313:440–448
47. Hashiba Y et al (1989) Capsaicin-sensitive nerves exert an inhibitory effect on the development of fibrin-induced pulmonary edema in rats. Am Rev Respir Dis 140:652–658
48. Baraldi PG et al (2010) Transient receptor potential ankyrin 1 (TRPA1) channel as emerging target for novel analgesics and anti-inflammatory agents. J Med Chem 53: 5085–5107
49. Szarka RJ et al (1997) A murine model of pulmonary damage induced by lipopolysaccharide via intranasal instillation. J Immunol Methods 202:49–57
50. Delclaux C et al (1997) Alveolar neutrophils in endotoxin-induced and bacteria-induced acute lung injury in rats. Am J Physiol 273:104–112
51. Yamada H et al (2000) Acid instillation enhances the inflammatory response to subsequent lipopolysaccharide challenge in rats. Am J Respir Crit Care Med 162: 1366–1371
52. Harkema JR, Hotchkiss JA (1992) In vivo effects of endotoxin on intraepithelial mucosubstances in rat pulmonary airways. Quantitative histochemistry. Am J Pathol 141:307–317
53. Yamamoto T et al (1998) The role of leukocyte emigration and IL-8 on the development of lipopolysaccharide-induced lung injury in rabbits. J Immunol 161:5704–5709
54. Chignard M, Balloy V (2000) Neutrophil recruitment and increased permeability during acute lung injury induced by lipopolysaccharide. Am J Physiol Lung Cell Mol Physiol 279:1083–1090
55. Rocksén D et al (2003) Vitamin E reduces transendothelial migration of neutrophils and prevents lung injury in endotoxin-induced airway inflammation. Am J Respir Cell Mol Biol 28:199–207
56. Okamoto T et al (2004) Multiple contributing roles for NOS2 in LPS-induced acute airway inflammation in mice. Am J Physiol Lung Cell Mol Physiol 286:198–209
57. Gatti S et al (1993) Role of tumour necrosis factor and reactive oxygen intermediates in lipopolysaccharide-induced pulmonary oedema and lethality. Clin Exp Immunol 91:456–461

58. Conti G et al (2010) Evaluation of lung inflammation induced by intratracheal administration of LPS in mice: comparison between MRI and histology. MAGMA 23:93–101
59. George CL et al (2001) Endotoxin responsiveness and subchronic grain dust-induced airway disease. Am J Physiol Lung Cell Mol Physiol 280:L203–L213
60. Brass DM et al (2003) Subchronic endotoxin inhalation causes persistent airway disease. Am J Physiol Lung Cell Mol Physiol 285:755–761
61. Brass DM, Savov JD, Schwartz DA (2003) Intercellular adhesion molecule-1 plays a pivotal role in endotoxin-induced airway disease. Chest 123:416S
62. Savov JD et al (2003) Fibrinolysis in LPS-induced chronic airway disease. Am J Physiol Lung Cell Mol Physiol 285:940–948
63. Brass DM et al (2008) Chronic LPS inhalation causes emphysema-like changes in mouse lung that are associated with apoptosis. Am J Respir Cell Mol Biol 39:584–590
64. O'Leary EC, Zuckerman SH (1997) Glucocorticoid-mediated inhibition of neutrophil emigration in an endotoxin-induced rat pulmonary inflammation model occurs without an effect on airways MIP-2 levels. Am J Respir Cell Mol Biol 16:267–274
65. Aimbire F et al (2005) Effect of LLLT Ga-Al-As (685 nm) on LPS-induced inflammation of the airway and lung in the rat. Lasers Med Sci 20:11–20
66. Mafra de Lima F et al (2010) Low intensity laser therapy (LILT) in vivo acts on the neutrophils recruitment and chemokines/cytokines levels in a model of acute pulmonary inflammation induced by aerosol of lipopolysaccharide from Escherichia coli in rat. J Photochem Photobiol B 101:271–278
67. Garat C et al (1995) Alveolar endotoxin increases alveolar liquid clearance in rats. J Appl Physiol 79:2021–2028
68. de Garavilla L et al (2005) A novel, potent dual inhibitor of the leukocyte proteases cathepsin G and chymase: molecular mechanisms and anti-inflammatory activity in vivo. J Biol Chem 280:18001–18007
69. Kaneko Y et al (2007) Effects of theophylline on chronic inflammatory lung injury induced by LPS exposure in guinea pigs. Allergol Int 56:445–456
70. De Castro CM et al (1995) Fenspiride: an anti-inflammatory drug with potential benefits in the treatment of endotoxemia. Eur J Pharmacol 294:669–676
71. Arbibe L et al (1998) Generation of lysophospholipids from surfactant in acute lung injury is mediated by type-II phospholipase A2 and inhibited by a direct surfactant protein A-phospholipase A2 protein interaction. J Clin Invest 102:1152–1160
72. Marek W et al (2008) Endotoxin-induced airway hyperresponsiveness in rabbits: contribution of neuropeptides. J Physiol Pharmacol 59:421–432
73. Jie Z et al (2003) Protective effects of alpha 1-antitrypsin on acute lung injury in rabbits induced by endotoxin. Chin Med J (Engl) 116:1678–1682
74. Mikawa K et al (2003) ONO-1714, a nitric oxide synthase inhibitor, attenuates endotoxin-induced acute lung injury in rabbits. Anesth Analg 97:1751–1755
75. Pacht ER, Kindt GC, Lykens MG (1992) Increased antioxidant activity in bronchoalveolar lavage fluid after acute lung injury in anesthetized sheep. Crit Care Med 20:1441–1447
76. Kirov MY et al (2004) Extravascular lung water assessed by transpulmonary single thermodilution and postmortem gravimetry in sheep. Crit Care 8:R451–R458
77. Nieman GF et al (1996) Surfactant replacement in the treatment of sepsis-induced adult respiratory distress syndrome in pigs. Crit Care Med 24:1025–1033
78. Lutz C et al (1998) Aerosolized surfactant improves pulmonary function in endotoxin-induced lung injury. Am J Respir Crit Care Med 158:840–845
79. Maurenbrecher H et al (2001) An animal model of response and nonresponse to inhaled nitric oxide in endotoxin-induced lung injury. Chest 120:573–581
80. Tabor DR, Kiel DP, Jacobs RF (1987) Receptor-mediated ingestion responses by lung macrophages from a canine model of ARDS. J Leukoc Biol 41:539–543
81. Welsh CH et al (1988) Pentoxifylline decreases endotoxin-induced pulmonary neutrophil sequestration and extravascular protein accumulation in the dog. Am Rev Respir Dis 138:1106–1114
82. Cao L et al (2010) Maternal endotoxin exposure attenuates allergic airway disease in infant rats. Am J Physiol Lung Cell Mol Physiol 298:670–677
83. Brackett DJ et al (1985) Evaluation of cardiac output, total peripheral vascular resistance, and plasma concentrations of vasopressin in the conscious, unrestrained rat during endotoxemia. Circ Shock 17:273–284
84. Natanson C et al (1986) Gram-negative bacteremia produces both severe systolic and diastolic cardiac dysfunction in a canine model

that simulates human septic shock. J Clin Invest 78:259–270
85. Law WR, Ferguson JL (1988) Naloxone alters organ perfusion during endotoxin shock in conscious rats. Am J Physiol 255: H1106–H1113
86. Denis M et al (1994) A mouse model of lung injury induced by microbial products: implication of tumor necrosis factor. Am J Respir Cell Mol Biol 10:658–664
87. Peristeris P et al (1992) N-acetylcysteine and glutathione as inhibitors of tumor necrosis factor production. Cell Immunol 140: 390–399
88. Lefort J et al (1998) Systemic administration of endotoxin induces bronchopulmonary hyperreactivity dissociated from TNF-alpha formation and neutrophil sequestration into the murine lungs. J Immunol 161:474–480
89. Kuo MY et al (2011) Luteolin attenuates the pulmonary inflammatory response involves abilities of antioxidation and inhibition of MAPK and NFκB pathways in mice with endotoxin-induced acute lung injury. Food Chem Toxicol 49:2660–2666
90. Yang KY et al (2011) IV delivery of induced pluripotent stem cells attenuates endotoxin-induced acute lung injury in mice. Chest 140:1243–1253
91. Arbibe L et al (1997) Endotoxin induces expression of type II phospholipase A2 in macrophages during acute lung injury in guinea pigs: involvement of TNF-alpha in lipopolysaccharide-induced type II phospholipase A2 synthesis. J Immunol 159:391–400
92. Larsson R et al (2000) Dose-dependent activation of lymphocytes in endotoxin-induced airway inflammation. Infect Immun 68:6962–6969
93. Ciencewicki J, Trivedi S, Kleeberger SR (2008) Oxidants and the pathogenesis of lung diseases. J Allergy Clin Immunol 122: 456–468
94. Balakrishna S et al (2011) Environmentally persistent free radicals induce airway hyperresponsiveness in neonatal rat lungs. Part Fibre Toxicol 8:11
95. Donaldson K et al (2005) Combustion-derived nanoparticles: a review of their toxicology following inhalation exposure. Part Fibre Toxicol 2:10
96. Sydlik U et al (2009) The compatible solute ectoine protects against nanoparticle-induced neutrophilic lung inflammation. Am J Respir Crit Care Med 180:29–35
97. Sydlik U et al (2006) Ultrafine carbon particles induce apoptosis and proliferation in rat lung epithelial cells via specific signaling pathways both using EGF-R. Am J Physiol Lung Cell Mol Physiol 291:725–733
98. Kim YM et al (2005) Ultrafine carbon particles induce interleukin-8 gene transcription and p38 MAPK activation in normal human bronchial epithelial cells. Am J Physiol Lung Cell Mol Physiol 288:432–441
99. Bolognin M et al (2009) Characterisation of the acute and reversible airway inflammation induced by cadmium chloride inhalation in healthy dogs and evaluation of the effects of salbutamol and prednisolone. Vet J 179: 443–450
100. Gavett SH, Oberdörster G (1994) Cadmium chloride and cadmium metallothionein-induced pulmonary injury and recruitment of polymorphonuclear leukocytes. Exp Lung Res 20:517–537
101. Wirth D et al (2004) Evidence for a role of heat shock factor 1 in inhibition of NF-kappaB pathway during heat shock response-mediated lung protection. Am J Physiol Lung Cell Mol Physiol 287:953–961
102. Zhang W et al (2010) Anti-inflammatory effects of formoterol and ipratropium bromide against acute cadmium-induced pulmonary inflammation in rats. Eur J Pharmacol 628:171–178
103. Mannino DM et al (2004) Urinary cadmium levels predict lower lung function in current and former smokers: data from the Third National Health and Nutrition Examination Survey. Thorax 59:194–198
104. Kirschvink N et al (2005) Repeated cadmium nebulizations induce pulmonary MMP-2 and MMP-9 production and emphysema in rats. Toxicology 211:36–48
105. Corbel M, Boichot E, Lagente V (2000) Role of gelatinases MMP-2 and MMP-9 in tissue remodeling following acute lung injury. Braz J Med Biol Res 33:749–754
106. Abboud RT, Vimalanathan S (2008) Pathogenesis of COPD. Part I. The role of protease-antiprotease imbalance in emphysema. Int J Tuberc Lung Dis 12:361–367
107. Hotchkiss JA et al (1989) Comparison of acute ozone-induced nasal and pulmonary inflammatory responses in rats. Toxicol Appl Pharmacol 98:289–302
108. Wagner JG et al (2003) Ozone exposure enhances endotoxin-induced mucous cell metaplasia in rat pulmonary airways. Toxicol Sci 74:437–446
109. Hatch GE et al (1994) Ozone dose and effect in humans and rats. A comparison using oxygen-18 labeling and bronchoalveolar lavage. Am J Respir Crit Care Med 150: 676–683

110. Harkema JR, Wagner JG (2002) Non-allergic models of mucous cell metaplasia and mucus hypersecretion in rat nasal and pulmonary airways. Novartis Found Symp 248:181–197, discussion 197-200, 277–282
111. Selig WM, Whalley ET, Ellis JL (2006) Asthma. In: Stevenson CS, Marshall LA, Morgan DW (eds) In vivo models of inflammation, 2nd edn. Birkhäuser Verlag, Basel, pp 1–28
112. Kucharewicz I, Bodzenta-Łukaszyk A, Buczko W (2008) Experimental asthma in rats. Pharmacol Rep 60:783–788
113. Nials AT, Uddin S (2008) Mouse models of allergic asthma: acute and chronic allergen challenge. Dis Model Mech 1:213–220
114. Stevenson CS, Birrell MA (2011) Moving towards a new generation of animal models for asthma and COPD with improved clinical relevance. Pharmacol Ther 130:93–105
115. Colsoul B, Nilius B, Vennekens R (2009) On the putative role of transient receptor potential cation channels in asthma. Clin Exp Allergy 39:1456–1466
116. Sel S et al (2008) Loss of classical transient receptor potential 6 channel reduces allergic airway response. Clin Exp Allergy 38: 1548–1558
117. Raemdonck K et al (2011) A role for sensory nerves in the late asthmatic response. Thorax 67(1):19–25
118. Watanabe N et al (2008) Immunohistochemical localization of transient receptor potential vanilloid subtype 1 in the trachea of ovalbumin-sensitized Guinea pigs. Int Arch Allergy Immunol 146:28–32
119. Dinh QT, Klapp BF, Fischer A (2006) Airway sensory nerve and tachykinins in asthma and COPD. Pneumologie 60:80–85
120. Li M et al (2011) The effect of substance P on asthmatic rat airway smooth muscle cell proliferation, migration, and cytoplasmic calcium concentration in vitro. J Inflamm 8:18
121. Kucharewicz I et al (2008) The concentration of kynurenine in rat model of asthma. Folia Histochem Cytobiol 46:199–203
122. Brewer JM et al (1999) Aluminium hydroxide adjuvant initiates strong antigen-specific Th2 responses in the absence of IL-4- or IL-13-mediated signaling. J Immunol 163: 6448–6454
123. Fedele G et al (2007) Lipooligosaccharide from Bordetella pertussis induces mature human monocyte-derived dendritic cells and drives a Th2 biased response. Microbes Infect 9:855–863
124. Nakagome K et al (2005) Antigen-sensitized CD4+CD62Llow memory/effector T helper 2 cells can induce airway hyperresponsiveness in an antigen free setting. Respir Res 6:46
125. Renz H et al (1992) Aerosolized antigen exposure without adjuvent causes increased IgE production and increased airway responsiveness in the mouse. J Allergy Clin Immunol 89:1127–1138
126. Lefort J et al (1996) Effect of antigen provocation of IL-5 transgenic mice on eosinophil mobilization and bronchial hyperresponsiveness. J Allergy Clin Immunol 97:788–799
127. Eum S et al (1995) Eosinophil recruitment into the respiratory epithelium following antigenic challenge in hyper-IgE mice is accomplished by interleukin-5-dependent bronchial hyperresponsiveness. Proc Natl Acad Sci USA 92:12290–12294
128. Zuany-Amorim C, Vargaftig BB, Pretolani M (1994) Strain-dependency of leukotriene C4 generation from isolated lungs of immunised mice. Br J Pharmacol 112:1230–1236
129. Zuany-Amorim C, Cordeiro RSB, Vargaftig BB (1993) Involvement of platelet-activating factor in death following anaphylactic shock in boosted and in unboosted mice. Eur J Pharmacol 235:17–22
130. Kool M et al (2008) Alum adjuvant boosts adaptive immunity by inducing uric acid and activating inflammatory dendritic cells. J Exp Med 205:869–882
131. Lloyd CM et al (2001) Resolution of bronchial hyperresponsiveness and pulmonary inflammation is associated with IL-3 and tissue leukocyte apoptosis. J Immunol 166:2033–2040
132. Hailé S et al (1999) Mucous-cell metaplasia and inflammatory-cell recruitment are dissociated in allergic mice after antibody- and drug-dependent cell depletion in a murine model of asthma. Am J Respir Cell Mol Biol 20:891–902
133. Hailé S et al (1999) Suppression of immediate and late responses to antigen by a non-anaphylactogenic anti-IgE antibody in a murine model of asthma. Eur Respir J 13:961–969
134. Trifilieff A, El-Hashim A, Bertrand C (2000) Time course of inflammatory and remodeling events in a murine model of asthma: effect of steroid treatment. Am J Physiol Lung Cell Mol Physiol 279:L1120–L1128
135. Trifilieff A et al (2003) PPAR-alpha and -gamma but not -delta agonists inhibit airway inflammation in a murine model of asthma: in vitro evidence for an NF-kappaB-independent effect. Br J Pharmacol 139: 163–171
136. Wyss D, Bonneau O, Trifilieff A (2005) Mast cell involvement in the adenosine mediated

airway hyper-reactivity in a murine model of ovalbumin-induced lung inflammation. Br J Pharmacol 145:845–852
137. Henderson WR Jr et al (1996) The importance of leukotrienes in airway inflammation in a mouse model of asthma. J Exp Med 184:1483–1494
138. Tomkinson A et al (2001) A murine IL-4 receptor antagonist that inhibits IL-4- and IL-13-induced responses prevents antigen-induced airway eosinophilia and airway hyperresponsiveness. J Immunol 166:5792–5800
139. Kanehiro A et al (2001) Inhibition of phosphodiesterase 4 attenuates airway hyperresponsiveness and airway inflammation in a model of secondary allergen challenge. Am J Respir Crit Care Med 163:173–184
140. Penn AL et al (2007) In utero exposure to environmental tobacco smoke potentiates adult responses to allergen in BALB/c mice. Environ Health Perspect 115:548–555
141. Rouse RL, Boudreaux MJ, Penn AL (2007) In utero environmental tobacco smoke exposure alters gene expression in lungs of adult BALB/c mice. Environ Health Perspect 115:1757–1766
142. Janssen EM et al (2000) The efficacy of immunotherapy in an experimental murine model of allergic asthma is related to the strength and site of T cell activation during immunotherapy. J Immunol 165:7207–7214
143. Zuany-Amorim C et al (1998) Requirement for gammadelta T cells in allergic airway inflammation. Science 280:1265–1267
144. Blyth DI et al (2000) Airway subepithelial fibrosis in a murine model of atopic asthma: suppression by dexamethasone or anti-interleukin-5 antibody. Am J Respir Cell Mol Biol 23:241–246
145. Nigo YI et al (2006) Regulation of allergic airway inflammation through Toll-like receptor 4-mediated modification of mast cell function. Proc Natl Acad Sci USA 103:2286–2291
146. Shibata Y (2002) Ras activation in T cells determines the development of antigen-induced airway hyperresponsiveness and eosinophilic inflammation. J Immunol 169:2134–2140
147. Kamata T (2003) src homology 2 domain-containing tyrosine phosphatase SHP-1 controls the development of allergic airway inflammation. J Clin Invest 111:109–119
148. Zosky GR et al (2008) Ovalbumin-sensitized mice are good models for airway hyperresponsiveness but not acute physiological responses to allergen inhalation. Clin Exp Allergy 38:829–838
149. Eidelman DH, Bellofiore S, Martin JG (1988) Late airway responses to antigen challenge in sensitized inbred rats. Am Rev Respir Dis 137:1033–1037
150. Hylkema MN et al (2002) The strength of the OVA-induced airway inflammation in rats is strain dependent. Clin Exp Immunol 129:390–396
151. Liu S, Chihara K, Maeyama K (2005) The contribution of mast cells to the late-phase of allergic asthma in rats. Inflamm Res 54: 221–228
152. Pauluhn J, Mohr U (2005) Experimental approaches to evaluate respiratory allergy in animal models. Exp Toxicol Pathol 56:203–234
153. Murphey SM et al (1974) Reagin synthesis in inbred strains of rats. Immunology 27:245–253
154. Panettieri RA Jr et al (1998) Repeated allergen inhalations induce DNA synthesis in airway smooth muscle and epithelial cells in vivo. Am J Physiol 274:L417–L424
155. Schneider T et al (1997) Kinetics and quantitation of eosinophil and neutrophil recruitment to allergic lung inflammation in a brown Norway rat model. Am J Respir Cell Mol Biol 17:702–712
156. Olivenstein R et al (1993) Depletion of OX-8 lymphocytes from the blood and airways using monoclonal antibodies enhances the late airway response in rats. J Clin Invest 92:1477–1482
157. Sirois J, Bissonnette EY (2001) Alveolar macrophages of allergic resistant and susceptible strains of rats show distinct cytokine profiles. Clin Exp Immunol 126:9–15
158. Ikawati Z, Nose M, Maeyama K (2001) Do mucosal mast cells contribute to the immediate asthma response? Jpn J Pharmacol 86:38–46
159. Zhou Y, Zhou X, Wang X (2008) 1,25-Dihydroxyvitamin D3 prevented allergic asthma in a rat model by suppressing the expression of inducible nitric oxide synthase. Allergy Asthma Proc 29:258–267
160. Han B et al (2011) Adverse effect of nano-silicon dioxide on lung function of rats with or without ovalbumin immunization. PLoS One 6:e17236
161. Ricciardolo FL et al (2008) The guinea pig as an animal model for asthma. Curr Drug Targets 9:452–465
162. Nemzek JA, Kim J (2009) Pulmonary inflammation and airway hyperresponsiveness in a mouse model of asthma complicated by acid aspiration. Comp Med 59: 321–330

163. Kim J et al (2004) Prevention and reversal of pulmonary inflammation and airway hyperresponsiveness by dexamethasone treatment in a murine model of asthma induced by house dust. Am J Physiol Lung Cell Mol Physiol 287:503–509
164. McKinley L et al (2004) Reproducibility of a novel model of murine asthma-like pulmonary inflammation. Clin Exp Immunol 136:224–231
165. Kim J et al (2011) Herbal medicine treatment reduces inflammation in a murine model of cockroach allergen-induced asthma. Ann Allergy Asthma Immunol 107:154–162
166. Coyle AJ et al (1996) Central role of immunoglobulin (Ig) E in the induction of lung eosinophil infiltration and T helper 2 cell cytokine production: inhibition by a non-anaphylactogenic anti-IgE antibody. J Exp Med 183:1303–1310
167. Mitchell VL, Van Winkle LS, Gershwin LJ (2011) Environmental tobacco smoke and progesterone alter lung inflammation and mucous metaplasia in a mouse model of allergic airway disease. Clin Rev Allergy Immunol (in press) DOI: 10.1007/s12016-011-8280-0
168. Campbell EM et al (1998) Temporal role of chemokines in a murine model of cockroach allergen-induced airway hyperreactivity and eosinophilia. J Immunol 161:7047–7053
169. McGee HS, Edwan JH, Agrawal DK (2010) Flt3-L increases CD4+CD25+Foxp3+ICOS+ cells in the lungs of cockroach-sensitized and -challenged mice. Am J Respir Cell Mol Biol 42:331–340
170. Arizmendi NG et al (2011) Mucosal allergic sensitization to cockroach allergens is dependent on proteinase activity and proteinase-activated receptor-2 activation. J Immunol 186:3164–3172
171. Tournoy KG et al (2000) Airway eosinophilia is not a requirement for allergen-induced airway hyperresponsiveness. Clin Exp Allergy 30:79–85
172. Sarpong SB, Zhang LY, Kleeberger SR (2003) A novel mouse model of experimental asthma. Int Arch Allergy Immunol 132:346–354
173. Kim J et al (2001) Eotaxin represents the principal eosinophil chemoattractant in a novel murine asthma model induced by house dust containing cockroach allergens. J Immunol 167:2808–2815
174. Lambert AL et al (2001) TNF-alpha enhanced allergic sensitization to house dust mite in brown Norway rats. Exp Lung Res 27:617–635
175. De Alba J et al (2010) House dust mite induces direct airway inflammation in vivo: implications for future disease therapy? Eur Respir J 35:1377–1387
176. Dong W, Selgrade MK, Gilmour MI (2003) Systemic administration of Bordetella pertussis enhances pulmonary sensitization to house dust mite in juvenile rats. Toxicol Sci 72: 113–121
177. Jobse BN et al (2009) Evaluation of allergic lung inflammation by computed tomography in a rat model in vivo. Eur Respir J 33:1437–1447
178. Singh P et al (2003) Phenotypic comparison of allergic airway responses to house dust mite in three rat strains. Am J Physiol Lung Cell Mol Physiol 284:588–598
179. Hsiue TR et al (1997) Mite-induced allergic airway inflammation in guinea pigs. Int Arch Allergy Immunol 112:295–302
180. Norris Reinero CR et al (2004) An experimental model of allergic asthma in cats sensitized to house dust mite or bermuda grass allergen. Int Arch Allergy Immunol 135:117–131
181. Bischof RJ et al (2003) Induction of allergic inflammation in the lungs of sensitized sheep after local challenge with house dust mite. Clin Exp Allergy 33:367–375
182. Sur S et al (1996) Immunomodulatory effects of IL-12 on allergic lung inflammation depend on timing of doses. J Immunol 157:4173–4180
183. Yadav UC et al (2009) Inhibition of aldose reductase prevents experimental allergic airway inflammation in mice. PLoS One 4:e6535
184. Misawa M et al (1987) Strain difference in an allergic asthma model in rats. Jpn J Pharmacol 45:63–68
185. Shampain MP et al (1982) An animal model of late pulmonary responses to Alternaria challenge. Am Rev Respir Dis 126:493–498
186. Minshall EM et al (1993) A novel animal model for investigating persistent airway hyperresponsiveness. J Pharmacol Toxicol Methods 30:177–188
187. Keir SD et al (2011) Airway responsiveness in an allergic rabbit model. J Pharmacol Toxicol Methods 64:187–195
188. Walters EH et al (2007) Nonpharmacological and pharmacological interventions to prevent or reduce airway remodelling. Eur Respir J 30:574–588
189. Moise NS et al (1989) Clinical, radiographic, and bronchial cytologic features of cats with bronchial disease: 65 cases (1980–1986). J Am Vet Med Assoc 194:1467–1473
190. Kirschvink N et al (2007) Functional, inflammatory and morphological characterisation

of a cat model of allergic airway inflammation. Vet J 174:541–553
191. Kirschvink N, Reinhold P (2008) Use of alternative animals as asthma models. Curr Drug Targets 9:470–484
192. Chapman RW (2008) Canine models of asthma and COPD. Pulm Pharmacol Ther 21:731–742
193. Woolley MJ et al (1995) Role of airway eosinophils in the development of allergen-induced airway hyperresponsiveness in dogs. Am J Respir Crit Care Med 152: 1508–1512
194. Baldwin F, Becker AB (1993) Bronchoalveolar eosinophilic cells in a canine model of asthma: two distinctive populations. Vet Pathol 30: 97–103
195. Fornhem C et al (1995) Allergen-induced late-phase airways obstruction in the pig: mediator release and eosinophil recruitment. Eur Respir J 8:1100–1109
196. Abraham WM et al (1983) Characterization of a late phase pulmonary response after antigen challenge in allergic sheep. Am Rev Respir Dis 128:839–844
197. Hein WR, Griebel PJ (2003) A road less travelled: large animal models in immunological research. Nat Rev Immunol 3:79–84
198. Abraham WM (2008) Modeling of asthma, COPD and cystic fibrosis in sheep. Pulm Pharmacol Ther 21:743–754
199. Ware LB (2008) Modeling human lung disease in animals. Am J Physiol Lung Cell Mol Physiol 294:L149–L150
200. Abraham WM et al (2005) Airway responses to aerosolized brevetoxins in an animal model of asthma. Am J Respir Crit Care Med 171:26–34
201. Kasaian MT et al (2007) Efficacy of IL-13 neutralization in a sheep model of experimental asthma. Am J Respir Cell Mol Biol 36:368–376
202. Maryanoff BE et al (2010) Dual inhibition of cathepsin G and chymase is effective in animal models of pulmonary inflammation. Am J Respir Crit Care Med 181:247–253
203. Michel O et al (1996) Severity of asthma is related to endotoxin in house dust. Am J Respir Crit Care Med 154:1641–1646
204. Yamashita M, Nakayama T (2008) Progress in allergy signal research on mast cells: regulation of allergic airway inflammation through toll-like receptor 4-mediated modification of mast cell function. J Pharmacol Sci 106:332–335
205. Kulhankova K et al (2009) Early-life co-administration of cockroach allergen and endotoxin augments pulmonary and systemic responses. Clin Exp Allergy 39:1069–1079

Chapter 17

Enhanced Cough, Animal Models

Raffaele Gatti, Pamela Pedretti, Romina Nassini, and Marcello Trevisani

Abstract

Under specific pathophysiological circumstances, cough becomes a chronic condition. Unfortunately, apart from drugs that treat the underlying diseases that cause the cough symptom, medicines directly oriented to reduce or abolish the cough reflex are limited to narcotic or narcotic-like drugs. The poor availability of antitussive medications is due to many causes but mainly to the poor understanding of the molecular mechanisms that cause and maintain chronic cough. Most of the mechanistic information on cough has been generated from animal models in which there is little or no airway pathology. It is essential then to develop animal models of chronic cough that reflect the disease in man. Here we described diverse animal models of enhanced coughing in guinea pigs, cats, dogs, pigs, and rabbits with airway inflammation produced by ovalbumin sensitization and challenge, exposure to cigarette smoke, sulfur dioxide, angiotensin-converting enzyme inhibitors, and ozone.

Key words: Ovalbumin, Chronic cough, Allergic cough, Capsaicin, Citric acid, Resiniferatoxin, Sulfur dioxide, Angiotensin-converting enzyme, Cigarette smoke, Ozone, Tracheitis

1. Introduction

Cough is an important defensive reflex mechanism and represents one of the most common symptoms of many inflammatory diseases of the airways such as postviral infections, pulmonary fibrosis, and bronchiectasis. Cough is also the first and most persistent symptom of lung chronic diseases such as asthma and chronic obstructive pulmonary disease (COPD) (1). Based on its duration, cough is generally defined as acute, subacute, and chronic (2).

Acute cough is an extremely common condition and develops to promote protection towards potentially harmful agents inhaled or produced (secretions) in the airways and lungs. It refers to a cough lasting for a maximum of 3 weeks and, in the majority of

Arpad Szallasi and Tamás Bíró (eds.), *TRP Channels in Drug Discovery: Volume I*, Methods in Pharmacology and Toxicology, DOI 10.1007/978-1-62703-077-9_17,

patients, it is mainly caused by viral or bacterial infections of the upper respiratory tract (3). Acute cough caused by such infections is usually self-limited with the clearing of the infection. Subacute cough has been defined as a cough of 3–8 week duration. Most cases of subacute cough refer to an increase in bronchial hyperresponsiveness mainly due to an airway postviral infection that has extended beyond 3 weeks. Inhaled corticosteroids or leukotriene receptor antagonists are often prescribed for this condition, but their use results self-limited to the infection (2).

Under specific pathophysiological circumstances cough becomes a chronic condition (4). Chronic cough maintains the protective role of the reflex mechanism, although occasionally its persistency may worsen the patient's quality of life by causing severe discomfort and requiring for an appropriate treatment.

Different studies have described the main etiologies to explain the causes promoting chronic cough. They comprise upper airway cough syndrome, due to a variety of rhinosinus conditions, as postnasal drip syndrome or other conditions such as asthma, chronic bronchitis, and gastroesophageal reflux (5–7).

The morbidity of chronic cough in humans is likely a product of the enhanced frequency and intensity that occurs as a result of increased excitability of this behavior. Clearly, the sensitivity of cough in response to inhaled irritants in patients with a variety of pulmonary disorders is enhanced (8, 9), and the frequency and intensity of cough can be elevated as well (10, 11).

The current poor availability of antitussive medications is due to many causes but mainly to the poor understanding of the molecular pathways that orchestrate the cough response in experimental animals and in man. Unfortunately, apart from drugs that treat the underlying diseases that cause the cough symptom, pharmacological treatment directly oriented to reduce or abolish cough are limited to narcotic or narcotic-like drugs (12).

The poor knowledge of the mechanisms of cough responses promotes the characterization of cough receptor(s) and the peripheral and central mechanisms for cough sensitization. Therefore, the identification of new therapeutic targets for the treatment of chronic cough will be of immense therapeutic benefit and will greatly enhance the quality of life of patients.

For these purposes, it is essential to develop animal models of chronic cough that reflect the disease in man. Indeed, a reliable, robust, and reproducible model of cough is essential to profile and establishes the efficacy of novel antitussive therapies under development before moving to man.

This chapter describes diverse animal models of enhanced coughing in the presence of airway inflammation in guinea pigs, cats, dogs, pigs, and rabbits. An enhanced cough response is described in animals sensitized to and challenged with ovalbumin and exposed to cigarette smoke, sulfur dioxide (SO_2), angiotensin-converting enzyme (ACE) inhibitors, and ozone.

2. Materials

2.1. Guinea Pig Models

2.1.1. Animals and General Equipment (Fig. 1)

1. Use male Dunkin–Hartley guinea pigs (250–700 g) (Note 1). Usually, animals are acclimatized in cages, (24 ± 0.5°C) for 1 week before the beginning of the experiments with free access to water and standard rodent diet.
2. Transparent 850–2,000 ml perspex chamber (Note 1), frequently custom-built, or body plethysmograph (type 855, Hugo Sachs Electronic, March-Hugstetten, Germany; model PMUA 1SAR, Buxco Electronics, Sharon, CT) (13, 14) with an air inlet and outlet, ventilated with a constant airflow of 400–1,000 ml/min. The chamber may be composed by a head chamber which is isolated from the body (14, 15). Changes in airflow can be measured by a pneumotachograph (Fleisch 0.6 V) connected to the outflow and recorded on a fast response recorder (Reba UK; Par 1000) (16, 17).
3. Jet (Pariprovocation test I, Pari Starneberg, Germany; Puritan Bennett, Lenexa, KS) or ultrasonic nebulizer (model 646, model 65, Ultra Neb 99, De Vilbiss Co., Somerset, PA; model AER5211, Buxco Electronics) (14, 17–20). The particle size mass median diameter should be 0.5–5 μm and the output of the nebuliser 0.037–0.4 ml/min (Note 1).

2.1.2. Agents to Induce Cough

1. Capsaicin (8-Methyl-*N*-vanillyl-*trans*-6-nonenamide, Sigma-Aldrich, St. Louis, MO), from *Capsicum* sp. Form and color: crystalline-white. Storage temperature: 2–8°C. Dissolve capsaicin

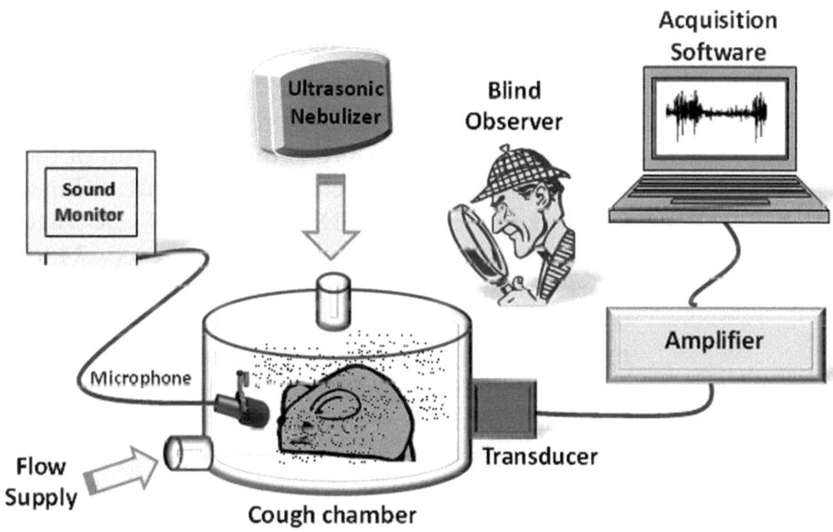

Fig. 1. General experimental setup for the evaluation of cough induced by aerosolized tussive stimuli in the conscious unrestrained guinea pig.

in pure Tween 80 and ethanol, and then add physiologic saline to obtain a 100-mM stock solution (final Tween 80 and ethanol concentration is 10%). Store the 100-mM stock solution at −20°C. This solution will be diluted with physiologic saline to obtain final protussive concentrations (21) (Notes 2 and 3).

2. Citric acid (Sigma-Aldrich). Form and color: crystalline-white. Dissolve citric acid in physiologic saline. Storage temperature: room temperature (Note 3).
3. Albumin, from chicken egg white (Ovalbumin, Sigma-Aldrich). Keep container tightly closed in a dry and well-ventilated place. Storage temperature: 2–8°C (Note 3).
4. Resiniferatoxin (Sigma-Aldrich). Form and color: powder and white with pungent odor. Storage temperature: −20°C. Prepare 1-mM stock concentration in 100% dimethylsulfoxide (DMSO) (Notes 2 and 3).

2.1.3. Allergic Cough

1. Aluminum hydroxide ($Al(OH)_3$; Sigma-Aldrich). Storage: room temperature.
2. Cyclophoslphamide (Sigma-Aldrich). Storage temperature: 2–8°C.
3. Heat-killed pertussis organisms (usually custom-prepared).

2.1.4. Cigarette Smoke Exposure

1. Standard 2Rl reference cigarettes (Tobacco and Health Research Institute, Lexington, KY). Physical Data: CRW (cigarette rod weight) = 1.184 g/cig, RTD (resistance to draw) = 8.9 cm of H_2O, static burning 660 s/40 mm, length = 85.5 mm/cig, circumference = 25.0 mm/cig, paper porosity = 47.6 s/50 cm^3 and paper additive type = phosphate 0.16%. Smoke Analyses: TPM (total particulate matter) = 44.6 mg/cig, nicotine = 2.45 mg/cig, water = 5.36 mg/cig, puff = 12.7 count/cig, CO = 25.1 mg/cig (22).
2. Standard 2R4F reference cigarettes (University of Kentucky, Louisville, KY). The 2R4F is the second run of the manufacturer of the 1R4F reference cigarettes. Physical Data: CRW = 1.055 g/cig, RTD = 13.40 cm of H_2O, static burning n/d, length = 83.90 mm/cig, circumference = 24.90 mm/cig, paper porosity = 24 s/50 ml and paper additive type = citrate 0.60%. Smoke Analyses: TPM = 11.70 mg/cig, nicotine = 0.85 mg/cig, water = 1.12 mg/cig, puff = 9.20 count/cig, CO = 13 mg/cig (22). To monitor carbon dioxide content of the exhaust air from the chamber: system Beckam LB-2 CO_2 monitor (Schiller Park, IL).

2.1.5. Angiotensin-Converting Enzyme Inhibitors

1. Osmotic pump (Model 2ML1, Alzet, Cupertino, CA) (see below).

2. ACE inhibitors: enalaprilat, lisinopril, imidapril, captopril, enalapril, zofenopril, and ramipril (Sigma-Aldrich).
3. Arabic gum (Sigma-Aldrich).

2.1.6. Sulfur Dioxide

1. SO_2 (Air gas, Inc., Radnor, Pennsylvania, PA).
2. The aerosolization chambers should have a single port for the introduction of either air (O_2=21%; N=79%) or SO_2 (1,000 ppm). An air outlet is also located on the bottom side corner of the exposure chamber which allows air to exit the chamber. Set the airflow rate at 3 L/min.

2.2. Cat Models

2.2.1. Enhanced Cough in Local Tracheitis Cats

1. Cats weighing 1.8–3.2 kg
2. Pentobarbital (Spofa, Praha, Czech Republic)
3. Tracheal cannula
4. Single silk suture
5. Nylon fiber
6. Polyethylene tubing
7. Electromanometer (Tesla, Valašské Meziříčí, Czech Republic)
8. Mingograph 81 (Elena-Schönandern, Solna, Sweden)

2.3. Dog Models

2.3.1. Allergic Animals

1. Male beagle dogs weighing 8–15 kg
2. Ragweed (Bayer, Elkhart, IN) (23)
3. Histamine dihydrochloride (0.3–1%, Sigma-Aldrich)
4. Alhydrogel ($Al(OH)_3$, 1.3%, Accurate Chemical, Westbury, NY)
5. Propofol (Diprivan; Zeneca Pharmaceuticals, Wilmington, DE)
6. Electric shaver
7. Surflo® catheter (Terumo Medical, Elkton, MD) and injection plug
8. Cuffed endotracheal tube (Rusch, Waiblingen, Germany; size 7.0 mm)
9. A balloon-tipped polyethylene catheter (i.d. = 2 mm)
10. Infusion pump (Gemini, PC-2TX; Imed, San Diego, CA)
11. Pneumotachograph
12. Low pressure transducer for pulmonary pressure and air flow measurement (model MP 45-14-871; Validyne, Northridge, CA)
13. Differential pressure transducer (model MP-45-24-87; Validyne)
14. Pulmonary function analyzer (model XA; Buxco Electronics)
15. Chart recorder (MFE Instruments, Beverly, MA)
16. Dinamap veterinary blood pressure monitor (model 8300; Critikon, Tampa, FL)

17. Digital handheld pulse oximeter (model 8500V, Nonin Medical, Minneapolis, MN)
18. Cardiocap monitor (Datex-Engstrom, Helsinki, Finland)
19. Fishing line (25-lb test, Berkley, Spring Lake, IA)
20. Jet nebulizer (Raindrop; Puritan Bennett, Lenexa, KS)
21. One-way respiratory valve (Hans-Rudolph, Shawnee, KS)

2.4. Pig Models

2.4.1. Enalapril-Enhanced Citric Acid-Induced Cough

1. Healthy Belgian Landrace piglets of both sexes weighing 13.3 ± 1.2 kg.
2. Closed stainless-steel and plastic 1.9-m^3 individual inhalation chamber designed for piglets.
3. Fiberglass filters (CM 295, Camfil, Brussels, Belgium) to remove dust and aerial bacteria from the chamber. Airflow rate: 10 m^3/h.
4. Ultrasonic nebulizer (ultra-neb 2000; DeVilbiss, Somerset, PA). The aerodynamic mass diameter of the droplets ranged from 0.5 to 5 μm (manufacturer's indications).
5. Citric acid (Merck, Darmstadt, Germany).
6. Enalapril (Renitec, Merck Sharp and Dohme, Brussels, Belgium).

2.5. Rabbit Models

2.5.1. Exposure to Ozone and Citric Acid

1. New Zealand White rabbits (2.4 ± 5.0 kg) of either sex
2. Purpose-built perspex exposure chamber for rabbits
3. Tie-clip microphone
4. Ozone generator (Ozonizer Model 25, Sander, Munich, Germany)
5. Nebulizer (De Vilbiss Ultraneb 2000 nebulizer, De Vilbiss)
6. Fleisch 00 pneumotachograph (OEM Medical, Richmond, VA)
7. Polygraph (Grass Instrument Division, West Warwick, RI)
8. Preamplifier and loudspeaker
9. Differential pressure transducer (PTS, Grass Instrument Division)
10. Computerized data acquisition system (Biopac Systems Inc, Santa Barbara, CA)

3. Methods

3.1. Guinea Pig Models

3.1.1. Allergic Cough: Ovalbumin Sensitization and Challenge

Most of the available data on enhanced coughing has been generated in allergic guinea pigs. Since many different sensitizing protocols have been disclosed so far, a list of the major and more standardized approaches is here provided. Dose-dependent increases in coughing have been produced by a passive sensitization paradigm (24). In other studies, cough was elicited in actively sensitized guinea pigs by acute exposure to antigen aerosols (see below). Sensitizing procedures here described induced an exacerbation of capsaicin-induced cough.

1. Inject ovalbumin (2 mg/ml/animal, i.p.) plus aluminum hydroxide (100 mg/ml) suspended in saline. Three weeks later, perform a boosting administration with ovalbumin (0.01 mg, i.p.) plus $Al(OH)_3$ (100 mg). To induce cough, perform a 60-s aerosol challenge, 3–4 weeks after the boosting, with ovalbumin (10 mg/ml) or saline solution. Inject cyclophosphamide (30 mg/kg, i.p.) 2 days prior to the ovalbumin i.p. administration (14, 15, 21, 25–27) (Note 4).
2. Inject ovalbumin (200 μg/0.5 ml/animal, i.p.) plus aluminum hydroxide (200 mg/ml) suspended in saline. In addition, administer 0.3 ml of 10×10^{10} heat-killed pertussis organisms (i.p.). To induce cough, challenge with aerosolized ovalbumin (0.1–1%) 4 weeks after the intraperitoneal ovalbumin administration (19) (Note 4).
3. Perform a 2×0.5 ml bilateral intraperitoneal injections with ovalbumin (20 μg/ml/animal) plus aluminum hydroxide (100 mg/ml) in distilled water. Two weeks later, challenge with aerosolized ovalbumin (0.1–1%) (Author's unpublished protocol) (Note 4).
4. After the ovalbumin sensitization, return the animal to the animal facility and put the naïve guinea pigs in the same surroundings as the ovalbumin-treated animals (Note 5).

A model of cough-variant asthma obtained by ovalbumin inhalation in actively sensitized guinea pigs has been described as well (28).

3.1.2. Tobacco Smoke Exposure

It is well documented that cigarette smoke exposure is one of the major causes of chronic cough in human smokers. In animal models (e.g., guinea pigs), chronic exposure of the airways to cigarette smoke induced cough hypersensitivity to various tussive inhalation challenges such as capsaicin and citric acid (18). Cough hyperresponsiveness following cigarettes smoke exposure is further enhanced in ovalbumin-sensitized guinea pigs (29). Here are described different guinea pig protocols of chronic tobacco smoke exposure.

1. Expose guinea pigs to mainstream tobacco smoke generated from four standard 2R1 reference cigarettes drawn into the exposure chamber (36 L in volume). Exposures are 7 days/week and 30 min/day. Maintained the concentration of tobacco smoke inside the chamber at 5 mg/L TPM during the exposure. Monitored carbon dioxide content of the exhaust air

from the chamber. CO_2 content should remain less than 1% of the exhaust air. Expose animals daily throughout the duration of the study. Induce cough between day 70 and day 100 of chronic exposure to tobacco smoke or compressed air. Induce cough by challenging guinea pigs with aerosolized capsaicin (0.001%, 30 s) (Notes 6 and 7) (18).

2. Expose guinea pigs to tobacco smoke (1–5 2R4F research cigarettes) daily, for 30–35 min exposure for up to 10 days. Deliver cigarette smoke into the exposure chamber (7 L in volume) through a tube connected to a puffer, to give a 3 s puff of smoke (25 ml) every 30 s. During the intervals between the generation of the puffs of smoke give to animals normal air pumped into the exposure unit (600 ml/min). 1–10 days after cigarette smoke exposure, place unrestrained animals in the transparent plastic whole body plethysmograph. To induce cough, aerosolize citric acid (0.3 M) for 10 min or capsaicin (10 mM) for 7 min (30).
3. Expose guinea pigs to filtered air or to aged and diluted side-stream cigarette smoke as a surrogate for environmental tobacco smoke (ETS) from 1 to 6 weeks of age. ETS exposures: 1 mg/m^3 of total suspended particulates for 6 h/day, 5 days/week (31–33). At 6 weeks of age, place animal in the whole-body plethysmograph to measure cough. Expose guinea pigs to citric acid aerosol (0.4 M) for 3 min, followed by a 12-min observation period (34).
4. After the chronic tobacco smoke exposure, return the animal to the animal facility and put the naïve guinea pigs in the same surroundings as the tobacco smoke exposed animals.

3.1.3. ACE Inhibitors

The subchronic systemic treatment with several different ACE inhibitors has been shown to potentiate capsaicin and citric acid induced and spontaneous cough in the guinea pig (see below) (35–37).

1. Treat guinea pigs daily at 13:00 hours for 14 days with 3 mg/kg enalapril (oral). On day 15, induce cough aerosolizing 30 μM capsaicin solution dissolved in 10% Tween 80 and ethanol 10% in physiologic saline (see above) (38).
2. Treat guinea pigs daily for 30 days with oral 2% arabic gum solution (vehicle group) or oral 3 mg/kg enalapril (treated group). Measure spontaneous coughing frequency for 30 min 2 h after the last administration on day 30 (39) (Note 8).
3. Treat guinea pigs with zofenopril (10 mg/kg, day), ramipril (3–10 mg/kg, day), or water solubilizing zofenopril and ramipril in animals drinking water for 14 days. To induce cough, at the end of the 14th day of oral treatment, place unanesthetized guinea pigs in the cough chamber and

expose to nebulized aqueous solution of 0.1 M citric acid for 10 min (40) (Note 9).

4. Add captopril to the drinking water of animals to be treated in a concentration of 0.3 mg/ml (approximately mean daily dose of 20 mg/animal) for 14 days. To induce cough, place unanesthetized guinea pigs in the cough chamber and expose to nebulized aqueous solution of 0.25 M citric acid for 10 min (41).
5. Insert the osmotic pump into the peritoneal cavity of the guinea pig 3 days before testing. Administer enalaprilat (0.1, 0.5, 1, 5, 10, 20 mg/ml), lisinopril (0.1, 0.5, 1, 5, 10, 20 mg/ml), imidapril (1, 5, 10, 20, 30 mg/ml) or 0.08 M Na_2HPO_4 (enalaprilat) and saline as control by using the osmotic pump for 3 days. To induce cough, 4 days after implantation of the osmotic pump, place unanesthetized guinea pigs in the cough chamber and expose to a nebulized solution of citric acid (5%) (42).

3.1.4. Subacute SO_2 Exposure

SO_2 is a major component of industrial smog that contributes significantly to the pathogenesis of obstructive airway diseases. SO_2 inhalation in dogs has been shown to produce mucus hypersecretion, airway obstruction, eosinophilia, and spontaneous productive coughing (43, 44). However, a methodical analysis of cough in dogs exposed to SO_2 is lacking.

In the guinea pig, exposure to SO_2 enhances coughing in response to capsaicin (37, 45).

1. Expose animals to air or SO_2 for 4 days for 3 h/day starting at 9:40 a.m. Between exposure sessions, return animals to their home cage and supply food and water ad libitum (45).
2. On day 5, place the animals in the cough chamber and aerosolize capsaicin (3, 10, 30 μM) for 4 min (see also: (46, 47)).
3. Expose animals to capsaicin once only.

3.1.5. Cough Measurement and Identification

1. Place unanesthetized guinea pigs in the aerosol chamber.
2. Induce cough by aerosolizing a protussive agent (ovalbumin, capsaicin, citric acid, resiniferatoxin) for 2–10 min from the nebulizer (see above).
3. Identification of a cough response could be achieved by the following different approaches (21, 25, 26, 48):
   (a) By observing/counting the typical cough posture of the guinea pig (by a trained and treatment-blind observer) during the challenge.
   (b) By a transient change in the box pressure (a rapid inspiration followed by rapid expiration).
   (c) By the presence of an investigator who confirmed the cough sounds during the challenge (transmitted from

the microphone in the cage to the recorder and to outside speakers).

(d) By the subsequent analysis of the sound waves recorded into a personal computer. The cough sounds can be recorded, digitally stored, and counted by a blind observer afterwards.

3.2. Cat Models

3.2.1. Enhanced Cough in Local Tracheitis Cats

It is well known that experimentally induced acute diffuse inflammation of the airways is accompanied by an increase in the intensity of the cough reflex (49–52). Hanacek and coworkers (51) have described a model of an experimentally induced local tracheitis in cats. Inflammation has been produced by a silk suture fixed in the intrathoracic trachea, and cough was elicited by mechanical stimulation of different parts of the respiratory tract mucosa (51). Within 1-week postsurgery the animals displayed spontaneous coughing and mucus hypersecretion. Authors have observed a significant increase only in cough elicited from the inflamed site in the trachea. At 2-week postsurgery, cough number and tracheal pressures during both the inspiratory and expiratory phases of cough were significantly increased relative to control animals when the mechanical stimulus was applied to the inflamed site in the trachea (51, 53).

1. Anesthetize the animals with pentobarbital (40 mg/kg, i.p.).
2. Cut the throat skin and discover the trachea.
3. Suture a chronic tracheal cannula ("control" animals).
4. Place a single silk suture in the dorsal wall of the cervical trachea as the means of producing inflammation ("treated" animals).
5. Suture the operation wound.
6. During healing, the cannula should be kept clean and changes in breathing determined by auscultation (Note 10).
7. After 10 days, elicit the cough reflex by mechanically stimulating the laryngopharyngeal mucosa, tracheal mucosa in the inflamed area, and the tracheobronchial mucosa with a nylon fiber slipped through a small hole in the polyethylene tubing connecting the chronic tracheal cannula with the electromanometer and along the tracheal cannula into the airways.
8. Record the cough with the electromanometer measuring the changes in the lateral tracheal pressure through the mingograph.

3.3. Dog Models

3.3.1. Allergic Animals

Dogs that have been neonatally sensitized to ragweed demonstrate many characteristic features of human asthma. Moreover, an altered cough reflex in allergic dogs after antigen challenge was reported by House and coworkers (54). This work shows that aerosolized ragweed challenge did not induce cough per se but increased the cough number, the respiratory rate, and pulmonary resistance and reduced tidal volume and dynamic lung compliance of coughs induced by mechanical stimulation of the intrathoracic trachea in dogs. Mechanical stimulation of the trachea did not produce enhanced cough in animals that were sensitized but not challenged. Both the number and amplitude of mechanically induced coughs returned to baseline values by 24–48 h after the ragweed challenge. Similar results were obtained after challenge with aerosolized histamine that did not induce spontaneous coughs but increased the number of mechanically induced cough.

1. To actively sensitized animals to ragweed use newborn dogs as described by Theodorou and coworkers (23). Briefly, inject (500 mg, i.p.) of short ragweed to puppies and 0.5 ml i.p. $Al(OH)_3$ within 24 h of birth. Repeat these injections weekly until the age of 8 week, every other week until age 14 week, and then monthly for the duration of the study. Do not inject ragweed in dogs in the control group.
2. Shave the dog (8–15 kg) front paw and insert the catheter into the cephalic vein. Connect the injection plug to the i.v. catheter to facilitate the injection of drugs.
3. Induce anesthesia by i.v. bolus injection of propofol at a dose of 8 mg/kg given over a 1-min period. Insert a cuffed endotracheal tube into the trachea with the aid of a laryngoscope, immediately following the propofol injection (Note 11).
4. Insert the balloon-tipped polyethylene catheter through the mouth into the esophagus and position it in the mid-thoracic region.
5. Connect the endotracheal tube to a heated pneumotachograph to calculate differential pressure across it and derivate pulmonary airflow.
6. Monitor the mean arterial blood pressure with the blood pressure monitor using an inflatable cuff around the hind paw.
7. Measure arterial oxygen saturation and heart rate with the pulse clipped onto the tongue.
8. Measure end-tidal CO_2 directly at the tip of the endorracheal tube using the cardiocap monitor.
9. Cough induction. Disconnect the endotracheal tube from the pneumotachograph and induce cough mechanically stimulating the intrathoracic trachea by inserting and withdrawing a determined length of fishing line through the endotracheal tube for 1 s. This procedure will induce a brief period of coughing followed by a period of apnea. When normal breathing had resumed, repeat this procedure by inserting and withdrawing the fishing line ten times over a 10-s period. At the end of this 10-s stimulation, when normal breathing had resumed, connect

the endotracheal tube to the pneumotachograph and measure pulmonary functions for 3 min. At the end of this 3-min period, discontinue the anesthesia and allow the dogs to recover and return to their cages. Count the number of coughs by observing the animal behavior. Measure the cough amplitude as the increase in mean expiratory pressure (cm H_2O) over the baseline values obtained during normal tidal breathing immediately before the induction of cough. Obtain these measurements from the chart recorder (55) (Note 12).

10. To evaluate the effect of antigen challenge on the cough reflex, expose dogs to aerosolized ragweed or saline. Generate the aerosols with the jet nebulizer at a flow of 150 ml/min at 40 pounds per square inch pressure (56). Measure pulmonary functions immediately before ten consecutive inhalations of either ragweed or saline aerosol delivered to the dogs via a one-way breathing valve.
11. Perform cough trial using a 5-s stimulation approximately 9 min after the ragweed or physiologic saline challenge to coincide with the peak ventilatory response to the ragweed. At the end of the cough trial, approximately 10 min after the ragweed or saline challenge, connect the dogs to the pneumotachograph and measure lung functions. Allow animals to recover from the anesthesia and return them to their cage.
12. 24 hour after challenge with ragweed or saline, anesthetize the dogs and measure the cough reflex. Repeat this procedure at 48 h after the ragweed or saline challenge.
13. In studies involving histamine aerosol challenge, measure baseline lung functions and perform a cough trial with 5-s stimulation. Challenge the dogs with five consecutive breaths of aerosolized histamine dihydrochloride (0.3–1% histamine). Perform a cough trial approximately 1 min after the histamine challenge and immediately measure lung functions (Note 13).

3.4. Pig Models

3.4.1. Enalapril-Enhanced Citric Acid-Induced Cough

It has been demonstrated that the treatment with ACE inhibitors induced enhancement of the cough reflex is the pig. Particularly, acute administration of enalapril produced an enhanced cough response to citric acid in pigs (57, 58).

1. Place individually the piglets in the inhalation chamber on a grating located 15 cm above the floor. Set the airflow rate at 10 m^3/h. The piglets were set on a grating located (Note 14).
2. Use the fiberglass filter to filter the air by removing dust and aerial bacteria (Note 15).
3. Connect the nebulizer to the air entry of the inhalation chamber and nebulize 0.8 M citric acid dissolved in saline into the chamber with the ultrasonic nebulizer at a rate of 2.5 ml/min for 15 min. Remember to switch off the chamber ventilator during this period.

4. After the 15 min of aerosolization, ventilate the inhalation chamber with fresh air for 15 min to remove the citric acid.
5. A trained observer counts the coughs during these two successive 15-min periods (Note 16).
6. To study drug effects, challenge the piglets twice with citric acid, with a 2-day interval between the two challenges (days 1 and 3). Take the values obtained during the first cough induction test (day 1) as control values.
7. Administer enalapril (7.5 and 15 μg/kg) intravenously 30 min before the second challenge. A control animal group should receive no drugs, apart from citric acid, will serve to assess the possible effect of repeated citric acid solution nebulizations (Note 17).

3.5. Rabbit Models

3.5.1. Exposure to Ozone and Citric Acid

Rabbits do not cough consistently to citric acid aerosol (59). However, when conscious rabbits are exposed to ozone at 3 ppm for 1 h, citric acid aerosol evokes a significant cough response (59, 60).

1. Connect the pneumotachograph to the differential pressure transducer and attach it to the outflow from the exposure chamber.
2. Connect the differential pressure transducer to the polygraph, for hard copy recording.
3. Input the output from the polygraph to the computerized data acquisition system for real-time recording of data.
4. Place the tie-clip microphone in the exposure chamber and connect it via a preamplifier to a loudspeaker output to provide the observer with an audio monitor of cough responses.
5. Place rabbits unrestrained into the perspex exposure chamber and allowed to acclimatize.
6. Expose animals to ozone at 3 ppm generated by passing cylinder air through the ozonizer at a flow rate of 5 L/min for 1 h.
7. Immediately after exposure to ozone, expose conscious animals to an aerosolized solution of citric acid (1.6 M) for 10 min in a flow rate of 5 L/min.
8. A trained observer should continuously monitor the animals, and the number of coughs should be counted over the 10-min period from commencement of the aerosol administration of the tussive agent.

3.6. Other Cough Exacerbating Protocols

3.6.1. Effect of Pressure Stress on Cough-Reflex Sensitivity

Hara and coworkers have recently demonstrated that negative mechanical pressure stress applied to the airway induced neutrophilic airway inflammation resulting in increased cough-reflex sensitivity in guinea pigs. Mechanical stress applied to the airway resulted in a self-perpetuating cough-reflex cycle (61).

3.6.2. Cough-Reflex Sensitivity in Guinea Pigs with Parainfluenza Virus Infection

Guinea pigs injected with parainfluenza virus type 3 (PIV3) exhibited an increased cough-reflex sensitivity to capsaicin (62).

3.6.3. Enhanced Cough and Protein Kinases

We have recently shown that protease-activated receptor-2 (PAR2) activators exaggerated transient receptor potential vanilloid, capsaicin receptor, type 1- (TRPV1) dependent cough through diverse mechanism(s), including PKC, PKA, and prostanoid release in the guinea pig (63).

Similarly, inhaled ethanol (1–3%) selectively exaggerated, via a PKC-dependent pathway, the cough response evoked by resiniferatoxin (another potent TRPV1 activator) in the guinea pig (20). Furthermore, it has been disclosed that ethanol (5 and 25%) inhalation enhanced the cough reaction to capsaicin in patients with airway sensory hyperreactivity (64).

4. Notes

1. The guinea pig body weight, the cough chamber dimensions, the aerosol particle size mass median diameter, and the output of the nebulizer are variable and depend on the model used.
2. Precautions for safe handling: avoid contact with skin and eyes (serious irritation), avoid formation of dust and aerosols, and provide appropriate exhaust ventilation at places where dust is formed.
3. To prevent the possible contribution of bronchoconstriction in the tussive response to the diverse stimuli, guinea pigs may be intraperitoneally administered with the β-adrenoceptor agonist, terbutaline (0.5 mg/kg), 5 min prior to the beginning of the cough induction.
4. Dissolve ovalbumin in distilled water and stir for 2 h then add aluminum hydroxide. The mix suspension should be stirred for additional 30 min prior to the administration.
5. Successful sensitization can be confirmed by an intradermal injection of ovalbumin (25 μl of 200 μg/ml) into the dorsal back surface.
6. Chronic tobacco smoke exposure may cause coughing during and for several minutes after its exposures (see also (29, 65)).
7. Set the bias flow in the chamber as 25 L/min (well above the estimated combined minimum volume of 3–5 L of eight guinea pigs housed in the chamber). This procedure will prevent the development of either a hypoxic or hypercapnic environment.

8. A single enalapril (10 mg/kg) oral administration 2 h before an exposure to capsaicin significantly increases the number of capsaicin-induced cough (66).
9. Put a constant volume of the "drinking solutions," exceeding the daily amount necessary, in the drinking bottles and change every day in order to adjust the drugs' concentration to the previous day's volume intake and to the guinea pigs' body weight variation (40).
10. At the end of the first week after surgery, treated animals may begin to display signs of respiratory distress and spontaneous cough may be observed. During spontaneous coughing, apathy, loss of appetite, and irritability may be observed as well. At 25-day postsurgery, "treated" animals exhibit an intense hyperemia at the site of the suture a diminished hyperemia is observed further down the trachea.
11. Infuse propofol continuously throughout the experiments (0.3–0.8 mg/kg/min) using the infusion pump. Perform the test with dogs in the supine position.
12. Measure tidal volume by electrical integration of the flow signal. Measure transpulmonary pressure with the differential pressure transducer as the pressure difference between the esophageal catheter and a port at the front of the endotracheal tube. Use pulmonary airflow, tidal volume, and transpulmonary pressure for the derivation of pulmonary resistance and dynamic lung compliance (67) using the pulmonary function analyzer.
13. Experiments with ragweed and histamine challenge were performed using a randomized, cross-over experimental design.
14. To avoid stress effects due to introduction of the pig into the chamber, do not perform the cough induction test until 2 days after introduction of the animals into the chamber.
15. As specified by the manufacturer, this filter retains 95% of all particles exceeding 1 μm in diameter.
16. Express the cough data as means ± SEM of the number of coughs counted over 30 min. Check that saline induces no coughing or any other respiratory clinical symptom.
17. Dissolve enalapril in water with NaCl, NaOH, and benzyl alcohol (Renitec, IV solution).

References

1. Widdicombe JG (1999) Advances in understanding and treatment of cough. Monaldi Arch Chest Dis 54:275–279
2. Irwin RS et al (2006) Diagnosis and management of cough executive summary: ACCP evidence-based clinical practice guidelines. Chest 129:1S–23S
3. Dicpinigaitis PV et al (2009) Acute cough: a diagnostic and therapeutic challenge. Cough 5:11

4. McGarvey LP, Morice AH (2006) Clinical cough and its mechanisms. Respir Physiol Neurobiol 152:363–371
5. Irwin RS et al (1981) Chronic persistent cough in the adult: the spectrum and frequency of causes and successful outcome of specific therapy. Am Rev Respir Dis 123:413–417
6. Poe RH et al (1989) Chronic persistent cough. Experience in diagnosis and outcome using an anatomic diagnostic protocol. Chest 95: 723–728
7. Dicpinigaitis PV (2011) Cough: an unmet clinical need. Br J Pharmacol 163:116–124
8. Doherty MJ et al (2000) Capsaicin responsiveness and cough in asthma and chronic obstructive pulmonary disease. Thorax 55:643–649
9. O'Connell F et al (1994) Capsaicin cough sensitivity decreases with successful treatment of chronic cough. Am J Respir Crit Care Med 150:374–380
10. Loudon RG, Brown LC (1967) Cough frequency in patients with respiratory disease. Am Rev Respir Dis 96:1137–1143
11. Sharpey-Schafer EP (1953) The mechanism of syncope after coughing. Br Med J 2:860–863
12. Birring SS (2011) New concepts in the management of chronic cough. Pulm Pharmacol Ther 24:334–338
13. Brozmanova M et al (2007) Experimental allergic rhinitis-related cough and airway eosinophilia in sensitized guinea pigs. J Physiol Pharmacol 58(Suppl 5):57–65
14. Liu Q et al (2001) Characterization of increased cough sensitivity after antigen challenge in guinea pigs. Clin Exp Allergy 31:474–484
15. Xiang A et al (1998) Effects of airway inflammation on cough response in the guinea pig. J Appl Physiol 85:1847–1854
16. Forsberg K et al (1988) Cough and bronchoconstriction mediated by capsaicin-sensitive sensory neurons in the guinea-pig. Pulm Pharmacol 1:33–39
17. Laude EA et al (1995) Antitussive and antibronchoconstriction actions of fenspiride in guinea-pigs. Eur Respir J 8:1699–1704
18. Bergren DR (2001) Chronic tobacco smoke exposure increases cough to capsaicin in awake guinea pigs. Respir Physiol 126:127–140
19. Bolser DC et al (1995) Pharmacological studies of allergic cough in the guinea pig. Eur J Pharmacol 277:159–164
20. Gatti R et al (2009) Ethanol potentiates the TRPV1-mediated cough in the guinea pig. Pulm Pharmacol Ther 22:33–36
21. Myou S et al (2001) Effects of suplatast tosilate, a new type of anti-allergic agent, on airway cough hypersensitivity induced by airway allergy in guinea-pigs. Clin Exp Allergy 31: 1939–1944
22. http://tobaccodocuments.org/ctr/CTRCONTRACTS024947-4963.html
23. Theodorou A et al (1997) Ragweed sensitization alters pulmonary vascular responses to bronchoprovocation in beagle dogs. J Appl Physiol 83:912–917
24. Winter CA, Flataker L (1955) The effects of drugs upon a graded cough response obtained in sensitized guinea pigs exposed to aerosol of specific antigen. J Exp Med 101:17–24
25. Katayama N et al (2001) Effects of carbocysteine on antigen-induced increases in cough sensitivity and bronchial responsiveness in guinea pigs. J Pharmacol Exp Ther 297: 975–980
26. Hara J et al (2005) Comparison of cough reflex sensitivity after an inhaled antigen challenge between actively and passively sensitized guinea pigs. Cough 1:6
27. Andersson P (1981) Antigen-induced bronchial anaphylaxis in actively sensitized guinea pigs. The effect of booster injection and cyclophosphamide treatment. Int Arch Allergy Appl Immunol 64:249–258
28. Nishitsuji M et al (2004) A guinea pig model for cough variant asthma and role of tachykinins. Exp Lung Res 30:723–737
29. Karlsson JA et al (1991) Hyperresponsiveness to tussive stimuli in cigarette smoke-exposed guinea-pigs: a role for capsaicin-sensitive, calcitonin gene-related peptide-containing nerves. Acta Physiol Scand 141:445–454
30. Lewis CA et al (2007) Animal models of cough: literature review and presentation of a novel cigarette smoke-enhanced cough model in the guinea-pig. Pulm Pharmacol Ther 20: 325–333
31. Bonham AC et al (1996) Sidestream smoke exposure enhances rapidly adapting receptor responses to substance P in young guinea pigs. J Appl Physiol 81:1715–1722
32. Mutoh T et al (1999) Chronic exposure to sidestream tobacco smoke augments lung C-fiber responsiveness in young guinea pigs. J Appl Physiol 87:757–768
33. Mutoh T et al (2000) Chronic passive cigarette smoke exposure augments bronchopulmonary C-fibre inputs to nucleus tractus solitarii neurones and reflex output in young guinea-pigs. J Physiol 523(Pt 1):223–233
34. Joad JP et al (2004) Passive smoke effects on cough and airways in young guinea pigs: role of brainstem substance P. Am J Respir Crit Care Med 169:499–504

35. Kamei J et al (1994) Cough-induced activity of (-)-(2S, 3aR, 7aS)-1-[(S)-N-[(S)-1-ethoxycarbonyl-3- phenylpropyl]alanyl]hexahydro-2-indolinecarboxylic acid (trandolapril) in guinea pigs. Res Commun Mol Pathol Pharmacol 85:105–108
36. Takahama K et al (1993) Differences in the mode of cough augmentation by four angiotensin-converting enzyme inhibitors in guinea-pigs. J Pharm Pharmacol 45:1003–1005
37. Takahama K et al (1995) Inhalation of phosphoramidon, a neutral endopeptidase inhibitor, induces cough in awake guinea-pigs. Arch Int Pharmacodyn Ther 330:241–250
38. Kamei J, Morita K (1996) Antitussive effect of moguisteine on the enhanced coughing associated with enalapril in guinea-pig. Eur J Pharmacol 312:235–239
39. Hirata R et al (2003) Augmentation of spontaneous cough by enalapril through up-regulation of bradykinin B1 receptors in guinea pigs. Eur J Pharmacol 474:255–260
40. Cialdai C et al (2010) Differences between zofenopril and ramipril, two ACE inhibitors, on cough induced by citric acid in guinea pigs: role of bradykinin and PGE2. Naunyn Schmiedebergs Arch Pharmacol 382:455–461
41. Fox AJ et al (1996) Bradykinin-evoked sensitization of airway sensory nerves: a mechanism for ACE-inhibitor cough. Nat Med 2: 814–817
42. Ito K et al (1995) Toxicodynamic analysis of cough and inflammatory reactions by angiotensin-converting enzyme inhibitors in guinea pig. J Pharmacol Exp Ther 275:920–925
43. Greene SA et al (1984) Sulfur dioxide-induced chronic bronchitis in beagle dogs. J Toxicol Environ Health 13:945–958
44. Shore SA et al (1987) Sulfur-dioxide-induced bronchitis in dogs. Effects on airway responsiveness to inhaled and intravenously administered methacholine. Am Rev Respir Dis 135: 840–847
45. McLeod RL et al (2007) Sulfur-dioxide exposure increases TRPV1-mediated responses in nodose ganglia cells and augments cough in guinea pigs. Pulm Pharmacol Ther 20:750–757
46. Bolser DC et al (1991) Ruthenium red decreases capsaicin and citric acid-induced cough in guinea pigs. Neurosci Lett 126:131–133
47. McLeod RL et al (2001) Nociceptin inhibits cough in the guinea-pig by activation of ORL(1) receptors. Br J Pharmacol 132: 1175–1178
48. Venkatasamy R et al (2010) Use of within-group designs to test anti-tussive drugs in conscious guinea-pigs. J Pharmacol Toxicol Methods 61:157–162
49. Korpas J et al (1965) Changes in coughing in cats with experimental or spontaneous inflammation of the respiratory tract. Physiol Bohemoslov 14:488–494
50. Korpas J (1972) Differentiation of the expiration and the cough reflex. Physiol Bohemoslov 21:677–680
51. Hanacek J et al (1979) Cough reflex changes in local tracheitis. Physiol Bohemoslov 28: 375–380
52. Friebel H (1963) Physiologie, Pathologie and Pharmakologie der Bronchialsekretion und des Hustens. In Müller A: Klinische Physiologie. G Thieme Verl., Stuttgart: 301–339
53. Bolser DC (2004) Experimental models and mechanisms of enhanced coughing. Pulm Pharmacol Ther 17:383–388
54. House A et al (2004) Cough reflex in allergic dogs. Eur J Pharmacol 492:251–258
55. Chapman RW et al (2001) A simple non-invasive method to measure the cough reflex in dogs. J Pharmacol Toxicol Methods 46: 21–26
56. Abraham WM et al (1986) Effect of an inhaled glucocorticosteroid (budesonide) on post-antigen induced increases in airway responsiveness. Bull Eur Physiopathol Respir 22: 387–392
57. Moreaux B et al (1999) Development of a cough induction test in pigs: effects of SR 48968 and enalapril. J Vet Pharmacol Ther 22:387–389
58. Moreaux B et al (2001) Role of bradykinin and tachykinins in the potentiation by enalapril of coughing induced by citric acid in pigs. Fundam Clin Pharmacol 15:23–29
59. Adcock JJ et al (2003) RSD931, a novel anti-tussive agent acting on airway sensory nerves. Br J Pharmacol 138:407–416
60. Gascoigne MH et al (1999) The effect of pre-exposure to ozone in the cough response to inhaled citric acid in conscious rabbits. Eur Resp J 14:157S
61. Hara J et al (2008) Effect of pressure stress applied to the airway on cough-reflex sensitivity in Guinea pigs. Am J Respir Crit Care Med 177:585–592
62. Ye XM et al (2011) Cough reflex sensitivity is increased in guinea pigs with parainfluenza virus infection. Exp Lung Res 37:186–194
63. Gatti R et al (2006) Protease-activated receptor-2 activation exaggerates TRPV1-mediated cough in guinea pigs. J Appl Physiol 101: 506–511

64. Millqvist E et al (2008) Inhaled ethanol potentiates the cough response to capsaicin in patients with airway sensory hyperreactivity. Pulm Pharmacol Ther 22:22

65. Kohrogi H et al (1988) Neutral endopeptidase inhibitors potentiate substance P- and capsaicin-induced cough in awake guinea pigs. J Clin Invest 82:2063–2068

66. Hatanaka M et al (1996) Effect of indapamide on capsaicin-induced cough by enalapril. Jpn Pharmacol Ther 24:1797–1804

67. Roy R et al (1974) Estimation of respiratory parameters by the method of covariance ratios. Comput Biomed Res 7:21–39

Chapter 18

Capsaicin Inhalation Test in Man

Eva Millqvist

Abstract

Stimulation of TRP receptors on sensory nerves is likely involved in chemically induced airway symptoms. A subgroup of patients has been identified with nonallergic, nonasthmatic symptoms from the upper and lower airway that are induced by environmental stimuli normally regarded as nontoxic. These patients have substantially increased cough response to inhaled capsaicin, known to initiate cough via TRPV1 receptors on airway sensory nerves. Hence, the condition has been termed sensory hyperreactivity (SHR). As a diagnostic tool of SHR, a standardized capsaicin inhalation cough test was developed. However, the capsaicin inhalation test can also be positive in other conditions which must be excluded before a diagnosis of SHR can be made. The capsaicin inhalation protocol described here could also be applied to testing TRP channels blockers as antitussive agents.

Key words: Airways, Capsaicin, Diagnostics, Sensory hyperreactivity, TRPV1

1. Introduction

The discovery of the transient receptor potential (TRP) superfamily has changed our understanding of how the human body reacts to its inner and outer environment. Regarding the airways, the expression "sensing the air around us" has got a new significance. The TRP system sets up limit values for airway reactions to temperature, humidity, osmolarity, noxious substances, and odorous substances turning into noxious. When these limits are exceeded, the airways respond with reflex avoidance and other physiological reactions. Though, in general, the TRP airway limits are evident, there are individual differences, sometimes due to medical airway conditions.

Trigeminal neural responses to inhaled irritants have been examined using carbon dioxide, mannitol, adenosine, hypertonic

Arpad Szallasi and Tamás Bíró (eds.), *TRP Channels in Drug Discovery: Volume I*, Methods in Pharmacology and Toxicology, DOI 10.1007/978-1-62703-077-9_18, © Springer Science+Business Media, LLC 2012

saline, and different nociceptive agents. The main targets of these agents in the airways are ion channels belonging to the TRPV and the TRPA families expressed on the C-fibers of sensory nerves (1–3). Nociceptive sensory neurons also take part in protective reflexes, including the cough and sneeze reflexes, and release inflammatory neuropeptides in the periphery upon stimulation by different environmental stimuli. Most trigeminal stimuli also have an odor but the odor threshold, not assessed by TRP channels, is generally lower. A key characteristic of many nociceptive neurons is their sensitivity to capsaicin, the pungent ingredient in chili peppers; this guided the cloning of the first molecular nociceptor, TRPV1, in the mammalian nervous system (4).

Besides capsaicin, TRPV1 can be excited by a variety of stimuli like high temperature, protons, lipids and noxious substances. The discovery that TRPV1 is activated by endogenous *N*-acylethanolamines (5, 6), and subsequent studies showing that other structurally related lipids also activate TRPV1 (7), indicate that TRPV1 also may serve as a sensor of the lipid environment. This is interesting because many of these lipids are generated during inflammation. The TRPV1 channel is viewed as a molecular integrator of chemical and physical stimuli that elicit pain (7). To open the channel, different stimuli lower the critical temperature threshold for activation. However, the activation and integration of different TRP channels in the airways is complex and not yet fully understood.

Cough is an essential protective physiological mechanism to prevent food, liquids, dust, and chemicals to reach the lower airways. Cough is also a symptom of many inflammatory diseases of the lungs for which patients most commonly consult a doctor in the western world (8, 9). When known causes of cough [such as various infections, cancer, foreign body aspiration, cystic fibrosis, alveolitis, asthma, chronic obstructive pulmonary disease (COPD), medication with angiotensin converting enzyme (ACE) inhibitors, gastroesophageal reflux disease (GERD), or postnasal drip syndrome) have been excluded, a group of patients with unexplained chronic cough remains. Cough is arbitrarily defined as chronic when it lasts for more than 8 weeks (10) and it is clearly associated with significant social and psychological impacts (11).

Epidemiological studies indicate that chronic cough is very prevalent in the community (up to 20% of the population) and may be increasing in relation to rising environmental pollution (9). Some cough patients can be labeled as having chronic refractory unexplained cough, a condition without medical explanation and resistant to different medications (12, 13). A novel paradigm has recently been introduced for understanding chronic cough (14). This paradigm, known as the "cough hypersensitivity syndrome," includes patients with symptoms that may indicate a reflux disease, as well as patients with a general hypersensitivity towards environmental irritants (15–18).

A similar group of patients with airway symptoms induced by environmental irritants like odorous chemicals and scents report problems with cough, chest discomfort and dyspnea (19, 20). The symptoms mimic allergy and asthma, but allergy and asthma-specific tests are negative. These patients have an increased cough reaction to inhaled capsaicin, reflecting sensory nerve reactivity and hence the symptoms are interpreted as airway sensory hyperreactivity (SHR). Cigarette smoke, car exhaust, and perfumed products in levels normally regarded as nontoxic are some of the known triggers for SHR symptoms (19, 20). Our earlier results showed changes in nasal lavage levels of nerve growth factor (NGF) after the capsaicin inhalation test (21) indicating a neuro-chemical imbalance, as do the findings that the capsaicin cough reaction can be blocked by inhalation of local anesthesia (22). As a tool to diagnose these patients, a standardized capsaicin inhalation cough test was developed, as well as a question-score system (23–25). The prevalence of SHR, often misdiagnosed as asthma or allergy, is estimated to be 6% in the Swedish population (23). However, the molecular mechanisms behind the SHR syndrome and hyperreactivity in other airway diseases remain obscure. Though there is so far no good method of treating the symptoms in SHR, future possibilities may come in the form of TRP channel blockers (26).

In airway diseases (e.g., asthma, rhinitis, and COPD), symptoms induced by environmental irritants are common. These patients often complain of symptoms induced by exposure to cold air, smoke, exhaust fumes, strong odorants, and exercise. Although the role of sensory nerves in airway inflammation and obstruction is controversial, there is a growing body of evidence that sensory nerves mediate many of the symptoms in these patients (1, 27). In line with this, patients with asthma and COPD exhibit exaggerated cough and secretory responses to inhaled capsaicin (28, 29). Despite extensive research, the link between the inflammatory cascade of asthma (causing bronchial hyperreactivity) and the axon reflex of bronchoconstriction is still not clarified (30). Increasing evidence points to a potential role of TRP channels on airway sensory nonadrenergic noncholinergic (e-NANC) nerves in the development of bronchial hyperresponsiveness and several features of asthmatic disease (1, 30, 31).

In their search for the "cough receptor," Canning et al. found a unique sensory nerve fiber subtype which may be responsible for initiating defensive coughing (32). The study was, however, performed in guinea pigs, a species known to have a different airway reaction when stimulated with capsaicin compared to humans. Furthermore, bronchial obstruction following capsaicin inhalation is evident in guinea pigs, but not in humans. Also, as Chung and Widdicombe pointed out, if there is a unique airway cough receptor then what is the role of all the other airway sensory receptors (33)? Morice and Geppetti claimed that TRPV1 is the true receptor of cough (34), but, as they also underlined, the final proof must await development of a specific antagonist for use in humans. Numerous studies have shown that the afferent fibers on sensory neurons are not static but can change in modulation by changes in the airway environment (35). Patients with chronic cough had an increase in TRPV1-staining nerve profiles, and a significant correlation between increased capsaicin tussive response and the number of TRPV1-positive nerves was reported (36, 37).

Inhaled capsaicin has long been used to induce cough in a safe and dose-dependent manner (38, 39) with good short- and long-term reproducibility (19, 20, 40–43). However, our earlier experiences demonstrated that different inhalation devices produce different numbers of cough (42). A number of capsaicin inhalation methods have been described to provoke cough (44). In the single-dose method, a single concentration of capsaicin is delivered. The dose–response challenge, on the other hand, involves either administration of incremental concentrations of capsaicin over a fixed time period, or administration of single breaths of incremental concentrations of capsaicin. Both methods produce reproducible results with good reciprocal agreement (43). In SHR patients, even inhalation of saline can induce coughing (19, 22). In early studies, using a single-dose provocation method to determine the capsaicin concentration threshold that evokes two, five or ten coughs caused significant problems. Therefore, we chose the dose–response method with incremental concentrations of capsaicin over a fixed time period. This method is stable and produces reproducible results (19, 20, 42). Using the Maxin MA3® device, a well-known equipment for nebulization (45, 46), we have developed a capsaicin inhalation test and set up limits for the expected number of coughs to identify patients with SHR (23, 25, 47).

2. Materials

2.1. Study Participants

1. SHR patients defined as nonatopic, nonallergic with a negative metacholine provocation test (48), normal lung function, and no pulmonary reversibility (Note 1). Smokers, pregnant, and breast-feeding women were excluded from the studies.
2. Control subjects without airway symptoms may be recruited from hospital workers, relatives, and friends. The control subjects may be matched by age and gender to the included patients.

2.2. Equipment

1. Maxin MA3® device (Clinova Medical AB, Malmö, Sweden) for capsaicin inhalation test (technical specifications, Table 1).

Table 1
Technical specifications of the Maxin MA3® device (Clinova Medical AB, Malmö, Sweden) for capsaicin inhalation test

Maxin MA3® (Clinova Medical AB, Malmö, Sweden)

Compressor	Maxin MA3®
Nebulizer	Maxin
Air flow rate	4.5 L/min
Nebulizer output	0.25 ml/min
Inhalation time/each concentration	4 min
Mass median particle diameter	2.27 μm

2. Spectrophotometer (Shimadzu UV 16, Kyoto, Japan) for capsaicin solution stability studies.

2.3. Capsaicin Solutions

1. Capsaicin stock solution: 100 μM capsaicin (Sigma, St. Louis, MO) in ethanol (99.5%), stored at −20°C.
2. Aqueous provocation solutions: dilute stock to 0.06 and 0.3 μM capsaicin with 0.9% saline containing 1% w/v ethanol; keep in glass bottles refrigerated and use within 1 week.

3. Methods

3.1. Capsaicin Stability

The stability of capsaicin in the stock solution was investigated (49). Three aliquots of the stock solution were stored at different conditions: one was placed in a refrigerator, one was kept in darkness at room temperature, and the third was exposed to daylight at room temperature. The stability of the stock solution stored at −20°C was also checked.

1. Prepare serial 3× dilutions of the capsaicin stock (0–100 μM) and measure absorbance at 281 nm (Shimadzu Spectrophotometer).
2. Determine the concentration of capsaicin aliquots kept at different conditions (Note 2).

3.2. Capsaicin Inhalation Provocation Method

1. Fill the Maxin MA3® nebulizer (Clinova) with (a) 2 ml saline or (b) 2 ml capsaicin, 0.06 μM or 0.3 μM (Notes 3 and 4).
2. Place the patient in a seated position in a semi-closed provocation box with separate ventilation to prevent any aerosol from reaching the outer room.

3. Instruct the patient to inhale with tidal volume breathing without a nose-clip for 4 min (the time estimated to inhale 1 ml of provocation solution).
4. Instruct the patient to inhale 1 ml of saline, followed by 1 ml of capsaicin aqueous provocation solution, 0.06 μM or 0.3 μM (Note 5).
5. Allow 6 min for rest between each dose.
6. Count the number of coughs for 10 min from the onset of each provocation step and registered manually (Notes 6 and 7).
7. Repeat above steps to determine the reproducibility of the results (Notes 8 and 9).

4. Notes

1. Indications for capsaicin inhalation test: (a) upper or/and lower airway symptoms induced by chemicals and scents like perfume, flower scents, car exhausts; (b) occupation-related airway sensitivity to chemical and scents; (c) an unclear diagnosis of asthma that cannot be defined by pulmonary tests and where the patients have little or no benefit from asthma medication; (d) chronic idiopathic cough.
2. Stock solution (100 μM) can be stored in the refrigerator for almost 2 months without significant loss of capsaicin, whereas room-temperature storage over an extended time period caused deterioration. The stock solution showed total stability after 3 months of storage in a freezer.
3. The Maxin MA3® device nebulizes continually; produces a fine aerosol; has a reservoir for nebulized mist in a closed inhalation system; and provides a fixed constant flow of 0.25 ml/min.
4. We have no indications of any risk for the staff performing the tests. Capsaicin is irritating for the skin and protecting gloves should be used when handling the substance. We use natural capsaicin (from chili fruits) and any remaining substance may be thrown in the drain.
5. After having performed more than 1,000 capsaicin inhalation tests, we have no indications of any serious side effects of the capsaicin inhalation test. Thirty-five patients who answered a questionnaire about late symptoms occurring on the days after capsaicin inhalation provocation had significantly more cough and feeling of "weight on the chest" on the same evening compared to before the provocation (49). No such relationship was found for other symptoms like phlegm, rhinorrhea, or general symptoms. On the following 2 days there were no

significant differences in recorded symptoms. This in agreement with the experiences of other research groups (41).

6. At each provocation, FEV_1 was measured before and after each completed capsaicin test. Many studies have shown no significant changes in FEV_1 after the capsaicin provocations among either the patients or the healthy controls (41), which is in line with our own experiences.
7. In a series of studies we have evaluated the effect of different cut-off values on sensitivity and specificity and have established the limit values for the capsaicin test to 35 coughs for 0.06 and 0.3 μM capsaicin, respectively, with a positive reaction already on the first capsaicin concentration step indicating a more severe form of increased sensitivity (23, 25, 47). When the patient coughs 35 times or more the provocation is stopped.
8. At the higher concentration of capsaicin (0.3 μM), the number of coughs correlated significantly between the two challenge occasions ($r=0.97$, $P<0.001$), but not at the lower capsaicin concentration (0.06 μM) ($r=0.63$, $P<0.09$). The SD from the mean difference between eight pairs of repeated measurements was 7.4 coughs for the capsaicin concentration of 0.06 μM and the coefficient of reproducibility was thus calculated as 14.8 coughs. For 0.3 μM capsaicin, the corresponding values were 5 and 10 coughs, respectively.
9. The ability to distinguish patients from control subjects of the capsaicin inhalation test was evaluated by constructing a

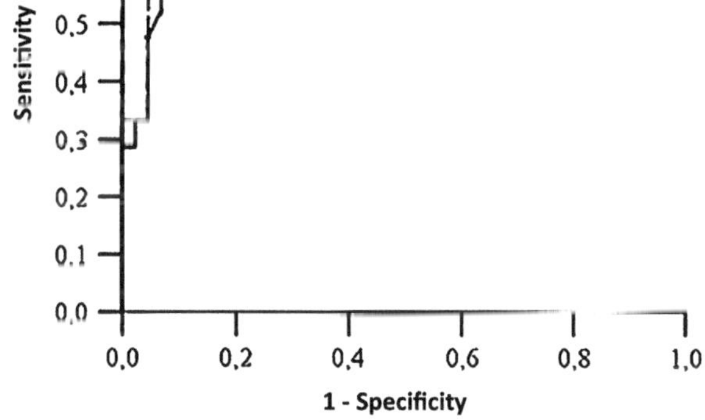

Fig. 1. ROC curves of optimal cough cut-off levels to distinguish patients with SHR from healthy controls with the Maxin MA3® device (capsaicin concentrations 0.06 and 0.3 μmol/L).

"receiver operator characteristic curve" (ROC curve) in which sensitivity versus 1-specificity was plotted for each possible cough cut-off level (50). The ROC curve allows measurement of a method's discriminative ability, for example, the ability to distinguish two groups from one another. An area under the curve of 0.5 suggests no discriminative ability, while an area under the curve of more than 0.90 indicates outstanding discriminative ability, more than 0.80 indicates excellent discriminative ability, and more than 0.7 indicates acceptable discriminative ability (51). The area under the curve with the Maxin MA3® device was 0.80 for 0.06 μM capsaicin and 0.94 for 0.3 μM capsaicin (Fig. 1).

References

1. Bessac BF, Jordt SE (2008) Breathtaking TRP channels: TRPA1 and TRPV1 in airway chemosensation and reflex control. Physiology (Bethesda) 23:360–370
2. Nilius B, Owsianik G, Voets T, Peters JA (2007) Transient receptor potential cation channels in disease. Physiol Rev 87(1):165–217
3. Banner KH, Igney F, Poll C (2011) TRP channels: emerging targets for respiratory disease. Pharmacol Ther 130(3):371–384
4. Caterina MJ, Schumacher MA, Tominaga M, Rosen TA, Levine JD, Julius D (1997) The capsaicin receptor: a heat-activated ion channel in the pain pathway. Nature 389(6653):816–824
5. Movahed P, Jonsson BA, Birnir B, Wingstrand JA, Jorgensen TD, Ermund A, Sterner O, Zygmunt PM, Hogestatt ED (2005) Endogenous unsaturated C18 N-acylethanolamines are vanilloid receptor (TRPV1) agonists. J Biol Chem 280(46):38496–38504
6. Zygmunt PM, Petersson J, Andersson DA, Chuang H, Sorgard M, Di Marzo V, Julius D, Hogestatt ED (1999) Vanilloid receptors on sensory nerves mediate the vasodilator action of anandamide. Nature 400(6743):452–457
7. Szallasi A, Cortright DN, Blum CA, Eid SR (2007) The vanilloid receptor TRPV1: 10 years from channel cloning to antagonist proof-of-concept. Nat Rev Drug Discov 6(5):357–372
8. Irwin RS, Curley FJ, French CL (1990) Chronic cough. The spectrum and frequency of causes, key components of the diagnostic evaluation, and outcome of specific therapy. Am Rev Respir Dis 141(3):640–647
9. Chung KF, Pavord ID (2008) Prevalence, pathogenesis, and causes of chronic cough. Lancet 371(9621):1364–1374
10. Morice AH, Fontana GA, Sovijarvi AR, Pistolesi M, Chung KF, Widdicombe J, O'Connell F, Geppetti P, Gronke L, De Jongste J, Belvisi M, Dicpinigaitis P, Fischer A, McGarvey L, Fokkens WJ, Kastelik J (2004) The diagnosis and management of chronic cough. Eur Respir J 24(3):481–492
11. Young EC, Smith JA (2010) Quality of life in patients with chronic cough. Ther Adv Respir Dis 4(1):49–55
12. Dicpinigaitis PV (2011) Cough: an unmet clinical need. Br J Pharmacol 163(1):116–124
13. Vertigan AE, Gibson PG (2011) Chronic refractory cough as a sensory neuropathy: evidence from a reinterpretation of cough triggers. J Voice 25(5):596–601
14. Morice AH, Faruqi S, Wright CE, Thompson R, Bland JM (2011) Cough hypersensitivity syndrome: a distinct clinical entity. Lung 189(1): 73–79
15. Morice AH (2009) The cough hypersensitivity syndrome: a novel paradigm for understanding cough. Lung 188:87–90
16. Chung KF (2011) Chronic 'cough hypersensitivity syndrome': a more precise label for chronic cough. Pulm Pharmacol Ther 24(3): 267–271
17. Millqvist E (2011) The airway sensory hyperreactivity syndrome. Pulm Pharmacol Ther 24(3):263–266
18. Birring SS (2011) New concepts in the management of chronic cough. Pulm Pharmacol Ther 24(3):334–338
19. Millqvist E, Bende M, Löwhagen O (1998) Sensory hyperreactivity – a possible mechanism underlying cough and asthma-like symptoms. Allergy 53(12):1208–1212
20. Ternesten-Hasseus E, Lowhagen O, Millqvist E (2007) Quality of life and capsaicin sensitivity in patients with airway symptoms induced by chemicals and scents: a longitudinal study. Environ Health Perspect 115(3): 425–429

21. Millqvist E, Ternesten-Hasseus E, Stahl A, Bende M (2005) Changes in levels of nerve growth factor in nasal secretions after capsaicin inhalation in patients with airway symptoms from scents and chemicals. Environ Health Perspect 113(7):849–852
22. Millqvist E (2000) Cough provocation with capsaicin is an objective way to test sensory hyperreactivity in patients with asthma-like symptoms. Allergy 55(6):546–550
23. Johansson A, Millqvist E, Nordin S, Bende M (2006) Relationship between self-reported odor intolerance and sensitivity to inhaled capsaicin: proposed definition of airway sensory hyperreactivity and estimation of its prevalence. Chest 129(6):1623–1628
24. Nordin S, Millqvist E, Lowhagen O, Bende M (2004) A short chemical sensitivity scale for assessment of airway sensory hyperreactivity. Int Arch Occup Environ Health 77(4):249–254
25. Johansson A, Löwhagen O, Millqvist E, Bende M (2002) Capsaicin inhalation test for identification of sensory hyperreactivity. Respir Med 96(9):731–735
26. Planells-Cases R, Valente P, Ferrer Montiel A, Qin F, Szallasi A (2011) Complex regulation of TRPV1 and related thermo-TRPs: implications for therapeutic intervention. Adv Exp Med Biol 704:491–515
27. Nilius B (2007) TRP channels in disease. Biochim Biophys Acta 1772(8):805–812
28. Doherty MJ, Mister R, Pearson MG, Calverley PM (2000) Capsaicin responsiveness and cough in asthma and chronic obstructive pulmonary disease. Thorax 55(8):643–649
29. Ekstrand Y, Ternesten-Hasseus E, Arvidsson M, Lofdahl K, Palmqvist M, Millqvist E (2011) Sensitivity to environmental irritants and capsaicin cough reaction in patients with a positive methacholine provocation test before and after treatment with inhaled corticosteroids. J Asthma 48(5):482–489
30. Veres TZ, Rochlitzer S, Braun A (2009) The role of neuro immune cross-talk in the regulation of inflammation and remodelling in asthma. Pharmacol Ther 122(2):203–214
31. Colsoul B, Nilius B, Vennekens R (2009) On the putative role of transient receptor potential cation channels in asthma. Clin Exp Allergy 39(10):1456–1466
32. Canning BJ, Mazzone SB, Meeker SN, Mori N, Reynolds SM, Undem BJ (2004) Identification of the tracheal and laryngeal afferent neurones mediating cough in anaesthetized guinea-pigs. J Physiol 557(Pt 2): 543–558
33. Chung KF, Widdicombe JG (2004) Cough as a symptom. Pulm Pharmacol Ther 17(6):329–332
34. Morice AH, Geppetti P (2004) Cough. 5: the type 1 vanilloid receptor: a sensory receptor for cough. Thorax 59(3):257–258
35. Carr MJ (2004) Plasticity of vagal afferent fibres mediating cough. Pulm Pharmacol Ther 17(6):447–451
36. Groneberg DA, Niimi A, Dinh QT, Cosio B, Hew M, Fischer A, Chung KF (2004) Increased expression of transient receptor potential vanilloid-1 in airway nerves of chronic cough. Am J Respir Crit Care Med 170(12):1276–1280
37. Mitchell JE, Campbell AP, New NE, Sadofsky LR, Kastelik JA, Mulrennan SA, Compton SJ, Morice AH (2005) Expression and characterization of the intracellular vanilloid receptor (TRPV1) in bronchi from patients with chronic cough. Exp Lung Res 31(3):295–306
38. Fuller RW, Dixon CM, Barnes PJ (1985) Bronchoconstrictor response to inhaled capsaicin in humans. J Appl Physiol 58(4):1080–1084
39. Midgren B, Hansson L, Karlsson J-A, Simonsson BG, Persson CGA (1992) Capsaicin-induced cough in humans. Am Rev Respir Dis 146:347–351
40. Dicpinigaitis PV (2003) Short- and long-term reproducibility of capsaicin cough challenge testing. Pulm Pharmacol Ther 16(1):61–65
41. Dicpinigaitis PV, Alva RV (2005) Safety of capsaicin cough challenge testing. Chest 128(1):196–202
42. Ternesten-Hasseus E, Johansson Å, Lowhagen O, Millqvist E (2006) Inhalation method determines outcome of capsaicin inhalation in patients with chronic cough due to sensory hyperreactivity. Pulm Pharmacol Ther 19(3):172–178
43. Nejla S, Fujimura M, Kamio Y (2000) Comparison between tidal breathing and dosimeter methods in assessing cough receptor sensitivity to capsaicin. Respirology 5(4):337–342
44. Morice AH, Fontana GA, Belvisi MG, Birring SS, Chung KF, Dicpinigaitis PV, Kastelik JA, McGarvey LP, Smith JA, Tatar M, Widdicombe J (2007) ERS guidelines on the assessment of cough. Eur Respir J 29(6):1256–1276
45. Dahlstrom K, Thorsson L, Larsson P, Nikander K (2003) Systemic availability and lung deposition of budesonide via three different nebulizers in adults. Ann Allergy Asthma Immunol 90(2):226–232
46. Jendle J, Karlberg BE, Persliden J, Franzen L, Arborelius M Jr (1995) Delivery and retention of an insulin aerosol produced by a new jet nebulizer. J Aerosol Med 8(3):243–254
47. Ternesten-Hasseus E, Larsson C, Bende M, Millqvist E (2008) Capsaicin provocation using two different inhalation devices. Respir Med 102(12):1784–1790

48. Sterk PJ, Fabbri LM, Quanjer PH, Cockcroft DW, O'Byrne PM, Anderson SD, Juniper EF, Malo JL (1993) Airway responsiveness. Standardized challenge testing with pharmacological, physical and sensitizing stimuli in adults. Report Working Party Standardization of Lung Function Tests, European Community for Steel and Coal. Official Statement of the European Respiratory Society. Eur Respir J Suppl 16:53–83
49. Ternesten-Hasseus E, Farbrot A, Löwhagen O, Millqvist E (2002) Sensitivity to methacholine and capsaicin in patients with unclear respiratory symptoms. Allergy 57(6):501–507
50. Hanley JA, McNeil BJ (1982) The meaning and use of the area under a receiver operating characteristic (ROC) curve. Radiology 143(1):29–36
51. Hosmer DW, Lemeshow S (2000) Applied logistic regression. Wiley, New York, NY, 162

Part IV

TRPs and the Genitourinary (GU) Tract

Chapter 19

TRP Channels in the Genitourinary Tract

Ana Charrua and Francisco Cruz

Abstract

Recently, TRP channels have been implicated in numerous pathologies of the genitourinary (GU) tract. TRP channels are differently expressed along the GU tract, and several lines of evidence suggest that they also have different roles in the pathophysiology of GU tract diseases. In this chapter, we focus on the expression and role of TRP in the urinary bladder and give an overall idea of TRP channel expression and function in the remaining GU tract.

Key words: TRPV, TRPM, TRPA1, TRPC, TRPP, TRPML, Urinary tract, Male genital tract, Female genital tract

1. TRP Channels in Bladder Afferents Coursing the Normal Urinary Bladder

TRP channels are abundantly expressed by bladder primary afferents (1). TRPV1 is the foremost studied TRP channel in the urinary bladder, where it can be found in varicose nerve plexus both in the mucosa and muscular layer (2, 3). In the mucosa, TRPV1-positive fibers are located in the suburothelial space, apposed to the urothelial basal cells or running between them up to the bladder lumen (2, 3). This receptor can also be found in nerve fibers surrounding blood vessels deeper in the *lamina propria* (2, 3). In what concerns the muscular layer, TRPV1-expressing fibers are located close to smooth muscle cells, with some terminals encroached in shallow grooves on the surface of muscle cells themselves (2, 3). In the rodents TRPV1-expressing fibers coursing the urinary bladder have been reported to decrease with age (4).

The secondly most well-studied TRP channel in the urinary bladder is TRPV4. Although TRPV4 could not be detected in nerve fibers that course the urinary bladder (5), its presence has been described in the cell bodies of neurons of the L4–L6 dorsal

Arpad Szallasi and Tamás Bíró (eds.), *TRP Channels in Drug Discovery: Volume I*, Methods in Pharmacology and Toxicology, DOI 10.1007/978-1-62703-077-9_19, © Springer Science+Business Media, LLC 2012

root ganglia and in nerve terminal in the dorsal horn of the spinal cord (6–8). In fact, 88% of L4–L6 dorsal root ganglia neurons express TRPV4 but only a small proportion of L4–L6 dorsal root ganglia neurons co-express TRPV4 and TRPV1 (8, 9).

Rat TRPA1 fibers were found throughout the bladder with higher prevalence in the outflow region (10). Interestingly, TRPA1 is expressed in the same fibers that express TRPV1, with a 1:1 overlap (10).

TRPM8 can be found in the dorsal root ganglia neurons that project to the bladder and in suburothelial nerve fibers (9, 11, 12). TRPM8 is highly expressed in CGRP-immunopositive fibers and also has a small co-expression with TRPV1-imunopositive fibers and even lesser co-expression with IB4 population of primary afferents (13).

2. TRP Channels in Nonneuronal Cells in the Urinary Bladder

Urothelial cells express TRPV1, TRPA1, TRPV4, TRPV2, and TRPM7, 8 (2, 10, 14–16). TRPV1 can be found in all urothelial layers, with higher prevalence in the superficial layers where large umbrella cells prevail (3, 10). Urothelial TRPV1 is functional. In vitro experiments performed with rodent urothelial cells have demonstrated that TRPV1 activation with capsaicin induces calcium influx and nitric oxide release (15). Consisting with these data, TRPV1 present in human urothelial (hu) cells is activated by capsaicin, heat (41 and 45°C) and protons, and also by inflammatory mediators (16). Furthermore, stimulation of huTRPV1 with capsaicin results in a substantial increase of ATP release from urothelial cells (16).

TRPV4 expression in urothelial cells is undisputable and was detected in the urothelial membrane of rat, mice, and guinea pig (5, 7, 14, 17, 18). In the mouse urothelium, TRPV4 is present in the basolateral membrane of the basal and intermediate layers (14, 19). In the human healthy urothelium TRPV4 was identified near the adherence junctions in close connection alpha-catenin that links E-cadherins to actin network (20). It is thought that around 20% of urothelial cells express both TRPV4 and TRPV1 and around 50% express both TRPV4 and TRPM8 (18).

The presence of TRPA1 in rat urothelium was detected both by fluorescence immunohistochemistry and RT-PCR (10). However, in contrary to what is observed in neurons, there is almost no co-localization between TRPA1 and TRPV1. While these TRP receptors prevail in the umbrella cells, TRPA1 is found almost exclusively in the basal layers of the urothelium (10).

TRPV2 and TRPM4 have been found in the superficial urothelial layer, in the apical membrane of umbrella cells (19). In fact, these cells express both the full-length TRPV2 and a short splice-variant (21). TRPC1, TRPM7, and TRPML1 were found in the cytoplasm and in cytoplasmic membrane of urothelial cells (17). TRPM8 was demonstrated to be present in human urothelium both by RT-PCR analysis (9) and immunohistochemistry (9, 12).

Immunohistochemical analysis revealed the presence of TRPV1 in interstitial cells, smooth muscle cells, and endothelial cells (3). In addition the human and rodent smooth muscle cells express TRPV4 mRNA and protein (22).

One report compared the density of TRPV1, TRPA1, and ASIC2 by gender (23). Apparently, these TRP receptors are more abundantly expressed in female than in male urinary bladder (23). It is temptive to relate this observation to the well known fact that painful urinary bladder pathologies such as interstitial cystitis are more common in women than men.

3. TRP Role in Bladder Function

Birder and coworkers showed that TRPV1 knockout mice have more nonvoiding contractions, an increased bladder capacity and their bladder distension was unable to induce an increase in spinal cord Fos expression (24). TRPV1 knockout mice produced more small-diameter urine spots that their wild-type littermates indicating a higher frequency of small-volume voidings (24). Furthermore, they also found that cultured urothelial cells from TRPV1 knockout mice release less ATP when stretched (24). Altogether these data indicated that TRPV1 participates in normal bladder function. However, these observations could not be partially reproduced by Charrua and coworkers (25). The latter authors performed cystometries in anesthetized TRPV1 knockout mice and did not observe nonvoiding contractions during bladder filling. Moreover, TRPV1 knockout mice had the same urinary frequency of to their wild-type littermate (25). In agreement with the last observation, TRPV1 antagonist in therapeutic doses was devoid of any relevant effect when administered to normal animals (26). Therefore, future experiments are needed to elucidate the role of TRPV1 in the normal bladder reflex activity. Eventually, the use of general anesthesia during cystometry may contribute for the differences between the two studies (24, 25).

TRPV4 knockout mice present an increased micturition threshold and a reduced stretch-induced ATP release (5). Awaken TRPV4 knockout mice have an incontinent phenotype, as shown by an increased number of small urine spots between micturitions (5). Cystometry under anesthesia revealed that TRPV4 knockout mice have lower frequency of voiding contractions but a higher number of nonvoiding urinary bladder contractions (5). The application of TRPV4 agonist to awake rats did not alter the frequency of intact bladders but increased the amplitude of contractions. This effect was not abolished by capsaicin treatment, indicating that it was mediated by the stimulation of capsaicin-insensitive bladder afferents (17), in accordance with the fact that TRPV4 and TRPV1 do not colocalize in bladder afferents (27). The functional role of TRPA1 is basically unclear. The finding that TRPA1 agonist induces detrusor overactivity in intact animals is strongly indicative that this receptor may have in sensory transduction (10). However the role TRPA1 expressed in urothelium and nerve fibers is totally unknown.

TRPM8 seems to be involved in the control of bladder micturition (28). The application of AMTB, a specific TRPM8 antagonist, to naïve animals reduced the bladder frequency in a dose-dependent manner (28). TRPM8 is a cold receptor that detects temperatures between 8 and 30°C (29, 30). Therefore, it has been suggested that this receptor is responsible for the urinary urgency evoke by cold sensation (31). The explanation may not reside in the capacity of TRPM8 to encode an eventual decrease of urine temperature but rather in the fact that the receptor is expressed in a subset of dichotomizing afferents that have their cell bodies in L6–S1 dorsal root ganglia and have termination both in the skin and in the urinary bladder (31).

As stated above, interstitial cells express TRPV1 (32). TRPV1 activation by capsaicin reduced the contractility of interstitial cells expressed in those cells (32). However, it remains uncertain in which way TRPV1 and TRPV4 expressed in interstitial cells, smooth muscle cells and endothelial cells (3) contribute to normal micturition reflex.

4. TRP Role During Micturition Dysfunction

TRPV1 role in urinary bladder micturition dysfunction was long foreseen before the receptor has been cloned. Capsaicin-sensitive fibers desensitization resulting from intravesical treatment with vanilloids, such as capsaicin and resiniferatoxin, reduced urinary incontinence by inhibiting involuntary detrusor contractions in patients with neurogenic (NDO) (33–50) and idiopathic detrusor overactivity (IDO) (47, 51–55). In addition, instillation of the two vanilloid solutions was shown to improve pain and frequency in patients with interstitial cystitis (IC) (54–58). These results should however be interpreted with caution since a large randomized clinical trial was unable to demonstrate the superiority of resiniferatoxin over placebo in IC patients (59).

The eventual effects of intravesical vanilloids on low urinary tract of patients with NDO, IDO and IC have, nevertheless, a strong molecular support. TRPV1 was shown to be increased in patients with NDO, IDO (49) as well as in patients with IC (60).

TRPV1 is desensitized by capsaicin and resiniferatoxin, decreasing its function and expression to further stimuli (61, 62).

Recent animal studies confirmed the importance of TRPV1 for micturition control during inflammation and NDO. TRPV1 knockout mice do not develop bladder overactivity during bladder inflammation (25). Moreover, TRPV1 antagonists reverse bladder overactivity caused by inflammation or by chronic spinal cord transection (63).

The mechanism by which TRPV1 increases in the bladder of IDO, NDO, and IC patients is being progressively unveiled. One mechanism might be the increase in TRPV1 levels due to an increase in TRPV1 translation (56, 64). The TRPV1 promoter is nerve growth factor-dependent (65, 66) and coincidentally there is an increase in nerve growth factor in the urine of the patients with NDO, IDO, and IC (67–72). In addition, the cleavage of phosphatidylinositol 4,5-bisphosphate by a nerve growth factor-mediated pathway demonstrated in in vitro experiments may also be operative in the bladder (73).

TRPV1 may be more active in the bladder of some of these patients due to a decrease in the amount of inhibitory splice variant of TRPV1, TRPV1b. These inhibitory splice variant have been suggested to make nonfunctional heterodimers with TRPV1 (74, 75). In the rat, cystitis was shown to decrease the expression of TRPV1b (76), therefore increasing the probability of formation of active homotetramers, composed exclusively by TRPV1 active channels.

TRPV1 can be directly activated by *N*-arachidonoyl-ethanolamine (anandamide) (77), *N*-arachidonoyl-dopamine (78), *N*-oleoyl-dopamine (79), eicosanoid acids and leucotrienes (80), among other molecules. Inflamed urinary bladders have an increase concentration of anandamide compared to normal ones (81). As these molecules are weak TRPV1 activators, it is also possible that the role of these endogenous agonists reside in the receptor sensitization (82–84). An increase of TRPV1 trafficking from the cytoplasm to the neuronal membrane is another mechanism that may, additionally, lead to an increase in TRPV1 expression in the neuronal surface during inflammatory situations (84).

An increase in TRPV1 expression is also detected in urothelial cells during bladder pathologies. In fact, TRPV1 mRNA is increased when human urothelial cells grow in culture media containing inflammatory mediators (15). Also, urothelial TRPV1 was also increased in biopsies of patients with overactive bladder (85, 86). Furthermore, intravesical vanilloid treatment decreased urothelial TRPV1 expression (85).

Altogether, these data prompt TRPV1 as a therapeutic target to overcome bladder disorders (87). However, the discovery of a specific, effective molecule that could block TRPV1 activity has suffered a few, but important, setbacks. For instances, it has been discovered that most TRPV1 antagonists promote vasoconstriction and launch into circulation the blood stored in the visceral bed, leading to an increase in body temperature (88–93). However, recent data showed that hyperthermia is attenuated after repeated application of TRPV1 antagonist, which renews the interest in already synthesized molecules (94, 95). Another setback is the potential risk of enhancing the ischemic area of the heart after acute coronary obstruction (96). In fact, TRPV1 knockout mice presented an increased postmyocardial infarction fibrosis and impaired myocardial contractile performance than their wild-type littermates indicating that TRPV1 has a protective role in postmyocardial infarction recovery (96).

The role of TRPV4 in pathological conditions has been studied in animal models and in vitro experiments. TRPV4 knockout mice do not develop bladder hyperactivity in the cystitis model of cyclophosphamide, suggesting a potential role during bladder inflammation (97). TRPV4 may also have a role bladder outlet obstruction. It was recently reported that patients with such condition have an increased expression of TRPV4 (98). Therefore, as observed for both TRPV1 and TRPA1, TRPV4 antagonists might be promising tools for the treatment of LUT in patients with prostatic enlargement.

TRPA1 activation by specific exogenous agonists, allyl isothiocyanate or cinnamaldehyde, leads to an increased frequency of reflex bladder contractions, in cystometries carried out in conscious rats (10, 99). Hydrogen sulfide, which may be produced in tissues during inflammation, can be an endogenous agonist of TRPA1 (10). Using a rat spinal cord injury (SCI) model, Andrade and coworkers have demonstrated that TRPA1 protein and mRNA levels were elevated both in the urinary bladder and in L6–S1 dorsal root ganglion of animals with SCI, but not at the corresponding spinal cord segments (100). Treating those animals with either a TRPA1 antagonist or with a TRPA1 antisense decreased the receptor expression, the number of nonvoiding contractions, and bladder overactivity (100). TRPA1 antagonists do not induce hyperthermia, nor have any locomotor or cardiovascular problems that have been reported for some TRPV1 antagonist (101).

Patients with IDO and IC present an increased expression of TRPM8 in thin, but not in thick, nerve fibers coursing the urinary bladder (12). TRPM8 immunoreactivity in the urothelium was not changed in IDO and IC patients (12). Interestingly, the increase in TRPM8 expression correlated with clinical symptoms, frequency and pain, but not with urgency (12). In contrast with IDO and IC patients, those with bladder outlet obstruction that developed detrusor overactivity showed no alteration on TRPM8 expression in the bladder mucosa (102, 103). Accordingly, the application of menthol, a TRPM8 agonist, in normal rats and in rats with bladder outlet obstruction produced similar effects (103). The reason why

TRPM8 seems to be involved only IDO and IC, and not in other bladder pathologies is not yet known.

5. Interactions Between Different TRPs and Between TRP and Non-TRP Receptors

The interaction between TRP channels is a new field of research. Recent data from Charrua et al. have shown that co-application of TRPV1 and TRPV4 antagonist is more effective in reversing frequency during cystitis than the application of these drugs per se (8).

Furthermore, TRPV1 and TRPA1 present cross-desensitization, which allows the use of much lower doses of antagonist molecules to counteract these receptors activity (99).

TRPV1 sensitization and activation results and is controlled by several molecules. TRPV1 can by phosphorylated with Ca^{2+}/calmodulin-dependent protein kinase, protein kinase A or protein kinase C (104). Protease-activated receptor 2 (105, 106) or 5-hydroxytryptamine 7 receptor (107) are known to be involved in protein kinase A or C-induced TRPV1 sensitization. Accordingly, protease-activated receptor 2 is expressed in TRPV1-expressing fibers and is upregulated during cystitis (108). Contrariwise, activation of group II metabotropic glutamate receptors (109) or mu opioid receptor (110, 111) inhibits TRPV1 activation, possibly by acting on the cAMP/PKA pathway (112).

The ureter and/or bladder distension induce ATP release from rodent (113) and human (114) urothelium that activates P2X3 receptors in suburothelial sensory fibers, thus contributing to modulation of bladder reflex activity (115). During chronic bladder pathologies such as interstitial cystitis or IDO, there is an increase in the urothelial ATP release (116, 117). Interestingly, the activation of TRPV1 by specific agonist like capsaicin induces release of ATP from human (16) and rodents' (118) urothelial cells, therefore bridging the two receptors pathway.

P2X3 co-localize both with CB1 (119–122) and TRPV1 receptor (123). CB1 and TRPV1 can be activated by anandamide. Anandamide levels increase in the urinary bladder tissues during cystitis, accompanying bladder hyperreflexia and hyperalgesia (81). Curiously, the co-application of anandamide and CB1 antagonist substantially increased the potency of anandamide (81). This indicates that during inflammation the net effect of anandamide results from a balance between the activation of TRPV1 which increases bladder activity and the activation of CB1 which tends to produce the opposite effect.

Bradykinin receptor activation sensitizes TRPV1 through a PKC-dependent mechanism (124–129). Activation of bradykinin receptor can also induce phospholipase C and phospholipase A2 activation, leading to the production of arachidonic acid metabolites which are TRPV1 agonists (128, 130, 131).

6. TRP Channels in the Kidney, Ureters, and Urethra

TRP channels are not exclusively found in the bladder. Several authors have described the presence of these receptors in the kidney, ureters, and urethra. TRPV1 has been found in nerve terminal that course both mucosa and muscular layer of the renal pelvis, ureters, and urethra (2). It is thought that TRPV1 has a protective role in situations of exaggerated renal function and structural injury, by attenuating the progression of renal fibrosis possibly through down-regulation of TGF-β/Smad2/3 signaling pathway (132). Recently experiments have demonstrated that TRPV1 attenuates renal inflammatory responses in mice subjected to DOCA-salt hypertension (133). Oral administration of SA13353, a TRPV1 agonist, prevents the ischemia/reperfusion-induced acute renal injury (134).

The activation of TRPV1 in the renal pelvis TRPV1 expressed in afferent pelvic renal nerve increases glomerular filtration rate and distal tubular delivery of sodium (135–139). Thus renal TRPV1 activation was shown to reverse salt-induced hypertension and its expression is reduced in hypertensive rats (138, 140, 141). This is thought to be achieved through activation of endothelin receptor-1B, via the protein kinase C pathway, and by the release of SP and CGRP in the renal pelvis (135, 142–145). One possible endogenous activator of renal pelvic TRPV1 is anandamide, since this compound increases urine volume excretion (132) although capsazepine was unable to blockade the anandamide-induced increase of diuresis (146). Moreover, anandamide-induced increase in urine volume excretion did not alter sodium excretion (146).

As observed in the urinary bladder, TRPV1 splice variants are also present in the kidney. VR1 5 splice variant (74, 147), TRPV1 (VAR) (148), TRPV1b (75), but not TRPV1β (149), can be found in the kidney, and all of them have a dominant negative activity over TRPV1 (75, 148, 150, 151). However, the precise function of these kidney splice variants is still unveiled.

TRPV4 is strongly expressed throughout the kidney (152, 153) in the basolateral membranes of epithelial cells of the ascending thin limb (152), thick ascending limb (152), the distal tubules (152, 154, 155), connecting tubule (152) and in intercalated cells (type A) (152). In other words, TRPV4 is present in water impermeant nephron segments in which a great transcellular osmotic gradient occurs (152). A weaker expression can be found in all collecting duct portions and in papillary epithelium (152) and no expression can be observed in the proximal tubule, in descending thin limb and in water-permeant cells of the macula densa (152). TRPV4 has similar expression as the renal exclusive sodium–potassium–chloride cotransporter 2 in thick ascending limb, although the latter being expressed apically (156) suggesting that TRPV4 might also be involved in the regulation of sodium and water balance.

Several other TRPs have been described in the kidney, such as TRPV5, 6 (157, 158), TRPC1, 3, 4, 6, TRPM3, 6, TRPP1, and TRPP2. Although rat kidney seems to express only TRPV5, human kidney expresses both TRPV5 and TRPV6 in cells of the distal convoluted tubule and connecting segment of the nephron (157–162). TRPV5 and TRPV6 control active reabsorption of calcium by limiting the rate of apical calcium entry into cells of the distal convoluted tubule and connecting segment of the nephron, during *trans*-cellular calcium reabsorption (157–162). It is known that TRPV5 inactivation leads to severe hypercalciuria (163). However, calcium reabsorption involves many other mechanism and pathways making unclear how important are TRP channel for calcium homeostasis.

TRPC1, 3, 4, and 6 were detected in different parts of the nephrons. TRPC1, TRPC4 and TRPC6 can be found in mesangial cell, and TRPC1 can also be found in the proximal tubule and in the thin descending limb, where it co-localizes with aquaporin-1 (164–167). TRPC1 is present in the cytoplasm, while TRPC4 is found mostly in the cytoplasmic membrane of mesangial cells (166). Using co-immunoprecipitation, it has been shown that TRPC1 directly interacts with TRPC4 and TRPC6 in human mesangial cells (168, 169).

TRPC1, TRPP1/TRPP2, and alpha-epithelial sodium channel are expressed in primary cilium of renal epithelial cells, indicating that these receptors might have sensory transducer properties, possibly responding to changes in fluid flow or having mechanical perception (170–172).

TRPC3 and 6 can be found in podocytes, and are also expressed in connecting tubule and collecting duct cells, since they co-localize with aquaporin-2 (164, 165, 173). However, both receptors do not co-localize with sodium/calcium exchanger or with renal peanut lectin binding site, meaning that they are expressed in principle cells of the collecting duct (164). Although TRPC3 is localized in apical domain, while TRPC6 can be found in both the basolateral and apical membranes, the expression of these receptors can also be found in the cytoplasm (164, 173, 174). In this cellular compartment only TRPC3 co-localizes with aquaporin-2 and their insertion in the membrane is stimulated by arginine-vasopressin stimuli (173).

TRPC1, 3, and 6 can also be found in afferent arteriolar vascular smooth muscle cells, where they are thought to mediate calcium influx upon noradrenalin stimulation (175). TRPC4 was found to be expressed in descending vasa recta, where it interacts with sodium/hydrogen exchanger regulatory factor-2, having a possible role in the regulation of vasoactivity and the release of NO (176).

Renal distal convoluted tubule cells express TRPM6 on their apical membrane (177). This receptor co-localizes with parvalbumin and calbindin-D28K on those cells, indicating that it may act as Mg^{2+} buffers (178).

As stated before, the urethra and ureters also express TRP channels. Similarly to what was observed in the urinary bladder, TRPV1 receptor expression was foreseen long before the receptor was cloned. Capsaicin produced an inhibitory effect on the nerve-mediated contractions of the rat proximal urethra (179, 180), showing that capsaicin-sensitive sensory nerves activation promotes urethra relaxation and thus facilitating urine voiding. In fact, CGRP released from capsaicin-sensitive fibers, possible after TRPV1 activation, acts as inhibitory transmitter, contributing to ureteral motility (179–183). Furthermore, capsaicin-induced urethra relaxation is nitric oxide-dependent (182). It is now known that the interaction between urethral capsaicin-sensitive primary afferents and the somatic efferents leads to a supraspinally-mediated activation of the external urethral sphincter (183). Similar observations were witnessed in the ureters (184). In vitro experiments showed that capsaicin inhibited neuropeptide neurokinin A-activated ureter rhythmic contractions (184).

Capsaicin also induces an inflammatory response, both in the urethra and ureters (184) indicating that capsaicin-sensitive fibers promote vasodilatation.

It is now known that TRPV1 is present in nerve terminal that course both mucosa and muscular layer of ureters and urethra (2). Curiously, urethra activity and sensitization is modulated by several factors, such as estrogen levels, and, therefore, cross-organ sensitization should be taken into account when studying this organ (185). In fact, TRPV1-mediated cross-organ sensitization between uterus and urethra varies during estrous cycle (185, 186).

Urethral TRPV1 expressing fibers that course the mucosa and the muscular layers also co-express TRPA1 (187). Therefore, it was not surprising that TRPA1 agonist were able to induce urethra relaxation (187, 188). TRPA1 immunoreactive fibers also express cannabinoid receptor 1 and cannabinoid receptor 2 (188). These two latter receptors might displace TRPV1/TRPA1 coupled activity (188). In the urethralurothelium, TRPV1 was more predominant in the outer layers and TRPA1 in the basal layers (187). Interestingly, urethral TRPA1 was also found in interstitial cells (187).

7. TRP Channels in the Male Genital Tract

The testicles express TRPV1, 4–6 and TRPM2–4, 7, 8 (9, 189–192). TRPV1 can be found in sertoli cells where it regulates the acidity of extracellular microenvironment which is crucial to maintain male fertility (193). In fact, TRPV1 knockout mice present testicular hyperthermia which results in massive germ cell depletion from the seminiferous tubules (194). However, TRPV1 knockout mice are not sterile, raising some questions about the relevance of these observations.

TRPM2, 4, 7, and 8 were found in spermatogenic cells (192). Activation of TRPM8 with menthol induces acrosome reaction, similar to what was observed upon progesterone or zona pellucida stimulation (192), indicating that TRPM8 is eventually involved in sperm activity during fertilization.

TRPM8 can be found in smooth muscle cells from the epididymal and prostatic portions of the rat vas deferens, both in the membrane and in sarcoplasmic reticulum, having a possible role in smooth muscle contraction (195, 196). TRPV1 presence in vas deferens was pharmacologically demonstrated in several experiments, where it is thought to inhibit vas deferens smooth muscle cells contractions, through the release of neuropeptides (197–203).

TRP channels such as TRPV1–4, TRPM1–4, and TRPM6–8 have also been described to be present in the prostate (204, 205). The most abundant TRP mRNA found in the prostate is from TRPM2–4, TRPM8, TRPV2, and TRPV4. Although low levels TRPV1 mRNA were found in the prostate, high levels of the protein were detected both in the epithelium and in smooth muscle cells (204). In fact, TRPV1 has been described to be present in primary afferents that course the urethral mucosa, verumontanum, ejaculatory ducts, and periurethral prostatic acini (205). Furthermore, this receptor has also been found in interstitial cells (206).

TRPA1 is also expressed in the prostate (102). Immunoreaction against TRPA1 showed that this receptor is present on nerve cells that were also positive for CB1 and CB2 receptors, for calcitonin gene-related peptide, for nitric oxide synthase, or for vesicular acetylcholine transporter, but not on adrenergic nerves (207). TRPA1 was also found on basal cells of glandular epithelium and on basal and subepithelial cells that also co-expressed vimentin and c-kit (102, 207). The application of TRPA1 agonist into the precontracted prostate resulted in relaxation (207), indicating that this receptor has a role on mechano afferent signaling of the gland.

Both TRPM8 RNA and protein have been found in the prostate, with more abundance in apical secretory epithelial cells (9, 208–210). However, only secretory mature differentiated prostate primary epithelial cells expressed functional cytoplasmic membrane TRPM8 (211). TRPM8 expression on prostate epithelial cells is androgen receptor-dependent (212, 213). In the prostate, TRPM8 activation might be induced by lysophospholipids, an end product of calcium-independent phospholipase A2 signaling pathway, which induces long TRPM8 channel opening (214).

In the prostate, TRPM8 can also be found in endoplasmic reticulum, where it controls calcium homeostasis and consequently cell survival, independently of cell differentiation (211, 215). This does not occur with TRPM8 expressed in cytoplasmic membrane of prostate epithelial cells since it is observed that TRPM8 in cytoplasmic membrane is downregulated during the loss of the apical differentiated phenotype (208, 215). Although altogether these data prompt TRPM8 as a switch-off of epithelial cell mobility and proliferation, care should be taken when analyzing data coming from experiments where common known TRPM8 agonist is used. Both menthol and icilin have demonstrated to induce effects similar to the ones attributed to TRPM8, in a TRPM8-independent process (216, 217).

The only TRP channel described in the penis is TRPM8 (218). TRPM8 expression decreases after castration and was recovered upon dihydrotestosterone treatment, indicating that TRPM8 is regulated by serum androgen (218). Although TRPV1 is not described to be present in the penis, intraurethral application of capsaicin induced penile erection possibly involving the activation of a urethra-corpora cavernosa reflex arc (212).

8. TRP Channels in the Female Genital Tract

Little is known about TRP channels expression and function in the ovaries, fallopian tube, the uterus and vagina. TRPP1 and TRPP2 are known to be in primary cilia of granulosa cells of antral follicles, in the motile oviduct cilia, where possible they have a role in granulosa cell differentiation and in follicle maturation (219). TRPP1/2 expression increases upon ovulatory gonadotropic stimulation (219). TRPV4 is present in a subpopulation of epithelial cells motile cilia of the ampulla and isthmus, with higher intensity in the proximal invaginations of the epithelial folds (219). TRPV4 role is thought to have a role in the oviduct properties upon ovulation (219) detecting viscosity changes, promoting cilia activity (220). TRP channels are thought to be involved in uterus contractility and physiology during labor (221). In what concern vagina, TRPV1 is thought to be expressed in nerve fibers that innervate the vagina (222), although these data need clarification.

9. TRP Channels and Cancer

In the human bladder, urothelial TRPV1 expression decreases in areas of transitional cell carcinoma (223). In vitro experiments have shown that TRPM8 activation on bladder cancer cells induces cell death, revealing that TRPM8 has anticancer properties (103). Similarly, TRPV2 activation on bladder cancer cells induces cell death (224). This might be related to the progressive lost of a TRPV2 splice variant concomitant with a great increase of full-length TRPV2 (21).

The use of vanilloids, such as capsaicin and resiniferatoxin, to treat prostate cancer has been valuable. It is thought that capsaicin may act as an inhibitor of electron transport chain, inducing an excess of reactive oxygen species (225, 226). The capsaicin direct action on TRPV1 will induce activation of caspase 1 and 3 (225). Furthermore, prostate TRPV1 activation by capsaicin also elicits a tumor necrosis factor-α-dependent production and secretion of interleukine-6, through a signal pathway involving the activation of protein kinase C-α, phosphoinositol-3 phosphate kinase, Akt and extracellular signal-regulated protein kinase (ERK) (227, 228). Capsaicin also activates c-Jun N-terminal kinases (JNK), which will induce ceramide production/accumulation, interfering with cell viability (228). Capsaicin increases the expression of p53, p21, and Bax, and downregulate the expression of prostate-specific antigen (PSA) and androgen receptor (229). Given orally, capsaicin has shown to reduce the rate of growth of prostate cancer xenografts (226). This might be mediated by endoplasmic reticulum stress-regulated gene, such as GADD153/CHOP (230). Interestingly, capsaicin may stabilize PSA (231). If all the described effects are a result of capsaicin direct action over TRPV1 is still unknown. In fact, in some experiments, incubation of PC3 prostate cancer cells with a moderate specific TRPV1 antagonist capsazepine did not have any effect or also resulted in cell death (226, 232). However, in other experiment capsazepine was able to reverse capsaicin-induced pro-apoptotic responses (227).

As for TRPV1, TRPM8 also seems to be involved in the control of mechanisms such as cell proliferation, apoptosis, and migration during prostate cancer (233–235). In fact, the expression of TRPM8 is enhanced during prostate cancer (210, 233). This seems to be a consequence of the increase of PSA during prostate cancer (233). In fact, PSA can activate TRPM8, inducing a decrease in cancer cell mobility, suggesting that the increased transport of TRPM8 to the cytoplasmic membrane has a protective role in prostate cancer progression (233). Importantly, TRPM8 was found to be more elevated in the urine and blood of patients with metastatic disease than in those of healthy men but it was equally found in prostate cancer patients with localized disease and healthy men, indicating that tissue TRPM8 levels can be used as biomarker of prostate cancer, while urine and blood TRPM8 levels may distinguish metastatic disease from clinically localized prostate cancer during diagnosis (213).

Acknowledgments

This work was partially supported by INComb FP7 HEALTH Project 223234. AC has been supported by SFRH/BPD/68716/2101 Fellowship from Fundação para a Ciência e a Tecnologia.

References

1. Andersson KE, Gratzke C, Hedlund P (2010) The role of the transient receptor potential (TRP) superfamily of cation-selective channels in the management of the overactive bladder. BJU Int 106:1114–1127
2. Avelino A, Cruz C, Nagy I et al (2002) Vanilloid receptor 1 expression in the rat urinary tract. Neuroscience 109:787–798
3. Lazzeri M, Vannucchi MG, Zardo C et al (2004) Immunohistochemical evidence of vanilloid receptor 1 in normal human urinary bladder. Eur Urol 46:792–798
4. Saleh HA (2006) Vanilloid receptor type 1-immunoreactive nerves in the rat urinary bladder and primary afferent neurones: the effects of age. Folia Morphol (Warsz) 65:213–220
5. Gevaert T, Vriens J, Segal A et al (2010) Deletion of the transient receptor potential cation channel TRPV4 impairs murine bladder voiding. J Clin Invest 117:3453–3462
6. Cao DS, Yu SQ, Premkumar LS (2009) Modulation of transient receptor potential vanilloid 4-mediated membrane currents and synaptic transmission by protein kinase C. Mol Pain 5:5–17
7. Yamada T, Ugawa S, Ueda T (2009) Differential localizations of the transient receptor potential channels TRPV4 and TRPV1 in the mouse urinary bladder. J Histochem Cytochem 57: 277–287
8. Charrua A, Cruz CD, Cruz F (2012) TRPV1 and TRPV4 antagonists have synergistic effect for treating bladder overactivity in rats. 27th Annual meeting EAU, Paris (accepted)
9. Stein RJ, Santos S, Nagatomi J et al (2004) Cool (TRPM8) and hot (TRPV1) receptors in the bladder and male genital tract. J Urol 172:1151–1178
10. Streng T, Axelsson HE, Hedlund P et al (2008) Distribution and function of the hydrogen sulfide-sensitive TRPA1 ion channel in rat urinary bladder. Eur Urol 53:391–399
11. Tsukimi Y, Mizuyachi K, Yamasaki T et al (2005) Cold response of the bladder in guinea pig: involvement of transient receptor potential channel, TRPM8. Urology 65:406–410
12. Mukerji G, Yiangou Y, Corcoran SL et al (2006) Cool and menthol receptor TRPM8 in human urinary bladder disorders and clinical correlations. BMC Urol 6:6–16
13. Hayashi T, Kondo T, Ishimatsu M et al (2009) Expression of the TRPM8-immunoreactivity in dorsal root ganglion neurons innervating the rat urinary bladder. Neurosci Res 65:245–251
14. Everaerts W, Vriens J, Owsianik G et al (2010) Functional characterization of transient receptor potential channels in mouse urothelial cells. Am J Physiol Renal Physiol 298:F692–F701
15. Birder LA, Kanai AJ, de Groat WC et al (2001) Vanilloid receptor expression suggests a sensory role for urinary bladder epithelial cells. Proc Natl Acad Sci USA 98:13396–13401
16. Charrua A, Reguenga C, Cordeiro JM et al (2009) Functional transient receptor potential vanilloid 1 is expressed in human urothelial cells. J Urol 182:2944–2950
17. Birder L, Kullmann FA, Lee H et al (2007) Activation of urothelial transient receptor potential vanilloid 4 by 4alpha-phorbol 12,13-didecanoate contributes to altered bladder reflexes in the rat. J Pharmacol Exp Ther 323:227–235
18. Kullmann FA, Shah MA, Birder LA et al (2009) Functional TRP and ASIC-like channels in cultured urothelial cells from the rat. Am J Physiol Renal Physiol 296:F892–F901
19. Yu W, Hill WG, Apodaca G et al (2010) Expression and distribution of transient receptor potential (TRP) channels in bladder epithelium. Am J Physiol Renal Physiol 300:F49–F59
20. Janssen DA, Hoenderop JG, Jansen KC et al (2011) The mechanoreceptor TRPV4 is localized in adherence junctions of the human bladder urothelium: a morphological study. J Urol 186:1121–1127
21. Caprodossi S, Lucciarini R, Amantini C et al (2008) Transient receptor potential vanilloid type 2 (TRPV2) expression in normal urothelium and in urothelial carcinoma of

human bladder: correlation with the pathologic stage. Eur Urol 54:612–620
22. Thorneloe KS, Sulpizio AC, Lin Z et al (2008) N-((1S)-1-{[4-((2S)-2-{[(2,4-dichlorophenyl) sulfonyl]amino}-3-hydroxypropanoyl)-1-piperazinyl]carbonyl}-3-methylbutyl)-1-benzothiophene-2-carboxamide (GSK1016790A), a novel and potent transient receptor potential vanilloid 4 channel agonist induces urinary bladder contraction and hyperactivity: part I. J Pharmacol Exp Ther 326:432–442
23. Kobayashi H, Yoshiyama M, Zakoji H et al (2009) Sex differences in the expression profile of acid-sensing ion channels in the mouse urinary bladder: a possible involvement in irritative bladder symptoms. BJU Int 104:1746–1751
24. Birder LA, Nakamura Y, Kiss S et al (2002) Altered urinary bladder function in mice lacking the vanilloid receptor TRPV1. Nat Neurosci 5:856–860
25. Charrua A, Cruz CD, Cruz F et al (2007) Transient receptor potential vanilloid subfamily 1 is essential for the generation of noxious bladder input and bladder overactivity in cystitis. J Urol 177:1537–1541
26. Charrua A, Cruz CD, Narayanan S et al (2009) GRC-6211, a new oral specific TRPV1 antagonist, decreases bladder overactivity and noxious bladder input in cystitis animal models. J Urol 181:379–386
27. Charrua A, Boudes M, De Ridder D, Cruz CD, Cruz F (2012) TRPV1 and TRPV4 expression in bladder neurons during normal condition and during cystitis. 27th Annual meeting EAU, Paris (accepted)
28. Lashinger ES, Steiginga MS, Hieble JP et al (2008) AMTB, a TRPM8 channel blocker: evidence in rats for activity in overactive bladder and painful bladder syndrome. Am J Physiol Renal Physiol 295:F803–F810
29. McKemy DD, Neuhausser WM, Julius D (2002) Identification of a cold receptor reveals a general role for TRP channels in thermosensation. Nature 416:52–58
30. Peier AM, Moqrich A, Hergarden AC et al (2002) A TRP channel that senses cold stimuli and menthol. Cell 108:705–715
31. Shibata Y, Ugawa S, Imura M et al (2011) TRPM8-expressing dorsal root ganglion neurons project dichotomizing axons to both skin and bladder in rats. Neuroreport 22:61–67
32. Sui GP, Wu C, Roosen A et al (2008) Modulation of bladder myofibroblast activity: implications for bladder function. Am J Physiol Renal Physiol 295:F688–F697
33. Fowler CJ, Jewkes D, McDonald WI et al (1992) Intravesical capsaicin for neurogenic bladder dysfunction. Lancet 339:1239
34. Cruz F, Guimarães M, Silva C et al (1997) Desensitization of bladder sensory fibers by intravesical capsaicin has long lasting clinical and urodynamic effects in patients with hyperactive or hypersensitive bladder dysfunction. J Urol 157:585–589
35. Geirsson G, Fall M, Sullivan L (1995) Clinical and urodynamic effects of intravesical capsaicin treatment in patients with chronic traumatic spinal detrusor hyperreflexia. J Urol 154:1825–1829
36. Das A, Chancellor MB, Watanabe T et al (1996) Intravesical capsaicin in neurologic impaired patients with detrusor hyperreflexia. J Spinal Cord Med 19:190–193
37. De Séze M, Wiart L, Joseph PA et al (1998) Capsaicin and neurogenic detrusor hyperreflexia. A double blind placebo controlled study in 20 patients with spinal cord lesions. Neurourol Urodyn 17:513–523
38. Lazzeri M, Spinelli M, Beneforti P et al (1998) Intravesical resiniferatoxin for the treatment of detrusor hyperreflexia refractory to capsaicin in patients with chronic spinal cord diseases. Scand J Urol Nephrol 32:331–334
39. Kuo HC (2003) Effectiveness of intravesical resiniferatoxin in treating detrusor hyper-reflexia and external sphincter dyssynergia in patients with chronic spinal cord lesions. BJU Int 92:597–601
40. Silva C, Silva J, Ribeiro MJ (2005) Urodynamic effect of intravesical resiniferatoxin in patients with neurogenic detrusor overactivity of spinal origin: results of a double-blind randomized placebo-controlled trial. Eur Urol 48:650–655
41. Kuo HC (2005) Multiple intravesical instillation of low-dose resiniferatoxin is effective in the treatment of detrusor overactivity refractory to anticholinergics. BJU Int 95:1023–1027
42. Chai TC, Gray ML, Steers W (1998) The incidence of a positive ice water test in bladder outlet obstructed patients: evidence for bladder neural plasticity. J Urol 160:34–38
43. Fowler CJ, Beck RO, Gerrard S et al (1994) Intravesical capsaicin for treatment of detrusor hyperreflexia. J Neurol Neurosurg Psychiatry 57:169–173
44. De Ridder D, Chandiramani V, Dasgupta P et al (1997) Intravesical capsaicin as a treatment for refractory detrusor hyperreflexia: a dual center study with long-term followup. J Urol 158:2087–2092
45. Cruz F, Guimarães M, Silva C et al (1997) Supression of bladder hyperreflexia by intravesical resiniferatoxin. Lancet 350:640–641
46. Lazzeri M, Beneforti P, Turini D (1997) Urodynamic effects of intravesical resinifera-

toxin in humans: preliminary results in stable and unstable detrusor. J Urol 158:2093–2096
47. Kuo HC, Liu HT, Yang WC (2006) Therapeutic effect of multiple resiniferatoxin intravesical instillations in patients with refractory detrusor overactivity: a randomized, double-blind, placebo controlled study. J Urol 176:641–645
48. Silva C, Rio ME, Cruz F (2000) Desensitization of bladder sensory fibers by intravesical resiniferatoxin, a capsaicin analog: long-term results for the treatment of detrusor hyperreflexia. Eur Urol 38:444–452
49. Apostolidis A, Popat R, Yiangou Y et al (2005) Decreased sensory receptors P2X3 and TRPV1 in suburothelial nerve fibers following intradetrusor injections of botulinum toxin for human detrusor overactivity. J Urol 174:977–982, discussion 982–983
50. Silva C, Silva J, Castro H et al (2007) Bladder sensory desensitization decreases urinary urgency. BMC Urol 11:7–9
51. Silva C, Ribeiro MJ, Cruz F (2002) The effect of intravesical resiniferatoxin in patients with idiopathic detrusor instability suggests that involuntary detrusor contractions are triggered by C-fiber input. J Urol 168:575–579
52. Kuo HC (2003) Effectiveness of intravesical resiniferatoxin for anticholinergic treatment refractory detrusor overactivity due to nonspinal cord lesions. J Urol 170:835–839
53. Dinis P, Silva J, Ribeiro MJ et al (2004) Bladder C-fiber desensitization induces a long-lasting improvement of BPH-associated storage LUTS: a pilot study. Eur Urol 46:88–93
54. Lazzeri M, Beneforti M, Spinelli A et al (2000) Intravesical resiniferatoxin for the treatment of hypersensitive disorder: a randomized placebo controlled study. J Urol 164:676–679
55. Apostolidis A, Gonzales GE, Fowler CJ (2006) Effect of intravesical resiniferatoxin (RTX) on lower urinary tract symptoms, urodynamic parameters, and quality of life of patients with urodynamic increased bladder sensation. Eur Urol 50:1299–1305
56. Mukerji G, Yiangou Y, Agarwal SK et al (2006) Transient receptor potential vanilloid receptor subtype 1 in painful bladder syndrome and its correlation with pain. J Urol 176:797–801
57. Chen TY, Corcos J, Camel M et al (2005) Prospective, randomized, double-blind study of safety and tolerability of intravesical resiniferatoxin (RTX) in interstitial cystitis (IC). Int Urogynecol J Pelvic Floor Dysfunct 16:293–297
58. Peng CH, Kuo HC (2007) Multiple intravesical instillations of low-dose resiniferatoxin in the treatment of refractory interstitial cystitis. Urol Int 78:78–81
59. Payne CK, Mosbaugh PG, Forrest JB et al (2005) Intravesical resiniferatoxin for the treatment of interstitial cystitis: a randomized, double-blind, placebo controlled trial. J Urol 173:1590–1594
60. Liu L, Mansfield KJ, Kristiana I et al (2007) The molecular basis of urgency: regional difference of vanilloid receptor expression in the human urinary bladder. Neurourol Urodyn 26:433–438
61. Caudle RM, Karai L, Mena N et al (2003) Resiniferatoxin-induced loss of plasma membrane in vanilloid receptor expressing cells. Neurotoxicology 24:895–908
62. Lee SY, Hong Y, Oh U (2004) Decreased pain sensitivity of capsaicin-treated rats results from decreased VR1 expression. Arch Pharm Res 27:1154–1160
63. Santos-Silva A, Charrua A, Cruz CD et al (2012) Rat detrusor overactivity induced by chronic spinalization can be abolished by a transient receptor potential vanilloid 1 (TRPV1) antagonist. Auton Neurosci 166:35–38
64. Brady CM, Apostolidis AN, Harper M et al (2004) Parallel changes in bladder suburothelial vanilloid receptor TRPV1 and pan-neuronal marker PGP9.5 immunoreactivity in patients with neurogenic detrusor overactivity after intravesical resiniferatoxin treatment. BJU Int 93:770–776
65. Xu X, Wang P, Zou X et al (2009) Increases in transient receptor potential vanilloid-1 mRNA and protein in primary afferent neurons stimulated by protein kinase C and their possible role in neurogenic inflammation. J Neurosci Res 87:482–494
66. Ji RR, Samad TA, Jin SX et al (2002) p38 MAPK activation by NGF in primary sensory neurons after inflammation increases TRPV1 levels and maintains heat hyperalgesia. Neuron 36:57–68
67. Xue Q, Jong B, Chen T et al (2007) Transcription of rat TRPV1 utilizes a dual promoter system that is positively regulated by nerve growth factor. J Neurochem 101:212–222
68. Okragly AJ, Niles AL, Saban R, Schmidt D, Hoffman RL, Warner TF, Moon TD, Uehling DT, Haak-Frendscho M (1999) Elevated tryptase, nerve growth factor, neurotrophin-3 and glial cell line-derived neurotrophic factor levels in the urine of interstitial cystitis and bladder cancer patients. J Urol 161:438–441
69. Kim JC, Park EY, Seo SI, Park YH, Hwang TK (2006) Nerve growth factor and prostaglandins in the urine of female patients with overactive bladder. J Urol 151:1773–1776

70. Giannantoni A, Di Stasi SM, Nardicchi V, Zucchi A, Macchioni L, Bini V, Goracci G, Porena M (2006) Botulinum-A toxin injections into the detrusor muscle decrease nerve growth factor bladder tissue levels in patients with neurogenic detrusor overactivity. J Urol 151:2341–2344
71. Liu HT, Chancellor MB, Kuo HC (2008) Urinary nerve growth factor level could be a biomarker in the differential diagnosis of mixed urinary incontinence in women. BJU Int 102:1440–1444
72. Liu HT, Chancellor MB, Kuo HC (2009) Urinary nerve growth factor levels are elevated in patients with detrusor overactivity and decreased in responders to detrusor botulinum toxin-A injection. Eur Urol 56:700–706
73. Stein AT, Ufret-Vincenty CA, Hua L et al (2006) Phosphoinositide 3-kinase binds to TRPV1 and mediates NGF-stimulated TRPV1 trafficking to the plasma membrane. J Gen Physiol 128:509–522
74. Schumacher MA, Moff I, Sudanagunta SP et al (2000) Molecular cloning of an N-terminal splice variant of the capsaicin receptor. Loss of N-terminal domain suggests functional divergence among capsaicin receptor subtypes. J Biol Chem 275:2756–2762
75. Lu G, Henderson D, Liu L, Reinhart PH, Simon SA (2005) TRPV1b, a functional human vanilloid receptor splice variant. Mol Pharmacol 67(4):1119–1127
76. Charrua A, Reguenga C, Paule CC et al (2008) Cystitis is associated with TRPV1b-downregulation in rat dorsal root ganglia. Neuroreport 19:1469–1472
77. Zygmunt PM, Petersson J, Andersson DA et al (1999) Vanilloid receptors on sensory nerves mediate the vasodilator action of anandamide. Nature 400:452–457
78. Huang SM, Bisogno T, Trevisani M et al (2002) An endogenous capsaicin-like substance with high potency at recombinant and native vanilloid VR1 receptors. Proc Natl Acad Sci USA 99:8400–8405
79. Chu CJ, Huang SM, De Petrocellis L et al (2003) N-oleoyldopamine, a novel endogenous capsaicin-like lipid that produces hyperalgesia. J Biol Chem 278:13633–13639
80. Hwang SW, Cho H, Kwak J et al (2000) Direct activation of capsaicin receptors by products of lipoxygenases: endogenous capsaicin-like substances. Proc Natl Acad Sci USA 97:6155–6160
81. Dinis P, Charrua A, Avelino A et al (2004) Anandamide-evoked activation of vanilloid receptor 1 contributes to the development of bladder hyperreflexia and nociceptive transmission to spinal dorsal horn neurons in cystitis. J Neurosci 24:11253–11263
82. Lee SY, Lee JH, Kang KK et al (2005) Sensitization of vanilloid receptor involves an increase in the phosphorylated form of the channel. Arch Pharm Res 28:405–412
83. Sculptoreanu A, de Groat WC, Buffington CA et al (2005) Protein kinase C contributes to abnormal capsaicin responses in DRG neurons from cats with feline interstitial cystitis. Neurosci Lett 381:42–46
84. Cortright DN, Szallasi A (2004) Biochemical pharmacology of the vanilloid receptor TRPV1. An update. Eur J Biochem 271:1814–1819
85. Li M, Sun Y, Simard JM et al (2011) Increased transient receptor potential vanilloid type 1 (TRPV1) signaling in idiopathic overactive bladder urothelial cells. Neurourol Urodyn 30:606–611
86. Apostolidis A, Brady CM, Yiangou Y et al (2005) Capsaicin receptor TRPV1 in urothelium of neurogenic human bladders and effect of intravesical resiniferatoxin. Urology 65:400–405
87. Szallasi A, Cruz F, Geppetti P (2006) TRPV1: a therapeutic target for novel analgesic drugs? Trends Mol Med 12:545–554
88. Jancsó-Gabor A, Szolcsanyi J, Jancso N (1970) Irreversible impairment of thermoregulation induced by capsaicin and similar pungent substances in rats and guinea-pigs. J Physiol 206:495–507
89. Swanson DM, Dubin AE, Shah C et al (2005) Identification and biological evaluation of 4-(3-trifluoromethylpyridin-2-yl)piperazine-1-carboxylic acid (5-trifluoromethylpyridin-2-yl)amide, a high affinity TRPV1 (VR1) vanilloid receptor antagonist. J Med Chem 48:1857–1872
90. Romanovsky AA (2007) Thermoregulation: some concepts have changed. Functional architecture of the thermoregulatory system. Am J Physiol Regul Integr Comp Physiol 292:R37–R46
91. Steiner AA, Turek VF, Almeida MC et al (2007) Nonthermal activation of transient receptor potential vanilloid-1 channels in abdominal viscera tonically inhibits autonomic cold-defense effectors. J Neurosci 27:7459–7468
92. Gavva NR (2008) Body-temperature maintenance as the predominant function of the vanilloid receptor TRPV1. Trends Pharmacol Sci 29:550–557
93. Gavva NR, Treanor JJ, Garami A et al (2008) Pharmacological blockade of the vanilloid receptor TRPV1 elicits marked hyperthermia in humans. Pain 136:202–210
94. Wong GY, Gavva NR (2008) Therapeutic potential of vanilloid receptor TRPV1 agonists

and antagonists as analgesics: recent advances and setbacks. Brain Res Rev 60:267–277
95. Szallasi A, Cortright DN, Blum CA et al (2007) The vanilloid receptor TRPV1: 10 years from channel cloning to antagonist proof-of-concept. Nat Rev Drug Discov 6:357–372
96. Huang W, Rubinstein J, Prieto AR et al (2010) Enhanced postmyocardial infarction fibrosis via stimulation of the transforming growth factor-beta-Smad2 signaling pathway: role of transient receptor potential vanilloid type 1 channels. J Hypertens 28:367–376
97. Everaerts W, Zhen X, Ghosh D et al (2010) Inhibition of the cation channel TRPV4 improves bladder function in mice and rats with cyclophosphamide-induced cystitis. Proc Natl Acad Sci USA 107:19084–19089
98. Miyamoto T, Mochizuki T, Zakohji H et al (2009) The expression of transient receptor potential (TRP) V4, A1, and V1 in the human bladder mucosa of normal and bladder outlet obstruction – a novel mechanism in the obstruction-induced bladder overactivity. Neurourol Urodyn 28:12
99. Lemos AE, Ferreira J, André E et al (2006) Contractile mechanisms coupled to TRPA1 receptor activation in rat urinary bladder. Biochem Pharmacol 72:104–114
100. Andrade EL, Forner S, Bento AF et al (2011) TRPA1 receptor modulation attenuates bladder overactivity induced by spinal cord injury. Am J Physiol Renal Physiol 300:F1223–F1234
101. Chen J, Joshi SK, DiDomenico S et al (2011) Selective blockade of TRPA1 channel attenuates pathological pain without altering noxious cold sensation or body temperature regulation. Pain 152:1165–1172
102. Du S, Akari I, Kobayashi H et al (2008) Differential expression profile of cold (TRPA1) and cool (TRPM8) receptors in human urogenital organs. Urology 72:450–455
103. Hayashi T, Kondo T, Ishimatsu M et al (2011) Function and expression pattern of TRPM8 in bladder afferent neurons associated with bladder outlet obstruction in rats. Auton Neurosci 164:27–33
104. Jung J, Shin JS, Lee SY et al (2004) Phosphorylation of vanilloid receptor 1 by Ca^{2+}/calmodulin-dependent kinase II regulates its vanilloid binding. J Biol Chem 279: 7048–7054
105. Amadesi S, Cottrell GS, Divino L et al (2006) Protease-activated receptor 2 sensitizes TRPV1 by protein kinase Cepsilon- and A-dependent mechanisms in rats and mice. J Physiol 575:555–571
106. Amadesi S, Nie J, Vergnolle N et al (2004) Protease-activated receptor 2 sensitizes the capsaicin receptor transient receptor potential vanilloid receptor 1 to induce hyperalgesia. J Neurosci 24:4300–4312
107. Ohta T, Ikemi Y, Murakami M et al (2006) Potentiation of transient receptor potential V1 functions by the activation of metabotropic 5-HT receptors in rat primary sensory neurons. J Physiol 576:809–822
108. Dattilio A, Vizzard MA (2005) Up-regulation of protease activated receptors in bladder after cyclophosphamide induced cystitis and colocalization with capsaicin receptor (VR1) in bladder nerve fibers. J Urol 173:635–639
109. Carlton SM, Du J, Zhou S (2009) Group II metabotropic glutamate receptor activation on peripheral nociceptors modulates TRPV1 function. Brain Res 1248:86–95
110. Vetter I, Wyse BD, Monteith GR et al (2006) The mu opioid agonist morphine modulates potentiation of capsaicin-evoked TRPV1 responses through a cyclic AMP-dependent protein kinase A pathway. Mol Pain 2:22–37
111. Vetter I, Cheng W, Peiris M et al (2008) Rapid, opioid-sensitive mechanisms involved in transient receptor potential vanilloid 1 sensitization. J Biol Chem 283:19540–19550
112. Yoshiyama M, de Groat WC (2007) Role of spinal metabotropic glutamate receptors in regulation of lower urinary tract function in the decerebrate unanesthetized rat. Neurosci Lett 420:18–22
113. Andersson KE (2002) Bladder activation: afferent mechanisms. Urology 59:43–50
114. Kumar V, Chapple CR, Rosario D et al (2010) In vitro release of adenosine triphosphate from the urothelium of human bladders with detrusor overactivity, both neurogenic and idiopathic. Eur Urol 57:1087–1092
115. Ford AP, Cockayne DA (2011) ATP and P2X purinoceptors in urinary tract disorders. Handb Exp Pharmacol 202:485–526
116. Sun Y, Keay S, de Deyne PG et al (2001) Augmented stretch activated adenosine triphosphate release from bladder uroepithelial cells in patients with interstitial cystitis. J Urol 166:1951–1956
117. Ford APDW, Gever JR, Nunn PA et al (2006) Purinoceptors as therapeutic targets for lower urinary tract dysfunction. Br J Pharmacol 147:S132–S143
118. Sadananda P, Shang F, Liu L et al (2009) Release of ATP from rat urinary bladder mucosa: role of acid, vanilloids and stretch. Br J Pharmacol 158:1655–1662
119. Elneil S, Skepper JN, Kidd EJ et al (2001) Distribution of P2X(1) and P2X(3) receptors in the rat and human urinary bladder. Pharmacology 63:120–128

120. Sun Y, Chai TC (2004) Up-regulation of P2X3 receptor during stretch of bladder urothelial cells from patients with interstitial cystitis. J Urol 171:448–452
121. Tempest HV, Dixon AK, Turner WH et al (2004) P2X and P2X receptor expression in human bladder urothelium and changes in interstitial cystitis. BJU Int 93:1344–1348
122. Walczak JS, Price TJ, Cervero F (2009) Cannabinoid CB1 receptors are expressed in the mouse urinary bladder and their activation modulates afferent bladder activity. Neuroscience 159:1154–1163
123. Walczak JS, Cervero F (2011) Local activation of cannabinoid CB1 receptors in the urinary bladder reduces the inflammation-induced sensitization of bladder afferents. Mol Pain 7:31
124. Premkumar LS, Ahern GP (2000) Induction of vanilloid receptor channel activity by protein kinase C. Nature 408:985–990
125. Vellani V, Mapplebeck S, Moriondo A et al (2001) Protein kinase C activation potentiates gating of the vanilloid receptor VR1 by capsaicin, protons, heat and anandamide. J Physiol 534:813–825
126. Sugiura T, Tominaga M, Katsuya H et al (2002) Bradykinin lowers the threshold temperature for heat activation of vanilloid receptor 1. J Neurophysiol 88:544–548
127. Carr MJ, Lollarik M, Meeker SN et al (2003) A role for TRPV1 in bradykinin-induced excitation of vagal airway afferent nerve terminals. J Pharmacol Exp Ther 304:1275–1279
128. Tang HB, Inoue A, Oshita K et al (2004) Sensitization of vanilloid receptor 1 induced by bradykinin via the activation of second messenger signaling cascades in rat primary afferent neurons. Eur J Pharmacol 498:37–43
129. Mizumura K, Sugiur T, Koda H et al (2005) Pain and bradykinin receptors-sensory transduction mechanism in the nociceptor terminals and expression change of bradykinin receptors in inflamed condition. Nihon Shinkei Seishin Yakurigaku Zasshi 25:33–38
130. Shin J, Cho H, Hwang SW et al (2002) Bradykinin-12-lipoxygenase-VR1 signalling pathway for inflammatory hyperalgesia. Proc Natl Acad Sci USA 99:10150–10155
131. Ferreira J, da Silva GL, Calixto JB (2004) Contribution of vanilloid receptors to the overt nociception induced by B2 kinin receptor activation in mice. Br J Pharmacol 141: 787–794
132. Wang Y, Wang DH (2011) Protective Effect of TRPV1 against Renal Fibrosis via Inhibition of TGF-β/Smad Signaling in DOCA-Salt Hypertension. Mol Med 17:1204–1212
133. Wang Y, Wang DH (2009) Aggravated renal inflammatory responses in TRPV1 gene knockout mice subjected to DOCA-salt hypertension. Am J Physiol Renal Physiol 297:F1550–F1559
134. Ueda K, Tsuji F, Hirata T et al (2009) Preventive effect of SA13353 [1-[2-(1-adamantyl)ethyl]-1-pentyl-3-[3-(4-pyridyl)propyl]urea], a novel transient receptor potential vanilloid 1 agonist, on ischemia/reperfusion-induced renal injury in rats. J Pharmacol Exp Ther 329:202–209
135. Xie C, Sachs JR, Wang DH (2008) Interdependent regulation of afferent renal nerve activity and renal function: role of transient receptor potential vanilloid type 1, neurokinin 1, and calcitonin gene-related peptide receptors. J Pharmacol Exp Ther 325:751–757
136. Zhu Y, Wang Y, Wang DH (2005) Diuresis and natriuresis caused by activation of VR1-positive sensory nerves in renal pelvis of rats. Hypertension 46:992–997
137. Li J, Wang DH (2008) Increased GFR and renal excretory function by activation of TRPV1 in the isolated perfused kidney. Pharmacol Res 57:239–246
138. Kawamata T, Ji W, Yamamoto J et al (2008) Contribution of transient receptor potential vanilloid subfamily 1 to endothelin-1-induced thermal hyperalgesia. Neuroscience 154: 1067–1076
139. Zhu Y, Wang DH (2008) Segmental regulation of sodium and water excretion by TRPV1 activation in the kidney. J Cardiovasc Pharmacol 51:437–442
140. Gao F, Sui D, Garavito RM et al (2009) Salt intake augments hypotensive effects of transient receptor potential vanilloid 4: functional significance and implication. Hypertension 53:228–235
141. Gao F, Wang DH (2010) Impairment in function and expression of transient receptor potential vanilloid type 4 in Dahl salt-sensitive rats: significance and mechanism. Hypertension 55:1018–1025
142. Xie C, Wang DH (2010) Effects of a high-salt diet on TRPV-1-dependent renal nerve activity in Dahl salt-sensitive rats. Am J Nephrol 32:194–200
143. Xie C, Wang DH (2009) Ablation of transient receptor potential vanilloid 1 abolishes endothelin-induced increases in afferent renal nerve activity: mechanisms and functional significance. Hypertension 54:1298–1305
144. Zhu Y, Xie C, Wang DH (2007) TRPV1-mediated diuresis and natriuresis induced by

hypertonic saline perfusion of the renal pelvis. Am J Nephrol 27:530–537
145. Feng NH, Lee HH, Shiang JC et al (2008) Transient receptor potential vanilloid type 1 channels act as mechanoreceptors and cause substance P release and sensory activation in rat kidneys. Am J Physiol Renal Physiol 294:F316–F325
146. Li J, Wang DH (2006) Differential mechanisms mediating depressor and diuretic effects of anandamide. J Hypertens 24:2271–2276
147. Sanchez JF, Krause JE, Cortright DN (2001) The distribution and regulation of vanilloid receptor VR1 and VR1 5′ splice variant RNA expression in rat. Neuroscience 107(3): 373–381
148. Vos MH, Neelands TR, McDonald HA et al (2006) TRPV1b overexpression negatively regulates TRPV1 responsiveness to capsaicin, heat and low pH in HEK293 cells. J Neurochem 99:1088–1102
149. Wang C, Hu HZ, Colton CK et al (2004) An alternative splicing product of the murine trpv1 gene dominant negatively modulates the activity of TRPV1 channels. J Biol Chem 279:37423–37430
150. Tian W, Fu Y, Wang DH et al (2006) Regulation of TRPV1 by a novel renally expressed rat TRPV1 splice variant. Am J Physiol Renal Physiol 290:F117–F126
151. Eilers H, Lee SY, Hau CW et al (2007) The rat vanilloid receptor splice variant VR.5′sv blocks TRPV1 activation. Neuroreport 18:969–973
152. Tian W, Salanova M, Xu H et al (2004) Renal expression of osmotically responsive cation channel TRPV4 is restricted to water-impermeant nephron segments. Am J Physiol Renal Physiol 287:F17–F24
153. Cohen DM (2005) TRPV4 and the mammalian kidney. Pflugers Arch 451:168–175
154. Delany NS, Hurle M, Facer P et al (2001) Identification and characterization of a novel human vanilloid receptor-like protein, VRL-2. Physiol Genomics 4:165–174
155. Taniguchi J, Tsuruoka S, Mizuno A et al (2006) TRPV4 as a flow sensor in flow-dependent K^+ secretion from the cortical collecting duct. Am J Physiol Renal Physiol 292:F667–F673
156. Haas M, Forbush B 3rd (1998) The Na-K-Cl cotransporters. J Bioenerg Biomembr 30: 161–172
157. Kim JA, Yang H, Hwang I et al (2011) Expression patterns and potential action of the calcium transport genes trpv5, trpv6, ncx1 and pmca1b in the canine duodenum, kidney and uterus. In Vivo 25:773–780
158. Hwang I, Jung EM, Yang H et al (2011) Tissue-specific expression of the calcium transporter genes TRPV5, TRPV6, NCX1, and PMCA1b in the duodenum, kidney and heart of *Equus caballus*. J Vet Med Sci 73(11):1437–1444
159. Peng JB, Brown EM, Hedigerb MA (2001) Structural conservation of the genes encoding CaT1, CaT2, and related cation channels. Genomics 76:99–109
160. Peng JB, Chen XZ, Berger UV et al (2000) A rat kidney-specific calcium transporter in the distal nephron. J Biol Chem 275:28186–28194
161. Hoenderop JGJ, Van der Kemp AW, Hartog A et al (1999) Molecular identification of the apical epithelial Ca^{2+} channel in 1, 25-vitamin D3 responsive epithelia. J Biol Chem 274: 8375–8378
162. Hoenderop JG, Vennekens R, Müller D et al (2001) Function and expression of the epithelial Ca(2+) channel family: comparison of mammalian ECaC1 and 2. J Physiol 537:747–761
163. Mensenkamp AR, Hoenderop JG, Bindels RJ (2007) TRPV5, the gateway to Ca^{2+} homeostasis. Handb Exp Pharmacol 179:207–220
164. Goel M, Sinkins WG, Zuo CD et al (2006) Identification and localization of TRPC channels in the rat kidney. Am J Physiol Renal Physiol 290:F1241–F1252
165. Ilatovskaya DV, Levchenko V, Ryan RP et al (2011) NSAIDs acutely inhibit TRPC channels in freshly isolated rat glomeruli. Biochem Biophys Res Commun 408:242–247
166. Wang X, Pluznick JL, Wei P et al (2004) TRPC4 forms store-operated Ca^{2+} channels in mouse mesangial cells. Am J Physiol Cell Physiol 287:C357–C364
167. Graham S, Ding M, Sours-Brothers S et al (2007) Downregulation of TRPC6 protein expression by high glucose, a possible mechanism for the impaired Ca^{2+} signaling in glomerular mesangial cells in diabetes. Am J Physiol Renal Physiol 293:F1381–F1390
168. Sours S, Du J, Chu S et al (2006) Expression of canonical transient receptor potential (TRPC) proteins in human glomerular mesangial cells. Am J Physiol Renal Physiol 290:F1507–F1515
169. Sours-Brothers S, Ding M, Graham S et al (2009) Interaction between TRPC1/TRPC4 assembly and STIM1 contributes to store-operated Ca^{2+} entry in mesangial cells. Exp Biol Med (Maywood) 234: 673–682
170. Raychowdhury MK, McLaughlin M, Ramos AJ et al (2005) Characterization of single channel currents from primary cilia of renal epithelial cells. J Biol Chem 280:34718–34722
171. Yoder BK, Hou X, Guay-Woodford LM (2002) The polycystic kidney disease proteins,

polycystin-1, polycystin-2, polaris, and cystin, are co-localized in renal cilia. J Am Soc Nephrol 13:2508–2516

172. Luo Y, Vassilev PM, Li X et al (2003) Native polycystin 2 functions as a plasma membrane Ca^{2+}-permeable cation channel in renal epithelia. Mol Cell Biol 23:2600–2607
173. Goel M, Sinkins WG, Zuo CD et al (2007) Vasopressin-induced membrane trafficking of TRPC3 and AQP2 channels in cells of the rat renal collecting duct. Am J Physiol Renal Physiol 293:F1476–F1488
174. Bandyopadhyay BC, Swaim WD, Liu X et al (2005) Apical localization of a functional TRPC3/TRPC6-Ca^{2+}-signaling complex in polarized epithelial cells. Role in apical Ca^{2+} influx. J Biol Chem 280:12908–12916
175. Salomonsson M, Braunstein TH, Holstein-Rathlou NH et al (2010) Na^{+}-independent, nifedipine-resistant rat afferent arteriolar Ca^{2+} responses to noradrenaline: possible role of TRPC channels. Acta Physiol (Oxf) 200:265–278
176. Lee-Kwon W, Wade JB, Zhang Z et al (2005) Expression of TRPC4 channel protein that interacts with NHERF-2 in rat descending vasa recta. Am J Physiol Cell Physiol 288: C942–C949
177. Muallem S, Moe OW (2007) When EGF is offside, magnesium is wasted. J Clin Invest 117:2086–2089
178. Voets T, Nilius B, Hoefs S et al (2004) TRPM6 forms the Mg^{2+} influx channel involved in intestinal and renal Mg^{2+} absorption. J Biol Chem 279:19–25
179. Maggi CA, Santicioli P, Manzini S et al (1989) Functional studies on the cholinergic and sympathetic innervation of the rat proximal urethra: effect of pelvic ganglionectomy or experimental diabetes. J Auton Pharmacol 9: 231–241
180. Andersson PO, Malmgren A, Uvelius B (1990) Functional responses of different muscle types of the female rat urethra in vitro. Acta Physiol Scand 140:365–372
181. Maggi CA, Giuliani S, Santicioli P et al (1987) Visceromotor responses to calcitonin gene-related peptide (CGRP) in the rat lower urinary tract: evidence for a transmitter role in the capsaicin-sensitive nerves of the ureter. Eur J Pharmacol 143:73–82
182. Nishizawa S, Igawa Y, Okada N et al (1997) Capsaicin-induced nitric-oxide-dependent relaxation in isolated dog urethra. Eur J Pharmacol 335:211–219
183. Conte B, Maggi CA, Giachetti A et al (1993) Intraurethral capsaicin produces reflex activation of the striated urethral sphincter in urethane-anesthetized male rats. J Urol 150: 1271–1277
184. Maggi CA, Santicioli P, Abelli L et al (1987) Regional differences in the effects of capsaicin and tachykinins on motor activity and vascular permeability of the rat lower urinary tract. Naunyn Schmiedebergs Arch Pharmacol 335:636–645
185. Peng HY, Huang PC, Liao JM et al (2008) Estrous cycle variation of TRPV1-mediated cross-organ sensitization between uterus and NMDA-dependent pelvic-urethra reflex activity. Am J Physiol Endocrinol Metab 295: E559–E568
186. Peng HY, Chang HM, Lee SD et al (2008) TRPV1 mediates the uterine capsaicin-induced NMDA NR2B-dependent cross-organ reflex sensitization in anesthetized rats. Am J Physiol Renal Physiol 295:F1324–F1335
187. Gratzke C, Streng T, Waldkirch E et al (2009) Transient receptor potential A1 (TRPA1) activity in the human urethra – evidence for a functional role for TRPA1 in the outflow region. Eur Urol 55:696–704
188. Weinhold P, Gratzke C, Streng T et al (2010) TRPA1 receptor induced relaxation of the human urethra involves TRPV1 and cannabinoid receptor mediated signals, and cyclooxygenase activation. J Urol 183:2070–2076
189. Xu ZP, Gao WC, Wang HP et al (2009) Expression of transient receptor potential subfamily mRNAs in rat testes. Nan Fang Yi Ke Da Xue Xue Bao 29:519–520
190. Mizrak SC, van Dissel-Emiliani FM (2008) Transient receptor potential vanilloid receptor-1 confers heat resistance to male germ cells. Fertil Steril 90:1290–1293
191. Li S, Wang X, Ye H et al (2010) Distribution profiles of transient receptor potential melastatin- and vanilloid-related channels in rat spermatogenic cells and sperm. Mol Biol Rep 37: 1287–1293
192. Martínez-López P, Treviño CL, de la Vega-Beltrán JL et al (2011) TRPM8 in mouse sperm detects temperature changes and may influence the acrosome reaction. J Cell Physiol 226:1620–1631
193. Auzanneau C, Norez C, Antigny F et al (2008) Transient receptor potential vanilloid 1 (TRPV1) channels in cultured rat Sertoli cells regulate an acid sensing chloride channel. Biochem Pharmacol 75:476–483
194. Mizrak SC, Gadella BM, Erdost H et al (2008) Spermatogonial stem cell sensitivity to capsaicin: an in vitro study. Reprod Biol Endocrinol 6:52
195. Boldyrev OI, Sotkis HV, IeM K et al (2009) Expression of the cold receptor TRPM8 in

the smooth muscles of the seminal ejaculatory ducts in rats. Fiziol Zh 55:17–27
196. Filippov IB, Vladymyrova IA, IeM K (2009) Modulation of the smooth muscle contractions of the rat vas deferens by TRPM8 channel agonist menthol. Fiziol Zh 55:30–40
197. Maggi CA, Giuliani S, Santicioli P et al (1987) Capsaicin-induced inhibition of motility of the rat isolated vas deferens: do multiple neuropeptides mediate the visceromotor effects of capsaicin? J Auton Pharmacol 7:243–255
198. Ellis JL, Burnstock G (1989) Modulation of neurotransmission in the guinea-pig vas deferens by capsaicin: involvement of calcitonin gene-related peptide and substance P. Br J Pharmacol 98:707–713
199. Wardle KA, Furey G, Sanger GJ (1996) Pharmacological characterization of the vanilloid receptor in the rat isolated vas deferens. J Pharm Pharmacol 48:285–291
200. Filippelli A, Falciani M, Piucci B (1999) Endothelin-1 affects capsaicin-evoked release of neuropeptides from rat vas deferens. Eur J Pharmacol 364:183–191
201. Ross RA, Gibson TM, Brockie HC et al (2001) Structure-activity relationship for the endogenous cannabinoid, anandamide, and certain of its analogues at vanilloid receptors in transfected cells and vas deferens. Br J Pharmacol 132:631–640
202. Thomas A, Pertwee RG (2006) The bioassay of cannabinoids using the mouse isolated vas deferens. Methods Mol Med 123:191–207
203. Sheykhzade M, Gupta S, Sørensen T et al (2011) Characterization of capsaicin induced responses in mice vas deferens: evidence of CGRP uptake. Eur J Pharmacol 667:375–382
204. Wang HP, Pu XY, Wang XH (2007) Distribution profiles of transient receptor potential melastatin-related and vanilloid-related channels in prostatic tissue in rat. Asian J Androl 9:634–640
205. Dinis P, Charrua A, Avelino A et al (2005) The distribution of sensory fibres immunoreactive for the TRPV1 (capsaicin) receptor in the human prostate. Eur Urol 48:162–167
206. Van der Aa F, Roskams T, Blyweert W et al (2003) Interstitial cells in the human prostate: a new therapeutic target? Prostate 56:250–255
207. Gratzke C, Weinhold P, Reich O et al (2009) Transient receptor potential A1 and cannabinoid receptor activity in human normal and hyperplastic prostate: relation to nerves and interstitial cells. Eur Urol 57:902–910
208. Bidaux G, Roudbaraki M, Merle C et al (2005) Evidence for specific TRPM8 expression in human prostate secretory epithelial cells: functional androgen receptor requirement. Endocr Relat Cancer 12:367–382
209. Kondrats'kyĭ AP, Sotkis HV, Boldyriev OI (2007) Functional identification of the TRPM8 cold receptor in rat prostate epithelial cells. Fiziol Zh 53:3–13
210. Tsavaler L, Shapero MH, Morkowski S et al (2001) Trp-p8, a novel prostate-specific gene, is up-regulated in prostate cancer and other malignancies and shares high homology with transient receptor potential calcium channel proteins. Cancer Res 61:3760–3769
211. Bidaux G, Flourakis M, Thebault S et al (2007) Prostate cell differentiation status determines transient receptor potential melastatin member 8 channel subcellular localization and function. J Clin Invest 117:1647–1657
212. Lazzeri M, Barbanti G, Beneforti P et al (1994) Intraurethrally infused capsaicin induces penile erection in humans. Scand J Urol Nephrol 28:409–412
213. Bai VU, Murthy S, Chinnakannu K et al (2010) Androgen regulated TRPM8 expression: a potential mRNA marker for metastatic prostate cancer detection in body fluids. Int J Oncol 36:443–450
214. Vanden AF, Zholos A, Bidaux G et al (2006) Ca^{2+}-independentphospholipaseA2-dependent gating of TRPM8 by lysophospholipids. J Biol Chem 281:40174–40182
215. Zhang L, Barritt GJ (2004) Evidence that TRPM8 is an androgen-dependent Ca^{2+} channel required for the survival of prostate cancer cells. Cancer Res 64:8365–8373
216. Kim SH, Nam JH, Park EJ et al (2008) Menthol regulates TRPM8-independent processes in PC-3 prostate cancer cells. Biochim Biophys Acta 1792:33–38
217. Kim SH, Kim SY, Park EJ et al (2011) Icilin induces G1 arrest through activating JNK and p38 kinase in a TRPM8-independent manner. Biochem Biophys Res Commun 406:30–35
218. Yang Z, Wang X, Zhu G et al (2012) Effect of surgical castration on expression of TRPM8 in urogenital tract of male rats. Mol Biol Rep 39(4):4797–4802
219. Teilmann SC, Byskov AG, Pedersen PA et al (2005) Localization of transient receptor potential ion channels in primary and motile cilia of the female murine reproductive organs. Mol Reprod Dev 71:444–452
220. Andrade YN, Fernandes J, Vázquez E et al (2005) TRPV4 channel is involved in the coupling of fluid viscosity changes to epithelial ciliary activity. J Cell Biol 168:869–874
221. Dörr J, Fecher-Trost C (2011) TRP channels in female reproductive organs and placenta. Adv Exp Med Biol 704:909–928

222. Meseguer V, Karashima Y, Talavera K et al (2008) Transient receptor potential channels in sensory neurons are targets of the antimycotic agent clotrimazole. J Neurosci 28(3):576–586
223. Kalogris C, Caprodossi S, Amantini C et al (2010) Expression of transient receptor potential vanilloid-1 (TRPV1) in urothelial cancers of human bladder: relation to clinicopathological and molecular parameters. Histopathology 57:744–752
224. Yamada T, Ueda T, Shibata Y et al (2010) TRPV2 activation induces apoptotic cell death in human T24 bladder cancer cells: a potential therapeutic target for bladder cancer. Urology 76:509.e1–509.e7
225. Ziglioli F, Frattini A, Maestroni U et al (2009) Vanilloid-mediated apoptosis in prostate cancer cells through a TRPV-1 dependent and a TRPV-1-independent mechanism. Acta Biomed 80:13–20
226. Sánchez AM, Sánchez MG, Malagarie-Cazenave S et al (2006) Induction of apoptosis in prostate tumor PC-3 cells and inhibition of xenograft prostate tumor growth by the vanilloid capsaicin. Apoptosis 11:89–99
227. Malagarie-Cazenave S, Olea-Herrero N, Vara D et al (2011) The vanilloid capsaicin induces IL-6 secretion in prostate PC-3 cancer cells. Cytokine 54:330–337
228. Sánchez AM, Malagarie-Cazenave S, Olea N et al (2007) Apoptosis induced by capsaicin in prostate PC-3 cells involves ceramide accumulation, neutral sphingomyelinase, and JNK activation. Apoptosis 12:2013–2024
229. Mori A, Lehmann S, O'Kelly J et al (2006) Capsaicin, a component of red peppers, inhibits the growth of androgen-independent, p53 mutant prostate cancer cells. Cancer Res 66:3222–3229
230. Sánchez AM, Martínez-Botas J, Malagarie-Cazenave S et al (2008) Induction of the endoplasmic reticulum stress protein GADD153/CHOP by capsaicin in prostate PC-3 cells: a microarray study. Biochem Biophys Res Commun 372:785–791
231. Jankovic B, Loblaw DA, Nam R (2010) Capsaicin may slow PSA doubling time: case report and literature review. Can Urol Assoc J 4:E9–E11
232. Huang JK, Cheng HH, Huang CJ et al (2005) Effect of capsazepine on cytosolic Ca(2+) levels and proliferation of human prostate cancer cells. Toxicol In Vitro 20:567–574
233. Gkika D, Flourakis M, Lemonnier L et al (2010) PSA reduces prostate cancer cell motility by stimulating TRPM8 activity and plasma membrane expression. Oncogene 29:4611–4616
234. Gkika D, Prevarskaya N (2011) TRP channels in prostate cancer: the good, the bad and the ugly? Asian J Androl 13(5):673–676
235. Yang ZH, Wang XH, Wang HP et al (2009) Effects of TRPM8 on the proliferation and motility of prostate cancer PC-3 cells. Asian J Androl 11:157–165

Chapter 20

Animal Models of Cystitis

Célia D. Cruz and António Avelino

Abstract

As in other areas of biomedical research, studies of bladder pathology rely heavily in animal models of cystitis. These can be basically divided in two types: induced by systemically administration and locally induced by challenge with immunogens or irritants. The choice between models is not easy, and several factors must be taken into account, namely side effects, intensity of the pathology, simplicity, and/or reproducibility of the method.

Key words: Bladder, Inflammation, Animal model, Cystitis

1. Introduction

Translational urological research strongly relies on modeling of lower urinary tract dysfunction in order to describe and understand the pathophysiological mechanisms and propose new therapeutic approaches. Computational models of bladder function are attractive but are still far from being useful due to the complexity of the regulation of lower urinary tract function and the current lack of integration of available data (1). In vitro studies are highly desirable as they avoid the use of experimental animals. However, they are unlikely to accurately reproduce the pathophysiological mechanisms of lower urinary tract dysfunction. In the other end of the spectrum, patient-based investigation also poses problems due to ethical concerns (2). Animal models assume, therefore, an exceptional importance for urological research and can be used for investigating the fundamental biological mechanisms regulating bladder function in normal and pathological conditions. One should, however, be aware that only rarely animal models will fully reflect the human state in normal and

Arpad Szallasi and Tamás Bíró (eds.), *TRP Channels in Drug Discovery: Volume I*, Methods in Pharmacology and Toxicology, DOI 10.1007/978-1-62703-077-9_20, © Springer Science+Business Media, LLC 2012

pathological conditions (3) but will still strongly contribute toward a better understanding of pathological states (4, 5).

Animal models can be divided into (3):

1. Induced animal models: The disease or disorder is induced either via a surgical approach or experimentally. Transgenic animal models also fall in this category.
2. Spontaneous animal models: In this case, the animals naturally exhibit genetic variants that results in the development of a given pathology.
3. Negative models: These models are in which a certain disease does not develop.
4. Orphan animal models: These models refer to diseases initially identified in animals but without identification of the human counterpart.

In urological research, rodents are the most commonly used animals and include rats, mice, and guinea pigs (6). Therefore, investigators should be fully aware of the physiological and morphological differences between these animals and human subjects in what concerns the regulation of bladder function. One major difference resides in the neurotransmitters that are used to control bladder contractions. Whereas in humans, bladder contractions depend on acetylcholine, and in rats, mice, and guinea pigs, ATP is the most prominent mediator (7–10). In rats, used in many studies, there are no intramural ganglia in their bladder wall (11), but they can be found both in mice and in guinea pigs. In most studies, female rodents are typically preferred for transurethral cystometry or for easy bladder instillation, but there may be small differences between males and females which may not be of importance in normal animals but may be exacerbated by genetic manipulation in the case of mice (7, 12).

A particularly difficult disease to model is bladder pain syndrome/interstitial cystitis (BPS/IC). BPS/IC is characterized by suprapubic pain related to bladder filling, usually accompanied by frequency and nocturia, in the absence of urinary infection or other pathology (13, 14). The prevalence in USA was suggested to reach 100/100,000 women (13). However, more recent European epidemiological surveys have reported a prevalence of about 200–300/100,000 women (15, 16). No cure exists yet for this disease and the etiology is unknown. This poses an immense difficulty in establishing accurate animal models for this syndrome. Nevertheless, translational urological research on phosphate buffer saline (PBS)/IC commonly utilizes induced animal models that rely in the administration of irritants or immune stimulants to the animal either systemically (such as cyclophosphamide (CYP), pseudorabies virus (PRV), or bladder homogenates) or directly to the bladder (bacterial products, acids, mustard oil, and croton oil).

There is also a natural animal model of PBS/IC, the feline interstitial cystitis (FIC). However, there is no perfect model and advantages and disadvantages differ. This has been reviewed recently (17), and here we will describe the procedures needed for the most commonly used models and pinpoint some of their advantages and weaknesses.

2. Materials

It should be mentioned in the first place that all the procedures that require urethral catheterization should be performed in females.

2.1. Feline Interstitial Cystitis

1. In this case, no special materials are necessary to induce BPS/IC.

2.2. Experimental Autoimmune Cystitis

1. Female mice or rats. Some studies have used Balb/cAN mice, C57BL/6, and C3H/HEN (18, 19) while other have reported the use of Lewis rats (VAF, Harlan Sprague–Dawley) (20, 21).
2. PBS with Triton.
3. BCA protein assay reagent or Bradford reagent.
4. Complete Freund's adjuvant (CFA).

2.3. Antigen-Induced Cystitis

1. Female guinea pigs or mice; we have not found any indication of the strains used.
2. Allergen: Ovalbumin (OVA) or dinitrophenyl $(DNP)_4$-human serum albumin (HSA).

2.4. Virus-Induced Cystitis

1. Adult female C57BL/6J mice or Sprague–Dawley rats
2. Samples of the attenuated Bartha strain of PRV
3. PK-15, ATCC CCL33 (pig kidney cells) cell line
4. DMEM with 10% fetal calf serum, l-glutamine, and antibiotics
5. Anesthetics: Xylazine/ketamine, volatile anesthesia
6. Hamilton syringe
7. Sutures

2.5. Cystitis Induced by Administration of Irritant Compounds

2.5.1. Cyclophosphamide

1. Cyclophosphamide.
2. Adult animals. Rats (Wistar and Sprague–Dawley) and mice have both been used.

2.5.2. Lipopolysaccharide

1. Lipopolysaccharide (LPS)
2. Adult rats or mice

2.5.3. Bacillus Calmette-Guérin

1. Bacillus Calmette-Guérin (BCG)
2. Adult mice

2.5.4. Chemical Irritants: Turpentine, Mustard Oil, Croton Oil, Acids, and Acrolein

1. Turpentine, mustard oil, and croton oil are dissolved in olive oil.
2. Acetic and hydrochloric acids as well as acrolein are dissolved in saline.

3. Methods

3.1. Feline Interstitial Cystitis

FIC is a spontaneously occurring disease affecting domestic cats (22). Therefore, in this case no animal manipulation is required. Indeed, if one allows for the differences between species, FIC is the closest resembling animal model of BPS/IC. The list of similarities comprises important features of both diseases, including increased urothelial permeability (23), decreased urinary excretion of glycosaminoglycan (24), alterations in the expression of purinergic receptors in the bladder (25), and higher levels of nerve growth factor (NGF) in the urine of FIC cats (26). The FIC model possesses, however, some important drawbacks. Like in human BPS/IC, the causes of FIC remain largely unknown. Whereas BPS/IC affects mostly women (13, 14), FIC each gender is equally affected. Finally, it is very difficult to have access to affected cats which depends on the willingness of owners and veterinarians (27).

3.2. Experimental Autoimmune Cystitis (See Note 1)

The contribution of autoimmune mechanisms to BPS/IC is still controversial. Several studies report the association between BPS/IC and several autoimmune disorders, including rheumatoid arthritis, ulcerative colitis or lupus erythematosus, and higher levels of autoantibodies (28, 29). Therefore, the development of an induced model of autoimmune cystitis has attracted attention over the years. The classical method is as follows (18–21):

1. Bladders from healthy animals are collected, fragmented with sterile scissors, and homogenized in sterile PBS, 0.1 M, pH 7.2, containing 0.5% Triton X-100. Homogenates may be centrifuged for 10 min at 1,000 × *g*. The supernatants are collected the protein content assayed either with the Bradford method or via the BCA protein microassay.
2. The homogenate is then diluted in PBS to a concentration of 10 mg/ml protein and then further diluted 1:2 in CFA.
3. The mixture obtained is injected subcutaneously at the base of the tail.
4. Twelve to 16 weeks after immunization, the establishment of cystitis can be detected by the presence of bladder hyperactivity

(either by cystometry or by ultraviolet analysis of urine spots in filter paper) and by analyzing the histology of the bladder to detect signs of inflammation such as alterations of the urothelial layer and infiltration of inflammatory cells in the bladder wall.

The main disadvantage of this experimental model resides in the fact that it is not clear if PBS/IC has an autoimmune origin or if it is merely accompanied by autoimmune disorders.

3.3. Antigen-Induced Cystitis (See Note 2)

In this experimental model of PBS/IC, cystitis is induced by sensitizing experimental animals by exposing them to known allergens. In most studies, 1 week after allergen administration, cystitis is established. This model involves no infectious agents or direct exposure of the bladder to noxious substances. However, despite many PBS/IC patients showing a higher prevalence of allergies than the general population, it is still controversial if allergic reactions may be part of the pathophysiological mechanisms of PBS/IC.

In the cases OVA is chosen as the allergen, this is the most commonly used approach (30–34):

1. Guinea pigs are sensitized by intraperitoneal three injections of OVA 10 mg/kg at 48-h intervals. The control group should receive saline injections at the same occasions. Twenty-one days after the last OVA injection, animals are either sacrificed for tissue collection for in vitro studies or anesthetized for in vivo studies (35).
2. In the cases using DNP_4-HSA, the most frequently used routine is described in (36–39): Adult mice are sensitized via intraperitoneal injections of 1 μg of DNP_4-HSA in 1 mg of alum on days 0, 7, 14, and 21. One week after the last administration of this allergen, cystitis is induced by intravesical instillation of DNP_4-OVA (1 μg/ml) under deep anesthesia. Infusion of the bladder was repeated twice at 30-min intervals to ensure even exposure of the bladder mucosa to DNP_4-OVA.

3.4. Virus-Induced Cystitis (See Note 3)

In this model of cystitis, bladder inflammation is induced by delivery of a modified strain of virus. In the late 1990s, neurotropic viruses, including herpes simplex virus type 1 and the attenuated Bartha strain of PRV, were being widely used to map neuronal circuits due to their ability to cross several synapses and infect neurons involved in specific circuit (40). It was observed that administration of the PRV caused hemorrhagic cystitis that was prevented by bladder denervation (41, 42). The main advantage of this model is that it demonstrates that cystitis can result from abnormal activity of the neuronal circuitry regulating bladder function. However, like in other models, it is not clear if IC/BPS results from viral infection. Another important drawback is that there is no evidence of increased frequency of bladder reflex contraction in virus-induced cystitis. This model is seldom used nowadays due to the specific care required for virus manipulation. Nevertheless, the described method is as follows:

1. In most studies (41–47), the Bartha strain of PRV was used and propagated in an appropriate cell line. Studies refer the PK-15, ATCC CCL33 (pig kidney cells). These cells are cultivated in DMEM with 10% fetal calf serum, L-glutamine, and antibiotics. Cells are then infected with the virus, and the presence of infectious PRV is determined by plaque assay.
2. When plaques are visible, cells are fixed with methanol and the number of plaques counted. The virus titers are expressed as plaque-forming units (pfu).
3. Once the virus is titrated, animals are deeply anesthetized either with a mixture of ketamine (50 mg/kg) and xylazine (12 mg/kg) via an intraperitoneal injection (47) or with adequate volatile anesthesia (41, 42, 44).
4. An incision is performed at the base of the tail to expose the abductor caudalis dorsalis (ACD) muscle. With a Hamilton syringe, animals are injected with an appropriate volume containing 2.29×10^6 pfu of Bartha's PRV. As negative controls, ultraviolet-irradiated/heat inactivated PRV stocks should be employed. Injection should be performed slowly, followed by compression of the injection site to prevent systemic diffusion of the virus. The incision is then sutured.
5. Behavioral signs indicating the presence of cystitis are observed as early as 48 h after PRV injection (41, 42, 46).

3.5. Cystitis Induced by Administration of Irritant Compounds

In this category fall the majority of studies addressing the pathophysiological mechanisms of cystitis and investigating new therapeutic strategies. Administration of irritants to experimental animals is accomplished by intravesical administration or simply by intraperitoneal injections. In either case, it is easy to follow the development of cystitis, its duration and severity and choose when and how to intervene. However, one should be careful as the mechanism of the animal's response to the irritant may not necessarily correspond to the natural mechanisms of IC/BPS. The most widely used irritants used to induce bladder inflammation are LPS, BCG, acid, turpentine, mustard oil, croton oil, and acrolein. While this group of substances is administered via intravesical route, CYP is given as intraperitoneal injection. While the former induce cystitis by direct contact with the bladder mucosa, CYP is metabolized in the liver into acrolein. Acrolein is excreted via the urinary system and accumulates in the bladder until excreted and acts as a local irritant to the bladder mucosa.

3.5.1. Cyclophosphamide (See Note 4)

Systemic administration of CYP as a means to induce cystitis has been used since 1970s (48). Because CYP administration is easily performed, numerous studies have used this model. A PubMed search using the terms "CYP, rat, cystitis" results in 124 papers. The dosages and protocol may vary slightly in what concerns the dosage and time of inflammation (49). Therefore, we will refer to the following two most widely used protocols:

1. CYP may be given as single intraperitoneal injection. In this case, the dosage may vary between 150 mg/kg (50–52) and 200 mg/kg (53–56). Adult rats are injected and followed for a period of 2–3 days. Behavioral tests and cystometry demonstrate that this approach induces chronic cystitis.
2. Another protocol that may be followed for CYP administration involves a higher number of intraperitoneal injections over time. In this case, the most frequently used protocol consists of three intraperitoneal injections of CYP, at a dosage of 75 mg/kg, in a period of 10 days (50–52, 57, 58).

3.5.2. Lipopolysaccharide

LPS, obtained from *Escherichia coli*, is an endotoxin thought to induce bladder inflammation by binding to the toll-like receptor 4 (59). In this case, LPS is administered via intravesical instillation (39, 60–64).

1. Animals are anesthetized using either volatile anesthesia or by injection of a mixture of xylazine/ketamine.
2. A polypropylene catheter, with a size appropriate to mice or rats, is lubricated and inserted in the urethra. Light pressure is applied to the abdomen to drain urine from the bladder.
3. An adequate volume of an LPS solution (1 mg/ml) (50–150 μl for mice; 350–500 μl for rats) is injected in the bladder via the catheter. This injection should be made slowly in order to prevent trauma and vesicoureteral reflux.
4. In some studies, authors have repeated LPS infusion to ensure full contact of LPS with bladder mucosa.
5. LPS is left in contact with the bladder for 30 min, after which the bladder is rinsed with sterile saline.
6. On the following day, cystitis is established.

3.5.3. Bacillus Calmette-Guérin

BCG is prepared from an attenuated form of *Mycobacterium bovis*. Its use as a means to induce cystitis is very recent (65–67). It has been used to study the inflammatory components of cystitis in mice, but there are no reports analyzing its effects on bladder function. The protocol described is as follows and is very similar to LPS administration (see above):

1. Mice are anesthetized and a lubricated catheter is placed in the urethra. BCG is then instilled (200 μl, total dose of 1.35 mg).

Mice are euthanized and bladder collected 7 days after BCG administration.

2. To induce chronic cystitis, BCG may be administered once every week for a period of 4 weeks (66).

3.5.4. Chemical Irritants: Turpentine, Mustard Oil, Croton Oil, Acids, and Acrolein

Intravesical administration of chemical irritants has long been used to induce bladder inflammation. For administration of these compounds, animals (rats or mice) are anesthetized either with volatile anesthesia or with i.v. anesthetics. A lubricated catheter with an appropriate gauge is inserted into the urethra. Solutions are slowly injected to avoid vesicouretral reflux and damage of the bladder wall. In most studies, irritants are left in contact with the bladder mucosa for 30 min to 1 h, after which the bladder is rinsed with sterile saline and the animals allowed awaking and recovering.

1. Turpentine: The use of turpentine was first described by McMahon and Abe (68). Turpentine is prepared in olive oil in concentrations that vary between 25% (68) and 50–100% (69). Turpentine induces bladder inflammation 1 h after instillation that lasts for at least 48 h. This irritant is known to induce both alterations in bladder function, visceral pain and alterations on the properties of bladder sensory afferents similar to what is known to happen during IC/BPS (68, 70–74).
2. Mustard and croton oils: Like turpentine, mustard and croton oils use was first reported by McMahon and Abe (68). Both are prepared in olive oil in a concentration of 2–2.5%. An adequate volume of oil is instilled via a previously positioned catheter and left in the bladder for 30 min (68, 69, 74–76).
3. Acetic acid: This acid has been used since the early 1990s as an irritant capable of inducing bladder inflammation (77). In this case, it is infused into the bladder and left in contact with the mucosa for 30 min to 1 h (77–80).
4. Hydrochloric acid: Like acetic acid, hydrochloric acid is slowly infused in the bladder (0.4 M; 500 μl) and left in contact with the bladder mucosa for 1 h. Animals should receive daily saline injections in the following 3 days to prevent blood clot formation in the urine. Cystitis is present soon after acid instillation and is long-lasting (81, 82).
5. Acrolein: Acrolein was identified many years ago as the irritative agent of CYP-induced cystitis (83). It is now well established that acrolein per se can induce cystitis if it is administered directly to the bladder of an intact experimental animal (84, 85). The most commonly used procedure describes one infusion (acute cystitis) or three (at 72-h intervals; subacute/chronic cystitis) of 1 mM of an acrolein solution (86–89). Effective dosages of acrolein were established in previous studies (90, 91).

4. Notes

1. To experimental autoimmune cystitis:

   So far, investigators have used either Lewis rats or Balb/cAN mice. Recently, one group has successfully induced autoimmune cystitis using the SWXJ strain of mice (92). This strain of mice has the advantage of increased susceptibility to several autoimmune diseases initiated by Th1-type response (93, 94). Although classically autoimmune cystitis is considered an induced animal model, more recently at least two transgenic models of autoimmune cystitis have been developed. Altuntas et al. (95) used a homogenate obtained from the bladders of mice stabling expressing uroplakin 2 under the control of the SV40 promoter. The authors extracted the protein, which was used to inject SWXJ mice. According to the authors, the resultant autoimmune cystitis is specific to the bladder tissue. Very recently, a novel transgenic model of autoimmune cystitis has been reported (96, 97). This strain of mice, designated as URO-OVA, expresses a membrane form of ovalbumin under the control of the uroplakin 2 promoter and develops inflammatory signs in the bladder that closely follow BPS/IC. The main drawback of these new models relies on the fact that they have not yet been characterized in terms of pain, afferent hypersensitivity and urothelium integrity and function.

2. To antigen-induced cystitis:

   In most studies addressing bladder inflammation using the antigen exposure strategy, mice and guinea pigs have been preferred. However, in a few studies, Sprague–Dawley (98) or Wistar (99) rats have been used.

3. To virus-induced cystitis:

   It was noted (43, 44) that PRV administration altered fluid intake. Therefore, PRV-infected animals should be hydrated daily by subcutaneous bolus of sterile saline.

4. To cystitis induced by administration of irritant compounds CYP:

   When purchasing CYP, one must be aware that there are several formulations available in the market. CYP should be administered as an isotonic solution. Therefore, it should be prepared either in sterile water if the chosen formulation includes sodium chloride or in sterile saline. Another important matter researchers should be aware refers to the harshness of secondary effects of CYP administration as it may induce blood loss through the urine. It is important to daily evaluate the animals, particularly in longer periods of testing. Loss of a high number of animals is common when repeated doses are used. This outcome seems to vary between different laboratories, without a known cause.

This study was supported by INComb FP7 HEALTH project no 223234

References

1. Fry CH, Sadananda P, Wood DN, Thiruchelvam N, Jabr RI, Clayton R (2011) Modeling the urinary tract-computational, physical, and biological methods. Neurourol Urodyn 30:692–699
2. Fry CH, Daneshgari F, Thor K et al (2010) Animal models and their use in understanding lower urinary tract dysfunction. Neurourol Urodyn 29:603–608
3. Beynen AC, Hau J (2001) Animal models. In: Van Zuthphen LFM, Baumans V, Beynen AC (eds) Principles of laboratory animal science: a contribution to the humane use and care of animals and to the quality of experimental results. Elsevier, Amesterdam, pp 197–205
4. Bird SJ, Parlee MB (2000) Of mice and men (and women and children): scientific and ethical implications of animal models. Prog Neuropsychopharmacol Biol Psychiatry 24: 1219–1227
5. van der Staay FJ, Arndt SS, Nordquist RE (2009) Evaluation of animal models of neurobehavioral disorders. Behav Brain Funct 5:11
6. McMurray G, Casey JH, Naylor AM (2006) Animal models in urological disease and sexual dysfunction. Br J Pharmacol 147(Suppl 2): S62–S79
7. Andersson KE, Soler R, Fullhase C (2011) Rodent models for urodynamic investigation. Neurourol Urodyn 30:636–646
8. Brading AF, Williams JH (1990) Contractile responses of smooth muscle strips from rat and guinea-pig urinary bladder to transmural stimulation: effects of atropine and alpha, beta-methylene ATP. Br J Pharmacol 99:493–498
9. Inoue R, Brading AF (1990) The properties of the ATP-induced depolarization and current in single cells isolated from the guinea-pig urinary bladder. Br J Pharmacol 100:619–625
10. Andersson KE, Arner A (2004) Urinary bladder contraction and relaxation: physiology and pathophysiology. Physiol Rev 84:935–986
11. Uvelius B, Gabella G (1995) Intramural neurones appear in the urinary bladder wall following excision of the pelvic ganglion in the rat. Neuroreport 6:2213–2216
12. Cornelissen LL, Misajet B, Brooks DP, Hicks A (2008) Influence of genetic background and gender on bladder function in the mouse. Auton Neurosci 140:53–58
13. Hanno P, Nordling J, Fall M (2010) Bladder pain syndrome. Med Clin North Am 95:55–73
14. Hanno PM, Chapple CR, Cardozo LD (2009) Bladder pain syndrome/interstitial cystitis: a sense of urgency. World J Urol 27:5
15. Leppilahti M, Sairanen J, Tammela TL, Aaltomaa S, Lehtoranta K, Auvinen A (2005) Prevalence of clinically confirmed interstitial cystitis in women: a population based study in Finland. J Urol 174:581–583
16. Temml C, Wehrberger C, Riedl C, Ponholzer A, Marszalek M, Madersbacher S (2007) Prevalence and correlates for interstitial cystitis symptoms in women participating in a health screening project. Eur Urol 51:803–808, discussion 9
17. Bjorling DE, Wang ZY, Bushman W (2011) Models of inflammation of the lower urinary tract. Neurourol Urodyn 30:673–682
18. Bullock AD, Becich MJ, Klutke CG, Ratliff TL (1992) Experimental autoimmune cystitis: a potential murine model for ulcerative interstitial cystitis. J Urol 148:1951–1956
19. Phull H, Salkini M, Purves T, Funk J, Copeland D, Comiter CV (2007) Angiotensin II plays a role in acute murine experimental autoimmune cystitis. BJU Int 100:664–667
20. Luber-Narod J, Austin-Ritchie T, Banner B et al (1996) Experimental autoimmune cystitis in the Lewis rat: a potential animal model for interstitial cystitis. Urol Res 24:367–373
21. Mitra S, Dagher A, Kage R, Dagher RK, Luber-Narod J (1999) Experimental autoimmune cystitis: further characterization and serum autoantibodies. Urol Res 27:351–356
22. Buffington CA, Chew DJ, Woodworth BE (1999) Feline interstitial cystitis. J Am Vet Med Assoc 215:682–687
23. Lavelle JP, Meyers SA, Ruiz WG, Buffington CA, Zeidel ML, Apodaca G (2000) Urothelial pathophysiological changes in feline interstitial cystitis: a human model. Am J Physiol Renal Physiol 278:F540–F553
24. Buffington CA, Blaisdell JL, Binns SP Jr, Woodworth BE (1996) Decreased urine glycosaminoglycan excretion in cats with interstitial cystitis. J Urol 155:1801–1804
25. Birder LA, Ruan HZ, Chopra B et al (2004) Alterations in P2X and P2Y purinergic receptor expression in urinary bladder from normal cats and cats with interstitial cystitis. Am J Physiol Renal Physiol 287:F1084–F1091
26. Birder LA, Wolf-Johnston AS, Chib MK, Buffington CA, Roppolo JR, Hanna-Mitchell AT (2010) Beyond neurons: involvement of urothelial and glial cells in bladder function. Neurourol Urodyn 29:88–96
27. Westropp JL, Buffington CA (2002) In vivo models of interstitial cystitis. J Urol 167:694–702

28. van de Merwe JP (2007) Interstitial cystitis and systemic autoimmune diseases. Nat Clin Pract Urol 4:484–491
29. Ochs RL (1997) Autoantibodies and interstitial cystitis. Clin Lab Med 17:571–579
30. Kim YS, Levin RM, Wein AJ, Longhurst PA (1992) Effects of sensitization on the permeability of urothelium in guinea pig urinary bladder. J Urol 147:270–273
31. Kim YS, Longhurst PA, Wein AJ, Levin RM (1991) Effects of sensitization on female guinea pig urinary bladder function: in vivo and in vitro studies. J Urol 146:454–457
32. Lee JG, Levin RM, Krasnopolsky L, Wein AJ, Longhurst PA (1995) Effects of the peptide leukotriene receptor antagonist ICI 198,615 on the in vivo and in vitro changes in guinea pig bladder function which occur after sensitization with ovalbumin. J Urol 154:1217–1221
33. Christensen MM, Keith I, Rhodes PR et al (1990) A guinea pig model for study of bladder mast cell function: histamine release and smooth muscle contraction. J Urol 144:1293–1300
34. Lavelle JP, Apodaca G, Meyers SA, Ruiz WG, Zeidel ML (1998) Disruption of guinea pig urinary bladder permeability barrier in noninfectious cystitis. Am J Physiol 274:F205–F214
35. Bjorling DE, Saban MR, Saban R (1994) Neurogenic inflammation of Guinea-pig bladder. Mediators Inflamm 3:189–197
36. Saban R, Saban MR, Nguyen NB, Hammond TG, Wershil BK (2001) Mast cell regulation of inflammation and gene expression during antigen-induced bladder inflammation in mice. Physiol Genomics 7:35–43
37. Saban R, Saban MR, Nguyen NB et al (2000) Neurokinin-1 (NK-1) receptor is required in antigen-induced cystitis. Am J Pathol 156:775–780
38. Dozmorov I, Saban MR, Gerard NP et al (2003) Neurokinin 1 receptors and neprilysin modulation of mouse bladder gene regulation. Physiol Genomics 12:239–250
39. Saban MR, Nguyen NB, Hammond TG, Saban R (2002) Gene expression profiling of mouse bladder inflammatory responses to LPS, substance P, and antigen-stimulation. Am J Pathol 160:2095–2110
40. Aston-Jones G, Card JP (2000) Use of pseudorabies virus to delineate multisynaptic circuits in brain: opportunities and limitations. J Neurosci Methods 103:51–61
41. Jasmin L, Carstens E, Basbaum AI (1997) Interneurons presynaptic to rat tail-flick motoneurons as mapped by transneuronal transport of pseudorabies virus: few have long ascending collaterals. Neuroscience 76:859–876
42. Jasmin L, Janni G, Manz HJ, Rabkin SD (1998) Activation of CNS circuits producing a neurogenic cystitis: evidence for centrally induced peripheral inflammation. J Neurosci 18:10016–10029
43. Rudick CN, Bryce PJ, Guichelaar LA, Berry RE, Klumpp DJ (2008) Mast cell-derived histamine mediates cystitis pain. PLoS One 3:e2096
44. Rudick CN, Chen MC, Mongiu AK, Klumpp DJ (2007) Organ cross talk modulates pelvic pain. Am J Physiol Regul Integr Comp Physiol 293:R1191–R1198
45. Rudick CN, Schaeffer AJ, Klumpp DJ (2009) Pharmacologic attenuation of pelvic pain in a murine model of interstitial cystitis. BMC Urol 9:16
46. Jasmin L, Janni G, Ohara PT, Rabkin SD (2000) CNS induced neurogenic cystitis is associated with bladder mast cell degranulation in the rat. J Urol 164:852–855
47. Doggweiler R, Jasmin L, Schmidt RA (1998) Neurogenically mediated cystitis in rats: an animal model. J Urol 160:1551–1556
48. Locher GW, Cooper EH (1970) Repair of rat urinary bladder epithelium following injury by cyclophosphamide. Invest Urol 8:116–123
49. Bon K, Lichtensteiger CA, Wilson SG, Mogil JS (2003) Characterization of cyclophosphamide cystitis, a model of visceral and referred pain, in the mouse: species and strain differences. J Urol 170:1008–1012
50. Hu VY, Zvara P, Dattilio A et al (2005) Decrease in bladder overactivity with REN1820 in rats with cyclophosphamide induced cystitis. J Urol 173:1016–1021
51. Qiao LY, Vizzard MA (2002) Cystitis-induced upregulation of tyrosine kinase (TrkA, TrkB) receptor expression and phosphorylation in rat micturition pathways. J Comp Neurol 454:200–211
52. Vizzard MA (2000) Changes in urinary bladder neurotrophic factor mRNA and NGF protein following urinary bladder dysfunction. Exp Neurol 161:273–284
53. Dinis P, Charrua A, Avelino A et al (2004) Anandamide-evoked activation of vanilloid receptor 1 contributes to the development of bladder hyperreflexia and nociceptive transmission to spinal dorsal horn neurons in cystitis. J Neurosci 24:11253–11263
54. Pinto R, Frias B, Allen S et al (2010) Sequestration of brain derived nerve factor by intravenous delivery of TrkB-Ig2 reduces

bladder overactivity and noxious input in animals with chronic cystitis. Neuroscience 166:907–916

55. Meen M, Coudore-Civiale MA, Eschalier A, Boucher M (2001) Involvement of hypogastric and pelvic nerves for conveying cystitis induced nociception in conscious rats. J Urol 166:318–322
56. Meen M, Coudore-Civiale MA, Parry L, Eschalier A, Boucher M (2002) Involvement of N-methyl-D-aspartate receptors in nociception in the cyclophosphamide-induced vesical pain model in the conscious rat. Eur J Pain 6:307–314
57. Dinis P, Charrua A, Avelino A, Cruz F (2004) Intravesical resiniferatoxin decreases spinal c-fos expression and increases bladder volume to reflex micturition in rats with chronic inflamed urinary bladders. BJU Int 94:153–157
58. Cruz CD, Avelino A, McMahon SB, Cruz F (2005) Increased spinal cord phosphorylation of extracellular signal-regulated kinases mediates micturition overactivity in rats with chronic bladder inflammation. Eur J Neurosci 21:773–781
59. Tapping RI, Akashi S, Miyake K, Godowski PJ, Tobias PS (2000) Toll-like receptor 4, but not toll-like receptor 2, is a signaling receptor for Escherichia and Salmonella lipopolysaccharides. J Immunol 165:5780–5787
60. Bjorling DE, Jacobsen HE, Blum JR et al (2001) Intravesical *Escherichia coli* lipopolysaccharide stimulates an increase in bladder nerve growth factor. BJU Int 87:697–702
61. Meyer-Siegler KL, Ordorica RC, Vera PL (2004) Macrophage migration inhibitory factor is upregulated in an endotoxin-induced model of bladder inflammation in rats. J Interferon Cytokine Res 24:55–63
62. Saban MR, Saban R, Hammond TG et al (2002) LPS-sensory peptide communication in experimental cystitis. Am J Physiol Renal Physiol 282:F202–F210
63. Saban R, D'Andrea MR, Andrade-Gordon P et al (2007) Regulatory network of inflammation downstream of proteinase-activated receptors. BMC Physiol 7:3
64. Charrua A, Cruz CD, Cruz F, Avelino A (2007) Transient receptor potential vanilloid subfamily 1 is essential for the generation of noxious bladder input and bladder overactivity in cystitis. J Urol 177:1537–1541
65. Saban MR, O'Donnell MA, Hurst RE et al (2008) Molecular networks discriminating mouse bladder responses to intravesical bacillus Calmette-Guerin (BCG), LPS, and TNF-alpha. BMC Immunol 9:4
66. Saban MR, Sferra TJ, Davis CA et al (2011) Neuropilin-VEGF signaling pathway acts as a key modulator of vascular, lymphatic, and inflammatory cell responses of the bladder to intravesical BCG treatment. Am J Physiol Renal Physiol 299:F1245–F1256
67. Saban MR, Simpson C, Davis C et al (2007) Discriminators of mouse bladder response to intravesical Bacillus Calmette-Guerin (BCG). BMC Immunol 8:6
68. McMahon SB, Abel C (1987) A model for the study of visceral pain states: chronic inflammation of the chronic decerebrate rat urinary bladder by irritant chemicals. Pain 28:109–127
69. Habler HJ, Janig W, Koltzenburg M (1988) A novel type of unmyelinated chemosensitive nociceptor in the acutely inflamed urinary bladder. Agents Actions 25:219–221
70. McMahon SB (1988) Neuronal and behavioural consequences of chemical inflammation of rat urinary bladder. Agents Actions 25:231–233
71. Jaggar SI, Habib S, Rice AS (1998) The modulatory effects of bradykinin B1 and B2 receptor antagonists upon viscero-visceral hyper-reflexia in a rat model of visceral hyperalgesia. Pain 75:169–176
72. Jaggar SI, Scott HC, Rice AS (1999) Inflammation of the rat urinary bladder is associated with a referred thermal hyperalgesia which is nerve growth factor dependent. Br J Anaesth 83:442–448
73. Rice AS (1995) Topical spinal administration of a nitric oxide synthase inhibitor prevents the hyper-reflexia associated with a rat model of persistent visceral pain. Neurosci Lett 187: 111–114
74. Habler HJ, Janig W, Koltzenburg M (1993) Receptive properties of myelinated primary afferents innervating the inflamed urinary bladder of the cat. J Neurophysiol 69: 395–405
75. Callsen-Cencic P, Mense S (1997) Expression of neuropeptides and nitric oxide synthase in neurones innervating the inflamed rat urinary bladder. J Auton Nerv Syst 65:33–44
76. DeBerry J, Randich A, Shaffer AD, Robbins MT, Ness TJ (2010) Neonatal bladder inflammation produces functional changes and alters neuropeptide content in bladders of adult female rats. J Pain 11:247–255
77. Birder LA, de Groat WC (1992) Increased c-fos expression in spinal neurons after irritation of the lower urinary tract in the rat. J Neurosci 12:4878–4889
78. Fraser MO, Chuang YC, Tyagi P et al (2003) Intravesical liposome administration – a novel

treatment for hyperactive bladder in the rat. Urology 61:656–663

79. Birder LA, Nakamura Y, Kiss S et al (2002) Altered urinary bladder function in mice lacking the vanilloid receptor TRPV1. Nat Neurosci 5:856–860
80. Charrua A, Cruz CD, Narayanan S et al (2009) GRC-6211, a new oral specific TRPV1 antagonist, decreases bladder overactivity and noxious bladder input in cystitis animal models. J Urol 181:379–386
81. Rivas DA, Chancellor MB, Shupp-Byrne S, Shenot PJ, McHugh K, McCue P (1997) Molecular marker for development of interstitial cystitis in rat model: isoactin gene expression. J Urol 157:1937–1940
82. Wang Y, Zhou Y, Mourad MS, Hassouna MM (2000) Neuromodulation reduces urinary frequency in rats with hydrochloric acid-induced cystitis. BJU Int 86:726–730
83. Cox PJ (1979) Cyclophosphamide cystitis – identification of acrolein as the causative agent. Biochem Pharmacol 28:2045–2049
84. Kalota SJ, Stein PC, Parsons CL (1992) Prevention of acrolein-induced bladder injury by pentosanpolysulfate. J Urol 148:163–166
85. Sakata T, Smith RA, Garland EM, Cohen SM (1989) Rat urinary bladder epithelial lesions induced by acrolein. J Environ Pathol Toxicol Oncol 9:159–169
86. Guerios SD, Wang ZY, Boldon K, Bushman W, Bjorling DE (2008) Blockade of NGF and trk receptors inhibits increased peripheral mechanical sensitivity accompanying cystitis in rats. Am J Physiol Regul Integr Comp Physiol 295:R111–R122
87. Guerios SD, Wang ZY, Boldon K, Bushman W, Bjorling DE (2009) Lidocaine prevents referred hyperalgesia associated with cystitis. Neurourol Urodyn 28:455–460
88. Merriam FV, Wang ZY, Guerios SD, Bjorling DE (2008) Cannabinoid receptor 2 is increased in acutely and chronically inflamed bladder of rats. Neurosci Lett 445:130–134
89. Merriam FV, Wang ZY, Hillard CJ, Stuhr KL, Bjorling DE (2011) Inhibition of fatty acid amide hydrolase suppresses referred hyperalgesia induced by bladder inflammation. BJU Int 108(7):1145–1149
90. Bjorling DE, Elkahwaji JE, Bushman W et al (2007) Acute acrolein-induced cystitis in mice. BJU Int 99:1523–1529
91. Wang ZY, Wang P, Merriam FV, Bjorling DE (2008) Lack of TRPV1 inhibits cystitis-induced increased mechanical sensitivity in mice. Pain 139:158–167
92. Lin YH, Liu G, Kavran M et al (2008) Lower urinary tract phenotype of experimental autoimmune cystitis in mouse: a potential animal model for interstitial cystitis. BJU Int 102:1724–1730
93. Jaini R, Hannaman D, Johnson JM et al (2006) Gene-based intramuscular interferon-beta therapy for experimental autoimmune encephalomyelitis. Mol Ther 14:416–422
94. Jane-wit D, Yu M, Edling AE et al (2002) A novel class II-binding motif selects peptides that mediate organ-specific autoimmune disease in SWXJ, SJL/J, and SWR/J mice. J Immunol 169:6507–6514
95. Altuntas CZ, Daneshgari F, Sakalar C et al (2012) Autoimmunity to uroplakin II causes cystitis in mice: a novel model of interstitial cystitis. Eur Urol 61(1):193–200
96. Kim R, Liu W, Chen X, Kreder KJ, Luo Y (2011) Intravesical dimethyl sulfoxide inhibits acute and chronic bladder inflammation in transgenic experimental autoimmune cystitis models. J Biomed Biotechnol 2011: 937061
97. Liu W, Evanoff DP, Chen X, Luo Y (2007) Urinary bladder epithelium antigen induces CD8+ T cell tolerance, activation, and autoimmune response. J Immunol 178:539–546
98. Rajasekaran M, Locke KW, Parsons CL (2006) MN-001, a novel oral anti-inflammatory agent, suppresses bladder hyperactivity in a rat model. BJU Int 98:430–434
99. Ahluwalia A, Giuliani S, Scotland R, Maggi CA (1998) Ovalbumin-induced neurogenic inflammation in the bladder of sensitized rats. Br J Pharmacol 124:190–196

Chapter 21

Overactive Bladder Models

Roberto Soler, Lysanne Campeau, Claudius Füllhase, and Karl-Erik Andersson

Abstract

The overactive bladder (OAB) can be defined in humans as a urodynamic observation (detrusor overactivity), or symptomatically (urgency, frequency, incontinence, nocturia) as the OAB syndrome. For obvious reasons, there are no animal models of the OAB syndrome. In humans, urinary incontinence can be due to *involuntary* bladder contractions demonstrable by cystometry during the filling phase. In animals, cystometric bladder hyperactivity (bladder contractions voluntary and/or involuntary) can be found in many animal models. It can occur spontaneously or be provoked, and the pathophysiology may include both peripheral and central mechanisms. To study bladder hyperactivity in animals, cystometry plays an important role. The present protocol describes the basic cystometry technique and its application in a few animal models specifically used for the study of bladder hyperactivity.

Key words: Overactive bladder, Animal models, Cystometry, Bladder hyperactivity, Lower urinary tract

1. Introduction

Urinary incontinence can be due to involuntary bladder contractions during the filling phase. Such contractions may be demonstrated by cystometry. According to the International Continence Society (ICS), "*detrusor overactivity* is a *urodynamic observation* characterized by involuntary contractions during the filling phase which may be spontaneous or provoked." Detrusor overactivity (DO) may be neurogenic when there is a relevant neurological condition, or idiopathic when there is no defined cause [1]. The "*overactive bladder" (OAB) syndrome* (urge syndrome or urgency-frequency syndrome) is a *symptom-based* clinical diagnosis defined by the ICS as "urgency, with or without urge incontinence, usually

Arpad Szallasi and Tamás Bíró (eds.), *TRP Channels in Drug Discovery: Volume I*, Methods in Pharmacology and Toxicology, DOI 10.1007/978-1-62703-077-9_21, © Springer Science+Business Media, LLC 2012

with frequency and nocturia" (1). DO may be associated with the OAB syndrome in many, but not all cases. For example, Hashim and Abrams (2) found that overall, 83% of patients with DO had symptoms of OAB and 64% of patients with OAB had DO.

Bladder hyperactivity can be found in many animal models. It can occur spontaneously or be provoked, and the pathophysiology may include both peripheral and central mechanisms. It should be emphasized that *there are no animal models of the OAB syndrome.* OAB can only be diagnosed in humans, since urgency is the central symptom, and this obviously cannot be assessed in animals. What can be studied in animal models is bladder hyperactivity, which is probably a better term to use for animal recordings than DO, since this term is defined and used in human cystometry. However, the terms DO and bladder hyperactivity are often used synonymously in animal cystometry.

For animal in vivo physiological studies on bladder dysfunction, cystometry plays an important role (3). The cystometry technique and a few of the models specifically used for the study of bladder hyperactivity are discussed later. Each animal model has a unique pathophysiology and is therefore described separately in different subheadings of this chapter.

2. Cystometry

Cystometry is widely used in animals, e.g., rodents, rabbits, and cats, to functionally study lower urinary tract function (3). A number of animal models utilize cystometric investigation to evaluate the effect of spontaneous, or chemically or surgically induced voiding dysfunction, as well as the effect of different therapies on bladder function. Different approaches may be used and the methodological details may vary depending on the species investigated; those below can be used for cystometry in rodents.

2.1. Materials

1. Polyethylene tubing (PE-50; PE90 or PE10) (Harvard Apparatus, Holliston, MA, USA)—for bladder catheterization and connections (Note 1)
2. Metabolic cage (Harvard Apparatus)
3. Swivel for infusion (optional) (Harvard Apparatus)
4. Force displacement transducer—for micturition volume measurement (FT-302, iWorx/CB Sciences, Dover, NH, USA)
5. Pressure transducer—for bladder pressure measurement (BP-100, iWorx/CB Sciences)
6. Syringe infusion pump (Harvard Apparatus)

7. Three-way stopcock—to connect the tubing from the bladder, the pressure transducer, and the tubing from the infusion pump
8. Transducer amplifier (ETH-401, iWorx/CB Sciences)
9. Data acquisition system (PowerLab, AD Instruments, Colorado Springs, CO, USA)
10. Computer and specific software
11. Room temperature saline
12. Surgical instruments
13. 5-0 and 4-0 polyglactin suture
14. Needles with appropriate size to the specific tubing
15. Syringe (the 60-mL syringe is preferred, since the solution does not have to be replaced during cystometry)

Cystometry setup. The animal is placed in a metabolic cage with a funnel to provide urine collection. The urine is collected into a cup connected to a force displacement transducer. The bladder catheter is connected through tubing to a three-way stopcock, which is also connected to the pressure transducer and to the infusion pump. Both transducers are connected to the transducer amplifier, which is then connected to the data acquisition system. A computer with specific software is in turn connected to the data acquisition system and displays the tracings in real time.

2.2. Methods

2.2.1. Bladder Catheterization: Conscious Animals

1. Prepare the bladder catheter (Note 2).
2. Anesthetize the animal, and use appropriate analgesia.
3. Place the animal on a heating pad to avoid cooling and take other supportive measures (eye ointment, subcutaneous saline injection for hydration).
4. The hair on the suprapubic abdominal area is cut short or shaved, and povidone/iodine is applied on the surgical field. All surgical procedures are performed in an aseptic fashion.
5. A midline longitudinal incision of approximately 2 cm is done on the suprapubic abdominal area. The rectus fascia is opened in the midline.
6. Identify the bladder.
7. Place a purse-string absorbable suture (such as polyglactin 5-0) in the dome of the bladder.
8. With micro scissors (or a large gauge needle) incise (or puncture) the bladder in the middle of the purse-string suture and insert the catheter with the cuff extremity (avoid touching the trigone with the catheter).
9. Pull the purse-string suture tight, and slightly pull the catheter back until the cuff gets in contact with the purse-string suture and test bladder leakage (Note 3).

10. Relocate the bladder in situ and use the median umbilical ligament as an orientation landmark to avoid any distortion/twisting of the bladder.
11. Make a small incision on the back of the neck of the animal and dissect the surrounding tissues.
12. Insert a piercing tunneled instrument subcutaneously through the neck incision to the ventral surface of the abdomen. The piercing instrument should enter the abdominal cavity at the level where the bladder lies anatomically. The bladder catheter is passed through the instrument until it is exteriorized. This will prevent the animals from reaching the catheter.
13. Anchor the catheter to the neck incision and thermally seal the free extremity. This will prevent leakage and infection.
14. Close the incisions with 4-0 polyglactin suture or clips.
15. The cystometry may be performed within 0–3 days after bladder catheterization. The choice of our group is 3 days after catheterization to allow proper recovery and healing.

2.2.2. Bladder Catheterization: Anesthetized Cystometry

1. For cystometry in anesthetized animals, the catheter can be inserted either via the urethra or via the bladder, and in this case, it can be externalized suprapubically.
2. The same type of tubing described earlier can be used as catheters, but alternatively, other types of cannulas or catheters may be used.
3. Urethral catheters can be secured with a hitch suture onto the lower abdomen, avoiding altering the normal urethral axis.
4. Urethane is the anesthetic of choice, since other anesthetics may affect bladder function (3).

2.2.3. Cystometry

1. Calibrate pressure and volume transducers. The pressure transducer must be at the level of the animal bladder.
2. Open the sealed extremity of the catheter.
3. Empty the animal's bladder either via the catheter or by manual abdominal compression.
4. Place the animal in the metabolic cage.
5. Connect the bladder catheter to the swivel, which is connected by tubing to the three-way stopcock, as are the pressure transducer and the tubing from the infusion pump (Note 4).
6. Start the continuous infusion of saline in room temperature (Note 5).
7. Allow the animals to accommodate, usually for 30 min, and record the micturition cycles for a period of 30–120 min. Bladder pressure and urine volume are continuously and synchronously recorded (Notes 6 and 7).

2.3. Notes

1. The most suitable catheter will depend on the size of the animal, catheter resistance, and the tendency to cause obstruction and twisting.
2. The length of the tubing must be some centimeters longer than the distance from the suprapubic area to the back of the neck of the animal. The inserted end of the catheter should be heated to dilate and create a cuff. This can be made by approximating a cautery or a flame.
3. To test bladder leakage, inject saline through the catheter. Any leakage point must be sutured. Fill the bladder until fluid starts leaking out of the urethral meatus.
4. The tubing connections are made by using needles that have the appropriate diameter to that specific tubing.
5. Infusions rates may vary depending on the animal species (e.g., 1.5–3 mL/h for mice, 2.4–11 mL/h for rats, and 9–12 mL/h for guinea pigs).
6. The procedure must be supervised until completion in order to avoid and correct any artifact or complications, such as catheter twisting or obstruction, catheter disconnection, or any technical issue.
7. Troubleshooting:
   (a) Sudden high amplitude rise in bladder pressure: check for causes of obstruction, such as bladder catheter twisting, blood clots in the bladder extremity of the catheter (gently infuse saline through the bladder catheter), or obstruction of any part of the tubing and connections
   (b) Sudden decrease in bladder pressure: check for leaking of any part of the tubing and connections
   (c) Negative pressure or volume: re-calibrate the system
   (d) Absence of voiding contractions for a longer period than expected: check the volume of saline in the syringe

3. Animal Models

3.1. Spontaneous Hyperactivity

Spontaneous hypertensive rat. The spontaneous hypertensive rat (SHR) is widely used as a genetic model of hypertension. However, the SHR also exhibits abnormal bladder function and hyperactive behavior (4–6). Compared with the normotensive Wistar Kyoto (WKY) rats, SHRs exhibit a reduced bladder capacity and micturition volume, increased urinary frequency, and a greater occurrence of nonvoiding contractions (NVCs: i.e., pressure increases during the filling phase not associated with fluid output). These rats have an increased bladder sympathetic innervation, and hypertrophy of both afferent and efferent neurons supplying the bladder, but the exact cause for their abnormal voiding function is not known. However, both peripheral and spinal mechanisms seem to be involved. The effects of intrathecal and intra-arterial doxazosin on cystometric parameters were more pronounced in SHR than in WKY rats. There was a marked reduction in NVCs after intrathecal (but not intra-arterial) doxazosin in SHRs (4), suggesting involvement of spinal α-adrenoceptors (AR).

Many reports have studied awake SHRs (and other rodents) cystometrically and diagnosed DO (bladder hyperactivity) subjectively only by the shape of the curve of intravesical pressure (IVP) changes. However, it seems unlikely that this approach can reliably discriminate between true DO (=NVCs) and IVP changes caused by abdominal straining. Lee et al. (7) measuring intraabdominal pressure, reported that among total events of IVP increases during the filling phase in SHRs, true DO represented up to 76% of the pressure changes, whereas the remaining 24% were caused by abdominal straining.

As a control for SHRs in studies of hypertension, most researchers use the genetically normotensive WKY rat. Less often, the Wistar-ST (Wistar) rat of their mother strain or the Sprague–Dawley (SD) rat has been used. Jin et al. (6) showed that in contrast to WKY rats, which showed DO, the Wistar and SD strains did not. Therefore, with respect to bladder hyperactivity, the Wistar and SD strains should be the most appropriate controls for SHRs (6).

Notably, development of bladder hyperactivity in SHRs is age dependent: young animals (e.g., less than 12 weeks) often show normal cystometry.

3.1.1. Intraabdominal Pressure Measurement During Cystometry

Materials

1. Polyethylene catheter (PE-50) with a 0.05-mL balloon on its tip cuff
2. Same material as described in the bladder catheterization

Methods

1. Fill the balloon and the catheter with saline or distilled water and seal the distal end of the catheter.
2. Fix the intraabdominal catheter with a silk tie to the posterior side of the bladder catheter to a distance of approximately 1 cm from the bladder (Note 1).
3. Tunnel the catheter subcutaneously together with the bladder pressure to the back of the neck of the animal and anchor it.
4. The detrusor pressure is defined as the bladder pressure subtracting the intraabdominal pressure.

Notes

1. This distance needs to be fixed. If too short, the filled bladder will affect the balloon pressure erroneously and if too long, the perivesical pressures will not be representative, apart from in the bladder

3.2. Inflammation Provoked Hyperactivity

Instillation of various agents into the bladder to stimulate specific receptors or to produce inflammation/hypersensitivity is commonly used as acute models. For example, PGE_2 instillation (8) is effective in rodents (as in humans; (9)). Stimulation of TRPV1 receptors with intravesical resiniferatoxin and capsaicin (10), and TRPA1 receptors by hydrogen sulfide (NaHS, donor of H_2S), allyl isothiocyanate, and cinnamaldehyde, are other examples (11). Other means of stimulation of bladder hyperactivity more nonspecifically include instillation of, e.g., acid (acetic, citric, or hydrochloric acid (12), protamine sulphate (13, 14), and acrolein (15). It is important to emphasize that even if the common denominator for this intervention is probably stimulation of nociceptive C-fibers and consequent increases in afferent activity, they most probably do not reflect the DO that may be associated with the OAB syndrome.

Stimuli inducing bladder inflammation and provoking bladder hyperactivity include, e.g., systemic administration of *cyclophosphamide* (16), intravesical instillation of *bacterial endotoxin* (17), and *bladder radiation* (18). Chronic chemical cystitis can be induced using different compounds and routes of administration. The animal model of cyclophosphamide-induced bladder inflammation was developed following the observation of hemorrhagic cystitis following the systemic treatment with this antineoplastic drug in humans. The bladder effects of the drug are caused by urinary excretion of the active metabolite acrolein. The intraperitoneal injection of cyclophosphamide in rodents has been used as a model of visceral pain because the animals demonstrate characteristic behavioral signs in parallel with the development of bladder lesions. However, the cystitis related changes are particularly complex involving local, peripheral, central modifications, as well as micturition abnormalities. The exact mechanism of the induced inflammation was attributed to increased levels of endogenous nitric oxide (NO) that causes urothelial damage and other inflammatory events. Cyclophosphamide causes a rise in inducible NO synthase (NOS) that may be mediated by platelet activating factors (19). The cyclophosphamide cystitis also increases c-Fos expression in the subnuclei of the dorsal vagal complex and of the ventrocaudal bulbar reticular formation (20). This model is achieved in female rats (200–300 g) by injecting four doses of cyclophosphamide (75 mg/kg) intraperitoneally every third day for 7 days.

Intravesical chemical irritants are also used to induce persistent inflammation related bladder hyperactivity such as hydrochloric acid and acetic acid (21, 22). A short duration of intravesical protamine sulfate instillation can induce irritative effects and shorten the intermicturition interval in a rat model (23).

All these models have limitations. For example, cyclophosphamide administration may be a good chemically induced bladder inflammation model, associated with alterations in neurochemical, electrophysiological, and functional properties of micturition pathways (24), and may also reflect the hemorrhagic cystitis encountered in patients treated with the drug. However, it may not be a good model for OAB or interstitial cystitis/painful bladder syndrome (IC/PBS), because there is no evidence for inflammation in OAB (which, as mentioned, can only be diagnosed in humans), and inflammatory changes are not prominent in IC/PBS. DO is also not a characteristic finding when cystometry is performed in IC/PBS patients. In addition, cystometric findings can be influenced in an unpredictable way with cyclophosphamide as it produces profound systemic effects in the animals.

3.2.1. Transient Bladder Hyperactivity During Intravesical Instillation

Bladder hyperactivity can be observed acutely during intravesical administration of several compounds during conscious cystometry. The technique for this acute model first involves the bladder catheter implantation and cystometry recording described in the materials and methods of the cystometry section. Each compound described in the material section should be used distinctly on its own in different animals.

Materials

1. Syringe (10, 20, or 60 mL) compatible with the infusion pump used for cystometry
2. Prostaglandin E_2 (Sigma-Aldrich, St-Louis, MO, USA): A stock solution is made of 0.01 M in ethanol. Concentrations as low as 10 μM in saline (0.9%w/v NaCl solution) have been shown to increase micturition pressure and bladder pressure, and to decrease micturition volume and bladder capacity (8). Concentrations between 20 and 60 μM have been demonstrated to cause bladder hyperactivity (25), but the most commonly used concentration is 50 μM in saline.
3. Capsaicin (Sigma-Aldrich): Capsaicin-induced bladder hyperactivity in conscious rats was observed to be reversible and repeatable (26). Capsaicin is first prepared as a stock solution as 0.1 M in ethanol. Concentrations of 10 to 30 μM in saline have been described in visceral pain models
4. Resiniferatoxin (Sigma-Aldrich: Urodynamic changes can be found with concentration of 30–100 nM (10)).
5. Sodium hydrogen sulfide (donor of H_2S) (Sigma-Aldrich): The intravesical instillation of hydrogen sulfide (1 mM) should be preceded by intravesical protamine sulfate (10 mg/mL) for 30 min to increase its penetration, and therefore changes in urodynamic parameters (11).
6. Allyl isothiocyanate (AI) and cinnamaldehyde (CA) (Sigma-Aldrich): Stock solutions (0.1 mol/l in ethanol) of AI and CA should be kept at –20°C until use. Different concentrations of AI (10 and 100 μM) and CA (100 and 500 μM) can produce urodynamic changes (11).

Methods

1. The animal is placed in a metabolic cage, and its bladder catheter (PE 50) is connected to the three-way stopcock, which is in turn connected to a pressure transducer and a syringe with room temperature saline as described in the cystometry method.
2. With the help of a syringe infusion pump, room temperature saline is infused at a rate of 10 mL per hour.
3. It is first important to obtain a stable and analyzable cystometry recording including several micturition cycles (3, 4) and lasting for approximately 20–30 min.
4. The compound should be prepared the same day as the cystometry recording and should be at room temperature.
5. The time of syringe change should be noted on the cystometry recording software. The pump is stopped, and the syringe is disconnected from the PE tubing.
6. The syringe with the compound is connected to the PE tubing. The syringe is then placed back in the pump, and infusion is started again.
7. There will be a certain time lapse between the start of the infu sion and the effect of the intravesical compound in the bladder, as the remaining saline is being flushed through the PE tubing.
8. Instillation of the compound should be continued until a stable and analyzable cystometry recording is obtained (Note 1).

Notes

1. It is recommended to flush the polyethylene tubing after each usage with ethanol, and then with saline, because certain compounds can adhere to the tube lining.

3.3. Hypercholesterolemia/Hyperlipidemia

Another way of provoking cystometric changes is to induce *hypercholesterolemia/hyperlipidemia* in rat (27, 28). Sprague–Dawley rats fed a high fat/cholesterol diet developed not only erectile dysfunction, and prostate enlargement, but also bladder hyperactivity (27, 28). The mechanisms behind these changes are unclear, but may be due to both vascular, detrusor, and neurogenic components, similar to what has previously been shown in rabbits (29, 30).

Animal models of atherosclerotic vascular disease have initially been developed to study vasculogenic erectile dysfunction. A high cholesterol diet would be the most physiological method of inducing vascular disease in rats. However, rats require very prolonged period of this special diet before presenting endothelial dysfunction because they seem to be "resistant" to the effects of cholesterol. Park et al. (31) accelerated the process by interrupting the nitric oxide synthase pathway early on and demonstrated diminished erectile function. The same model was then found to decrease voiding intervals and bladder capacity on cystometrography (28).

The hyperlipidemia model developed by Rahman et al. (27) caused more bladder overactivity in rats.

Materials

3.3.1. Hypercholesterolemia

1. Male Sprague–Dawley rats (250–300 g)
2. 1% Cholesterol diet (CRF-1; Oriental Yeast Co. Ltd, Osaka, Japan)
3. NG-nitro-l-arginine methyl ester (L-NAME, 3 mg/ml) (Sigma-Aldrich)

Methods

1. The animals are fed a daily 1% cholesterol diet for 8 weeks.
2. The animals are also given a daily dose of L-NAME in their drinking water early on for 2 weeks.
3. Cystometry is performed as previously described (Note 1).

Notes

1. Cystometry should be done at the end of the required time frame.

3.3.2. Hyperlipidemia

Materials

1. Male Sprague–Dawley rats (250–300 g)
2. 2% cholesterol and 10% lard (Zeigler Brothers, Gardner, PA, USA)

Methods

1. The animals are fed a daily 2% cholesterol and 10% lard diet for 6 months.
2. Cystometry is performed as previously described (Note 1).

Notes

1. Cystometry should be done at the end of the required time frame.

3.4. Diabetes

Models of *diabetes* both type 1 and 2 have been successfully established in rats and mice (32). Destruction of pancreatic β-cells using streptozotocin, leading to type 1 diabetes is widely used and accepted as a valid model for type 1 diabetes complications, including diabetic cystopathy. Rat models of type 2 diabetes often have comorbidities that can influence urodynamic parameters, e.g., nephropathy and obesity (33, 34).

3.4.1. Materials

Type 1 Diabetes: Chemically Induced

1. Streptozocin diluted in 0.1 M citrate buffer solution, pH 4.5 (Sigma-Aldrich)
2. Alloxan monohydrate diluted in saline (Sigma-Aldrich)

Type 1 Diabetes: Genetic Models (Note 1)

The most common models are:

1. Nonobese diabetic mouse
2. Akita mouse
3. Diabetes prone BioBreeding rat

Type 2 Diabetes and Obesity: Genetic Models (Note 1)

The most common models are:

1. *ob/ob* mice
2. *db/dB* mice
3. Zucker diabetic fatty (ZDF) rats
4. Spontaneously diabetic Torii (SDT) rats

For all models, a blood glucose monitoring system (Roche Diagnostics, Indianapolis, IN) is required.

3.4.2. Methods for the Chemically Induced Type 1 Diabetes

Streptozocin-Induced Diabetes

1. Deprive the animals of food for 12–24 h.
2. Inject streptozocin (50–65 mg/kg body weight) intraperitoneally.
3. After 3 days collect blood from the tail vein to estimate the blood glucose level and select the diabetic ones (blood glucose equal or greater than 200–350 mg/dL) (Note 2).

Alloxan-Induced Diabetes

1. Deprive the animals of food for 12–24 h.
2. Inject alloxan (120–150 mg/kg body weight) intraperitoneally.
3. After 7–14 days collect blood from the tail vein to estimate the blood glucose level and select the diabetic ones (blood glucose equal or greater than 200–350 mg/dL) (Note 2).

3.4.3. Notes

1. Genetic models exhibit the chronological evolution of diabetic bladder dysfunction.
2. Changes in bladder function were seen 3 weeks after chemically inducing diabetes and a shift from compensated to decompensated bladder function was observed around 9–12 weeks after the induction.

3.5. Outflow Obstruction

Men with benign prostatic hyperplasia (BPH) often experience symptoms of OAB, and DO is a common cystometric finding in these patients (35). Bladder outlet obstruction (BOO) has been successfully employed in multiple animal species, including rats, mice, and guinea pigs (36–38). The obstruction can be acute (hours) (39) or chronic (days to weeks) (36). Generally, a metal rod with a fixed diameter (in rats usually around 1 mm) is applied close to the urethra, partially obstructing it. The approach can be transabdominal or transperineal (see below; (40)). A ligature is placed around the urethra and the metal rod, which is then removed. Even though BOO is a well established procedure in rodents, the resulting urodynamic changes are not always uniform, which may depend on several factors, including the degree and time of obstruction. Animals may have an overactive or underactive voiding pattern (41, 42). In overactivity, higher residual volume, threshold and maximum pressures, and the occurrence of NVCs are often seen. An underactive pattern may represent the development of decompensation. Increased bladder weight is an important parameter in this model and represents both muscle hypertrophy and hyperplasia. Besides being used to study the pathophysiology and therapeutic approaches to outflow obstruction, this model is also used to study bladder hyperactivity, since NVCs are typical cystometric findings. It is important to note though that in contrast to what happens in the majority of OAB cases, the changes in this model are the result of a specific lesion. It should also be noted that BOO in rodents is a dynamic process, both morphology and bladder function changing with time of obstruction. As a result, urodynamic findings change at different time points after obstruction, and comparisons between different studies cannot be made if this is not taken into consideration.

Below are described some of the techniques used for rats. This protocol can also in principle also be used for rabbits, guinea pigs, and mice, but has to be modified according to the species used and the size of the animal

3.5.1. Materials

1. Surgical instruments
2. Plastic or metal rod with 1 mm outer diameter (or a 19 gauge needle)
3. 3-0 polypropylene suture and 4-0 polyglactin suture

3.5.2. Methods

Partial Urethral Obstruction: Retropubic Approach (Note 1)

1. Anesthetize the animal.
2. The hair on the suprapubic abdominal area is cut short or shaved, and povidone/iodine is applied on the surgical field. All surgical procedures are performed in an aseptic fashion.
3. A midline longitudinal incision of approximately 2 cm is done on the suprapubic abdominal area. The rectus fascia is opened in the midline.
4. Identify the bladder.
5. Reach and expose the bladder neck area. Use cotton swabs to blunt separation of the bladder from surrounding connective tissue (Note 2).
6. Starting from the dorsal side, place a tie (e.g., 3-0 polypropylene) around the urethra right below the bladder neck and the uretrovesical junction, then move the bladder so that the anterior wall can be seen, and close the tie loosely (Note 3).
7. Place the sterile 1 mm metal bar under the tie on the proximal urethral surface, then pull the tie tight. When the suture is secured, remove the metal bar, leaving the urethra partially obstructed.
8. Close the incision with 4-0 polyglactin suture (Notes 4 and 5).

Partial Urethral Obstruction: Perineal Approach

1. Anesthetize the animal.
2. The hair on the perineal area is cut short or shaved, and povidone/iodine is applied on the surgical field. All surgical procedures are performed in an aseptic fashion.

3. Incise longitudinally the area from the penoscrotal junction to the midscrotum.
4. Bluntly dissect the surrounding connective tissue to isolate the bulbous penis with a string.
5. Identify the urethra and separate it from the cavernous bodies.
6. Place the metal bar on the outer surface of the urethra and proceed as described earlier (Notes 4 and 5).

3.5.3. Notes

1. The use of an operation microscope is highly recommended for the retropubic approach.
2. When dissecting the bladder, it is important to identify the ureters. In male rats, carefully dissect the prostatic urethra to avoid harming the seminal vesicles and the ureters at the ureterovesical junction.
3. In this step, be careful not to include the distal ureters in the tie.
4. The expected mortality rate after partial urethral obstruction procedures can reach up to 20%.
5. Two weeks after PUO changes in bladder function can be seen and the severity of the obstruction usually increases with time.

3.6. Neurological Models

Normal voiding function relies on the integrity of a complex neuronal control system. Important regulatory centers are located in the cortex, pons, and spinal cord. Depending on the level of the lesion in the CNS specific alterations in the bladder function will take place. Several models of CNS injury have been used to study the pathophysiology of the neurogenic bladder dysfunction and some of them are discussed below.

3.6.1. Spinal Cord Transection/Injury

Animal models of spinal cord injury (SCI) have greatly added to the understanding of the neuronal spinal control of bladder function (42). After SCI the bladder is initially arreflexic and the bladder has to be emptied manually at least twice a day. When a spinal micturition reflex pathway emerges, dependent in part on plasticity of neuronal connections within the spinal cord, the bladder often becomes overactive. Moreover, detrusor-sphincter dyssynergia may also occur (42). The development of neurogenic bladder dysfunction following SCI is due in part to changes in C-fibers afferent neurons. Nerve growth factor may play a role in those changes in afferent neurons and remodeling of spinal synaptic connections, since its level is increased in the spinal cord of SCI rats (43).

LUT dysfunction in spinal injury caused by cord contusion has a time course different from that in spinal cord transection (44–46). Since a majority of SCI patients are those with partial cord injury, animal models of spinal contusion-caused LUT dysfunction are important.

Materials

1. Female Sprague–Dawley rats (200–250 g)
2. Povidone/iodine
3. Scalpel and surgical instruments (scissors, needle-holder, graspers)
4. Small retractors
5. Cotton swabs and gauze pads
6. 4-0 vicryl and 4-0 PDS
7. Gelfoam (Baxter, Deerfield, IL, USA)

Methods

1. The animal is anesthetized and given proper analgesia.
2. The animals should be kept warm during the procedure with a heating pad.
3. The hair on the back is cut short or shaved, and povidone/iodine is applied on the surgical field. All surgical procedures are performed in an aseptic fashion.
4. A midline sagittal incision of approximately 2 cm is done on the spine from the level of the last rib caudally. The incision is extended through the spinal processes.
5. The muscle on both sides of the spinous processes from T9 to T11 is incised, and the vertebral bodies are exposed with the help of small retractors. Bleeding can also be avoided if the incision is done as close as possible to the spinous process in the tendinous region.
6. A laminectomy is performed at T10 by using a heavy scissors on the lateral lamellae of the vertebral body.
7. The spinal cord is exposed. If bleeding persists, pressure can be applied with cotton swabs (Note 1).
8. The spinal cord is completely severed under magnification (Note 2). A small square of gelfoam (1–2 mm thick) is placed at the site of transaction, preventing any spinal cord contact.
9. The paravertebral muscles are approximated using interrupted 4-0 vicryl sutures.
10. The skin is closed with interrupted 4-0 PDS sutures.
11. A dose of antibiotic is given pre- and postoperatively.
12. Following the procedure, the rats are housed singly in separate cages containing wood shaving bedding to prevent the formation of decubitus ulcers.
13. Upon awakening, the animals will be paraplegic. Their bladders should be evacuated 3 times daily using the Crede maneuver. This involves holding the animal with both hands and applying pressure to the lower abdomen (over the bladder) to empty it. This maneuver is performed until the reflex voiding recovers. It will become apparent as the volume expelled will become progressively smaller. They are subsequently emptied twice daily.

Notes

1. Exposure and visualization is the most important component of this procedure. It allows a clean and straight transection of the spinal at the adequate level.
2. The transection can be done with a No 11 Blade or a pair of sharp scissors.

3.6.2. Suprapontine Models

Parkinson's Disease

Lower urinary tract symptoms are frequently reported in PD patients, specially increased daytime frequency, nocturia, and urgency. Hesitancy and slow stream are also symptoms reported by PD patients. The most frequent urodynamic abnormality is detrusor overactivity, which may be accompanied by impaired contractility. Underactivity, acontractility, and detrusor-sphincter dyssynergia may also be observed (47–49). The dopaminergic pathway influences micturition depending on the type of dopamine receptors that are activated. Activation of D1-like dopamine receptors causes inhibition, whereas D2-like receptors are involved in facilitation of the micturition reflex (50, 51). Animal models of PD induced by 1-methyl-4-phenyl-1,2,3,6-tetrahydropyridine (MPTP) or by injection of 6-hydroxidopamine into the striatum, medial forebrain bundle (MFB), or the substantia nigra (SN) (51–53) have been used to study bladder dysfunction. Soler et al. (53) evaluated the development of bladder dysfunction after injection of 6-OHDA into the MFB. Changes were seen as early as 3 days after the injury. A defined pattern of bladder hyperactivity was established 14 days postlesion and persisted for 28 days.

Materials

1. Female Sprague–Dawley rats (150–250 g)
2. Povidone/iodine
3. Scalpel and surgical instruments (scissors, needle-holder, graspers)
4. Cotton swabs and gauze pads
5. 5-0 Vicryl suture and 4-0 PDS suture
6. Stereotactic frame (Stoelting, Wood Dale, IL, USA)
7. Small electrical drill (Ram Products Inc, Forth Worth, TX, USA)
8. Hamilton glass syringe (10 or 20 μL) (Hamilton, Reno, NV, USA)
9. 26 gauge blunt needle (flat end) (Hamilton)
10. Manual injector (Stoelting)
11. Stereotactic syringe minipump (optional) (Stoelting)
12. 6-hydroxydopamine (Sigma-Aldrich)

Methods

1. The compound 6-hydroxydopamine is supplied by Sigma at a concentration of 2.5 mg/mL in 0.1% ascorbic acid. A solution of 2 mg/mL is made by diluting the supplied compound with saline or Phosphate Buffered Saline (PBS). The volume of this solution injected in each animal will be 4 μL.

2. The animal is anesthetized with an intraperitoneal injection and given proper analgesia.
3. The hair on the head from behind the ears to above the eyes is cut short or shaved, and povidone/iodine is applied on the surgical field. All surgical procedures are performed in an aseptic fashion.
4. A midline sagittal incision of approximately 2 cm is done on the dorsal surface of the head. The incision is extended through the subcutaneous fascia.
5. The cranium is exposed with the help of cotton swabs and gauze pads, and the bregma is identified. The bregma is the point of reference used for stereotactic localization.
6. The animal is placed in the stereotactic frame by first symmetrically pressing the lateral rods to the temporal bone of the animal's head. The mouth piece is then gently inserted by opening the inferior jaw of the animal.
7. The dorsal surface of the cranium should then be leveled parallel to the horizon using the adjustable mouthpiece height, by ensuring that both the bregma and the dorsal sutures are located at the same height.
8. The solution of 6-OHDA is aspirated in the Hamilton syringe, and all air bubbles are removed (Note 1).
9. The syringe is placed in the injector. The stereotactic frame allows the movement of the syringe in three dimensions (antero-posteriorly (AP), medio-laterally (ML), and dorso-ventrally (DV)). The tip of the needle is placed directly in the center of the bregma. The corresponding coordinates (AP and ML) are noted and are used as references (Notes 2 and 3).
10. The 6-OHDA lesion is performed unilaterally, consistently on the same side for all animals. The coordinates to inject in the medial forebrain bundle (MFB) of a female Sprague–Dawley (150–250 g) are based on The Rat Brain in Stereotaxic Coordinates by Paxinos and Watson (54), and are: AP: −0.4; ML: −1.6 (for the right side); DV: −7 from bregma. Therefore, the AP and ML coordinates are added to the reference coordinates measured at the bregma. The syringe is elevated and displaced to the calculated coordinates. The needle is then descended to touch the cranium, and the corresponding site is marked. The needle and syringe is again elevated.
11. With the help of the drill, a small burr hole (less than 1 mm diameter) is done at the marked site. The hole should remain within the bony cranium and superficial to the dura membrane.
12. The needle and syringe is descended until the tip of the blunt needle touches the dura. This corresponds to the appropriate DV coordinates of reference (Note 3).

13. After ensuring that the head of the animal is immobile and stable, the needle is descended to the corresponding calculated DV coordinate.
14. Injection is performed at a rate of 0.5–1 μL per minute until a total of 8 μg (4 μL) is delivered. This can be done manually or with the help of a stereotactic syringe minipump (Note 4).
15. The needle is left in place for 5 min after injection, prior to slowly retracting it back. If bleeding occurs, slight pressure can be applied with a cotton swab until bleeding resolves.
16. The subcutaneous fascia is closed with a 5-0 vicryl suture in a running fashion. The skin is closed with interrupted 4-0 PDS suture.
17. The operative site is cleaned with saline of any excess blood or povidone/iodine.

Notes

1. The diluted solution of 6-OHDA should be made fresh on the day of the surgery and kept on ice covered from the light.
2. Good exposure and identification of the bregma is critical to the stereotactic injection.
3. The AP and ML coordinates are in reference to the bregma, while the DV coordinate is in reference to the dura at the injection site.
4. A slow and regular injection rate renders a more consistent lesion. Keeping the needle in place for 5 min following the injection allows the intracerebral pressure to equilibrate.

Medial Cerebral Artery Occlusion

Unilateral occlusion of the medial cerebral artery (MCA) is used to create an animal model of stroke. Cystometry in these animals mainly reveals low bladder capacity. This model of bladder hyperactivity has been utilized to study pathophysiology and therapeutic approaches. When dizocilpine or MK-801, NMDA glutamate receptor antagonists, was given to rats at the time of the MCA occlusion, bladder hyperactivity was prevented, suggesting a role for glutamatergic stimulation in the initial stage of the development of bladder dysfunction (52, 55). There is evidence of involvement of COX-2 and nitric oxide in bladder hyperactivity following MCA occlusion (55, 56).

Materials

1. Microsurgical instruments
2. Microsurgery microscope
3. 4-0 nylon suture (for rats)
4. 6-0 silk suture
5. 7-0 nylon suture
6. 4-0 polyglactin suture

Methods

1. Anesthetize the animal.
2. The hair on the anterior neck area is cut short or shaved, and povidone/iodine is applied on the surgical field. All surgical procedures are performed in an aseptic fashion. Place the animal under the microsurgery microscope.
3. Make a midline incision in the anterior neck.
4. Expose the common carotid artery.
5. Position a retractor between the digastrics and sternomastoid muscles and divide the omohyoid muscle.
6. Identify and coagulate the branches of the external carotid artery: the occipital, the superior thyroid, and the ascending pharyngeal arteries.
7. Dissect the external carotid artery further distally and coagulate it along with the terminal lingual and maxillary artery branches.
8. Isolate and carefully separate the internal carotid artery from the adjacent vagus nerve.
9. Ligate the pterygopalatine artery (a branch of the internal carotid artery) close to its origin with 7-0 nylon suture.
10. At this point, the internal carotid artery is the only remaining branch of the common carotid artery.
11. Place a loose 6-0 silk suture around the external carotid artery stump.
12. Place a curved microvascular clip across both the common carotid artery and the internal carotid artery adjacent to the external carotid artery origin.
13. Heat the tip of a 5-cm length 4-0 monofilament nylon suture (suitable for rats) to round it.
14. Puncture the external carotid artery and introduce the rounded tip nylon suture through it.
15. Tie the silk suture around the external carotid stump around the intraluminal nylon suture to prevent bleeding.
16. Remove the microvascular clip.
17. Gently advance the nylon suture from the external carotid artery to the internal carotid artery lumen.
18. After the suture passes the base of the skull and it is no longer visible, continue inserting it until a resistance is felt and a slight curving of the suture or stretching of the internal carotid artery is observed (Note 1).
19. The origin of the middle cerebral artery is then blocked and all sources of blood flow from the internal carotid, anterior cerebral, and posterior cerebral arteries are occluded.

20. Close the incision with 4-0 polyglactin suture, leaving 1 cm of the nylon suture protruding it, so it could be withdrawn if further reperfusion is planned in the experiments.
21. To allow reperfusion, pull back the suture until resistance is felt, indicating that the suture has cleared the anterior cerebral-internal carotid arteries lumens and is in the external carotid artery stump. The suture can be trimmed and this procedure does not require anesthesia (Note 2).

Notes

1. At this point the blunted tip of the suture had passed the middle cerebral artery origin and reached the proximal segment of the anterior cerebral artery, which has a smaller diameter (for rats weighing 300–400 g, from the origin of the external carotid artery, the suture is introduced exactly 17 mm into the internal carotid artery).
2. The model is suitable for studying bladder dysfunction as early as 1 h after the procedure.

References

1. Abrams P, Cardozo L, Fall M et al (2002) The standardisation of terminology of lower urinary tract function: report from the Standardisation Sub-committee of the International Continence Society. Neurourol Urodyn 21:167–178
2. Hashim H, Abrams P (2006) Is the bladder a reliable witness for predicting detrusor overactivity? J Urol 175(1):191–194
3. Andersson KE, Soler R, Füllhase C (2011) Rodent models for urodynamic investigation. Neurourol Urodyn 30(5):636–646
4. Persson K, Pandita RK, Spitsbergen JM et al (1998) Spinal and peripheral mechanisms contributing to hyperactive voiding in spontaneously hypertensive rats. Am J Physiol 275: R1366–R1373
5. Jin LH, Andersson KE, Kwon YH et al (2009) Selection of a control rat for conscious spontaneous hypertensive rats in studies of detrusor overactivity on the basis of measurement of intra-abdominal pressures. Neurourol Urodyn 29:1338–1343
6. Jin LH, Andersson KE, Kwon YH, Yoon SM, Lee T (2010) Selection of a control rat for conscious spontaneous hypertensive rats in studies of detrusor overactivity on the basis of measurement of intra-abdominal pressures. Neurourol Urodyn 29(7):1338–1343
7. Lee T, Andersson KE, Streng T, Hedlund P (2008) Simultaneous registration of intraabdominal and intravesical pressures during cystometry in conscious rats–effects of bladder outlet obstruction and intravesical PGE2. Neurourol Urodyn 27(1):88–95
8. Ishizuka O, Mattiasson A, Andersson KE (1995) Prostaglandin E2-induced bladder hyperactivity in normal, conscious rats: involvement of tachykinins? J Urol 153(6):2034–2038
9. Schussler B (1990) Comparison of the mode of action of prostaglandin E2 (PGE2) and sulprostone, a PGE2-derivative, on the lower urinary tract in healthy women. A urodynamic study. Urol Res 18:349–352
10. Ishizuka O, Mattiasson A, Andersson KE (1995) Urodynamic effects of intravesical resiniferatoxin and capsaicin in conscious rats with and without outflow obstruction. J Urol 154:611–616
11. Streng T, Axelsson HE, Hedlund P et al (2008) Distribution and function of the hydrogen sulfide-sensitive TRPA1 ion channel in rat urinary bladder. Eur Urol 53:391–399
12. Mitobe M, Inoue H, Westfall TD et al (2008) A new method for producing urinary bladder hyperactivity using a non-invasive transient intravesical infusion of acetic acid in conscious rats. J Pharmacol Toxicol Methods 57:188–193
13. Chuang YC, Chancellor MB, Seki S et al (2003) Intravesical protamine sulfate and potassium chloride as a model for bladder hyperactivity. Urology 61:664–670

14. Soler R, Bruschini H, Freire MP et al (2008) Urine is necessary to provoke bladder inflammation in protamine sulfate induced urothelial injury. J Urol 180:1527–1531
15. Bjorling DE, Elkahwaji JE, Bushman W et al (2007) Acute acrolein-induced cystitis in mice. BJU Int 99:1523–1529
16. Juszczak K, Ziomber A, Wyczolkowski M et al (2009) Urodynamic effects of the bladder C-fiber afferent activity modulation in chronic model of overactive bladder in rats. J Physiol Pharmacol 60:85–91
17. Stein PC, Pham H, Ito T et al (1996) Bladder injury model induced in rats by exposure to protamine sulfate followed by bacterial endotoxin. J Urol 155:1133–1138
18. Vale JA, Bowsher WG, Liu K et al (1993) Post-irradiation bladder dysfunction: development of a rat model. Urol Res 21:383–388
19. Souza-Fiho MV, Lima MV, Pompeu MM, Ballejo G, Cunha FQ, Ribeiro Rde A (1997) Involvement of nitric oxide in the pathogenesis of cyclophosphamide-induced hemorrhagic cystitis. Am J Pathol 150(1):247–256
20. Lantéri-Minet M, Bon K, de Pommery J, Michiels JF, Menétrey D (1995) Cyclophosphamide cystitis as a model of visceral pain in rats: model elaboration and spinal structures involved as revealed by the expression of c-Fos and Krox-24 proteins. Exp Brain Res 105(2):220–232
21. Birder LA, de Groat WC (1992) Increased c-fos expression in spinal neurons after irritation of the lower urinary tract in the rat. J Neurosci 12(12):4878–4889
22. Yu Y, Fraser MO, de Groat WC (2004) Effects of ZD6169, a K ATP channel opener, on neurally- mediated plasma extravasation in the rat urinary bladder induced by chemical or electrical stimulation of nerves. Brain Res 996(1):41–46
23. Chuang YC, Chancellor MB, Seki S, Yoshimura N, Tyagi P, Huang L, Lavelle JP, De Groat WC, Fraser MO (2003) Intravesical protamine sulfate and potassium chloride as a model for bladder hyperactivity. Urology 61(3):664–670
24. LaBerge J, Malley SE, Zvarova K et al (2006) Expression of corticotropin-releasing factor and CRF receptors in micturition pathways after cyclophosphamide-induced cystitis. Am J Physiol Regul Integr Comp Physiol 291:R692–R703
25. Takeda H, Yamazaki Y, Igawa Y, Kaidoh K, Akahane S, Miyata H, Nishizawa O, Akahane M, Andersson KE (2002) Effects of beta(3)-adrenoceptor stimulation on prostaglandin E(2)- induced bladder hyperactivity and on the cardiovascular system in conscious rats. Neurourol Urodyn 21(6):558–565
26. Ishizuka O, Igawa Y, Mattiasson A, Andersson KE (1994) Capsaicin-induced bladder hyperactivity in normal conscious rats. J Urol 152(2 Pt 1):525–530
27. Rahman NU, Phonsombat S, Bochinski D, Carrion RE, Nunes L, Lue TF (2007) An animal model to study lower urinary tract symptoms and erectile dysfunction: the hyperlipidaemic rat. BJU Int 100(3):658–663
28. Son H, Lee SL, Park WH, Park K, Park S, Kang MS, Kim DY, Kim SW, Paick JS (2007) New unstable bladder model in hypercholesterolemia rats. Urology 69(1):186–190
29. Azadzoi KM, Tarcan T, Kozlowski R, Krane RJ, Siroky MB (1999) Overactivity and structural changes in the chronically ischemic bladder. J Urol 162(5):1768–1778
30. Azadzoi KM, Yalla SV, Siroky MB (2007) Oxidative stress and neurodegeneration in the ischemic overactive bladder. J Urol 178(2):710–715
31. Park K, Son H, Kim SW, Paick JS (2005) Initial validation of a novel rat model of vasculogenic erectile dysfunction with generalized atherosclerosis. Int J Impot Res 17(5):424–430
32. Daneshgari F, Leiter EH, Liu G, Reeder J (2009) Animal models of diabetic uropathy. J Urol 182(6 Suppl):S8–S13
33. Janssen U, Phillips AO, Floege J (1999) Rodent models of nephropathy associated with type II diabetes. J Nephrol 12(3):159–172
34. Gasbarro G, Lin DL, Vurbic D, Quisno A, Kinley B, Daneshgari F, Damaser MS (2010) Voiding function in obese and type 2 diabetic female rats. Am J Physiol Renal Physiol 298(1):F72–F77
35. Oelke M, Baard J, Wijkstra H et al (2008) Age and bladder outlet obstruction are independently associated with detrusor overactivity in patients with benign prostatic hyperplasia. Eur Urol 54:419–426
36. Malmgren A, Sjögren C, Uvelius B, Mattiasson A, Andersson KE, Andersson PO (1987) Cystometrical evaluation of bladder instability in rats with infravesical outflow obstruction. J Urol 137(6):1291–1294
37. Mostwin JL, Karim OM, Van Koeveringe G, Seki N (1994) Guinea pig as an animal model for the study of urinary bladder function in the normal and obstructed state. Neurourol Urodyn 13(2):137–145
38. Pandita RK, Fujiwara M, Alm P, Andersson KE (2000) Cystometric evaluation of bladder function in non-anesthetized mice with and without bladder outlet obstruction. J Urol 164(4):1385–1389

39. Saito M, Miyagawa I (2001) Bladder dysfunction after acute urinary retention in rats. J Urol 165:1745–1747
40. Melman A, Tar M, Boczko J et al (2005) Evaluation of two techniques of partial urethral obstruction in the male rat model of bladder outlet obstruction. Urology 66:1127–1133
41. Schroder A, Uvelius B, Newgreen D et al (2003) Bladder overactivity in mice after 1 week of outlet obstruction. Mainly afferent dysfunction? J Urol 170:1017–1021
42. de Groat WC, Yoshimura N (2010) Changes in afferent activity after spinal cord injury. Neurourol Urodyn 29:63–76
43. Noto H, Roppolo JR, Steers WD et al (1991) Electrophysiological analysis of the ascending and descending components of the micturition reflex pathway in the rat. Brain Res 549: 95–105
44. Pikov V, Gillis RA, Jasmin L et al (1998) Assessment of lower urinary tract functional deficit in rats with contusive spinal cord injury. J Neurotrauma 15:375–386
45. Pikov V, Wrathall JR (2002) Altered glutamate receptor function during recovery of bladder detrusor-external urethral sphincter coordination in a rat ,model of spinal cord injury. J Pharmacol Exp Ther 300:421–427
46. Pikov V, Wrathall JR (2001) Coordination of the bladder detrusor and the external urethral sphincter in a rat model of spinal cord injury: effect of injury severity. J Neurosci 21:559–569
47. Araki I, Kitahara M, Oida T et al (2000) Voiding dysfunction and Parkinson's disease: urodynamic abnormalities and urinary symptoms. J Urol 164:1640–1643
48. Fowler CJ (2007) Update on the neurology of Parkinson's disease. Neurourol Urodyn 26:103–109
49. Sammour ZM, Gomes CM, Barbosa ER et al (2009) Voiding dysfunction in patients with Parkinson's disease: impact of neurological impairment and clinical parameters. Neurourol Urodyn 28:510–515
50. Seki S, Igawa Y, Kaidoh K et al (2001) Role of dopamine D1 and D2 receptors in the micturition reflex in conscious rats. Neurourol Urodyn 20:105–113
51. Yoshimura N, Kuno S, Chancellor MB, De Groat WC, Seki S (2003) Dopaminergic mechanisms underlying bladder hyperactivity in rats with a unilateral 6-hydroxydopamine (6-OHDA) lesion of the nigrostriatal pathway. Br J Pharmacol 139(8):1425–1432
52. Yokoyama O, Mizuno H, Komatsu K et al (2004) Role of glutamate receptors in the development and maintenance of bladder overactivity after cerebral infarction in the rat. J Urol 171:1709–1714
53. Soler R, Füllhase C, Santos C, Andersson KE (2011) Development of bladder dysfunction in a rat model of dopaminergic brain lesion. Neurourol Urodyn 30(1):188–193
54. Paxinos G, Watson C (2005) The rat brain in stereotaxic coordinates – the new coronal set, 5th edn. Academic, New York
55. Yotsuyanagi S, Yokoyama O, Komatsu K et al (2005) Role of cyclooxygenase-2 in the development of bladder overactivity after cerebral infarction in the rat. J Urol 174:365–369
56. Kodama K, Yokoyama O, Komatsu K et al (2002) Contribution of cerebral nitric oxide to bladder overactivity after cerebral infarction in rats. J Urol 167:391–396

Concluding Remarks

As highlighted in the introduction to this book, much remains to be discovered about the Transient Receptor Potential (TRP) superfamily of ion channels. Even the most thoroughly characterized member of the family continues to provide controversy and surprises. Our goal is to summarize the current state of the art in regard to the literature as well as methods. This comprehensive collection provides up-to-date information on targets that are being pursued in human clinical trials (TRPV1 and TRPA1) as well as less-explored family members that may serve as the next generation of "hot" drug targets. Of particular interests are the TRP channels that human genetic data implicate in disease. These include TRPML1 (mucolipidosis type IV), TRPC6 (Focal Segmental Glomerular Sclerosis), TRPP2 (Polycystic kidney disease), TRPV4 (varied peripheral neuropathies and spinal dysplasias), and TRPM4 (progressive familial heart block type 1). We expect the material presented here will help investigators as they think about targeting TRP channels to improve human health and will promote discussion of this exciting and diverse family of proteins.

Cambridge, MA, USA *Magdalene M. Moran*

Arpad Szallasi and Tamás Bíró (eds.), *TRP Channels in Drug Discovery: Volume I*, Methods in Pharmacology and Toxicology, DOI 10.1007/978-1-62703-077-9, © Springer Science+Business Media, LLC 2012

Index

A

Arpad Szallasi and Tamás Bíró (eds.), *TRP Channels in Drug Discovery: Volume I*, Methods in Pharmacology and Toxicology, DOI 10.1007/978-1-62703-077-9,

B

C

D

F

G

H

I

K

R

S

U

V

W

MIX
Papier aus verantwortungsvollen Quellen
Paper from responsible sources
FSC® C105338

If you have any concerns about our products,
you can contact us on
ProductSafety@springernature.com

In case Publisher is established outside the EU,
the EU authorized representative is:
Springer Nature Customer Service Center GmbH
Europaplatz 3, 69115 Heidelberg, Germany

Printed by Libri Plureos GmbH
in Hamburg, Germany